An Interplay of Cellular and Molecular Components of Immunology

Our immune system defends us against infection by employing multiple lines of defense. The relevance of the immune response in human health, disease prevention, and vaccinations becomes evident when the immune system is compromised as in the case of pathogenic infections or autoimmune diseases. The reader will gain a fundamental understanding of the essential principles of immunology, such as how our immune system recognizes/fights infectious agents, how our body differentiates between foreign and self-cells/molecules, and how the memory from previous infections aids in a faster and more effective immune response. The book is divided into 17 chapters, providing an overview of the immune system and its components, including its organs and cells. Chapters on the major histocompatibility complex, the complement system, hypersensitivity and tolerance, antibody diversity through DNA rearrangements, and autoimmune diseases are included in the book which further broadens the understanding of this very complex system of our body. Chapters on transplantation immunology and vaccines provide a perspective on the application of these immunological concepts and will be of great interest to readers.

Key features of the book:

- Simple, direct, and lucid language
- Comprehensive coverage of concepts for better understanding
- Well-labeled illustrations, flowcharts, and tables for enhanced learning
- Every chapter is followed up with a detailed summary and questionnaire
- A detailed glossary for users to know the right words
- Chapters contributed/reviewed by experienced experts in this field

The book provides broad, accessible, and up-to-date information about immunological perspectives to biotechnologists, biomedical scientists, biochemists, molecular biologists, and students from various streams of life sciences, including zoology, biotechnology, and microbiology, as well as instant access to a wealth of information.

An Interplay of Cellular and Molecular Components of Immunology

Edited by
Indrakant K. Singh
Poonam Sharma

CRC Press
Taylor & Francis Group
Boca Raton London New York

CRC Press is an imprint of the
Taylor & Francis Group, an **informa** business

First edition published 2023
by CRC Press
6000 Broken Sound Parkway NW, Suite 300, Boca Raton, FL 33487-2742

and by CRC Press
4 Park Square, Milton Park, Abingdon, Oxon, OX14 4RN

CRC Press is an imprint of Taylor & Francis Group, LLC

Library of Congress Cataloging-in-Publication Data
Names: Singh, Indrakant K., editor. | Sharma, Poonam, 1978- editor.
Title: An interplay of cellular and molecular components of immunology / edited by Indrakant K. Singh, Poonam Sharma.
Description: First edition. | Boca Raton : CRC Press, 2023. | Includes bibliographical references and index.
Identifiers: LCCN 2022025727 (print) | LCCN 2022025728 (ebook) | ISBN 9781032260730 (hardback) | ISBN 9781032260747 (paperback) | ISBN 9781003286424 (ebook)
Subjects: MESH: Immune System Phenomena–physiology | Immune System Diseases–physiopathology | Autoimmunity
Classification: LCC QR185.2 (print) | LCC QR185.2 (ebook) | NLM QW 504 | DDC 571.9/6–dc23/eng/20220830
LC record available at https://lccn.loc.gov/2022025727
LC ebook record available at https://lccn.loc.gov/2022025728

ISBN: 9781032260730 (hbk)
ISBN: 9781032260747 (pbk)
ISBN: 9781003286424 (ebk)

DOI: 10.1201/9781003286424

Typeset in Times
by codeMantra

Contents

Preface

Immunology is one of biology's most fascinating and challenging fields. It is a multidisciplinary field that draws students from zoology, biomedical science, biochemistry, microbiology, biotechnology, genetics, and medicine. Immunology deals with molecular and cellular components that constitute our immune system. Our immune system is incredibly complex and safeguards us from infection through various lines of defense. The relevance of the immune response in human health, disease prevention, and vaccines has become more apparent in the fight against the Covid-19 epidemic, which shows no signs of abating and is resulting in an increasingly terrible death toll. This textbook provides the reader with a holistic understanding of the molecular and cellular basis of the immune system and identification of its biological, clinical and therapeutic implications in a logical and easy-to-understand manner. This book comprises 17 chapters and begins with an overview of the immune system, including the organs and cells of immune system, before moving on to innate and adaptive immunity. Topics including the major histocompatibility complex, the complement system, hypersensitivity and tolerance, and autoimmunity are thoroughly covered, broadening the scope of the book. Furthermore, transplantation immunology, vaccines, and the principles of DNA rearrangements for antibody diversification are some of the book's other features, which will draw the interest of reader. The book also introduces various latest immunological techniques that are used in immunology. It provides biotechnologists, biomedical scientists, biochemists, molecular biologists, and students from various streams of zoology, biotechnology, and microbiology with comprehensive, accessible, and up-to-date information about immunological perspectives, as well as instant access to a wealth of data. Our heartfelt acknowledgement goes to Prof. Ashok K. Singh and Prof. Rup Lal, our mentors, for their blessing, motivation, and timely guidance as we explored the world of teaching and research. We are incredibly thankful to Prof. Rajiv Aggarwal, Principal, Deshbandhu College, University of Delhi, and Prof. Sangeeta Bhatia, Principal, Gargi College, University of Delhi, for providing overall support for our research and academic pursuits. We would also like to convey our gratitude to Almighty and Our Parents for their love, care, unconditional support, and blessings. Personal debts can never be adequately acknowledged. I, Dr. Indrakant Singh, owe an unquantifiable debt to my wife, Dr. Archana Singh, for her wholehearted support and my little angels, Saumya and Kimaya, for creating a beautiful ambiance. I owe a special thanks to my research team for their constant support. I would like to extend our sincere thanks to Dr. V.K. Kawatra, Dr. P K Singh, and Dr. Shalendra S Chauhan, for their encouragement and support. I, Dr. Poonam Sharma, am eternally grateful to my dear husband, Dr. Mandeep Dadhwal, and beloved daughters, Lavanya and Pragya, for their unwavering support. Last but not least, our sincere thanks to handling editors and publishers. We are optimistic that this book keeps pace with the latest knowledge and presents comprehensive information about the modern concepts of immunology.

Indrakant K Singh

Poonam Sharma

Contributors

Shailly Anand
Department of Zoology
Deen Dayal Upadhyaya College
University of Delhi
Delhi, India

Shanuja Beri
Department of Zoology
Kalindi College
University of Delhi
Delhi, India

Nidhi Garg
Department of Zoology
Shivaji College
University of Delhi
Delhi, India

Princy Hira
Department of Zoology
Maitreyi College
University of Delhi
Delhi, India

Monisha Khanna
Microbial Technology Laboratory
Department of Zoology
Acharya Narendra Dev College
University of Delhi
Delhi, India

Pooja Mittal
Molecular Biology Research Lab
Department of Zoology
Deshbandhu College
University of Delhi
Delhi, India

Garima Nagar
Molecular Biology Research Lab
Department of Zoology
Deshbandhu College
University of Delhi
Delhi, India

Gouri Nandi
Department of Zoology
Deshbandhu College
University of Delhi
Delhi, India

Neha Niharika
Department of Zoology
Daulat Ram College
University of Delhi
Delhi, India

Ritu Rai
Department of Zoology
Dyal Singh College
University of Delhi
Delhi, India

Archana Singh
Plant Protection & Human Health Research Lab.
Department of Botany
Hansraj College
University of Delhi
Delhi, India

Jyoti Singh
Department of Zoology
Maitreyi College
University of Delhi
Delhi, India

Jyoti Singh
Department of Zoology
Daulat Ram College
University of Delhi
Delhi, India

Priya Singh
Department of Zoology
Maitreyi College
University of Delhi
Delhi, India

Renu Solanki
Department of Zoology
Deen Dayal Upadhyaya College
University of Delhi
Delhi, India

Charu Tripathi
Department of Zoology
CMP College
University of Allahabad
Prayagraj, India

Vatsala Tripathi
Department of Zoology
Dyal Singh College
University of Delhi
Delhi, India

Meena Yadav
Department of Zoology
Maitreyi College
University of Delhi
Delhi, India

Sunita K. Yadav
Department of Zoology
Daulat Ram College
University of Delhi
Delhi, India

Glossary

CHAPTER 1

Active Immunity: Immune system of one's own body produces antibodies against pathogen.

Acute Phase Protein: Serum proteins like interferons that are formed in response to an inflammation.

Adaptive Immunity: It is acquired after birth and provides second line of defense against broad spectrum of pathogens due to high diversity, specificity and immunological memory.

Allergy: A type of hypersensitivity reaction caused by allergens and characterized by fever, asthma and anaphylaxis.

Allograft: Transplantation of any tissue or organ between two different individuals of same species.

Antibody: Immunoglobulin molecule expressed on membrane on B cells acting as B cell receptor and secreted by effector plasma cells in response to an antigenic challenge. Two heavy and two light chains with constant and variable regions form the structure.

Antigen Presenting Cell (APC): A cell that processes and presents foreign antigens in complex with MHC class II molecules to T cells.

Antigen: A foreign molecule that binds to the antibody or T cell receptor for immune response.

Attenuation: Pathogen with decreased virulence so that it does not cause the disease but elicits an immune response.

Autograft: Transplantation from one part of the body to other in the same individual.

Autoimmunity: Improper immune response against self-antigens.

Bone Marrow: Connective tissue present in medullary cavities of bones.

Cell-Mediated Immunity: Immune response mediated by T cells against intracellular pathogens.

Complement proteins: Serum proteins that are activated in response to foreign molecules ad carry out various effector functions.

Effector Cell: An immune cell capable to mediating an immune function. E.g. plasma cell or CTLs.

Graft versus host disease (GvHD): A condition where graft cells start attacking the recipient's body by recognizing it as a foreign component.

Haematopoiesis: Formation of all blood cells from a pluripotent stem cell in the bone marrow.

Humoral Immunity: Immune response mediated by B cells by producing antibodies against extracellular pathogens.

Hypersensitivity: Undesirable and inflated immune response produced by immune system leading to allergies and autoimmune diseases.

Immunity: A protected state of body against diseases.

Immunodeficiency: Compromised or deficient state of immune system leading to reduce ability to fight infections and cancer suppression. Caused by defect in the humoral response, phagocytosis or cell-mediated responses.

Innate Immunity: Fist line of defense present since the birth against common infections.

Lipopolysaccharide (LPS): An oligomer of lipid and polysaccharide composed of O-antigen, that constitutes the endotoxins and lipoglycans, found in outer membrane of Gram-negative bacteria.

Lymphocyte: A mononuclear white blood cell (leukocyte) that mediates humoral or cell-mediated immunity. Include natural killer cells, B cells and T cells.

Memory cells: Long-lived lymphocyte generated following encounters with a particular antigen and capable of responding on its reintroduction, readily stimulated than naïve lymphocytes and mediate a secondary response.

Naive: Un-primed and virgin mature B and T cells that have not encountered any antigen.

Opsonization: A process by which an antibody or complement protein binds or coats the antigen and makes it more susceptible for phagocytosis by macrophages.

PAMPs: Pathogen-associated molecular patterns are the cell surface markers present commonly on pathogenic cell membrane that can be identified as non-self-molecules by body.

Passive Immunity: Preformed antibodies are administered from a recovered patient to a healthy individual to gain immunity against a disease.

Pattern recognition receptors (PRRs): Innate immune system receptors that recognize motifs or molecular patterns present on pathogens but absent in host.

Phagocytosis: The ingestion or cellular uptake of particulate material by engulfment.

Pluripotent: A stem cell that can differentiate into multiple lineages of cells. For example, hematopoietic stem cell.

Primary response: Initial immune response following first exposure to the pathogen/antigen. Reinfection or previously encountered antigen.

Secondary response: Rapid and heightened immune response against a pathogen at time of repeated infected.

Stem cell: An undifferentiated cell of a multicellular organism capable of giving rise to indefinitely more cells of the same type, and from which certain other kinds of cell arise by differentiation.

TLR: Toll-like receptors are present on the membrane of innate immune cells that are capable of recognizing common molecular markers on pathogens.

Vaccination: Intentional introduction of harmless or less harmful pathogen to produce immunity against a disease.

Vaccine: Immunogenic material use to induce immunity against pathogenic organisms.

Xenograft: Tissue or graft transplanted from a donor of a different species to the recipient of another species.

CHAPTER 2

Acute phase proteins (APP): Acute phase proteins are the proteins primarily synthesized by hepatocytes during acute phase response. Example: C-reactive protein.

Acute phase response (APR): APR is a core component of innate immune response. It is a complex and systemic early response induced by trauma, infection, stress, neoplasia and inflammation.

Anergy: When an antigen is presented to a T cell and there is no costimulation, it results in tolerance to that antigen. This is known as anergy.

Antimicrobial peptides: Peptides secreted by cells that can kill microbes.

Antiserum: Serum containing antibodies, obtained from pathogen exposed individual.

Autoimmunity: When the immune system breaks down and it starts attacking the self-cells, the condition results in autoimmunity.

Basophils: Basophils are non-phagocytic cells and contain large granules which are filled with basophilic proteins. When circulating antibodies bind to basophils, they release the contents of their granules by a process called **degranulation**. Together with eosinophils, basophils respond to parasitic infections.

Cell-adhesion molecules (CAMs): Cell-adhesion molecules are a group of transmembrane proteins that are involved in cell–cell interactions. In immune system, CAMs are especially involved in helping the leucocytes to adhere to vascular endothelium before extravasation or diapedesis.

Cell-mediated immunity: The immunity mediated by cell-cell interaction is known as cell-mediated immunity. T lymphocytes are the primary cells involved in cell-mediated immunity.

Cellular innate response: A response triggered due to interaction between cell surface or intracellular receptors with conserved molecular components on the cell surface of the pathogen.

Central tolerance: In central tolerance, clones of T- and B-cells that recognize self-antigens are destroyed before maturation in bone marrow and thymus. Central tolerance is not completely successful.

C-reactive protein (CRP): It belongs to family of pentameric proteins called **pentraxins (PTX)**, which bind to ligands in a calcium-dependent reaction and is an important acute phase response protein.

Crohn's disease: It is an autoimmune disease. There is an attack on intestinal tissues that destroys the epithelial layer in the gut resulting in poor absorption of food.

C-type lectin receptor (CLR): CLRs are a type of pattern recognition receptors which activate innate and inflammatory responses.

Cytotoxic T lymphocytes (CTLs): CTLs are effector cells and they eliminate the cells which express MHC class I-peptide complex such as virus-infected cells, tumor cells or foreign graft cells.

Damage-associated molecular patterns (DAMPs): The components are present on the dead and damaged cells and are recognized by PRRs are called DAMPs.

Defensins and Cathelicidins: Antimicrobial peptides that are continuously secreted by epithelia in many tissues in humans and also stored in neutrophil granules.

Dendritic cells (DCs): DCs are non-lymphoid cells that can recognize an antigen as well as act as antigen-presenting cell.

Eosinophils: Eosinophils are motile phagocytes that can migrate from blood into tissue spaces.

Extravasation: Extravasation or diapedesis is a process of leakage of blood, lymph or other substances like drugs from a blood vessel to the surrounding tissue. It is a common phenomenon during the inflammatory reaction.

Goodpasture's syndrome: It is a type II hypersensitivity reaction with autoantibodies produced against collagen in glomeruli and pulmonary alveoli. The autoantibodies can also cause local activation of complement proteins, aggravating the tissue damage.

Granulocytes: Polymorphonuclear leucocytes, are diverse and have characteristic granules in the cytoplasm.

Humoral immunity: The immunity mediated through the blood (in the form of antibodies) is known as humoral immunity. B lymphocytes are the primary cells involved in humoral immunity.

Immune dysfunction: Immune dysfunction is a condition when the immune system of a person is compromised and results in failures and subsequent diseases.

Immunoglobulin-superfamily Cell Adhesion Molecules (ICAM): ICAMs are members of immunoglobulin superfamily. They play important roles in inflammation, immunity and intracellular signaling pathways. They are ligands for beta2 integrin molecules present on leukocytes. ICAM family has five members, i.e. ICAM-1 to ICAM-5.

Immunomodulatory molecules: The molecules that can stimulate or inhibit T cells are called immunomodulatory molecules. Example: CTLA-4.

Inducible nitric oxide synthase (iNOS or NOS_2): This enzyme leads to the production of reactive nitrogen species (RNS) and it is activated after binding of microbial components to various PRRs.

Inflammation: A process when the infected site becomes swollen and shows other signs like redness, heat and pain.

Innate immunity: Primitive form of protection which is spontaneous or in-built.

Langerhans cells: Langerhans cells are skin-resident and specialized dendritic cells that ingest antigen by phagocytosis or endocytosis and upon maturation in lymph nodes they present the processed antigens to cells like naïve T cells.

Leucine-rich repeats (LRRs): These are repeating segments of 24–29 amino acids having the sequence LxxLxLxx (L- leucine and x is any other amino acid) in the exterior domain of Toll-like receptors.

Macrophages: Monocytes can migrate into tissues as a result of infection and may get differentiated into tissue-specific **macrophages.** During innate immune response, macrophages may undergo several significant changes after encountering pathogens. Such macrophages are called as **inflammatory macrophages** and they can act as phagocytes.

Magainins: Antimicrobial peptides secreted from skin glands in frog. They have strong activity against bacteria, yeast and protozoans.

Mast cells: The precursors of mast cells are released from the bone marrow into the blood and when they leave blood, they mature into mast cells. Mast cells are involved in various allergic reactions.

Multiple schlerosis: The disease is characterized by production of self-reactive T cells that form inflammatory lesions in the myelin sheath of the nerve cells in brain and spinal cord resulting in numerous neurologic dysfunctions including loss of vision, paralysis or numbness in limbs.

Myasthenia gravis: In myasthenia gravis autoantibodies are formed against the α chain of nicotinic acetylcholine receptor and thus block the neuromuscular transmission. People suffering from myasthenia gravis become progressively weaker which may prove to be fatal.

Myeloid differentiation factor 88 (MyD88) and TIR domain containing adaptor-inducing IFN-β factor (TRIF): These are initial adaptor proteins for the downstream signaling of TLRs.

Natural killer cells (NK cells): NK cells are a population of lymphocytes which are activated during innate response and they release effector proteins from the preformed secretory granules that kill the altered cells by inducing apoptosis.

Neutrophils: They are the first major cells that respond to an infection and are recruited at the site of infection by chemokines secreted by other cells at the site of infection that have interacted with pathogens.

Nod-like receptors or NLRs: NLRs are a large family of proteins found in the cytoplasm. NLRs are activated by intracellular PAMPs, DAMPs or other harmful substances. They activate innate as well as inflammatory responses.

Non-professional antigen-presenting cells: The cells that can present the antigen at certain times only are known as non-professional antigen-presenting cells. They can be induced to express class II MHC or a costimulatory signal. They function only for a short duration such as during a sustained inflammatory response. Example: fibroblasts, glial cells

Opsonins: The proteins that enhance the process of phagocytosis.

Pattern associated molecular patterns (PAMPs): PAMPs are the conserved molecular components that are usually present in several copies on the surface of pathogens like bacteria, fungi, parasite or virus.

Pattern recognition receptors (PRRs): The receptors that recognize molecular patterns on microbes like PAMPs and DAMPs are called pattern recognition receptors.

Peripheral tolerance: The lymphocytes that recognize self-antigens and escape destruction during central tolerance are further destroyed by additional mechanisms that together constitute the peripheral tolerance. The peripheral tolerance also generates lymphocytes that inhibit immune reaction against self-cells and tissues.

Phagocytosis: It is a process by which specialized cells take up the bacteria or other material from the surroundings.

Phagosomes: Phagosomes are the endosomes formed during phagocytosis.

Primary immune response: The kind of immune response produced when an antigen enters the body for the first time.

Primary Immunodeficiency: Primary immunodeficiency (PID) is caused due to genetic factors. Example: X-linked agammaglobulinemia (XLA), Autoimmune Lymphoproliferative Syndrome (ALS) etc.

Professional antigen-presenting cells: The cells that are specialized in presenting the antigen to the T cells or B cells. Example: dendritic cells, macrophage.

Respiratory burst: It is a process by which the intake of oxygen by phagocytes increases many fold to produce ROS by NADPH oxidase.

Rheumatoid arthritis: Rheumatoid arthritis is an example of secondary immunodeficiency. It is caused by self-reactive antibodies called **rheumatoid factors,** which cause chronic inflammation of the joints of hands, feet, arms and legs. Other than the joints, hematologic, cardiovascular and respiratory systems are affected.

Secondary immune response: The kind of immune response produced when an antigen enters the body for the second or subsequent time.

Secondary Immunodeficiency: Secondary immunodeficiency (SID) is caused due to environmental factors such as chemical, physical or biological agents.

Sepsis: Sepsis is a condition which is a systemic response to infection with symptoms like fever, increased heartbeat and respiratory rate, low blood pressure and inappropriate functioning of organs due to defects in circulatory system as a result of inappropriate immune response.

Systemic lupus erythematosis (SLE): The individuals having SLE produce auto-antibodies (IgG) for several antigens like DNA, histones, RBCs, platelets, leucocytes, clotting factors and nucleoprotein particles. The symptoms of SLE include fever, weakness, arthritis, skin rashes as well as kidney dysfunction.

T regulatory (T_{REG}) cells: The T cells that down-regulate the immune response are called as T_{REG} cells. They are active in secondary lymphoid tissues and at the site of inflammation.

The Retinoic acid-inducible gene-I-like receptors (RLRs): These are soluble PRRs which are found in cytoplasm of cells and play critical role in cells infected with viruses. They are *Caspase* Recruitment Domains (CARD)-containing RNA helicases which recognize viral RNAs.

Tolerogens: Tolerogens are antigens that induce tolerance.

Wegener's granulomatosis: An autoimmune disease characterized by severe necrotizing vasculitis

CHAPTER 3

Anergy: When an antigen is presented to a T cell and there is no costimulation, it results in tolerance to that antigen. This is known as anergy.

Antiserum: Serum-containing antibodies, obtained from pathogen-exposed individual.

Autoimmunity: When the immune system breaks down and it starts attacking the self-cells, the condition results in autoimmunity.

Cell-mediated immunity: The immunity mediated by cell–cell interaction is known as cell-mediated immunity. T lymphocytes are the primary cells involved in cell-mediated immunity.

Central tolerance: In central tolerance, clones of T- and B-cells that recognize self-antigens are destroyed before maturation in bone marrow and thymus. Central tolerance is not completely successful.

Crohn's disease: It is an autoimmune disease. There is an attack on intestinal tissues that destroys the epithelial layer in the gut resulting in poor absorption of food.

Cytotoxic T lymphocytes (CTLs): CTLs are effector cells and they eliminate the cells which express MHC class I-peptide complex such as virus-infected cells, tumor cells or foreign graft cells.

Dendritic cells (DCs): DCs are non-lymphoid cells that can recognize an antigen as well as act as antigen-presenting cell.

Goodpasture's syndrome: It is a type II hypersensitivity reaction with autoantibodies produced against collagen in glomeruli and pulmonary alveoli. The autoantibodies can also cause local activation of complement proteins, aggravating the tissue damage.

Humoral immunity: The immunity mediated through the blood (in the form of antibodies) is known as humoral immunity. B lymphocytes are the primary cells involved in humoral immunity.

Immune dysfunction: Immune dysfunction is a condition when the immune system of a person is compromised and results in failures and subsequent diseases.

Immunomodulatory molecules: The molecules that can stimulate or inhibit T-cells are called immunomodulatory molecules. Example: CTLA-4.

Langerhans cells: Langerhans cells are skin-resident and specialized dendritic cells that ingest antigen by phagocytosis or endocytosis and upon maturation in lymph nodes they present the processed antigens to cells like naïve T cells.

Multiple schlerosis: The disease is characterized by production of self-reactive T cells that form inflammatory lesions in the myelin sheath of the nerve cells in brain and spinal cord resulting in numerous neurologic dysfunctions including loss of vision, paralysis or numbness in limbs.

Myasthenia gravis: In myasthenia gravis autoantibodies are formed against the α chain of nicotinic acetylcholine receptor and thus block the neuromuscular transmission. People suffering from myasthenia gravis become progressively weaker which may prove to be fatal.

Natural killer cells (NK cells): NK cells are a population of lymphocytes which are activated during innate response and they release effector proteins from the preformed secretory granules that kill the altered cells by inducing apoptosis.

Non-professional antigen-presenting cells: The cells that can present the antigen at certain times only are known as non-professional antigen-presenting cells. They can be induced to express class II MHC or a costimulatory signal. They function only for a short duration such as during a sustained inflammatory response. Example: fibroblasts, glial cells.

Opsonins: The proteins that enhance the process of phagocytosis.

Pattern recognition receptors (PRRs): The receptors that recognize molecular patterns on microbes like PAMPs and DAMPs are called pattern recognition receptors.

Peripheral tolerance: The lymphocytes that recognize self-antigens and escape destruction during central tolerance are further destroyed by additional mechanisms that together constitute the peripheral tolerance. The peripheral tolerance also generates lymphocytes that inhibit immune-reaction against self-cells and tissues.

Phagocytosis: It is a process by which specialized cells take up the bacteria or other material from the surroundings.

Phagosomes: Phagosomes are the endosomes formed during phagocytosis.

Primary immune response: The kind of immune response produced when an antigen enters the body for the first time.

Primary Immunodeficiency: Primary immunodeficiency (PID) is caused due to genetic factors. Example: X-linked agammaglobulinemia (XLA), Autoimmune Lymphoproliferative Syndrome (ALS) etc.

Professional antigen-presenting cells: The cells that are specialized in presenting the antigen to the T cells or B cells. Example: dendritic cells, macrophage.

Rheumatoid arthritis: Rheumatoid arthritis is an example of secondary immunodeficiency. It is caused by self-reactive antibodies called **rheumatoid factors,** which cause chronic inflammation of the joints of hands, feet, arms and legs. Other than the joints, hematologic, cardiovascular and respiratory systems are affected.

Secondary immune response: The kind of immune response produced when an antigen enters the body for the second or subsequent time.

Secondary Immunodeficiency: Secondary immunodeficiency (SID) is caused due to environmental factors such as chemical, physical or biological agents.

Systemic lupus erythematosis (SLE): The individuals having SLE produce auto-antibodies (IgG) for several antigens like DNA, histones, RBCs, platelets, leucocytes, clotting factors and nucleoprotein particles. The symptoms of SLE include fever, weakness, arthritis, skin rashes as well as kidney dysfunction.

T regulatory (T_{REG}) cells: The T cells that downregulate the immune response are called as T_{REG} cells. They are active in secondary lymphoid tissues and at the site of inflammation.

Tolerogens: Tolerogens are antigens that induce tolerance.

Wegener's granulomatosis: An autoimmune disease characterized by severe necrotizing vasculitis.

CHAPTER 4

Antigen: Any substance (usually foreign or recognized as foreign) that binds precisely to an antibody or T-cell receptor; often used as alternative word for **immunogen**.

Antigenicity: Ability of a foreign particle to bind to the products of the immune response.

Autoantigens or autologous antigens: A type of self-antigens that are endogenous. These are tolerated otherwise but become immunogenic upon certain changes in host.

Allogenic: Denotes members of same species that differ genetically for certain gene locus.

Auto-immune disease: An abnormal immune response against self-molecules of the host resulting in inflammation and some disorders.

Conformational epitope: Portion of a protein that is composed of amino acids that are close together in the three-dimensional structure making it immunogenic but may not be near each other in linear amino acid sequence impeding immunogenicity of molecule. They are also termed as conformational determinant and/or non-sequential epitope.

Epitope: The portion of the antigen that binds with antibody or TCR-MHC complex; often used as synonym of antigenic determinant.

Haptens: Small molecules that are antigenic but they themselves are not capable of inducing a specific immune response as they lack immunogenicity.

Immunogen: Any substance that can elicit an immune response upon it's binding to antibody or T-cell receptor and results in generation of monoclonal cell population specific for it. Antigens (*e.g.* haptens) are not always immunogens but all immunogens are certainly antigens.

Immunogenicity: Ability to induce the immune system to respond to the binding of certain foreign substance.

Neoantigenic epitopes: Newly formed antigens that have not been previously recognised by the immune system.

Superantigens: Class of antigens (viral or bacterial proteins) that bind simultaneously to TCR and class II MHC molecules, and it result in excessive activation of the immune system (Non-specific activation of T-cells resulting in polyclonal T-lymphocyte activation and massive cytokine release. Potent T-cell mitogen and can cause food poisoning and other disorders.

Tolerance: Immunologic unresponsiveness towards some antigens or group of antigens. Usually an organism is tolerant/unresponsive to self-antigens.

CHAPTER 5

Agretope- The portion of the processed antigen which interacts with class I or class II MHC molecule.

Aliquots: Dividing a total volume of certain components into the required number of tubes/ vials.

Antigenic determinant: The portion of the antigen that is recognized by the antibody molecule.

Antiserum: It is that serum that has antibodies for the desired antigen the serum containing antibodies for the desired antigen is called antiserum.

Beta sheets: One of the secondary structural organizations found in proteins.

Complement: The set of proteins synthesized by the liver or within plasma and are involved in opsonization, lysis and neutralization of pathogen.

Dimer: Two units joined to each other.

Domain: Homologous units present in antibody light and heavy chains, the number of which may vary among classes.

Electrophoresis: The process of migration of charged molecules under the effect of electric current.

Epitope: Immunologically discreet sites present on an antigen molecule that binds with the antibody.

Fab: Fragment which has antigen-binding ability.

Fc: Fragment which can be crystallized.

Hybridoma: A hybrid cell produced by fusion of cancerous cell with antibody-secreting B-cell, used for the production of monoclonal antibody.

Immunoglobulin family: One of the divisions or families where different proteins have been grouped according to presence of a few common structural features.

Immunization: Giving injection or inoculating a particular animal with a certain antigen.

J- chain: Polypeptide chain present in Ig A and Ig M which facilitates joining of two or more chains to form a dimer or pentamer.

Monoclonal antibody: The antibody produced by a single type of hybrid clone and having specificity for a single epitope in a culture, used as immunodiagnostic tools.

Opsonins: The molecules (antibody or C3b) which bind to the antigen and render it more susceptible to phagocytosis by macrophages through the process of opsonization.

Paratope: Region of antibody which interacts with an epitope of an antigen.

Passive immunization: It is the acquisition of immunity by receipt of preformed antibodies rather than by active production of antibodies after exposure to antigen.

Serum: It is pale, clear fluid collected when the blood of an animal is subjected to coagulation.

Titer: It is the reciprocal of the last dilution of an antiserum that can mediate antibody interaction such as precipitation etc.

T-Lymphocytes: The lymphocytes getting matured in the thymus are designated as T-lymphocytes.

CHAPTER 6

Affinity: The binding strength of a single receptor to its legend i.e. sum total of interactions involved between single epitope and single antibody binding site of an antibody.

Avidity: The functional binding strength between two molecules that is representative of interactive forces on all the binding sites.

Cross-reactivity: It is the ability of an antibody or T-cell receptor to react with two or more antigens that possess a common epitope.

ELISA: Enzyme-linked immunosorbent assay, a technique to quantify antigen or antibody-based enzyme-catalyzed reaction on an adsorbed medium.

Equilibrium dialysis: A technique that is used to determine the valency and affinity of antibody molecules.

Fluorochrome: It is a fluorescent dye that can be conjugated to a protein or an antibody, e.g., FITC, fluorescein isothiocyanate and rhodamine.

Hemagglutinin: An antibody that causes clumping of RBCs.

Immunoadsorption: Process or technique where antigen or antibody is adsorbed on a suitable solid medium to which complementary antibody or antigen can bind.

Immunofluorescence: The technique to visualize antigen/antibody through fluorescent microscope wherein cells or tissues are stained with fluorescent antibody.

Ligand: The molecule that can be recognized by the receptor.

ODD: Ouchterlony double immunodiffusion, is a method for detection of purity of antigen or antibody. It involves diffusion of both antigen and antibody in a semi-solid medium that can be visualized in the zone of equivalence.

Precipitin: The antibodies that precipitate or aggregate soluble antigen to give visible arcs of the antigen-antibody complex in form of precipitates.

RIA: Radioimmunoassay is a technique to quantify analyte (antigen or antibody) using radiolabeled antigen/antibody, the basis being competitive binding between labelled and unlabeled antigen/antibody.

Valency: Valency is equivalent to the number of binding sites available on an antibody.

CHAPTER 7

Allograft: It is the transplantation of either cells, or tissues, or even organs from a genetically non-identical individual to another of the same species. The transplanted cells, tissues, or organs are called as allograft.

Ankylosing spondylitis: It is a type of arthritis characterized by inflammation of the vertebral column with the disease peaking at around the age of 20 years. The reason is the presence of haplotype B27.

Antibody: These are big sized Y-shaped immunoglobulins, a class of proteins synthesized by plasma cells which are components of the immune system that help in counteract pathogens like bacteria and viruses.

Antigen: It is defined as any substance which upon entering the body is recognized as foreign and elicits the antibody production.

Celiac disease: It is a disease characterized by individuals showing intolerance to gluten thus intestinal damage takes place and leads to abdominal bloating, chronic diarrhoea and abdominal pain. It is associated with the presence of DR3 marker and a gluten-free diet is a must to avoid trouble.

Graft versus host disease (GVHD): It is defined as a condition where the allograft is unsuccessful because the tissue transplanted recognizes the body of the recipient as foreign and elicit an immune response. It is of two types Acute graft versus host disease (aGvHD) and Chronic graft versus host disease (cGvHD).

Graft versus leukaemia (GVL): It is defined as the ability of transplanted tissue cells like bone marrow cells when transplanted in an individual suffering from leukaemia to kill leukaemia cells. Therefore, it is used as a therapy to prevent a relapse.

HLA typing: It is defined as the identification of the HLA antigens of both donor and recipients for the purpose of transplantation to avoid graft rejection.

Immunodominance: Immunodominance is defined as the ability of some H antigens to elicit a stronger response from the T cells as compared to other H antigens even though they are present on the same cell.

Minor histocompatibility antigens: These are defined as proteins other than the MHC that are capable of acting as alloantigens capable of eliciting a response from the immune system. They play a significant role in hematopoietic cell transplant (HCT) with subdued effects.

Major Histocompatibility Complex (MHC): The MHC is actually a tightly linked cluster of genes encoding for a set of glycoproteins that are expressed on cell surface and behave as antigens upon introduction into the body of another organism. It is responsible for graft rejection upon transplantation in another individual belonging even to the same species.

Reiter's disease: It is an auto immune disorder affecting the genital mucosa and is associated with the marker HLA B27 and. These patients show receptors foe pathogens which cause diseases.

Transplantation: It is defined as a surgical procedure which requires the removal of cells, tissues or organs from an individual of same or different genus or species for being placed in another individual for improving his/her health status.

CHAPTER 8

Agonist: Molecule that enhances the activity of other molecules.

Antagonist: Molecule that interferes and inhibits the physiological effect of other molecules.

Biomarker: A naturally occurring molecule, gene, or anything that can be used as an indicator of a particular disease state or some other physiological state of an organism.

Biphasic: Having two phases.

Crohn's disease: A chronic inflammatory disease of the intestines, especially the colon and ileum, associated with ulcers and fistulae.

Hematopoiesis: It is a process by which all blood cells are formed, developed and differentiate. It occurs inside bone marrow in humans (adult).

Lymphocytes: White blood cells that protects body against disease.

Macrophage: A large phagocytic cell found in stationary form in the tissues or as a mobile white blood cell, especially at sites of infection.

Mast cells: Type of granulocyte that are present in connective tissue that mediate inflammatory response.

Neutropenia: The presence of abnormally few neutrophils in the blood, leading to increased susceptibility to infection.

Receptor: In immunological term, these are cell membrane glycoproteins which bind to cytokines, receive their signal and transduce the signaling pathway.

Rheumatoid arthritis: A chronic progressive disease causing inflammation in the joints and resulting in painful deformity and immobility, especially in the fingers, wrists, feet, and ankles.

Therapeutic: The branch of medicine concerned with the treatment of disease and the action of remedial agents or a treatment, therapy, or drug.

CHAPTER 9

Anaphylatoxin: Small cationic peptides generated by the complement cascade, that are capable of binding to the receptors on the basophiles and the mast cells causing release of the histamine and many other pharmacologically active mediators which causes inflammatory responses.

Antigen sequestration: One of the mechanisms of tolerance in which a self-antigen is sequestered and never encounters the immune system in normal conditions, it is protected by default from the immune system. E.g., blood-brain and blood-testes barriers.

C1 inhibitor (C1Inh): It is a heavily glycosylated plasma protein which causes regulation of C1 protein of classical complement pathway. C1Inh is a serine protease inhibitor.

Complement cascade system: The function of the serum that completes the action of the antibody by forming membrane attack complex to cause the lysis of the pathogen.

Complement fixation: Activation of classical complement cascade pathway due to binding of a secreted pentameric IgM to a red blood cell.

Complement receptor 1 (CR1): Receptors present on erythrocyte surfaces which binds to immune complexes and clears them from the blood.

Decay Accelerating Factor (DAF): A C3 convertase regulatory glycoprotein. DAF accelerates the decay of C3 convertase. Also known as CD55.

Hereditary Angioneurotic Edema (HANE): An autosomal dominant autoimmune disease caused due to defective C1Inh regulatory proteins. The patients suffer from extensive swelling of laryngeal, facial and intestinal mucosa, abdominal pain and trachea choking.

Innocent –bystander lysis: Killing of healthy host cells present nearby pathogenic surfaces. The lysis of host cells is caused by binding of C5b67 complex formed during MAC assembly.

Lectin: These proteins functions to recognize and bind the specific carbohydrate residues. Mannose binding lectins play essential role in activation of complement cascade.

Leukocyte Adhesion Deficiency (LAD): An autoimmune disease caused by CR3 and CR4 deficiencies. It leads to frequent pyogenic infections in patients.

Membrane attack complex (MAC): A macromolecular cytocidal structure formed in last few steps of complement cascade, due to interaction of proteins namely C5b, C6, C7, C8, and C9 in a sequential manner. MAC functions to cause lysis of pathogenic cells.

Opsonisation: The increase in phagocytosis rate due to binding of complement cascade activation generated molecules like C, C and iC(opsonins) to antigens and immune complexes.

Paraoxysmal Nocturnal Haemoglobulinuria (PNH): An autoimmune disorder caused by deficiency of regulatory protein CD59 or DAF. The patients suffer from red blood cell lysis by complement system.

Regulators of Complement Activation gene cluster (RCA): Gene cluster encoding the C3 convertase regulatory proteins in humans. It is located at chromosome number 1.

Short Consensus Repeats (SCRs): Characteristic repeated motifs spanning about 60 amino acids, present in a family of C3 regulatory proteins.

Systemic Lupus Erythematosus (SLE): An autoimmune disorder caused due to formation of immune complexes in large amounts in the body. The effected individuals suffer from damage caused by complement mediated tissue damage and hypersensitivity.

Zymogens: The inactive protein precursors of the functional complement proteins. Zymogens are activated by proteolytic removal of the inhibitory fragment.

CHAPTER 10

Auto-antibodies: Antibodies that recognize self-cells or self-components with more than a low threshold.

Autoimmune hemolytic anemia: Auto-antibodies formed against a self-RBC antigen, cause RBC agglutination and complement mediated lysis.

Autoimmunity: The failure of mechanisms in place to check the proliferation and activity of self-reactive immune components, leading to an immune attack on the host itself.

Central Tolerance: Mechanisms for the removal of self-reactive T and B cells that take place in the primary lymphoid organs, namely, thymus and bone marrow respectively comprise central tolerance.

CTLA-4: CTLA-4 or CD152 is an inhibitory receptor expressed on activated T cells. It interacts with B7 molecule on APCs and inhibits further proliferation of T cells.

Cytotoxic T lymphocyte (CTL): An effector T cell (usually $CD8^+$) that has been activated after exposure of naïve T_C cells to antigenic peptide presented on MHC molecules. CTLs bring about direct killing of cells by means of perforins/granzymes and by activating the Fas/FasL pathway.

F_C receptor: Receptor present on macrophages, NK cells, mast cells, eosinophils that can bind to the F_C portion of immunoglobins.

Goodpasture's syndrome: In this syndrome, auto-antibodies are formed against basement membrane antigens. Kidney glomeruli and lung alveoli are the sites whose basement membranes are targeted, causing damage and deterioration.

Graves' disease: Auto-antibodies are produced which bind to TSH receptors on thyroid cells, stimulating them to produce thyroid hormones. This continued and unregulated stimulation of thyroid cells leads to overproduction of thyroid hormones, causing hyperthyroidism.

Hashimoto's thyroiditis: Auto-antibodies and sensitized T_H1 lymphocytes are directed against thyroid gland proteins, thyroglobulin and thyroid peroxidase, resulting in hypothyroidism.

IDDM (Insulin-dependent diabetes mellitus): An autoimmune disease in which activated CTLs and/or auto-antibodies targeting the β cells of the Islets of Langerhans damage insulin producing cells, leading to rise in blood-glucose levels.

Multiple sclerosis: In multiple sclerosis, activated T cells destroy the myelin sheath of nerves. As myelin sheath of nerve fibers is responsible for its insulation, therefore the insulation is lost leading to neurologic disorders.

Myasthenia Gravis: Myasthenia gravis results due to the formation of antagonistic antibodies that bind to acetylcholine receptors which are present on the motor end plate of muscles. These antibodies, when bound to acetylcholine receptors, block the binding of acetylcholine and thus inhibit the stimulation of skeletal muscles when needed. Additionally, complement mediated lysis leads to deterioration of the membrane and receptors, causing weakening of skeletal muscles.

Negative selection: Lymphocytes having self-reactive receptors are destroyed in the bone marrow and thymus. It involves clonal deletion of self-reactive clones by apoptosis.

Organ-specific autoimmune diseases: In these diseases, cell-mediated or humoral immune response is directed against antigens of a single organ, affecting the normal functioning of that organ and causing physical damage to the organ.

Peripheral tolerance: The process occurring in peripheral lymph organs, responsible for rendering anergic or destroying self-reactive lymphocytes escaping central tolerance.

Pernicious anemia: Auto-antibodies are formed against intrinsic factor, a protein present in gastric parietal cells responsible for the absorption of vitamin B12 which is required for hematopoiesis.

Rheumatoid Arthritis: Auto IgM antibodies (rheumatoid factor) are generated against the Fc region of IgG antibodies, forming IgM-IgG complexes which get deposited in joints causing chronic inflammation of joints.

Self-reactive T cells: T cells bearing TCRs that recognize self-cells or self-components with more than a low threshold.

Systemic autoimmunity: In systemic autoimmune diseases, autoimmunity targets not only a single antigen or organ, but multiple antigens present throughout many organs.

Systemic lupus erythematosus: A systemic autoimmune disease in which a number of tissue antigens like red blood cells, platelets, DNA, histones, clotting factors, leucocytes and many others are targeted by auto-antibodies leading to different kinds of responses.

Tolerance: Mechanisms that ensure continuous removal of self-reactive T and B cells and thus prevent auto-immune responses are known as tolerance.

Tolerogen: An antigen which fails to evoke an immune response such that the host is tolerant to its presence is known as a tolerogen.

T_{reg} cells: T_{reg} (T regulatory) cells are $CD4^+$ T cells, expressing IL-2R α chain (CD25) on their surface. They downregulate autoimmune reactions in secondary lymphoid organs.

CHAPTER 11

Allergen: An allergen is any substance that can cause allergic reactions.

Anaphylaxis: *Anaphylaxis* is a severe, life-threatening allergic reaction to an allergen.

Granuloma: It is an area of inflammation formed by numerous macrophages.

Hypersensitivity: It refers to exaggerated response of immune system resulting in allergies and autoimmune disorders.

Prausnitz–Küstner test: PK test is a sensitive immunological test used to check allergic reactions to a specific allergen.

Tuberculin Reaction: It is a skin test to detect tuberculosis infection.

Wheal and flare reaction: It refers to the swelling and reddening of the skin.

CHAPTER 12

Acute graft rejection: Graft rejection in first six months after the transplant.

Allograft: Grafting of tissue for transplant in one individual from genetically different individual of the same species.

Autograft: Grafting of tissue from one part of the body to another in the same individual.
Chronic graft rejection: Graft rejection beyond six months of transplant procedure.
Donors: Individual from whom the tissue or organ is taken for transplant into another individual.
Graft: Piece of a living tissue (sometime organ) that is transplanted surgically.
Graft rejection: Immune response against the transplanted tissue/organ leading to its necrosis and failure.
Graft versus host disease (GvHD): It is defined as a condition where the allograft is unsuccessful because the tissue transplanted recognizes the body of the recipient as foreign and elicit an immune response. It is of two types Acute graft versus host disease (aGvHD) and Chronic graft versus host disease (cGvHD).
Hyperacute rejection: Exaggerated and quick graft rejection with minutes or hours of transplant.
Immunosuppression: Reducing the immune system efficacy and activity.
Immunosuppressive drugs: Pharmaceutical agents used to suppress the immune system function and activity.
Isograft: Graft between identical twins similar genetically.
Major Histocompatibility Complex (MHC): The MHC is actually a tightly linked cluster of genes encoding for a set of glycoproteins that are expressed on cell surface and behave as antigens upon introduction into the body of another organism. It is responsible for graft rejection upon transplantation in another individual belonging even to the same species.
Transplantation: Surgical placement of tissue or organ from one individual into another.
Xenograft: Graft (usually tissue or organ) transplanted from individual of one species into individual of another genetically unrelated species.

CHAPTER 13

Active immunity: This type of immunity develops by exposure of individual to the antigen such that an immune response is evoked thus, the immunity attained lasts for the life time.
Antibiotics: *It is a chemical* substance produced by organisms which either kills or inhibits the bacterial growth. In the manufacture of certain vaccines minor amounts of antibiotics are added to avoid bacterial contamination of viruses in tissue culture.
Antigen: It is defined as any substance which upon entering the body is recognized as foreign and elicits the antibody production.
Cholera: It is a communicable disease caused by *Vibrio cholerae* affecting the small intestine. The symptoms are copious watery diarrhea, nausea, muscular cramps and extreme dehydration, leading to diminution of electrolytes.
Diphtheria: It is a vaccine preventable disease caused by *Corynebacterium diphtheria.* The throat of the infected individuals shows presence of a false membrane.
Eradication: It refers to both complete as well as permanent universal reduction of new cases caused by an infectious disease to zero by continuous efforts. Additional control methods aren't required.
Haemophilus influenzae type b (Hib): Infection by this bacterium causes pneumonia and meningitis, commonly in children and immunocompromised individuals.
Herd immunity: Vaccination of large number of people in a population result in acquisition of immunity known as herd community.
Immunization: It is defined as the process which protects humans and animals from a particular disease through heightened immune response.
Inoculation: It is the intentional exposure of an individual to a small amount of smallpox pustules which starts a minor, protective response towards smallpox.
Live attenuated vaccine (LAV): The LAV's consists of laboratory weakened pathogenic viral or bacterial strains which replicate in the body of the vaccinated child, arousing an immune response equivalent to the one initiated by wild-type strain. But since the pathogen is weak, it would lead to either very mild or no disease at all.

Measles: It is a highly infectious disease caused by measles virus characterized by fever, red circular spots on skin proving fatal to young and debilitated people.

Microorganisms: These include organisms such as bacteria and viruses, which are visualized only through a microscope.

Passive immunity: It is attained by transferring the antibodies or activated T-cells to the individual leading to a short-term immunity.

Passive immunization: It is conferred through administration of external antibodies, which for a short term fortify body's response to a particular antigen without inducing immunological memory to the antigen.

Pathogen: These are disease-causing organisms such as bacteria, viruses, PPLO etc. or their biological products like toxins.

Pertussis (whooping cough): A communicable disease whose causative agent is the bacterium *Bordetella pertussis*. The disease is characterized by vehement and irregular coughing.

Plague: It's a fatal infectious disease caused by the biting of rodents to humans.

Rabies: It is a lethal infection caused by the virus which enters the body by the bite of animals like dogs. The virus affects the central nervous system.

Rotavirus: It refers to a group of viruses leading to gastroenteritis or diarrhoea commonly affecting children.

Rubella: (German measles): It is a potentially fatal disease caused by viral infection which is generally milder than measles. If a pregnant mother catches the infection, it can either cause serious damage or even death of the foetus.

Smallpox: It is an acute, extremely contagious and deadly disease. Its causative agent is the variola virus. The symptoms are high fever along with body aches and eruption of pimples forming blisters, producing pus, and scars.

Subunit vaccine: It consists of inactivated antigenic parts of the pathogen rather than a complete pathogen which are necessary for provoking a suitable immune response.

Tetanus: It is caused by the bacterium *Clostridium tetani* which produces the toxins that upon entry into muscles induces extremely painful contractions.

Tetanus toxoid (TT) vaccine: It is a vaccine containing tetanus toxoid protecting against tetanus. Expecting women are vaccinated protecting them against tetanus, and their newborns against neonatal tetanus.

Tuberculosis (TB): It is an infectious disease affecting the lungs the most although, other body parts such as bone, kidney, spine and even brain can be affected. The causative agent is the bacterium *Mycobacterium tuberculosis*.

Typhoid: It is a vaccine preventable disease caused by different strains of the bacterium *Salmonella typhi*. The characteristic features of the disease are high temperature, general body weakness, pains in stomach and head and loss of appetite.

Vaccine: It comprises of the antigen which may be a complete bacterium or virus or it's portion capable of initiating an immune response for developing long term immunity.

Vaccine-preventable diseases: These are diseases which are preventable through vaccination.

Vaccination: It is defined as a process of injecting an antigen into the body of an individual, so its immune system can be stimulated which leads to the development of adaptive immunity towards that antigen.

Virus: It is in the borderline of living and non-living and can be visualized through a microscope. Its genetic material can either be DNA or RNA enclosed by a capsid. For replication the virus needs a living host.

Yellow fever: A viral disease transmitted through mosquitoes and common in the tropics. Common symptoms are high fever, jaundice, and gastrointestinal bleeding.

CHAPTER 14

12/23 Rule: States that the recombination only occurs when one RSS has 12 bp spacer, and another has 23 bp spacer.

AID: During the event of somatic hypermutation in variable chain region, this enzyme deaminates cytosine into Uracil and creates Mutation in the DNA.

Antigen receptor: B cells and T cells have unique and specific multi-complex proteins that bind to a specific antigenic determinant and generate signals for acquired immune response.

Antigen: Foreign molecules that bind with an antigen receptor stimulate an immune response and activate lymphocytes.

Burkitt lymphoma: A type of non-Hodgkin lymphoma, i.e., lymphatic system cancer caused due to chromosomal translocation in the c-Myc gene. Three variants - endemic, which most common with t(8,14) Mutation, sporadic with t(2,8), and HIV associated with t(8,22) are present.

Chemoimmunotherapy: It's a combination therapy in which chemotherapy drugs are used for killing cancer cell, and immunotherapy cells are used for stimulating the fighting ability of the immune system.

DSB: Normal cellular processes result in the formation of reactive oxygen species, which causes damage in DNA bases, which block the replication. In B cells, the V gene segment is rearranged, and its secondary diversification is linked with the error-prone Non-homologous DSB repair.

Exon: A part of a gene that contains the information for coding a protein.

IGHV mutation: Mutation in a gene encoding for the variable region of Immunoglobulin heavy chain of B cell receptors. This mutational analysis is a prognostic biomarker in CLL.

Immunological memory: Some specific B and T memory cell clones are formed when our immune system.

Isotype switching: It is a mechanism that results in switching one class of Immunoglobulin to another. In this process, the heavy chain constant region is replaced, and there is no change in the variable region. Due to this, no change occurs in antigen specificity.

PARP 1: Poly ADP Ribosyl transferase does aspartate and glutamate ribosylation of the target protein and plays an essential role in replication fork stabilization in DNA repair.

RSS: Noncoding Heptamer or nanomer conserved DNA sequences which guide VDJ recombination by recognizing RAG ½ enzyme.

Southern blot: It is used to detect a specific DNA of interest in the sample DNA mixture.

VDJ recombination: Adaptive immune system cells cut a small DNA region and re-join it in an error-prone manner.

CHAPTER 15: BIOLOGY OF B LYMPHOCYTES

Affinity maturation: The affinity of an antibody increases for a particular antigen during an immune response.

B cell receptor: It is a complex of membrane-bound immunoglobulin along with Ig-α/Ig-β signal transducing molecules.

B lymphocyte: It is a type of lymphocyte which has membrane-bound immunoglobulin. These cells are produced and matured in bone marrow.

Centroblast: Activated and enlarged B cell found in the dark zone of proliferating germinal centre.

Centrocyte: B cell having cleaved nucleus which is found in the light zone of non-proliferating germinal centre.

Class switching: It is a change in the class of antibodies from one type to other which are produced by B cell.

Dark zone: It is the site of rapid cell division present in the germinal centre and having centroblasts.

Effector cell: After antigen interaction resting lymphocytes produce effector cells and memory cells. Effector B cells are antibody-secreting plasma cells and effector T cells are activated T helper cells and cytotoxic T cells.

Germinal centre: It is the site of extensive B cell somatic mutation and selection. It is present in the secondary follicles in lymph nodes and spleen.

Hematopoiesis: It is the formation and development of all blood cells including RBCs, WBCs and platelets.

Hematopoietic-inducing microenvironment (HIM): In the bone marrow hematopoietic-inducing microenvironment is provided by the stromal cells and various growth factors which are required for the formation and development of all types of blood cells.

Humoral immune response: It is the immune response which is mediated by the antibodies and provides protection against extracellular pathogens and foreign particles.

Light Zone: It is region of the germinal centre having follicular dendritic cells.

Memory cells: After the interaction of antigen, lymphocytes differentiate into effector cells and memory cells. Memory cells are long-lived and responsible for secondary immune response.

Plasma cells: Antibody-secreting B cells are called as plasma cells.

Precursor B cell: It is a stage of B cell development that produce cytoplasmic μ heavy chains and most of the precursor B cell express the pre-B-cell receptor.

Somatic hypermutation: It is an increased rate of mutation within the rearranged immunoglobulin genes regions and in adjacent area.

Thymus-dependent antigen: It is a soluble antigen capable of inducing antibody production by plasma cells with the interaction of T_H cell with B cell. It involves memory cell production, isotype switching and affinity maturation.

Thymus-independent (TI) antigen: It is the response of B cells without involvement of T cells. These antigens are further divided into two types one is TI-1 which are polyclonal B cell activators and another one is TI-2 which activate B cells by crosslinking the mIg receptor.

V pre B: The surrogate light chain of pre B cell receptor has Vpre B which is a polypeptide chain and λ5.

CHAPTER 16: BIOLOGY OF T LYMPHOCYTES

Cell-mediated immune response: The immune response which is mediated by antigen-specific T cells along with various nonspecific cells like macrophages. It targets intracellular bacteria and self-altered cells including virus infected cells and cancerous cells.

CD4: It is a single chain glycoprotein which act a coreceptor on Class II MHC restricted T cells and present on most of the T helper cells.

CD8: It is a dimeric glycoprotein which act a coreceptor on Class I MHC restricted T cells and present on most of the cytotoxic T cells.

Cytotoxic T lymphocyte: It is a type of T lymphocyte which can mediate the lysis of target cells having antigenic peptides complexed with Class I MHC molecule.

Negative Selection: Removal of developing lymphocytes which show reactivity against self-antigens.

Pre T cell receptor: It is a complex of CD3 with a structure having T cell receptor β chain.

Self-tolerance: No response against self-antigens.

Single positive CD4$^+$: Thymocyte having coreceptor CD4.

Single positive CD8$^+$: Thymocyte having coreceptor CD8.

T cell receptor: It is heterodimer (having αβ or γδchains) which is expressed on the surface of T cells along with CD3 for binding antigen.

T helper cell: A type of T lymphocytes which are MHC class II restricted and also producer of cytokines. Most of the T helper cells are $CD4^+$.

Thymocyte: These are developing T cells in the thymus.

T_H1 response: This response produces a cytokine profile which supports inflammatory responses and activates primarily T lymphocytes and macrophages.

T_H2 response: This response produces a cytokine profile which activates B lymphocytes and immune responses that depend on the antibodies.

CHAPTER 17

Antibody-dependent cell-mediated cytotoxicity (ADCC): A cell-mediated cytotoxicity mechanism in which cells (NK cells, macrophages, eosinophils, monocytes, neutrophils) having Fc receptors on their surfaces that can bind to the Fc region of antibodies, thus capturing antigen-antibody complexes and destroying them.

Antigen presenting cells (APCs): Cells that can process and present antigenic peptides on class II MHC molecules on their surface and deliver a co-stimulatory signal for the activation of T cells. Professional APCs include macrophages, dendritic cells and B cells. Thymic epithelial cells and vascular endothelial cells constitute non-professional APCs.

CD3: A molecule present on T cells in association with T-cell receptor that plays a role in signal transduction. It is a polypeptide made up of 2 heterodimers: $\gamma\varepsilon$ and $\varepsilon\delta$ and a homodimer $\xi\xi$.

CD4: A glycoprotein present on T_H cells which serves as a co-receptor during interaction of T_H cells with MHC class II bearing APCs.

Cell adhesion molecules (CAMs): Proteins that function in intercellular adhesion. These include integrins, selectins, immunoglobin superfamily and mucin-like proteins.

Cytotoxic T lymphocyte (CTL): An effector T cell (usually $CD8^+$) that has been activated after exposure of naïve T_C cells to antigenic peptide presented on MHC molecules. CTLs bring about direct killing of cells by means of perforins/granzymes and by activating the Fas/FasL pathway.

F_C receptor: Receptor present on macrophages, NK cells, mast cells, eosinophils that can bind to the F_C portion of immunoglobins.

GM-CSF: Granulocyte-macrophage colony-stimulating factor secreted by T cells, macrophages, fibroblasts and endothelial cells.

Granzymes: Protease enzymes released by CTLs that initiate apoptosis in target cells.

Interleukins (ILs): Cytokines secreted by leukocytes affecting the growth and differentiation of immune cells.

Interferon (IFN): A group of glycoprotein cytokines produced by some cells that induce an antiviral state in other cells. IFN-α is secreted by lymphocytes, dendritic cells and macrophages. IFN-β is secreted by fibroblasts, dendritic cells and some epithelial cells. IFN-γ is secreted by $CD4^+$, $CD8^+$ and NK cells.

Killer Immunoglobin-like Receptors (KIRs): Glycoprotein molecules belonging to the immunoglobin superfamily, present on the surface of NK cells.

MHC restriction: The process whereby an antigen is processed by APCs or target cells and its peptides are presented on surface molecules known as MHC class I or MHC class II to T lymphocytes.

Natural Killer Cells (NK Cells): Innate immunity cells bearing lectin-like or immunoglobin-like receptors on their surface. They bring about non-specific killing of infected cells and tumor cells by release of perforins and granzymes.

Natural Killer T cells (NKT cells): Innate immunity cells similar to NK cells, but bearing T-cell receptors that are non-specific and non-MHC restricted.

Perforin: Lytic proteins produced by CTLs and NK cells that polymerize in the presence of Ca^{2+} to form transmembrane pores in target cells causing their lysis.

T cell receptor (TCR): Molecule present on T lymphocytes in association with CD3 molecules which binds antigenic peptides presented in association with MHC molecules. They are heterodimers made up of α or β or γ and δ chains.

Tumor Necrosis Factor β (TNF-β): Proinflammatory cytokine secreted by activated T cells, B cells, fibroblasts, endothelial cells and epithelial cells.

Editors

Dr. Indrakant K. Singh received his PhD in Zoology from the Department of Zoology, University of Delhi (DU). Dr. Singh is currently working as an Associate Professor in the Department of Zoology, Deshbandhu College, University of Delhi, New Delhi, where he enjoys teaching Immunology, Genomics, Biotechnology & Computation Biology. He is a recipient of the *Young Scientist Award/ Fellowship* (2013–2016) by *Science & Engineering Research Board (SERB), Raman-Post Doctoral Fellowship* (2016–2017), *Max-Planck Society Visiting Post-Doctoral Fellowship* (2019); *ICMR-DHR Long Term International Fellowship for Young Biomedical Scientists* (2019–2020), *Prestigious Delhi University Excellence Award for Teacher in Service 2021* and *DHR Long Term Fellowship for Training in Indian Institute* (2022). *He has also been awarded with Best Researcher, 2017 and Best Teacher, 2018 by Grace India Trust India.* Recently, Dr. Singh selected as Fellow, Delhi School of Public Health, Institute of Eminence, University of Delhi, India. Dr. Singh also filed two patents (India & USA) on *Mycobacterium tuberculosis.* Dr. Singh is a life-member/Fellow/member of many prestigious academic societies including the AMCA, IBS, and AIMI, ESI, BSI. Dr. Singh has published his quality research in peer-reviewed journals of international repute like *Scientific Reports, International Journal of Biological Macromolecules, Critical Reviews in Biotechnology, Clinica Chimica Acta, International Journal of Molecular Sciences and BBA Molecular Basis of Disease, Computational Biology and Chemistry, Anti-Cancer Agents in Medicinal Chemistry.* He is an editorial board member in many international and national journals like Scientific Reports etc. He has also edited four books and authored many book chapters published by Springer, Taylor & Francis and Elsevier. Dr. Singh has served as a reviewer for several international and national refereed journals. Dr. Singh has trained many post-doctoral, doctoral, post-graduate and graduate students, and currently supervises five doctoral and two post-doctoral students. Over the years, he has pursued his research interests in the areas of Molecular biology, Cancer Biology, and Functional Genomics.

Dr. Poonam Sharma is an Associate Professor at Department of Zoology, Gargi College, University of Delhi, India and has more than 15 years of teaching and research experience. She received her PhD in Cell and Molecular Biology from the Department of Zoology, University of Delhi (DU). Dr. Sharma received the prestigious "BOYSCAST FELLOW" award from the Department of Science and Technology of the Government of India for outstanding research in the field of Environmental Engineering at LIEBE, Universitė Paul Verlaine-Metz, France. Dr. Sharma was also a Visiting fellow at Swiss Federal Institute for Environmental Science and Technology (EAWAG), Dubendorf, Switzerland. She has also been awarded with "Best Women Faculty Award-2021" by Novel Research Academy, India under the Ministry of MSME, Government of India. She has edited and written several chapters and modules for postgraduate, undergraduate, and senior secondary school students. Dr. Sharma has a number of research publications in peer-reviewed international and national journals to her credit. Dr. Sharma is actively involved in research and supervises graduate students in many different projects based on computational biology and molecular biology.

Abbreviations

CHAPTER 2

AChRs	Acetylcholine Receptors
ADCC	Antibody-Dependent Cell-mediated Cytotoxicity
AIDS	Acquired Immune Deficiency Syndrome
AMPs	Anti-Microbial Proteins
ANCA	Anti-Neutrophil Cytoplasm Antibodies
APCs	Antigen Presenting Cells
APR	Acute Phase Response
BCG	Bacillus Calmette-Guérin
BCR	B-Cell Receptor
CAM	Cell Adhesion Molecules
CARD	Caspase Recruitment Domains
CD	Cluster Differentiation
CGD	Chronic Granulomatous Disease
CLRs	C-type lectin receptors
COX2	Cyclooxegenase-2
CRP	C-reactive protein
CTLs	Cytotoxic T-Lymphocytes
CVID	Common Variable Immunodefieiency
DAMPs	Damage-associated Molecular Patterns
DCs	Dendritic Cells
DPT	Diptheria-Tetanus-Pertussis
EBV	Epstein-Barr Virus
FNG	Fibrinogen
GM-CSF	Granulocyte-Macrophage Colony-Stimulating Factor
HGB	Haptoglobulin
HLA	Human Leukocyte Antigen
HPV	Human Papilloma Virus
ICAM	Immunoglobulin-superfamily Cell Adhesion Molecules
ICOS	Inducible Costimulator
IFN	Interferon
IL	Interleukin
iNOS	Inducible Nitric Oxide Synthase
IPV	Inactivated Polio Vaccine
IRFs	Interferon Regulatory Factors
iT_{REG} cells	Induced T-Regulatory Cells
LCs	Langerhans Cells
LPS	Lipopolysaccharide
LRRs	Leucine Rich Repeats
MAC	Membrane Attack Complex
MAMPs	Microbe Associated Molecular Patterns
MASPs	MBL-associated serine proteases
MBK	Membrane Binding Lectin
MBP	Mannan Binding Protein
MHC	Major Histocompatibility Complex
MS	Multiple Sclerosis

MyD88 Myeloid Differentiation factor 88
NK cells Natural Killer cells
NLRs NOD Like Receptors
NSAIDs Non-Steroidal Anti-Inflammatory Drugs
PAMPs Pattern associated Molecular Patterns
PCV Pneumococcal conjugate vaccine
PRR Pattern Recognition Receptor
PTX Pentraxin
RA Rheumatoid Arthritis
RLRs Retinoic acid-inducible gene-I-Like Receptors
RNS Reactive Nitrogen Species
ROS Reactive Oxygen Species
RSV Respiratory Syncytial Virus
SCID Severe Combined Immunodeficiency
SLE Systemic lupus erythematosus
SP Surfactant Proteins
Tc cells T-cytotoxic cells
TCR T-Cell Receptor
T_{FHCs} T Follicular Helper Cells
TGF Transforming Growth Factor
T_H cells T-Helper cells
TIR Toll-Interleukin Receptor
TIRAP Toll/interleukin-1 receptor (TIR) domain containing adaptor protein
TLRs Tol-like Receptors
TNF Tumor Necrosis Factor
TRAM TIRF Related Adaptor Molecule
T_{REG} cells T-Regulatory Cells
TRIF TIR domain containing adaptor-inducing IFN-β Factor

CHAPTER 3

ADCC Antibody-Dependent Cell-mediated Cytotoxicity
AIDS Acquired Immune Deficiency Syndrome
APCs Antigen Presenting Cells
BCG Bacillus Calmette-Guérin
BCR B-Cell Receptor
CD Cluster Differentiation
CGD Chronic Granulomatous Disease
CTLs Cytotoxic T-Lymphocytes
CVID Common Variable Immunodeficiency
DCs Dendritic Cells
DPT Diptheria-Tetanus-Pertussis
EBV Epstein-Barr Virus
FNG Fibrinogen
GM-CSF Granulocyte-Macrophage Colony-Stimulating Factor
HGB Haptoglobulin
HLA Human Leukocyte Antigen
HPV Human Papilloma Virus
ICOS Inducible Costimulator
IPV Inactivated Polio Vaccine
IRFs Interferon Regulatory Factors

iT_{REG} cells	Induced T-Regulatory Cells
LCs	Langerhans Cells
MHC	Major Histocompatibility Complex
MS	Multiple Sclerosis
NK cells	Natural Killer cells
PCV	Pneumococcal conjugate vaccine
PRR	Pattern Recognition Receptor
RA	Rheumatoid Arthritis
SCID	Severe Combined Immunodeficiency
SLE	Systemic lupus erythematosus
Tc cells	T-cytotoxic cells
TCR	T-Cell Receptor
T_{FHCs}	T Follicular Helper Cells
TGF	Transforming Growth Factor
T_H cells	T-Helper cells
TNF	Tumor Necrosis Factor
T_{REG} cells	T-Regulatory Cells
IL	Interleukin
IFN	Interferon

1 Overview of Immune System

Princy Hira
University of Delhi

CONTENTS

1.1 INTRODUCTION AND BACKGROUND

Immunology deals with the study of the immune system and its responses against various infections. The state of body's defense against foreign challenges is Immunity (Latin word *immunis* meaning to exempt). Immunity evolved in multicellular organisms to protect them from invader disease-causing

DOI: 10.1201/9781003286424-1

micro-organisms commonly referred to as pathogens. The immune system is highly adaptable and diverse with respect to its response to various infectious agents (see Table 1.1) and involves a wide variety of cells, molecules and tissues that recognize and generate an immune response to a pathogen. Therefore, the two most important attributes are Recognition and Response. Recognition means discriminating self from non-self i.e. differentiating between the body's own cells and foreign cells based on molecular markers or patterns. Recognition also involves identifying the altered cells of one's own body, mainly cancerous or virus-infected cells. Recognition effectively leads to a particular effector response that eliminates or neutralizes the infection. Additionally, immunological memory is also generated that leads to a quick and heightened response upon subsequent exposures to the same pathogen. Generally, two different arms of immunity work independently as well as in association dynamically to fight against infections namely, innate and acquired immunity. Innate immunity provides the first line of defense with recognition and response mechanisms pre-deployed before exposure to infection. Innate immunity recognizes only a broad spectrum of foreign pathogens with the help of markers like pathogen-associated molecular markers (PAMPs). Conversely, acquired immunity provides a second line of defense in response to exposure to infection and is highly adaptable also known as adaptive immunity. Immunological memory and specificity are additional attributes of acquired immunity that provides the ability to recognize a variety of pathogens with even slight differences in their structure and generate more memory response for later exposures. This is why we don't get some of the diseases the second time. The molecules and cells involved in the two types of immunity will be discussed in subsequent sections.

History of immunology traces back to 430 BC during the times of the Peloponnesian war when historian Thucydides wrote that individuals who had recovered from the plague could only treat the sick people because they will not catch the disease for the second time. This concept came into practice only after 2,000 years in the 15th century wherein the Chinese and Turkish trying to prevent smallpox (a deadly disease caused by the virus *Variola major*) used dried crusts from the smallpox pustules from diseased (either inhaled or inserted into the skin, variolation) to provide immunity to healthy. Same attempts were made by Lady Mary Montagu on her own children in 1718, and positive results were seen. Major advancement in the field of immunology came from the work of Edward Jenner (English Physician, designated as the father of immunology) in the 18th century where inoculation of fluid from cowpox pustule (mild disease) could confer protection from fatal smallpox disease. This was tested on a healthy 8 years old boy who was inoculated with cowpox fluid and later infected by smallpox artificially and to his prediction, the boy did not develop smallpox disease and this procedure he termed as vaccination, which is presently in use to describe the inoculation of attenuated (weak) or killed pathogens into healthy individuals to provide immunity from several diseases. However, the vaccine for smallpox took around two centuries to become universal and smallpox was eradicated as announced by WHO in 1979. This technique was further extended by Louis Pasteur to develop vaccines for other diseases like cholera in fowl, anthrax in sheep and rabies in dogs. He used weak or aged strains of *Vibrio* and *Bacillus* strains (causative organisms of cholera and anthrax respectively) which are no more virulent for inoculation and called this attenuated strain as vaccine. Similarly, Emil von Behring and Shibasaburo Kitasato used antitoxic activity in the serum of animals immune to tetanus or diphtheria as short-term protection against tetani or diphtheria toxin. Later it was discovered that this antitoxic activity was provided by protein molecules called antibodies generated in response to the toxin. This is how the science of immunology and vaccination started and boomed during the past years. However, there are certain challenges to vaccine development and its universal delivery worldwide. Presently, a lot of research is in progress to solve various unanswered questions in immunology and vaccination. In this unit, an overall overview of immunology is outlined with an idea to deliver basic understanding and knowledge about the immune system. The purpose of this introductory unit is to provide preliminary information about various headings that are discussed in detail in subsequent units.

Table 1.1
Different Categories of Pathogens and their Associated Diseases

	Causative Organism	Disease Caused
	Category: Bacteria	
Gram-Negative	*Pseudomonas aeruginosa*	Pneumonia, Sepsis
	Neisseria gonorrhoeae	Gonorrhoea, sexually transmitted disease
	Salmonella typhimurium	Typhoid
	Shigella dysenteriae	Dysentery
	Escherichia coli	Gastroenteritis, Urinary tract infections
	Brucella spp.	Brucellosis
	Klebsiella spp.	Meningitis, blood stream infections
	Vibrio cholerae	Cholera
	Treponema pallidium	Syphilis
	Bordetella pertussis	Pertussis (Whooping Cough)
Gram-Positive	*Mycobacterium tuberculosis*	Tuberculosis
	Bacillus anthracis	Anthrax
	Clostridium tetani	Tetanus
	Corynebacterium diphtheriae	Diphtheriae
	Staphylococcus aureus	Skin and bone infections
	Enterococcus spp.	UTI, Prostatitis, Endocarditis
	Erysipelothrix rhusiopathiae	Erysipelothricosis; skin infection
	Category: Viruses	
RNA Viruses	Orthomyxoviruses	Influenza
	Hepatitis C virus (HCV); flavivirus	Hepatitis C
	Hepatitis E virus (HEV); flavivirus	Hepatitis E
	West Nile Virus (WNV); flavivirus	West Nile Fever
	Rhabdovirus	Rabies
	Enterovirus	Polio; Hepatitis A
	Paramyxoviruses	Measles
	Filovirus	Ebola
	Retroviruses	Leukemia, AIDS
DNA Viruses	Herpes Simplex Virus (HSV)	Herpes
	Hepadnavirus	Hepatitis B
	Poxvirus	Smallpox
	Papillomavirus	Cervical carcinoma
	Category: Fungi	
	Candida spp.	Candidiasis
	Sporothrix	Sporotrichosis
	Trichophyton	Ringworm, athlete's foot
	Claviceps purpurea	Ergotism (St. Anthony's fire)
	Aspergillus	Allergies, Mild Pneumonia
	Category: Protozoa	
	Trypanosoma gambliense	African Sleeping Sickness
	Schizotryanum cruzi	Chagas Disease
	Giardia intestinalis	Giardiasis
	Plasmodium falciparum	Malaria
	Leishmania donovanii	Kala-azar
	Entamoeba histolytica	Amoebiasis
	Toxoplasma gondii	Toxoplasmosis
	Balantidium coli	Balantidiasis

1.2 HUMAN PATHOGENS: PLETHORA OF ANTIGENS

Micro-organisms or invaders causing diseases in humans and animals are commonly referred to as pathogens as already mentioned in introduction. When Edward Jenner discovered the vaccine for deadly smallpox, there was no knowledge of the infectious agents that cause the disease, until proven by Robert Koch that micro-organisms are responsible for causing these infections and specific for a particular disease. The world of pathogen is quite vast and enormous, and to our surprise body's immune system is capable of discriminating a wide range of these pathogens and eliciting specific responses against them. As stated earlier, this specificity is a property of the adaptive immune system, while the innate system generates a non-specific immune response to a broad spectrum of molecules expressed by foreign invaders. Innate immunity cannot distinguish minor differences in the pathogens. To our knowledge, pathogens are divided into four broad categories: bacteria, viruses, fungi and protozoans. Some of the most common diseases and their causative micro-organisms are listed in Table 1.1. These pathogens can be recognized in our body and even a small molecular structure on its cell membrane has the ability to elicit an immune response mainly the formation of antibodies. These molecules that trigger the immune response are known as antigens and it includes whole organisms like viruses and bacteria, toxins, chemicals, pollen grains or any other foreign substances (Table 1.1).

1.3 HEMATOPOIESIS

The blood cells are derived from the stem cells in the bone marrow in a process termed as Haematopoiesis. Hematopoietic stem cell (HSC) differentiates and gives rise to all red and white blood cell types therefore it is also referred to as pluripotent stem cell. These stem cells are self-renewing but their study is difficult due to struggle in in-vitro culture and their scarcity (one HSC per 5×10^4 cells in bone marrow). During the first few weeks of development, haematopoiesis takes place in the yolk sac. Later from the third to seventh month of gestation, these stem cells differentiate into different cells in the fetal liver and spleen. Subsequently, in the last gestational month and from birth, the bone marrow becomes the primary site for the development of blood cells. The process starts with the differentiation of pluripotent HSC to the lymphoid and myeloid progenitor cells. Progenitor cells are committed to develop into specific cell lineages and have lost the property of self-renewal. In brief, lymphoid lineage differentiates into three types of cells namely B lymphocytes, T lymphocytes (commonly referred to as B and T cells) and natural killer (NK) cells. NK cells are innate immune cells. Myeloid lineage mostly gives rise to cells of the innate immunity like macrophages, granulocytes (neutrophils, eosinophils, basophils), dendritic cells and mast cells (Figure 1.1). Erythrocytes and platelets are also formed from erythroid progenitor and megakaryocyte (myeloid progenitor lineage) but they are of no concern in immunity. Additionally, stromal cells in bone marrow provide a conducive environment i.e., haematopoietic-inducing microenvironment (HIM) for these cells to develop and mature. Endothelial cells, fat cells, macrophages and fibroblasts form the stromal cell networks.

Haematopoiesis is a genetically regulated process, specific genes and transcription factors control the differentiation of specific lineages which were initially identified by using 'Knock-out' mice techniques. Some of the major transcription factors are as follows:

- **GATA-2:** Transcription factor GATA-2 (*GATA-2* gene) influences multiple lineages. It is necessary for the formation of myeloid, erythroid and lymphoid lineages.
- **GATA-1:** it regulates only erythroid lineage i.e., development of red blood cells.
- **Ikaros:** it regulates only lymphoid lineage i.e., B and T cells and NK cells
- **Bmi-1:** affects all lineages as it is necessary for self-renewal of HSCs
- **Oct-2:** regulates differentiation of plasma cells from B cells

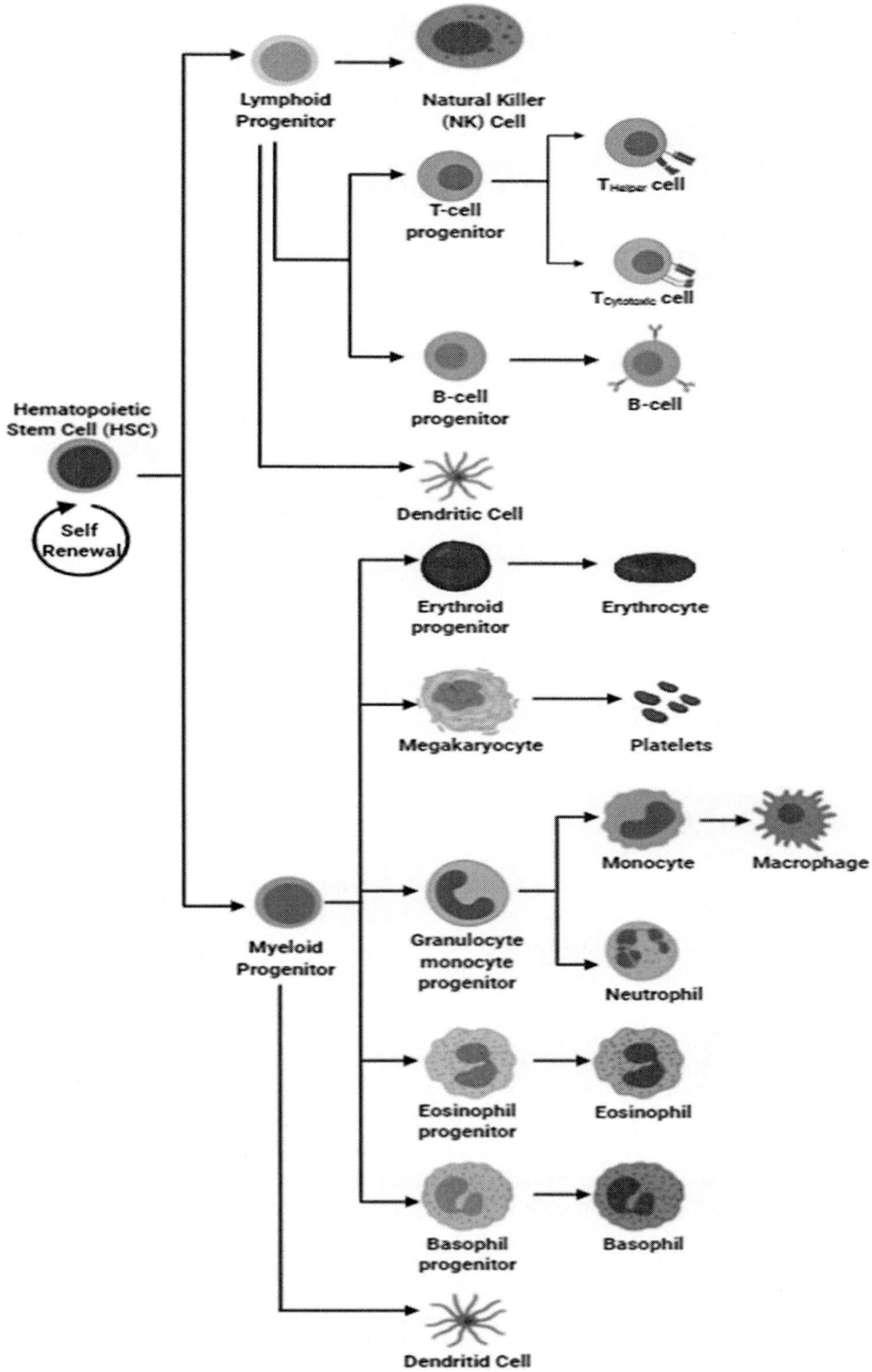

FIGURE 1.1 Process of haematopoiesis showing formation of blood cells in myeloid and lymphoid lineages by haematopoietic stem cell in bone marrow. (Created using Biorender.com.)

Haematopoiesis maintains homeostasis wherein the rate of formation of blood cells is equivalent to the rate at which blood cells are removed from the body due to aging or a limited life span. Therefore, a steady state balance is maintained between the number of cells produced and the number lost. Each blood cell has a definite life span like RBCs may live for 120 days and neutrophils only for 1 day. This homeostasis is maintained by programmed cell death of aged blood cells i.e., apoptosis. Apoptosis is a well-ordered process where a cell exhibits specific morphological changes that eventually leads to its own death. These profound changes include membrane blebbing, rearrangement of cytoskeleton, reduction in cell volume, DNA fragmentation and chromatin condensation. Subsequently, small bodies are pinched off from the cell having whole organelles known as apoptotic bodies. Finally, macrophages engulf and phagocytose these apoptotic bodies. All these changes are induced by different genes and signals. *Fas* gene induces apoptosis and *caspase* and *bax* genes promote apoptosis. On contrary, genes *bcl-2* and $bcl\text{-}X_L$ prevent apoptosis.

1.4 CELLS OF THE IMMUNE SYSTEM

White blood cells (leukocytes) are the major immune cells. These cells bear receptors to recognise foreign antigens and generate an appropriate immune response. We will discuss cells of adaptive immunity first *i.e.,* lymphocytes and antigen-presenting cells (APCs) followed by granulocytes and other cells.

1.4.1 Lymphoid Lineage Cells

Lymphocytes (B and T cells) are differentiated from lymphoid progenitor mainly in bone marrow in mammals but maturation takes place at different sites. Natural killer (NK) cells are also produced in lymphoid lineage.

- **B lymphocytes**

  The development and maturation of B cells take place in the bone marrow of long bones in most adult mammals and prenatally in the foetal liver. In birds, the site of B cell maturation is the bursa of fabricius. B lymphocytes are the major cells of adaptive immunity responsible for antibody (immunoglobulins Ig) production against antigens (pathogens). Maturation of B cells involves two types of selection: Positive and negative selection. Positive selection operates to hand-pick only those B cells that display antigen binding receptors on their surface. Subsequently, negative selection operates to eliminate the cells that are capable of recognizing self-antigens (molecules from our own body) to diminish the chances of autoimmune disorders. B cells after clearing the selection process complete the final stages of maturation and leave the bone marrow. These mature B cells are referred to as naïve B cells or unprimed as they have not interacted with any antigen yet. Once the B cell receptor (membrane-bound antibody) recognizes any particular antigen, it gets activated or primed and clonally expands to create more copies of the cell expressing the same receptor. Finally, clone of daughter cells differentiates into effector cells and memory cells. Effector cells are plasma cells that no longer express antibodies on the cell surface, instead they are specialized for secreting large amounts of antibodies in response to the same antigen. These are the only cells in the body that produce antibodies at a rate of about thousands of molecules per second.

  B cell activation can take place via T-cell independent or dependent manner. In the T-cell independent pathway, signals for B cell activation come from T-independent antigens like LPS and the interaction of TLRs with PAMPs or from the complement system proteins. However, this response does not result in the formation of memory cells and is short-lived. On the other hand, activation with dependency on T cells is more strong, complex and results in the formation of memory cells. Briefly, B cell recognizes and presents the antigens to T helper cells (one of the types of T cells) that gets activated and release cytokines which in turn activates B cells leading to differentiation into antibody-secreting plasma cells as well as memory cells for secondary immune response. T-dependent antigens are usually protein antigens and T-independent antigens are non-proteinaceous and characterized by repetitive units. This process is explained in detail in subsequent units. Therefore, along with antibody production B cells also act as antigen-presenting cells.

- **T lymphocytes**

  T lymphocytes are also derived from a lymphoid progenitor in none marrow but unlike B cells, lymphoblasts committed to differentiate into T cells migrate to a bilobed organ thymus for maturation. Immature T lymphocytes in the thymus are known as thymocytes. During the maturation, these thymocytes first develop receptors on their cell surface known as T cell receptors (TCR) and then undergo thymic selection. Negative selection eliminates thymocytes with defective TCRs. This is followed by a positive selection of cells capable of interacting with MHC molecules (Major-histocompatibility Complex). MHC is a cluster of genes located on chromosome 6 in humans referred to as HLA (Human leukocyte antigen). These are peptides that play a significant role in self and non-self recognition and discrimination between different cells, essentially, they are antigen-presenting molecules expressed on different cells. Negative selection also operates to eliminate self-reacting thymocytes to minimize the risk of autoimmune diseases.

  Mature T cells express TCR specific for antigen presented with MHC molecule. T cells cannot recognize antigen alone rather they recognize antigen complexed with MHC molecules presented by an APC. Along with TCR, certain markers are also present on T cell surface that differentiate them into three subtypes:

  - **T_H cells:** T helper cells express CD4 (Cluster of differentiation) marker and can only recognize antigens complexed with MHC II molecules expressed on the surface of APCs. Upon activation, effector helper cells secrete cytokines that stimulate B cells to produce antibodies. Hence, these cells help in the elimination of extracellular pathogens majorly. They also help macrophages and NK cells by enhancing their killing functions so also crucial in defense against intracellular pathogens.
  - **T_C cells:** Cytotoxic T cells express CD8 marker and recognize antigens complexed with MHC I molecules expressed on the cell surface of all nucleated cells of the body. As MHC I is expressed in all the cells, generally virus-infected or altered (tumour) host cells present antigens to cytotoxic cells. In response, they differentiate into effector cells known as CTLs (Cytotoxic T lymphocytes) and destroy target cells with cytotoxicity. Therefore, they act mainly on intracellular pathogens that enter the host cells
  - **T_{reg} cells:** Regulatory T cells have CD4 and CD25 markers. Their exact role is not well understood. But they are involved in regulating immune responses and tolerance.

- **Natural killer cells**

  They are large granular cells capable of killing target cells. These are derived from a common lymphoid progenitor with antigen-specific lymphocytes but unlike lymphocytes, NK cells don't possess antigen-specific receptors. These cells are best known for controlling tumours and viral infections as a first line of defense. They can kill cells without any activation naturally hence named so. But they possess activation and inhibition receptors on their surface that recognize altered host cells (cancerous or virus infected). Further, they also secrete cytokines (IFN-γ and TNF-α) that activate dendritic cells and macrophages to come into play.

1.4.2 Myeloid Lineage

These cells have granular appearance and are mainly involved in innate immunity.

- Neutrophils are usually the first cells to migrate and arrive at the site of infection. They have multilobed nucleus (polymorphonuclear leukocytes) and specialise in phagocytosis of pathogens like bacteria and fungi. Both acid and basic dyes can be used for staining of neutrophils.
- Eosinophils have bilobed nucleus and are chiefly phagocytose parasitic pathogens. Eosin red dye is used to stain these cells (acidic).
- Basophils are granulocytes containing granules with pharmacologically active substances inducing inflammation and allergic reactions to many parasitic infections. A basic dye methylene blue is used to stain basophils.
- Monocytes after formation in bone marrow circulate in blood for about 8 hours known as circulating phagocytes. Phagocytosis is engulfing the pathogen inside the cells into a vesicle with the help of the cell membrane. This vesicle is known as phagosome. There are two pathways for killing the internalized pathogen: Oxidative attack and non-oxidative attack. In oxidative attack, the uptake of oxygen in cells increases many fold and reactive oxidative species (ROS) are formed with the help of an enzyme NADPH phagosome complex (phox) in phagosome as microbicides. ROS includes superoxide, hydrogen peroxide, hypochlorous acid etc. Superoxide further reacts with nitric oxide to form reactive nitrogen species (RNS). On contrary, in non-oxidative attacks (for pathogens like *Candida* and *Staphylococcus aureus*) cellular granules fuse with phagosome to release antimicrobial proteases or enzymes like lysozymes and proteins like LPS-binding proteins and bactericidal permeability-increasing protein (BPI).
- Macrophages are differentiated from monocytes when they enter specific tissues. Macrophages are usually long-lived phagocytic cells that can be circulating as well as fixed in a particular tissue. Examples: alveolar macrophages (lung), intestinal (gut), osteoclasts (bone), microglial cells (brain), mesangial cells (kidney), and Kupffer cells (liver). Along with phagocytosis, macrophages also play other important roles in the innate and adaptive immune system. They secrete inflammatory mediators like cytokines (IL-1, IL-6 and TNF-α) and help induce inflammation. They play an important role in antigen presentation to the T lymphocytes with the help of MHC-II molecules. Lastly, they act as scavengers for clearing cell debris, dead cells and immune complexes.
- Mast cells are involved in protection against parasitic worms. Their progenitors differentiate only when they enter tissues and not in blood. Also, they possess large cytoplasmic granules which when activated release their contents inducing allergic responses. They are mainly present in connective tissues of organs, skin and epithelial tissues.
- Dendritic cells (DC) are named so because of their appearance like dendrites of a neuron. Like other phagocytic cells, they phagocytose pathogens and clear them with either oxidative attack (ROS, RNS) or non-oxidatively by antimicrobial peptides. However, their major role is to link innate and adaptive immunity. Immature DC circulate in the blood and enter tissue spaces. Receptors on dendritic cells recognize and internalize antigens. Following the encounter with antigen, DC mature into cells that further activate the T cells by presenting the antigens to them. There are four types of dendritic cells: Langerhans DC, monocyte-derived DC, plasmacytoid-derived DC, and Interstitial DC. They express MHC-II and co-stimulatory molecules on their membranes for interaction and activation of T cells. Attributed to the antigen presentation function, these cells are also known as antigen-presenting cells (APC).

1.4.3 Antigen Presenting Cells (APCs)

As already mentioned above, T lymphocytes only recognize antigens complexed with MHC I or MHC II molecules. Cells that express these MHC molecules and present antigens complexed with them to T cells are referred to as Antigen Presenting Cells (APC). Examples include macrophages, dendritic cells, B cells (before activation), virus-infected cells and altered host cells. Generally, APC engulfs the detected pathogen and processes it or cleaves it in different fragments which we call it as antigen. These antigenic fragments are then expressed on the cell surface complexed with MHC molecules and presented to T cells. This is known as antigen processing and presentation by APCs. Classical MHC molecules fall under two classes and have distinct pathways of antigen processing and presentation:

- **Class I MHC (endogenous antigens):** expressed on all nucleated cells and process intracellular (endogenous) pathogens by cytosolic pathway. So, pathogens that have infected the cells like virus, parasite or bacteria are degraded by proteasome into smaller peptides in cytoplasm of the cell. These peptides move to the endoplasmic reticulum and associate with MHC-I complex. Finally, MHC along with peptide move to golgi apparatus and passes to the plasma membrane of the cell. Therefore, the processed antigen-MHC-I complex is ready to be presented to cytotoxic T cells.
- **Class II MHC (exogenous antigens):** expressed only on specialized APCs like macrophages, dendritic cells and B cells. They present exogenous antigens via endocytic pathway *i.e.;* extracellular pathogens are internalized into endosomes via endocytosis and MHC-II molecules move from Golgi to endocytic compartment simultaneously. Inside the endosomes, the antigen is processed into peptides and gets associated with MHC-II. Finally, they move to the plasma membrane of APC and are ready to be presented to the helper T cell.

1.5 LYMPHOID ORGANS

Immune system comprises various organs and tissues having distinct functions. These are majorly classified into primary and secondary lymphoid organs also referred to as central and peripheral lymphoid organs respectively.

1.5.1 Primary Lymphoid Organs

They are the sites where lymphocyte development and maturation take place as they provide favorable environment for their growth. Therefore, the immature lymphocytes formed in hematopoiesis migrate to primary lymphoid organs for maturation. Maturation means that the lymphocytes attain immunocompetency (ability to elicit an immune response) and those recognizing self-antigens are negatively selected. There are two primary lymphoid organs, red bone marrow and thymus where B and T lymphocytes mature respectively. Although lymphocytes mature in primary lymphoid organs, they are still naïve (not interacted with the antigen) hence maturation of lymphocytes in primary organs is an antigen-independent process.

1.5.1.1 Bone Marrow

Bone marrow is gelatinous soft tissue present in the central medullary cavity of long bones where stromal cell niches secrete various growth factors and cytokines for the growth and proliferation of blood cells. It is a site for haematopoiesis where hematopoietic stem cell (HSC) gives rise to all the blood cell types in the human body. Stem cells differentiate into the myeloid and lymphoid lineage of cells in a hematopoietic inducing microenvironment (HIM) provided by stromal cells. Different cells after differentiation migrate to distinct tissues and organs for maturation but immature B cells remain in bone marrow only for the maturation process. Two chief mechanisms operate during B cell maturation i.e., receptor editing (expression of surface antibody) and negative selection. B cells that recognize self-antigens undergo programmed cell death (apoptosis) whereas that express appropriate receptors are proliferated. Finally, mature and naïve B cells migrate to secondary lymphoid organs for activation. Bone marrow also serves as a repository for immune effector cells, i.e., antibody-producing plasma cells and contributes to 90% of antibodies IgG and IgA in blood plasma. Bone marrow is also the site for fat depot and with age, around 50% of the area gets filled with fat (yellow marrow). Some physiological conditions like anaemia or blood loss lead to extra production of blood cells under the action of the hormone erythropoietin secreted by kidneys. However, the site of B cell maturation in birds is bursa of fabricius. In sheep and cattle, the fetal spleen is the site for B cell maturation and later Peyer's patches in the intestine.

1.5.1.2 Thymus

Thymus is a lobed organ (bilobed in mice and multilobed in humans) situated just behind the sternum and above the heart. It grows in size from infancy until puberty and then regresses with age in adulthood. Eventually, it fills with fat and shrinks to only 5 g in elders. Thymus, as a lymphoid organ is very important during the early development years as T lymphocytes formed in bone marrow undergo maturation in the thymus. Maturation process involves both negative selection and positive selection (commonly referred to as thymic selection). It also releases a hormone called thymosin that regulates T cell development. Fully mature T cells leave the thymus and enter into secondary or peripheral lymphoid organs where they encounter antigens. Anatomically thymus gland is pinkish and covered with a thick capsule. Each lobe is further divided into lobules by trabeculae (connective tissue strands). Centrally medulla region is presently surrounded by a peripheral cortex (Figure 1.2). Cortex contains dense populations of dividing thymocytes (T cells) as well as cortical epithelial cells and nurse cells that provide nutrition to developing thymocytes. T cell maturation and differentiation take place in the cortex. Medulla has a comparatively less number of thymocytes. Degenerating epithelial cells are organized into concentric layers as Hassall's corpuscles are also seen in the medulla region. After maturation naive T cells migrate to peripheral organs for antigen encounter and activation.

1.5.2 Secondary Lymphoid Organs

Secondary or peripheral lymphoid organs are spread throughout the body and monitor the pathogens entering body fluids, i.e., tissue fluid, blood and lymph. They are the sites for activation of lymphocytes, interaction with antigens and induction of immune response. Therefore, the proliferation of lymphocytes in secondary lymphoid organs is antigen-dependent. Antigen-activated lymphocytes proliferate and differentiate into effector and memory cells. Hence, these organs are densely populated with lymphocytes. Examples include spleen, lymph nodes, Peyer's patches, tonsils and Mucosal Associated Lymphoid Tissue (MALT).

1.5.2.1 Lymph Node

Lymph nodes are bean-shaped glandular in appearance and are present all along the lymphatic system in the human body. Lymph nodes are abundant in regions like armpits, inguinal, perihilar, cervical etc. The lymphatic system consists of lymph, lymphatic vessels and nodes. In the human body, fluid from blood plasma seeps into the tissues. If this fluid is not returned back to central circulation, then a condition called edema will occur, hence this lost fluid from the blood returns back by flowing into thin-walled lymphatic vessels that finally drain into veins. Inside these vessels, the fluid is called lymph. Each lymph node is covered with a dense connective tissue layer called capsule that extends inside as trabeculae. The surface of each lymph node is invaded by afferent lymphatic vessels that carry lymph containing antigens/pathogens, APCs and a few B-cells into the subcapsular sinus of the node. Further, the node is divided into an outer cortex, paracortex and inner

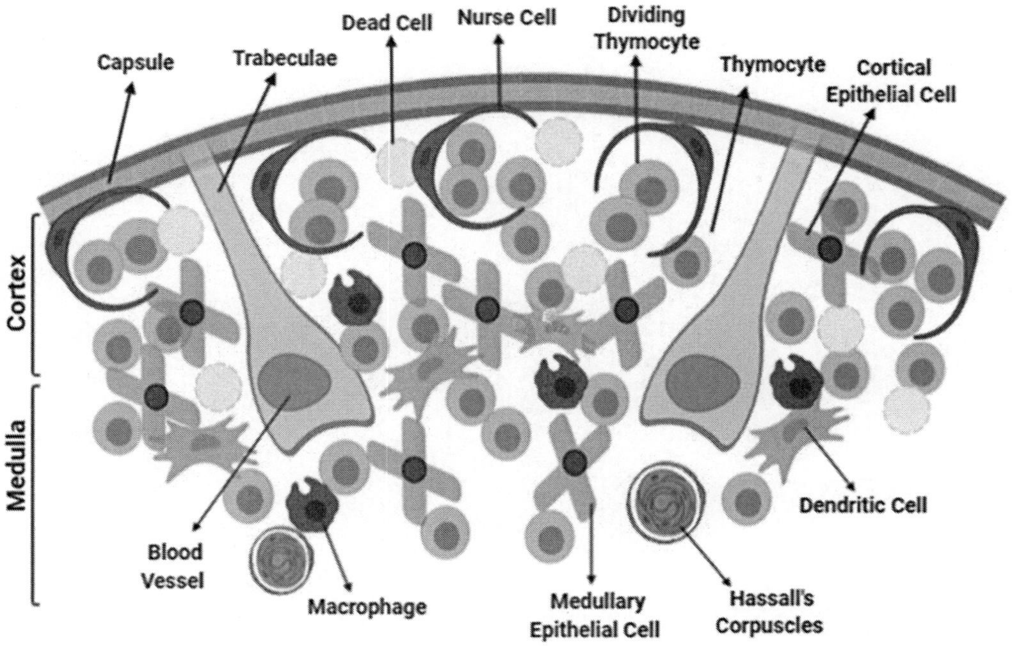

FIGURE 1.2 Diagrammatic view of Thymus showing a thick capsule covering and strands of connective tissue diving lobes into lobules. Outer cortex with dividing thymocytes and medulla with Hassall's corpuscles can also be seen. (Adapted from Kindt, T. J., R. A. Goldsby, B. A. Osborne and J. Kuby. 2007. *Kuby Immunology*. Macmillan. W.H. Freeman and Company, New York (6th Ed). Created using Biorender.com.)

medulla (Figure 1.3). Lymph from subcapsular sinuses enters the cortical sinuses. Cortex is filled with lymphoid follicles containing clusters of proliferating B cells and T cells with or without a germinal centre termed secondary and primary follicles, respectively. Cortex is also referred to as B cell zone. Also, macrophages trap antigens in the cortex and present them to B- cells. Paracortex is dominated by T-cells and here the antigens are presented to T cells by dendritic cells. Inner medulla mainly contains antibody-forming plasma cells. Plasma cells in lymph nodes generally survive for 3 days and produce IgG antibody. Finally, lymphocytes and antibody-rich lymph come out of the node *via* efferent lymphatic vessels through a depression called hilum from where the blood vessels leave. Therefore, lymph nodes fight against antigens arising in lymph and tissue spaces.

1.5.2.2 Spleen

Spleen is the largest lymphoid tissue specialized to trap blood-borne antigens. It is situated between the diaphragm and stomach on the left side of the abdomen. Spleen has two major functions, production of antibodies against blood-borne antigens and removal of aged and malfunctional red blood cells and platelets from the blood. Spleen is covered with a connective tissue capsule and trabeculae run into the interior of the organ called spleen parenchyma. Two main functions of spleen are performed by functionally and morphologically distinct regions red pulp and white pulp separated by a marginal zone. Red pulp is highly vascular and contains splenic cords (cords of Billroth) and sinusoids. Blood enters through the splenic artery that finally terminates into central arterioles in red pulp. Defective RBC or platelets are phagocytosed here by the macrophages, therefore red pulp is also known as the graveyard for RBC. White pulp consists of Periarterial lymphatic sheath (PALS) and marginal zone (Figure 1.4). PALS is a layer of lymphoid tissue surrounding the splenic arterioles. It is dominated by T cells and lymphoid follicles, whereas the marginal zone consists of macrophages and dendritic cells that function to trap and present antigens. Eventually upon activation, B cells in the primary follicles start proliferating and differentiating into plasma cells.

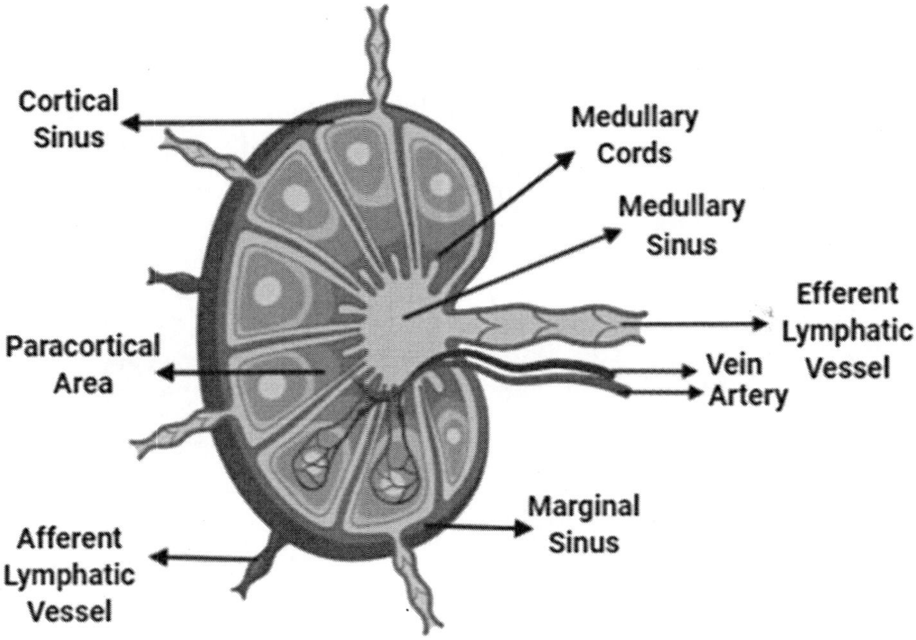

FIGURE 1.3 Diagrammatic representation of lymph node showing the covering capsule and three regions: outer cortex, deep paracortex and inner medulla. (Adapted from Janeway, C. A., P. Travers, M. Walport, and M. J. Shlomchik. 2005. *Immunobiology: The Immune System in Health and Disease*. Garland Science Publishing (6th Ed). Created using Biorender.com.)

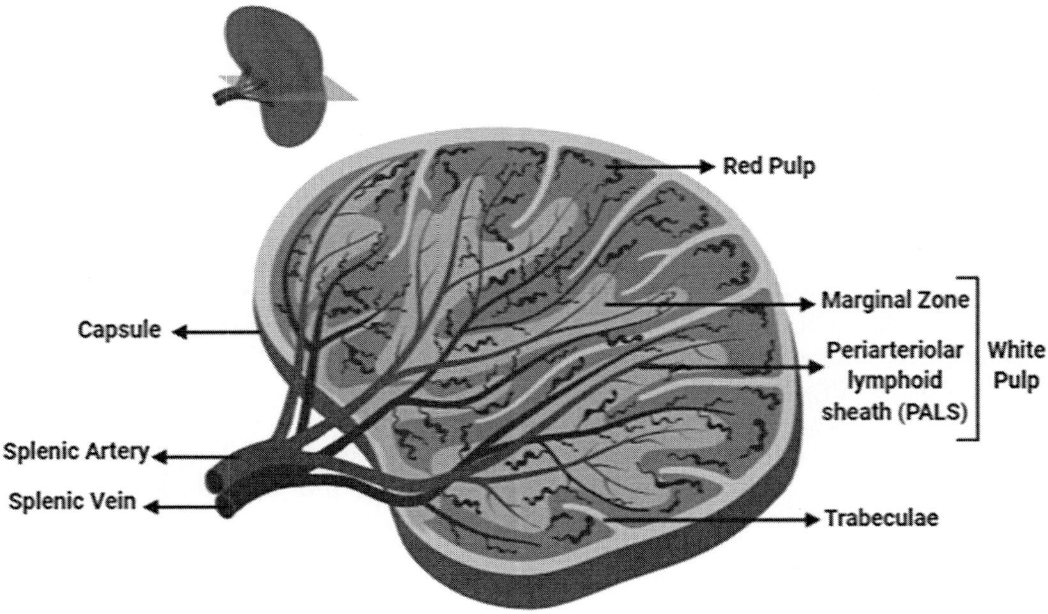

FIGURE 1.4 Secondary lymphoid organ spleen showing the red and white pulp. (Adapted from Kindt, T. J., R. A. Goldsby, B. A. Osborne and J. Kuby. 2007. *Kuby Immunology*. Macmillan. W.H. Freeman and Company, New York (6th Ed). Created using Biorender.com.)

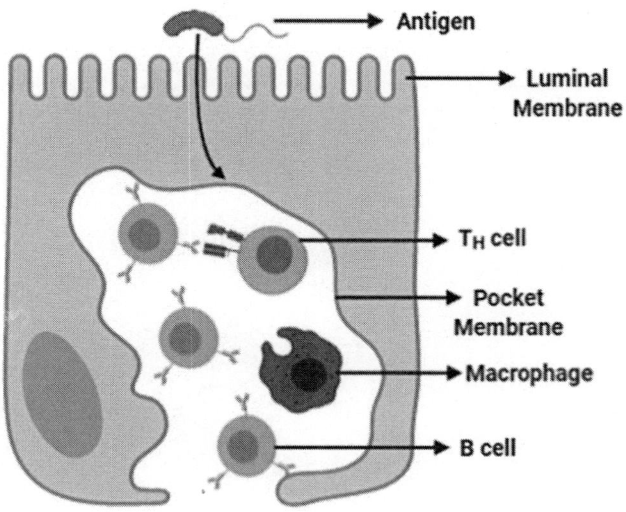

FIGURE 1.5 Structure of an M cell lacking microvilli and presence of pocket in mucosal lining of digestive tract. (Adapted from Kindt, T. J., R. A. Goldsby, B. A. Osborne and J. Kuby. 2007. *Kuby Immunology*. Macmillan. W.H. Freeman and Company, New York (6th Ed). Created using Biorender.com.)

1.5.2.3 MALT

Apart from well-organized organs, lymphoid tissue is also present as loose aggregations associated with the epithelial or mucosal lining of the digestive (Gut Associated Lymphoid Tissue: GALT), reproductive and respiratory tract (Bronchus Associated Lymphoid Tissue: BALT). However, some of these lymphoid tissues are clustered into much larger structures like Peyer's patches in the small intestine, tonsils and appendix. In MALT, many T cells in the epithelial layers as IELs (Intraepithelial lymphocytes) produce cytokines for targeted killing. In the gut, the mucosa contains special cells called M cells that lack microvilli and possess a deep pocket or invagination basolaterally (Figure 1.5). Antigen is taken up by the M cell from gut lumen *via* endocytosis, then it is transported to the pocket where antigen presentation takes place. As a result, B cells start secreting large amounts of IgA antibodies into the gut lumen. Similarly, our skin epithelium also contains Intraepidermal lymphocytes, keratinocytes and Langerhans dendritic cells that function as Cutaneous Associated Lymphoid Tissue (CALT).

1.6 TYPES OF IMMUNE RESPONSES

Body's defense mechanisms generate several types of immune responses and in the above section, two types of immune responses innate and acquired were already briefly introduced. Each type of immune response has a certain type of cells, molecules and barrier mechanisms involved which will be discussed in detail in later sections. All the immune responses elicited against different antigens are commonly known as effector responses with the difference in immune cells under action and type of molecules produced. Amongst the formed elements of blood, white blood cells (WBC) descending from the lymphoid lineage called lymphocytes have the capability to recognize and produce these effector molecules against any infection. Here we only mention about the different categories of immune responses.

1.6.1 Primary and Secondary Immune Response

The primary immune response is observed when an individual is exposed to the antigen for the first time and it takes a particular time for the formation of antibodies. During the primary response,

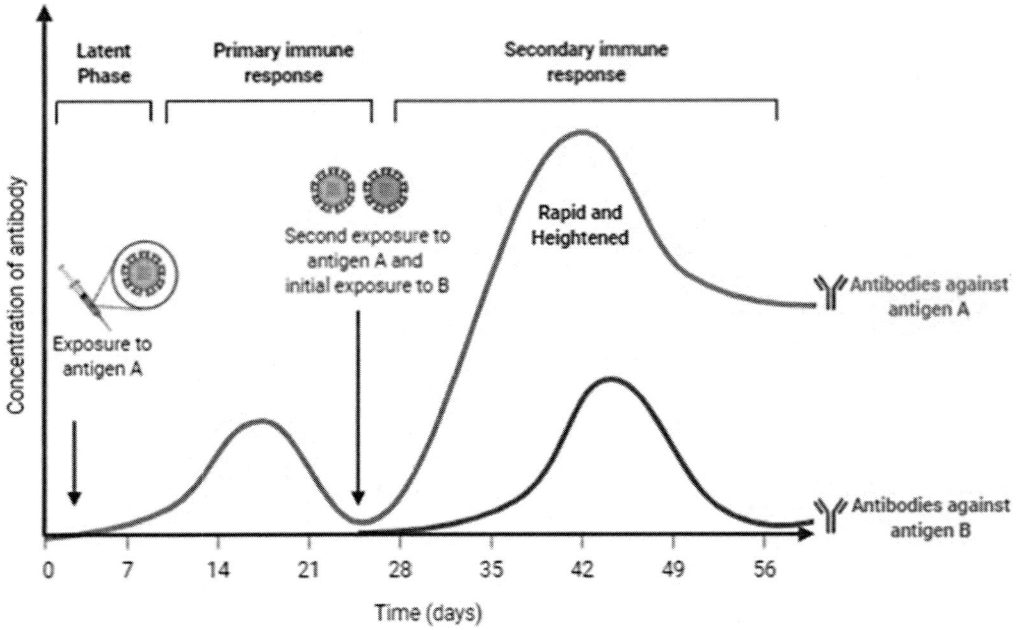

FIGURE 1.6 Primary and secondary immune response showing immune reaction against antigen A and B. Second exposure to antigen A gives heightened and rapid immune response. Created using Biorender.com.

IgM is the first antibody produced followed by IgG or other antibody classes (IgA, IgE). There are five classes of antibodies produced in the body depending on the type of polypeptide chain of heavy chain. Primary immune response can be understood more clearly with its four distinct phases (Figure 1.6):

- **Latent phase (lag):** It is the immediate phase after antigen exposure where no antibody against the antigen is detected in serum and it generally lasts for 1–2 weeks.
- **Exponential phase (log):** This is the stage of antibody production and it rises exponentially in concentration.
- **Steady state (stationary):** Antibodies formed to interact with an antigen and finally these complexes are cleared or degraded from the circulation. In the stationary stage, the amount of antibody formed and degraded are equal *i.e.* it's an equilibrium phase.
- **Declining phase:** This is the stage during which the concentration of antibody declines rapidly. This is attained when the body has fought and cleared the antigen that entered the body.

Further, secondary immune response results from second exposure to the same antigen. It is a more rapid and heightened action against the infection due to immunological memory. As compared to primary response which has a latent phase, low concentration of antibodies and short-lived; secondary immune response is quick, long-lived and more powerful. It is marked by no or very brief latent period and finally results in greater concentrations of antibodies produced in the body.

1.6.2 Active and Passive Immunity

The technique of vaccination discovered by Edward Jenner and Louis Pasteur is actually active immunity as the attenuated or weakened pathogen initiates an immune response or production of antibodies in one's own body. The aim is to provide immunity as well as immunological memory.

Table 1.2
A Comparison between Active and Passive Immunity

Active Immunity	Passive Immunity
• Healthy individuals are inoculated with attenuated or killed pathogens (or its antigenic determinants) capable of including immunity but do not cause disease	• Healthy individuals are directly given pre-formed antibodies from recovered humans or animals.
• Scientists: Edward Jenner and Loius Pasteur are recognized for their contribution for the discovery of vaccination and induction of active immunity.	• Scientist: Emil von Behring and Hidesaburo Kitasato discovered immunity can be transferred from diseased individuals to a healthy by injecting serum containing antibodies or effector immune cells, providing passive immunity.
• Exposure to antigen is required for immune response.	• Exposure to antigen is not required.
• Immunity starts developing after few days and provides long-term protection as immunological memory is there.	• Immunity is developed immediately but gives only short term/transient protection as no memory is there.

It can be achieved naturally by pathogen infection or artificially by using vaccines. Further, work done by Emil Von Behring and Hidesaburo Kitasato revealed immunity can also be attained by transferring serum-containing antibodies. As the antibodies are not produced in an individual's own body it's a form of passive immunity. Passive immunity i.e. maternal antibodies to streptococci, rubella, mumps, tetanus, diphtheria and poliovirus are naturally transferred from mother to the developing fetus via the placenta. Additionally, colostrum also contains antibodies which provide passive immunity to the infant naturally. It can be done artificially by taking serum from one diseased person having pre-formed antibodies and injecting it into another person. But passive immunization has several limitations; it provides only short-term protection, no immunological memory, pre-formed antibodies in different species than horse can also mount a response against foreign antibody and cause complications. Refer to Table 1.2 for differences between active and passive immunity.

1.6.3 Innate and Adaptive Immunity

Non-specific defenses of the body against infections present since birth is known as the innate immune response. It is the first line of defense and it is not specific to any particular pathogen rather common to any foreign molecule or antigens. Also, it is immediately functional to fight against the broad range of pathogens and does not require prior exposure to the pathogens. In contrast, adaptive immune responses are specific to recognize and eliminate antigens selectively providing second line of defense and generating memory. Subsequent exposures to the same antigen generate more rapid and powerful response (Table 1.3). As already mentioned, response and self/non-self-recognition are the standard attributes of any immunity, but adaptive immunity is characterized by its four hallmarks:

- **Self/non-self-recognition:** property of distinguishing foreign molecules from body's own molecules and generating immune response against non-self and avoiding against self, termed as self-tolerance.
- **Immunological memory:** initial contact with the pathogen generates an array of memory cells that are responsible for quick and robust response during second exposure to the same pathogen.
- **Specificity:** it is specific to particular antigens
- **Diversity:** it is the ability to respond to diverse and vast variety of antigens even if the body has not encountered them before.

Table 1.3
Difference between Innate and Adaptive Immunity

Property	Innate Immunity	Adaptive Immunity
Response time	Rapid within hours	Slow, may take few days to weeks
Second exposure	Same intensity response (primary)	Heightened and rapid response (secondary)
Specificity	No specificity	Highly specific to antigens
Diversity	Inadequate or limited	Extreme high
Memory	Absent	Present
Distribution	Many invertebrates and vertebrates	Evolved in Gnathostomes (jawed vertebrates)
Cellular barriers	Natural killer cells (NK), macrophages, neutrophils	B and T lymphocytes, antigen presenting cells (APC)

1.6.3.1 Components of Innate Immunity

Innate immunity possesses several barriers to fight against the infections even in the absence of encounter with antigen. Here we will discuss different types of barriers and cells that are involved in the non-specific defenses against microbes. Initially, physical barriers play an important role in blocking the entry of pathogens into tissues where they can cause infection. Skin epithelium prevents the entry of microbes; however, if a portion of skin is damaged, pathogens may gain entry from the injured site. The epidermal layer of skin contains keratinocytes keratin that makes the surface of the skin resistant and tough to degrade by pathogens. Sweat and sebaceous glands provide an acidic and salty environment that inhibits the growth of microbes. The mucous membrane lining the digestive and respiratory tract act as a non-specific barrier as they trap the microorganisms that gain entry into the body. This is accomplished by their sticky and moist secretions called mucus. Table 1.4 summarizes several physical barriers and their mode of actions.

Physical barriers like skin are composed of cells like macrophages that release various antimicrobial peptides against pathogens. Chemical mediators of innate immunity include various enzymes, cytokines, antimicrobial proteins, pattern recognition receptors (PRRs). Enzymes like lysozyme present in secretions like tears, saliva ad sweat degrade the bacterial cell wall. Cytokines are intercellular immunoregulatory molecules and include lymphokines, chemokines, interleukins, interferons etc. Cytokines bind to the cells adjacent to the cells that produce and secrete them and initiate a signal transduction pathway for an effective response. Antimicrobials involve proteins that act both locally and at distance. Local antimicrobials include defensins (α and β), cathelicidins, magninins, drosomycin etc. Proteins acting at distance generally involve complement and acute phase proteins. Complement proteins are glycoproteins that circulate in the blood and are produced by the liver. These plasma proteins can be triggered by pathogens directly as well as indirectly by

Table 1.4
Different Physical Barriers and their Mode of Action Involved in Innate Immunity

Barrier	Functional Component/Defense	Mode of Action
Skin	Keratinocytes	Prevent entry of microbes
Mucous membrane	Mucus and endothelial cells	
Ciliary movement	Cilia	Removal of pathogens from infection sites
Shedding of skin and Peristalsis	Mechanical defense	
Microbiome	Resident microbes of digestive, respiratory, urinogenital tract and skin	Competition with invaders for nutrition and colonization

Cell	Morphology	Lineage and Function
Granulocyte Neutrophil		Myeloid Progenitor; Phagocytosis of pathogens by bactericidal mechanisms
Granulocyte Eosinophil	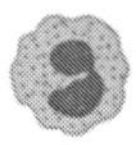	Myeloid Progenitor; Phagocytosis of parasites
Granulocyte Basophil		Myeloid Progenitor; Promotion of allergic reactions
Macrophage		Myeloid Progenitor; antigen presenting cells and phagocytosis
Mast Cells		Myeloid Progenitor; Release of histamine granules, allergic responses
Natural Killer (NK) cells		Lymphoid Progenitor; natural killing of target cells
Dendritic Cells		Myeloid and Lymphoid Progenitor; antigen uptake and presentation

FIGURE 1.7 Types of cells involved in innate Immunity and their activated functions.

antibody bound with antigen (pathogenic determinants) finally resulting in cascade of reactions and several effector functions for the elimination of the pathogens. Therefore, the complement system bridges innate and adaptive immunity.

Another set of proteins synthesized by the liver as an innate body defense is highly conserved acute phase proteins (APP). Their production is induced by different proinflammatory cytokines (immunoregulatory molecules) like IL-6, INF-γ and TNF in response to acute phase response. Two of the most important and extensively studied APPs are mannose-binding lectin (MBL) and C-reactive protein (CRP). These proteins influence phagocytosis of antigens by opsonization (a process by which a molecule known as opsonin binds to the pathogen and marks it more susceptible for phagocytosis). Therefore, these proteins function as opsonin.

If a pathogen escapes any physical or chemical barrier then specialized cells of innate immunity play an important role in body's defense. Innate immunity cells primarily function as phagocytic cells, secreting inflammatory mediators and killing (Figure 1.7). Phagocytosis is carried out largely by neutrophils, eosinophils, blood monocytes, tissue macrophages and dendritic cells. The next class of cells are basophils, mast cells and macrophages that induce inflammation. Finally, NK cells provide immunity against viral infections.

1.6.3.2 Receptors of Innate Immunity

As already mentioned above, innate immunity recognizes molecular patterns called PAMPs shared by a group of pathogens required for their survival in the host and cannot distinguish every single probable antigen. PAMPs include components of bacterial and fungal cell walls, flagellar proteins, viral RNA and many more. Further, molecules exclusively present on damaged or dead human cells can also be recognized as a part of innate immunity often referred to as damage-associated molecular patterns or DAMPs. PAMPs or DAMPs act as ligands and are recognized by pattern recognition receptors (PRR). PRRs are broadly classified into three types:

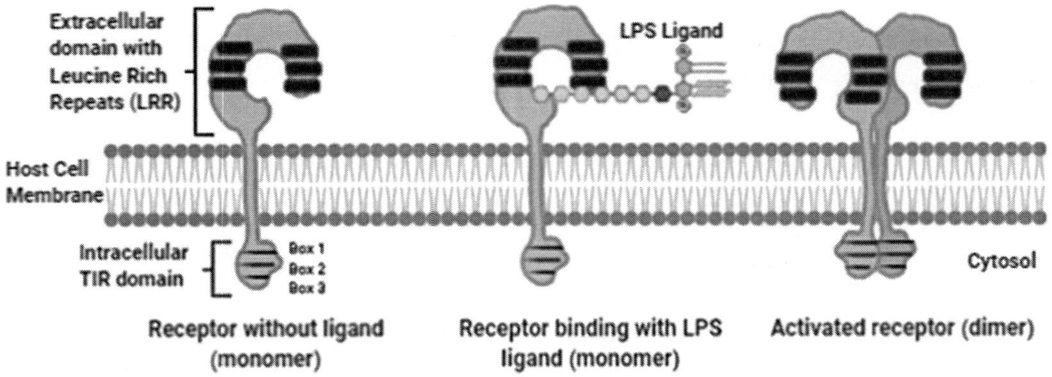

FIGURE 1.8 Structure of a toll-like receptor (TLR) on innate immune cell membrane showing extracellular and intracellular domains. Receptor binding with the appropriate ligand (PAMPs on pathogens) leads to receptor dimerization and activation. Activated receptor transduces the signal to the inside of the cell and effector response is initiated. (Adapted from Kindt, T. J., R. A. Goldsby, B. A. Osborne and J. Kuby. 2007. *Kuby Immunology*. Macmillan. W.H. Freeman and Company, New York (6th Ed). Created using Biorender.com)

- **Soluble or circulating PRRs:** They are serum proteins circulating in the body that bind to components on the cell surface of invading microbes. They are major of three types collectins, lectins and pentraxins. Acute phase proteins like MBL and CRP mentioned above also serve as soluble PRRs.
- **Membrane bound/cellular PRRs:** They are membrane-bound receptors associated with various immune cells. Of these most extensively studied are Toll-like receptors (TLRs). TLRs are typical transmembrane signal receptor proteins with extracellular N-terminal and intracellular C-terminal domains (Figure 1.8). The extracellular domain is constituted by leucine-rich repeats (LRRs) with formula xLxxLxLxx serving as ligand binding site for PAMPs. Intracellular domain commonly known as TIR (Toll/IL-1 receptor) comprises three conserved boxes that transduce the signal from the receptor to the inside of the cell. Several classes of TLRs are known that recognize different PAMPs and at least 11 are known in humans. Some of the major PAMPs and their corresponding TLRs are mentioned in Table 1.5. Another class of membrane-bound receptors are the scavenger receptors on the surface of macrophages that bind to microbes and facilitates phagocytosis.
- **Intracellular PRRs:** Some receptors are located inside the cells either in cytosol or endolysosomes. Examples include a few TLRs (TLR7/8/9/10) and NOD-like receptors (NLRs).

1.6.3.3 Components of Adaptive Immunity

As mentioned in types of immunity, adaptive immunity serves as the second line of defense and elicits a specific immune response with four hallmarks specificity, diversity, memory and self/non-self recognition. Adaptive immunity is named so because it is acquired as an adaptation during the lifetime of an individual. Like innate immunity, it has specific cells and its associated receptors that recognize different antigens. Adaptive immunity comes into play only after interaction with antigens, it can be active or passive as mentioned earlier. There are specialized cells, antigen-specific receptors and different mechanisms like gene rearrangements for diversity and clonal expansion associated with adaptive immunity. Adaptive and innate immunity work together in coordination and are interdependent in many ways. Broadly three cell types govern an acquired immune response. These are B lymphocytes and T lymphocytes derived from a lymphoid progenitor in haematopoiesis (Figure 1.9a). Another important class of cells involved in adaptive immunity are antigen-presenting cells (APC) like macrophages and dendritic cells derived from a myeloid progenitor cell.

Table 1.5
Major PAMPs and their Recognition by Different TLRs (Modified from Carmody and Chen 2007)

Pathogen	PAMP	Cellular Component	TLR
Gram-positive bacteria	Peptidoglycan	Cell wall	TLR2
	Lipoteichoic acids	Cell wall	TLR6/TLR2
Gram-negative bacteria	Lipopolysaccharide	Cell wall	TLR4
Bacteria	Unmethylated CpG DNA	Intracellular (Genome)	TLR9
	Flagellin	Cell surface	TLR5
	Porins	Cell surface	TLR2
	N-formyl methionine	Intracellular	
Virus	ssRNA	Intracellular	TLR8, TLR7
	dsRNA	Intracellular	TLR3
	DNA	Intracellular	TLR9
Fungi	Glycans	Cell wall	TLR4/2
	Zymosan	Cell wall	TLR6/2
Protozoa	Profilin-like	Cell surface	TLR11
	Hemozoin	Intracellular	TLR9
	Glyco-inositol phospholipids	Cell surface	TLR4
	GPI anchor	Cell surface	TLR2, TLR6

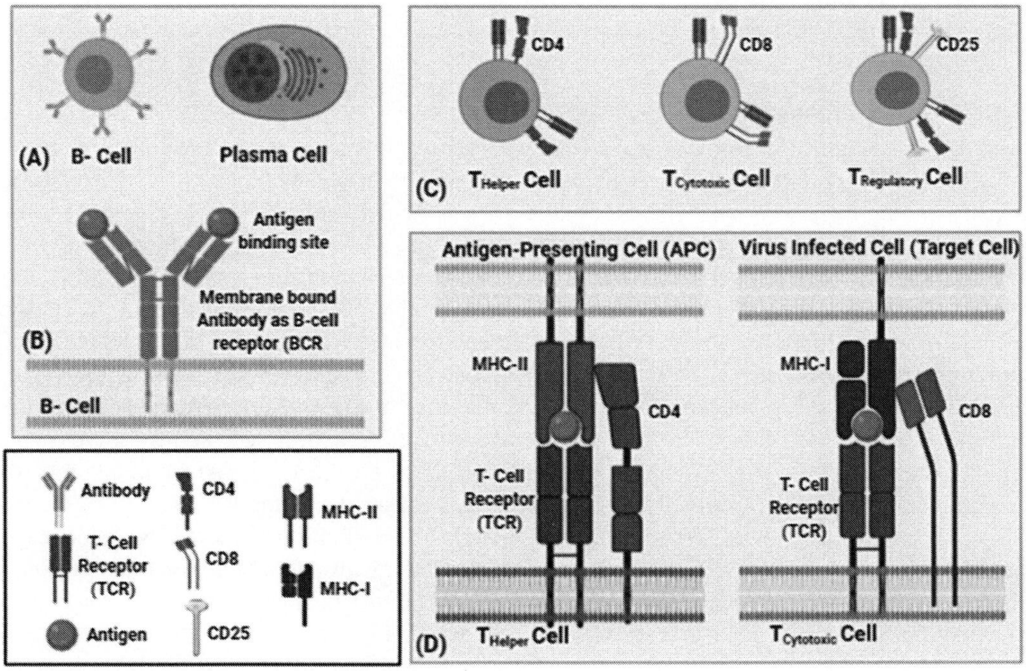

FIGURE 1.9 Cells and receptors adaptive immunity. (A) B lymphocyte with membrane bound antibody as BCR and effector plasma cell (B) Structure of BCR (Immunoglobulin) showing antigen binding site (C) Different types of T lymphocytes with the specific surface marker (D) Antigen presentation to the $T_{cytotoxic}$ and T_{helper} cells showing interaction of TCR and CD molecules with MHC-antigen complex. Target cells with MHC I present antigen to $T_{cytotoxic}$ cells bearing CD8 marker and specialized antigen presenting cells with MHC-II present antigens to T_{helper} cells expressing CD4 marker. (Kindt, T. J., R. A. Goldsby, B. A. Osborne and J. Kuby. 2007. *Kuby Immunology*. Macmillan. W.H. Freeman and Company, New York (6th Ed). Created using Biorender.com.)

1.6.3.4 Receptors of Adaptive Immunity

Like TLR receptors of innate immunity, B and T cells also possess receptors for pathogen recognition known as BCR (B-cell receptor) and TCR (T-cell receptor), respectively. However, unlike TLRs, BCR and TCR are antigen-specific and display enormous diversity and specificity. They can recognize a vast variety of antigens and this diversity arises primarily due to gene rearrangements at the developmental stage.

B-cell receptors (BCRs) are actually membrane-bound antibodies (monomeric forms of IgD and IgM). An antibody has a Y-shaped structure with two identical heavy and two identical light chains joined by disulphide bonds. Each chain consists of a constant and variable region, variable region has antigen binding sites on the exterior. Constant regions of heavy chain are embedded in the plasma membrane (Figure 1.9b). Typically, each B cell membrane has about 100,000 receptors and all of the BCRs on a single cell have the same specificity i.e. they bind to the same antigen. BCR also contains signal transducing molecules like CD79a and CD79b associated with immunoglobulins. These molecules don't bind to the antigens but function to transduce signals to the cell nucleus. Other molecules that help in signal transmission are present as B cell co-receptor and consist of CD81, CD19 and CD21. It also expresses MHC-II molecules as it acts as an APC. Also, receptors for complement proteins are present for activation of the complement pathway and opsonization.

T-cell receptors (TCRs) are also antigen-specific and structurally it is a heterodimer of two glycoprotein transmembrane chains α and β connected by disulphide bonds. The extracellular domains resemble the variable and constant domains of immunoglobulin. Variable regions possess the antigen binding sites. Transmembrane regions have positively charged amino acid residues. Co-receptor molecules on the T cell membrane like CD4 (T helper cells) and CD8 (cytotoxic T cells) bind to the MHC II and MHC I molecules respectively (Figure 1.9c and d). Again, the vast diversity of TCRs comes from genetic rearrangements. Like BCR, all TCRs on a single cell have the same antigen specificity.

1.6.3.5 Clonal Expansion

During the development of B and T cells, selection operates to eliminate cells that possess defective BCRs and TCRs. Cells with functional receptors undergo extensive genetic rearrangements in the genes coding for heavy and light chains (B cells) or α and β chains (T cells). These genes are divided into several segments commonly known as variable (V), diversity (D) and joining (J). It is rearrangements of these segments in all possible combinations that result in a repertoire of diverse receptors with different specificities. Therefore, a great number of lymphocytes develop from each lymphoid progenitor cell. Organs that provide microenvironment for lymphocyte development and maturation are known as primary lymphoid organs (bone marrow and thymus). Naive lymphocytes interact with antigens in secondary lymphoid organs (lymph nodes, spleen). After interaction with a foreign molecule, the lymphocyte gets primed and it starts to divide into the progeny of cells. Each daughter cell of progeny is a clone of the parent cell *i.e.*, genetically identical, in other words it displays the same antigen specificity. Hence, the antigen selects the lymphocyte-specific for it and leads to its clonal proliferation. This antigen-driven selection and division of specific lymphocytes is termed as clonal expansion. Further, the clone of cells differentiates into effector cells and memory cells. While effector cells generate immune response for clearance of antigen, memory cells remain in the body for subsequent exposures to the same antigen. Clonal Expansion for a B cell is diagrammatically explained in Figure 1.10. There are four basic principles of clonal selection and expansion:

- Receptors on a single lymphocyte have the same specificity.
- Binding of a specific antigen with the corresponding lymphocyte with high-affinity results in activation and clonal proliferation of that lymphocyte.
- Clones of lymphocytes bear the same receptor specificity as that of the parent cell.
- This clone of cells differentiates into effector and memory cells for primary and secondary immune responses, respectively.

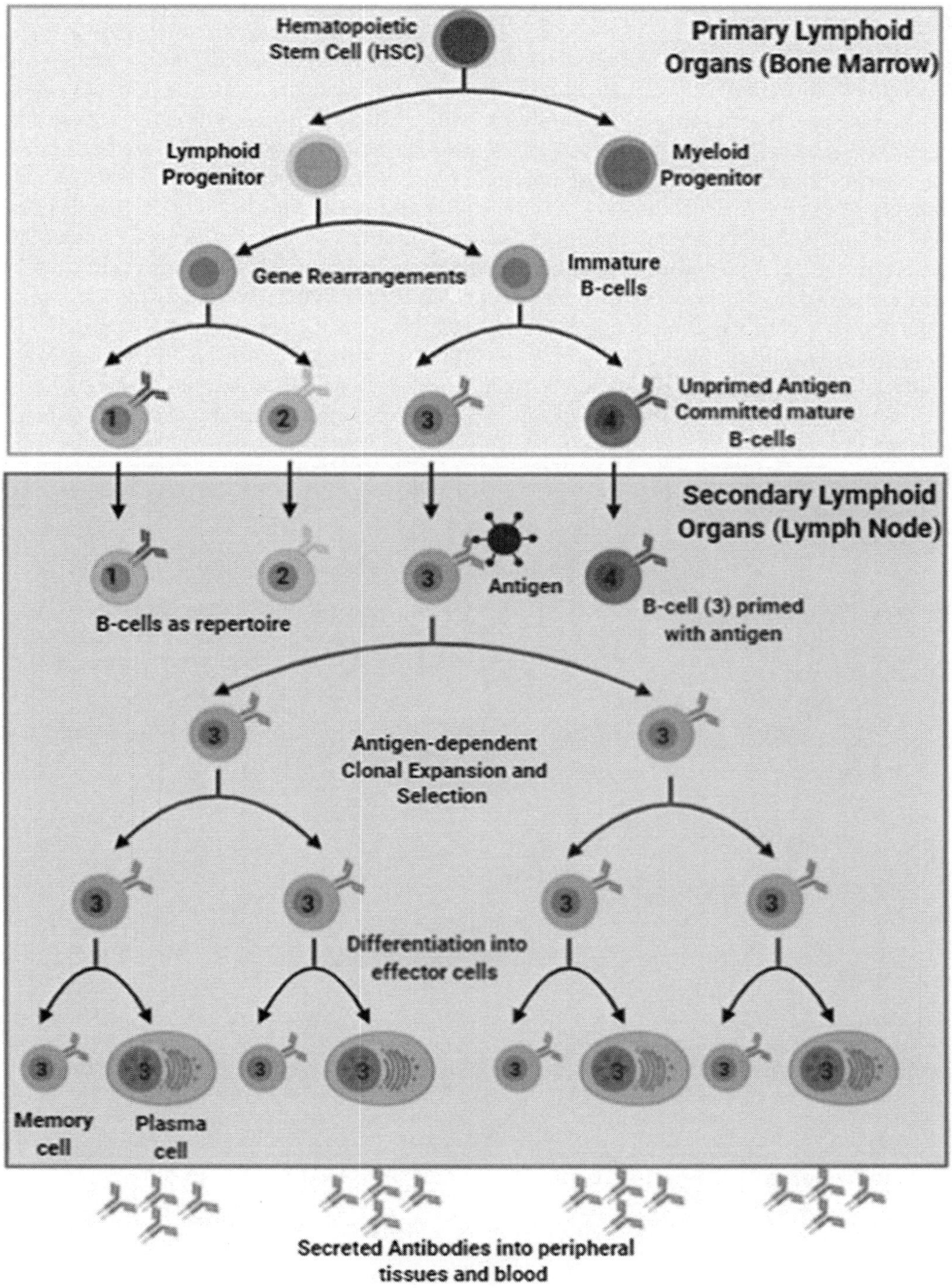

FIGURE 1.10 Antigen mediated clonal selection and proliferation of B lymphocytes. In this figure, B cells shown in different colours are specific for four antigens. For instance, if antigen 3 comes into contact with its complementary B cell receptor, it binds with high affinity and selects that lineage of cells leading to its clonal proliferation and generation of effector response against antigen 3. Rest of the B cells circulate in body as repertoire for diverse range of infections. (Adapted from Kindt, T. J., R. A. Goldsby, B. A. Osborne and J. Kuby. 2007. *Kuby Immunology*. Macmillan. W.H. Freeman and Company, New York (6th Ed). Created using Biorender.com.)

1.6.4 Humoral and Cell-Mediated Immunity

Adaptive immunity is best understood as two branches functioning differently with a different set of cells to eliminate the pathogen. However, they collaborate to achieve the aim of removing the antigen. One arm is referred to as humoral immunity mediated mainly by B cells (activated by helper T cells) by producing and secreting antibodies. The humoral word refers to the body fluids, as antibodies are secreted into the blood plasma, hence the name humoral immunity. Humoral immunity deals with freely circulating or extracellular pathogens. Antibodies in the blood plasma or lymph bind to these pathogens and either destroy, neutralize, or opsonize them. The other arm is referred to as cell-mediated or cellular immunity mediated by T cells against intracellular pathogens. Antigens displayed by APC on their cell surface or infected cells activate the T cells. Helper T cells secrete cytokines that perform several effector functions including the differentiation of cytotoxic T cells into CTLs (effectors). CTL then kills the target cell with cytotoxicity (Figure 1.11). Any cell that displays an abnormal expression of MHC molecules including virus-infected host cells, tumour cells or transplants is recognized by the cellular branch and a response is generated.

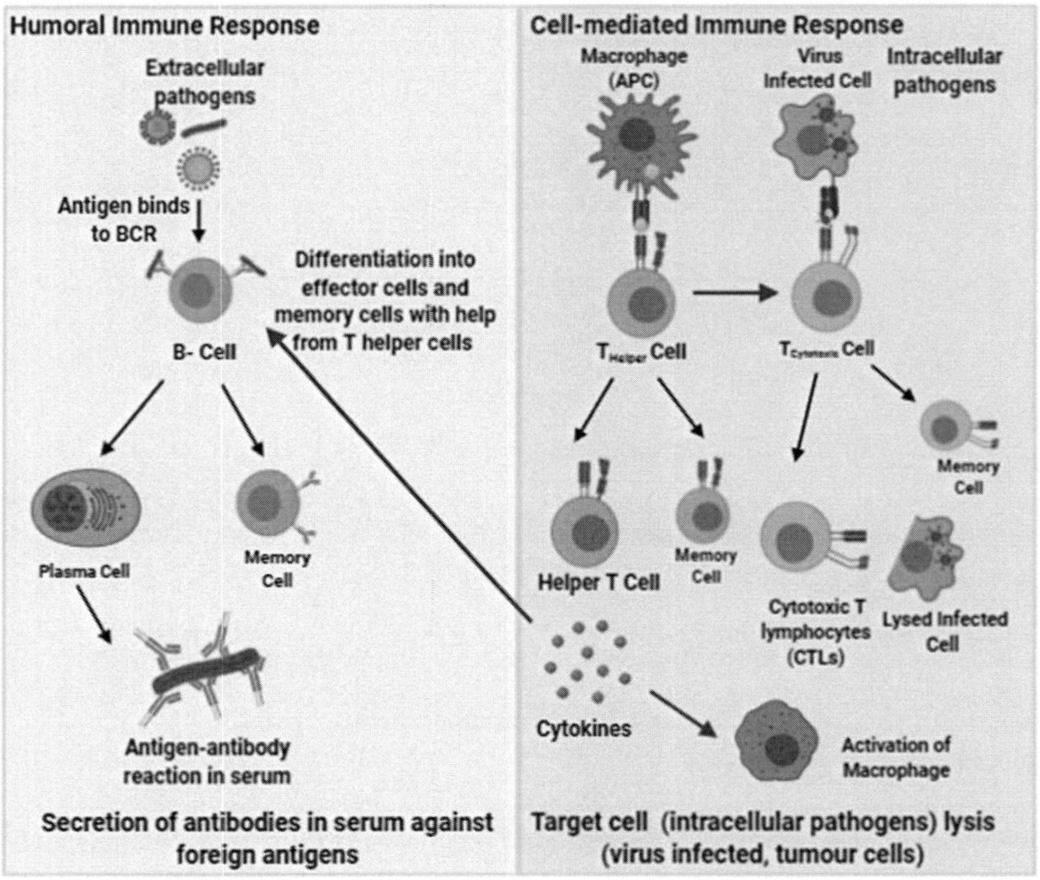

FIGURE 1.11 Humoral and Cell-mediated immune responses mediated by B cells and T cells respectively. (Adapted from Kindt, T. J., R. A. Goldsby, B. A. Osborne and J. Kuby. 2007. *Kuby immunology*. Macmillan. W.H. Freeman and Company, New York (6th Ed). Created using Biorender.com.)

1.7 IMMUNE DYSFUNCTIONS

We have studied briefly in the above overview that the immune system functions in various ways to protect the host from infections and other harmful pathogens. But what happens when the defense mechanisms do not function the way it should optimally? Improper functioning of the immune system leads to a variety of disorders arising primarily due to deficiency or over active immunity or autoimmunity. The dysfunctions can be acquired or congenital since birth. There are four chief categories of immune dysfunctions:

1.7.1 Immunodeficiency

An individual may have a fully or partially impaired immune system, and therefore less efficient in the fight against infections. These individuals are immunocompromised with a high risk of infections. The vulnerability of such individuals to certain conditions is more than normal individuals. Therefore, an individual having an immunodeficiency falls victim to recurrent infections by the same or different pathogens. Immunodeficiency can be of two types:

- **Primary immunodeficiency (PID)/congenital:** Individual has a weak immune system since birth i.e. congenital (resulting from genetic disorders) or inherited. Examples include B cell and T cell immunodeficiencies, SCID (severe combined immune deficiencies), phagocyte and complement defects. The survival and life span of affected individual depends on the condition and may be corrected during the initial stages by bone marrow transplant (BMT)
- **Secondary immunodeficiency (SID):** Individual acquires immunodeficiency due to certain environmental factors like chemotherapy or diseases like HIV. These are more prevalent than PIDs and develop as a result of a disease or other factors that can be treated by resolving the primary cause of it. Examples include malnutrition, chronic infections, acquired immunodeficiency syndrome (AIDS) and result of application of cytotoxic drugs.

1.7.2 Autoimmune Disorders

It is a condition where the immune system responds inappropriately to self-antigens. Under ideal conditions, the immune system can recognize between self and non-self components and does not elicit a response against self, a mechanism commonly known as self-tolerance. Autoimmunity arises due to breakdown or suppression of self-tolerance as a result of which the immune system starts attacking your own body i.e. injury to self. The disorders caused by autoimmunity are known as autoimmune disorders. It is mediated by releasing autoantibodies and self-reactive T cells (against self-antigens). The development of an autoimmune disorder is influenced by several factors like genetic, hormonal, viral, environmental and neuro-immunological. Therefore, it is essential to first understand the underlying factor or mechanism of autoimmune disorder in an individual before deciding a corrective therapy so that the normal immune functions are not hampered. Autoimmunity was first described as self-toxicity by Paul Ehrlich by using the term "Horror autotoxicus". There are over 100 different types of autoimmune disorders associated with various organs and tissues like blood, bones, muscles, thyroid gland, lungs, skin, intestinal tract and nervous system. Autoimmune diseases can be antibody-mediated (Graves' disease, Myasthenia gravis, Systemic lupus erythematous (SLE) etc.) or T cell-mediated (Diabetes mellitus type I, Multiple sclerosis, Rheumatoid arthritis etc.). Autoimmune disorders can be classified as:

- **Organ-specific:** immune response to target self-antigen is limited to a single gland or organ by direct cellular damage or blocking normal functions of the target organ by autoantibodies. Examples include Multiple sclerosis, Graves' disease, Hashimoto's thyroiditis, Pernicious anaemia, and Myasthenia gravis.

- **Systemic:** Immune response is against a vast range of antigens affecting multiple systems in the body. Examples include Rheumatoid arthritis, SLE, and Scleroderma. One of the most widely studied autoimmune disorder is SLE in which autoantibodies to several nuclear antigens develop and leads to widespread damage in tissues like skin, heart, blood, muscles, lungs and kidneys.

1.7.3 Allergies and Hypersensitivity

The immune system normally responds to antigens and generates an inflammatory response. However, under certain conditions, the immune system can produce unsuitable and an exaggerated response results in damage to the tissues. This inappropriate response of the immune system is termed as hypersensitivity or allergy and the antigens responsible are known as allergens. It has two components; the initial dose for priming the immune system and the shocking dose (second dose) that results in damaging effects. Hypersensitivity can result from immune response within minutes or hours after exposure to antigen termed as immediate type or, response takes few days termed as delayed-type hypersensitivity. Gell and Coombs (1963) proposed a classification system for hypersensitivity into four major types based on mediator molecules and arms of immune system involved.

- **Type I hypersensitivity (humoral branch):** commonly called as allergy and antibody mediated (IgE). It occurs in two forms anaphylaxis and atopy. Effector cells are mast cells and basophils. E.g. Food and drug allergy, asthma
- **Type II hypersensitivity (humoral branch):** Antibody mediated cytotoxic hypersensitivity (IgG and IgM). E.g. Hemolytic disease of the new born
- **Type III hypersensitivity (humoral branch):** Immune complexes lead to tissue damage by activating complement system. E.g. SLE and Arthus reaction
- **Type IV hypersensitivity (cell-mediated branch):** Delayed type hypersensitivity (DTH) mediated by T cells and macrophages. E.g. Graves' disease

1.7.4 Graft Rejection and Graft-versus-Host Disease

Graft refers to a cell, tissue or organ transplant from one organism to another of the same species (allograft) or from one site to the other in the same organism (autograft) as a replacement therapy for damaged tissues. Transplantation of cells or tissues between two different species known as xenograft. The immune system of the body generates a response against these transplants from other organisms as it is treated as foreign tissue. This immune response of the recipient is directed toward the destruction of the transplanted organ from the donor leading ultimately to Graft rejection. Therefore, many parameters need to be matched between recipient and donor before transplantation including blood group, tissue typing and recipient's serum responses. Additionally, immunosuppressive drugs are given to the recipient to alleviate the chance of graft rejection and promote its long-term survival. As these drugs are nonspecific, it compromises with the normal immune system of the recipient and the susceptibility to several other infections increases substantially as compared to healthy individuals. The basic mechanism of graft rejection is the generation of antibodies against foreign antigens in transplanted tissue that react with blood cells of the transplant leading to clotting and cutting off blood supply to the transplant eventually leading to its rejection. Generally, there are three clinical stages of graft rejection

- **Hyperacute rejection:** Immediate rejection within minutes to hours after transplantation
- **Acute rejection:** Rejection within first 3–6 months of transplantation.
- **Chronic rejection:** Years after acute rejection leading to complete transplantation failure.

Graft-versus-host disease (GvHD) is literally opposite reaction as that of graft rejection. In GvHD, stem cells or bone marrow of the donor's transplant recognizes the recipient body as foreign and eventually it starts generating an immune response against the body. It occurs in two forms, acute and chronic with a complex pathophysiology involving multiple interactions between immune cells of recipient and donor transplant. GvHD might affect various organs of the recipient body. Again, immunosuppressants are the primary therapy used to reduce the risk.

1.8 VACCINES AND FUTURE CHALLENGES

The above overview explains several important components, mechanisms and effectors of immunity. If we look at the history of immunological studies, vaccination has proved to be the most effective weapon for the prevention of several infections. Edward Jenner and Louis Pasteur are considered as the pioneers of vaccination for developing vaccines of smallpox, cholera, rabies etc. Ideally, a vaccine is a formulation of inactivated or attenuated pathogen given to an individual that induces active immunity. There are many types of vaccines based on the type of antigens used in preparation such as whole organism vaccines, purified antigen vaccines, recombinant vector vaccines, DNA vaccines and multivalent subunit vaccines.

Worldwide, immunologists are working toward vaccine development as there is an utmost for a vaccine against deadly diseases like AIDS, malaria and other viral diseases. Also, existing vaccines for certain diseases need to be improved in terms of cost, efficiency, safety and delivery method. However, the process of vaccine design and development is costly, cumbersome, and very time taking with many scrutinizing steps and ethical guidelines. Therefore, it is very challenging and massive research is being carried out in various laboratories for vaccine design and development. There are two major critical steps in vaccine design. First, which arm of the immune system will be activated with the vaccine and secondly would it lead to immunological memory. After a vaccine gets developed and passes all the trials, the method of delivery and its worldwide distribution also imposes a challenge especially reaching to the developing countries at low costs. Even the most successful vaccines can lead to unforeseen side effects when administered to humans.

Although vaccination during childhood has substantially alleviated the fatality from various infectious diseases in the entire world but finding vaccines against viral diseases is still challenging. Hence, immunologists linger and explore new possibilities for immunization. The future depends on the success of therapeutic research using advanced molecular biology tools for promising vaccine development and delivery.

1.9 SUMMARY

- Immunology is the study of the body's defense system against various diseases. Infectious agents causing the diseases are known as pathogens that include bacteria, fungi, virus and protozoans.
- The defense mechanism comprises primarily two components, innate and adaptive immunity. Innate immunity provides the first line of defense present since birth. It is capable of recognizing self and self-components with the help of certain markers on foreign particles referred as Pathogen associated molecular patterns (PAMPs).
- Innate immunity involves various physical, chemical and cellular barriers to fight against pathogens. Macrophages, granulocytes and natural killer cells are the major innate immune cells that function like phagocytosis, antimicrobial activity and natural killing.
- Adaptive immunity provides a second line of defense that operates after interacting with an antigen and exhibits attributes of diversity, specificity, immunological memory and self/non-self recognition. B and T Lymphocytes are the central cells that govern two arms of adaptive immunity, humoral and cell-mediated.

- Humoral immunity refers to the production of antibodies by effector B cells known as plasma cells, antibodies further bind to extracellular antigens.
- On the other hand, cell-mediated immunity is mediated by T lymphocytes that function to kill the intracellular pathogens that have infected the host cells by cytotoxicity. Along with effector response cells, memory cells are also formed in adaptive immunity which are long-lived and stay in the body for subsequent exposures to the same antigen.
- First exposure leads to a primary immune response which takes a few days to develop after a latent period, however second exposure leads to more rapid and heightened response because of memory cells (secondary immune response).
- Capability of adaptive immunity to recognize vast variety of antigens by naive B and T cells is due to the gene rearrangements during their development stage in primary lymphoid organs like bone marrow and thymus. Later, after interaction with antigens, cells undergo clonal selection and proliferation.
- Knowing the mechanisms and components of the immune system, vaccination is used as a most effective way to prevent diseases. Typically, a vaccine consists of inactivated or attenuated pathogen that induces active immunity after delivery into the human body.
- In certain conditions, the immune system may have improper functioning leading to immune dysfunctions like autoimmunity, immunodeficiency, hypersensitivity and graft rejections.

QUESTIONS

1. Differentiate between
   I. Innate and Adaptive Immunity
   II. Primary and Secondary Immune Response
   III. Humoral and Cell Mediated Immunity
   IV. Active and Passive Immunity
   V. Natural and Artificial Immunization
   VI. Lymphoid and Myeloid progenitor cell
2. What are the different types of immune dysfunctions?
3. How is vaccine designed and developed? What are the probable challenges and risks associated?
4. How does clonal selection help in generating immune response?
5. What are the different receptors of immune cells and how do they recognize the antigens?
6. Explain the role of MHC molecules in antigen presentation?

BIBLIOGRAPHY

1. Carmody, R. J. and Y. H. Chen. 2007. Nuclear factor-kappaB: activation and regulation during toll-like receptor signalling. *Cellular and Molecular Immunology* 4: 31–41.
2. Kindt, T. J., R. A. Goldsby, B. A. Osborne, and J. Kuby. 2007. *Kuby Immunology*. Macmillan. W.H. Freeman and Company, New York (6th Ed).
3. Murphy K, Weaver C. 2016. *Janeway's Immunobiology*, 9th edn. Garland Science, Taylor & Francis Group, New York, NY.
4. Janeway, C. A., P. Travers, M. Walport, and M. J. Shlomchik. 2005. *Immunobiology: The Immune System in Health and Disease*. Garland Science, Taylor & Francis Group, New York, NY. (6th Ed).
5. Parham, P. 2014. *The Immune System*. Garland Science, Taylor & Francis Group, New York, NY.

2 Innate Immunity

Meena Yadav
University of Delhi

CONTENTS

DOI: 10.1201/9781003286424-2

2.1 INTRODUCTION

Immunity refers to the ability of body to protect against invading pathogens or antigens which have the potential to harm the body. The immune system protects us from any invading pathogen, as small as ~30 nm in size to as big as ~100 cm or more in length. Immunology as a branch of biology started in the late 19th century. The two pioneering works in the field were: Elias Metchnikff's discovery of phagocytic cells, the cells which engulf and destroy invading pathogens, and Paul Ehrlich's discovery of antibodies which could neutralize microbial toxins. While the former's discovery paved way for innate immunity, the latter's work formed the basis of acquired immunity. In 1908, Metchnikff and Ehrlich were jointly awarded Nobel Prize in Physiology or Medicine for their work in the field of immunity.

Based on the mechanism of providing immunity to the body, immunity can be broadly classified into two types: innate immunity and adaptive immunity. *Innate immunity* or innate immune response is the mechanism of protecting the body from the harmful influence of pathogens and is found in all multicellular animals and plants. It is a primitive mechanism and is spontaneous. On the other hand, *adaptive immunity* is the acquired immunity and is present only in vertebrates as it is more specialized and was probably acquired during evolution. It is discussed in detail in Chapter 3. This chapter discusses the concept and mechanisms involved in innate immune response; cells and molecules involved in innate immune response; various types of barriers that are part of innate immune response; process of inflammation; harm caused due to dysregulated innate immune responses, and regulation of immune responses.

2.2 UNDERSTANDING THE CONCEPT OF INNATE IMMUNE RESPONSE

The concept of innate immunity was put forth by Russian immunologist Elie Metchnikoff. Metchnikoff discovered that several microbes entering our body could be engulfed and digested by specific cells called macrophages, a type of phagocytic cell. He said that these cells are readily and straightaway available, without any prior exposure to pathogens. Macrophages fight a diverse and wide range of pathogens and thus, are a vital part of the innate immune response. The components of the innate immune system act very fast, may even start acting within seconds of the entry of pathogen, but these components are not very specific. Other than macrophages, innate immunity involves cells such as mast cells, dendritic cells, granulocytes etc. Granulocytes, also called as polymorphonuclear leukocytes, are diverse and have characteristic granules in their cytoplasm, and include neutrophils, which work similarly to phagocytes. Metchnikoff hypothesized that the macrophages in humans and other vertebrates have been derived from similar phagocytic cells as observed in simpler animals such as sea stars. Before cells come into action, our body has several barriers which hinder the entry of pathogens. These barriers act as the '*first gate*' to inhibit the pathogens to enter the body and include physical barriers, physiological barriers, inflammatory barriers and phagocytic barriers. Table 2.1 lists various barriers which prevent infection, and constitute an integral part of the innate immune system.

Table 2.1
Different Types of Barriers Involved in Innate Immune Response

Barrier	Function
Physical Barriers (Discussed in Detail in the Section 2.2.1)	
Skin	• Skin is the largest physical barrier • Epidermis of skin is the mechanical barrier that hampers the entry of pathogens.
Mucous membrane	• Present in respiratory, gastrointestinal and urinary tract • The symbiotic microflora present in body compete with the pathogens for the attachment sites and nutrients • Mucous lining on the inner membranes entraps the foreign microorganisms • Cilia of the surface epithelium push the foreign microorganisms out of the body
Bronchial cilia	• Help in clearing pathogens from respiratory tract
Physiological Barriers	
Temperature	• The normal body temperature in humans retards the growth of some foreign microorganisms • The heightened temperature during fever retards the growth of some other pathogens
Low pH	• The low pH in stomach (pH ~1.5–3.5) kills most of the ingested microorganisms • The acidity of skin (pH 3–5) retards the growth and colonization of microorganisms • Low pH of vagina (~pH 4.5) prevents pathogen growth
Chemical mediators (discussed in detail in Section 2.2.2)	• Lysozyme, present in tears, breaks down the cell wall in bacteria • Interferons produce an antiviral state in the uninfected cells in the body • Complement system either lyses the foreign microorganisms or opsonizes and facilitates their destruction by phagocytosis
Inflammatory Barrier (Discussed in Detail in Section 2.3.1)	
Damage of tissue and infection	• Tissue damage stimulates the vascular fluid leakage, containing phagocytic cells and antibacterial proteins, from the capillaries to the site of injury
Phagocytic/Endocytic Barriers (Discussed in Detail in Section 2.3.2)	
Various types of cells	• The cells such as neutrophils, monocytes, macrophages and mast cells are specialized to endocytose or internalize the foreign microorganisms and then break them down

2.2.1 Physical Barriers

The physical barriers are tough and intact barriers that inhibit the entry and colonization of various types of microorganisms which can possibly be pathogens. They include:

i. **Epithelial layer of the skin:** The skin consists of two layers *i.e.* a thin outer epidermis and a thicker dermis underneath. Epidermis consists of several layers of tightly packed epithelial cells and its outer layer consists of mostly dead cells which are packed with a waterproof protein keratin. On the other hand, dermis consists of connective tissue which contains blood vessels, hair follicles, sebaceous glands, sweat glands and scattered leucocytes like dendritic cells, macrophages and mast cells. The sebaceous glands produce a secretion called as 'sebum' which is rich in lactic acid and fatty acids and maintains the pH of the skin between 3 and 5. This low pH inhibits the growth of many microorganisms. However, if the skin breaks due to wounds or abrasion, such regions become the routes of infection.

ii. **Epithelial layer of mucosal and glandular tissues present near the openings in the body:** The mucosal epithelial layer lines the respiratory, gastrointestinal and urogenital tracts, ducts of secretory glands like salivary glands, lacrimal glands and mammary glands. These epithelial cells are joined together by tight junctions and do not allow any pathogen to squeeze through them.

The skin and mucosal epithelial layer protect the underlying cells and organs from pathogens. The secretions of these tissues like mucus, urine, saliva, tears and milk not only wash away the invading pathogens but also contain antimicrobial and antiviral substances. Specialized cells in the mucosal epithelial layers secrete a viscous fluid called *mucus* which traps the foreign microorganisms. The epithelial cells in the lower respiratory tract are provided with hair-like protrusions of the cell membranes called *cilia*. These cilia move in a synchronized way and propel the mucus-entrapped microorganisms from the respiratory tract. In case of any respiratory infection, coughing – a mechanical response – helps us to get rid of the excess mucus and along with that the mucus-entrapped microorganisms are also removed. The flow of urine also removes several bacteria from the urinary tract. When we eat, with every meal we ingest large numbers of microorganisms but in order to survive these bacteria have to evade several lines of defense like antimicrobial compounds in saliva and epithelia of the mouth and, digestive enzymes and acid in the stomach. Similarly, the vaginal secretions have acidic pH which provides protection against bacterial and fungal infections.

2.2.2 Chemical Barriers

Along with the physical barriers, there exist soluble substances which have antibacterial activity or are acidic and hinder/inhibit the proliferation of pathogens. The common chemical mediators that are involved in innate immunity can be grouped into two types i.e. *antimicrobial proteins* and *antimicrobial peptides*. They act against a wide range of bacteria, mycobacteria, enveloped viruses and fungi.

2.2.2.1 Antimicrobial Proteins

If the pathogens breach the physical barrier, the epithelial layer and mucous membrane, the epithelial cells secrete a plethora of proteins and peptides which give protection against the pathogens. This capability to produce antimicrobial substances helps in providing a protective cover to the surface of the skin, against microbes, as any break in the physical barrier may allow the entry of pathogens inside the body which can cause infection in the absence of a chemical defense system.

There are several antimicrobial proteins that are produced by cells, which include enzymes and binding proteins, that either kill or inhibit the growth of pathogens. For example, lysozyme is an enzyme that is found in saliva, tears and fluids of respiratory tracts which cleaves the peptidoglycan component in the bacterial cell walls. Lactoferrin and calprotectin (a member of S-100 family) are two proteins that bind to and sequester metal ions which are needed for the growth of bacteria and fungi and hence, this sequestration hampers their growth. Human skin produces several antibacterial substances including psoriasin, a small protein of S-100 family, which has a strong antibacterial activity against *Escherichia coli*, an enteric bacterial species. Psoriasin prevents the colonization of *E.coli* on skin but is not effective against *Staphylococcus aureus*, a major causative agent of food poisoning and skin infections. On the other hand, calprotectin prevents colonization of *S. aureus* but not of *E. coli*.

2.2.2.2 Antimicrobial Peptides

In addition to antimicrobial proteins, the skin and other epithelia also secrete antimicrobial peptides which are less than 100 amino acids in length and are the primary components of the innate immunity in vertebrates, invertebrates, plants and even some fungi. The first evidence that vertebrate skin can produce antimicrobial peptides came from frogs, whose skin glands secrete peptides called *magainins* which have strong antimicrobial activity against bacteria, yeast and protozoans. In general, the antimicrobial peptides are rich in cysteine, are cationic and amphipathic *i.e.* having both hydrophilic and hydrophobic regions. Due to the overall positive charge on them, they interact with the acidic phospholipids on the lipid bilayer, which are negatively charged. This interaction disturbs the integrity of the lipid bilayer, thereby disrupting the plasma membrane of bacteria, fungi, viruses and parasites. This makes the entry of peptide inside the pathogen easier where the peptide shows toxic effects like inhibiting the synthesis of RNA, DNA or proteins and activating antimicrobial enzymes that cause apoptosis.

In humans, the principal antimicrobial peptides are – *defensins* (α and β) and *cathelicidin.* The human defensins kill several bacterial species like *E. coli, S. aureus, Streptococcus pneumoniae, Pseudomonas aeruginosa* and *Hemophilus influenzae.* Antimicrobial peptides can also disrupt the lipoprotein envelope of viruses like influenza virus and some herpes viruses. Defensins and cathelicidin LL-37 are continuously secreted by epithelial cells of many tissues and are stored in granules of neutrophil which are used during phagocytosis to kill the phagocytosed microbes. The human α-defensin is secreted into the gut by intestinal epithelial Paneth cells, located in crypts between the villi, and are important for maintaining the beneficial bacterial flora which are required for normal intestinal immune functions.

Further, the respiratory tract epithelia secrete several lubricating lipids and proteins known as surfactants. The two *surfactant proteins i.e.* SP-A and SP-D, present in lungs and other mucosal epithelial secretions are members of a class of microbe binding proteins called collectins. These surfactant proteins bind to different protein, carbohydrate and lipid components on the surface of pathogens, and thus block or modify surface components to clear the pathogens and prevent infection. For example, the lung pathogen *Klebsiella pneumoniae* exists in two states – one, coated with a thick polysaccharide capsule and second, with no coating. SP-A binds to the capsulated form of the pathogen while SP-D binds to the non-encapsulated form of the pathogen and promote phagocytosis. Some of the major antimicrobial proteins and peptides are presented in Table 2.2 (Box 2.1).

BOX 2.1 NATURAL ANTIMICROBIAL PEPTIDES (AMPs)

Antimicrobial Peptides (AMPs) are naturally occurring small cationic molecules which possess antimicrobial activity against several species of bacteria, viruses and fungi and more than 2,000 AMPs have been identified so far. They are important players in innate immune response and act as antimicrobial agents, immunomodulating molecules and act as a link between innate and adaptive immune response. Some of the common AMPs and their effects are given below:

AMPs	Effects
Defensins	The α-defensins are found in mammalian leucocytes and certain epithelial cells while β-defensins are commonly found in mammals and birds and are expressed by various epithelial cells at all mucosal surfaces. *The θ-defensins are found in neutrophils of macaques and Old World monkeys.* They inhibit microbial growth as well as recruit leucocytes with the help of chemokines and cytokines. The primary cells that can be recruited by defensins are monocytes, macrophages, neutrophils, immature dendritic cells, mast cells and T cells.
Cathelicidins	Cathelicidins are primarily found in azurophilic granules of neutrophils and also in mucosal surfaces and keratinocytes during inflammation. Cathelicidins kill bacteria by increasing the permeability of bacterial cell surfaces and by binding to LPS. Only cathelicidin LL-37 is expressed in humans.

Secretory Leucocyte Protease Inhibitor (SLPI)	SLPI is a serine protease inhibitor of several enzymes like trypsin, chymotrypsin, cathepsin G and neutrophil elastase. SLPI was initially isolated from bronchial secretions but later on, it was also found in the keratinocytes, neutrophils, macrophages and mucosal surfaces. SLPI has antimicrobial properties *in vitro* as well as *in vivo* and in humans, this activity has been seen against several bacterial species like *E.coli, P. aeruginosa, Stapylococcus epidermidis, Stapylococcus aureus, Streptococcus sp., Aspergillus fumigatus, Candida albicans* and HIV. In mouse and human, SLPI acts against mycobacteria and kills them. It also facilitates the phagocytosis of mycobacteria by murine and human macrophages.

Source: Sánchez ML, Martínez MMB, Maffia PC (2013). Natural Antimicrobial Peptides: Pleiotropic Molecules in Host Defense. *Cell Bio*, 2: 200–210.

Table 2.2
Some Antimicrobial Proteins and Peptides secreted by the Epithelium

Antimicrobial Peptides and Proteins	Present in	Function
Lysozyme	Mucosal and/or glandular secretions like saliva, tears, secretions of respiratory tract	Destroys peptidoglycan on bacterial cell walls by breaking glycosidic bonds leading to bacterial cell lysis
Lactoferrin (LF)	Mucosal and/or glandular secretions, for example, milk, mucus from intestine, respiratory and urogenital tracts	Binds to iron and sequesters it; limits the growth of bacteria and fungi;
Leucocyte protease inhibitor (Secretory)	Skin, mucus and/or glandular secretions like intestine, milk, respiratory tract and urogenital tract	Prevents infection of epithelium by bacteria, viruses, fungi as it has antimicrobial and anti-inflammatory properties
S-100 family proteins: • psoriasin • Calprotectin	Skin, mucosal and/or glandular secretions like tears, saliva, intestine, respiratory tract and urogenital tract	• Membrane lysis, kills cells • Binds to divalent metal ions and sequesters them, thus killing the pathogens (bacteria and fungi)
Defensins: α defensins (HNP-1, -2, -3; HD-5); β defensins (HBD-1 & HBD-2)	Skin; mucosal epithelia, for example, in intestine, mouth, respiratory and urogenital tracts; blood, breast	Membranes of bacteria, fungi, viruses and protozoan parasites disrupted; shows intracellular toxic effects; kill cells; disables viruses
BPI	Blood, lungs	
Cathelicidin (LL-37: present only in humans)	Mucosal epithelia like respiratory and urogenital tract; blood, breast, gut, skin	Disrupts the plasma membranes of bacteria and kills them; has intracellular toxic effects
Surfactant proteins (SP-A, SP-D)	Primarily secretions of respiratory tract; other mucosal epithelia	Blocks components of the bacterial surface; promotes phagocytosis

BPI, bactericidal/permeability-increasing protein; HBD, human β defensins; HD, human defensins; HNP, human neutrophil peptides.

Apart from antimicrobial proteins and peptides, there are other soluble chemical mediators, like cytokines and serum proteins, which are important players in innate immunity. Table 2.3 summarizes some of these soluble chemical mediators.

Table 2.3
Some Other Soluble Chemical Mediators Active in Innate Immunity

Soluble Mediators	Functions
Cytokines	
TNF-α, IL-1, chemokines	Involved in immune response and inflammation
IFN-α	Confers resistance to viruses
IFN-γ	Confers resistance to intracellular pathogen infection and macrophage activation
IL-6	Pleiotropic cytokine; Synthesis of acute phase proteins
IL-8	Attracts neutrophils and induces degranulation
IL-12	Induces production of IFN-γ by T cells and NK cells; increases cytolytic activity
IL-10	Regulates inflammatory response
IL-15	Activates and induces proliferation of NK cells
IL-18	Induces synthesis of IFN-γ; activates NK and T cells
TGF-β	Regulates inflammatory response
Serum Proteins	
Complement proteins	Opsonization, pathogen clearance and T cell activation
Collectins	Opsonization and complement system activation
C-reactive proteins (CRPs)	Opsonization and complement system activation
Coagulatory proteins	Localization of infected or damaged tissue

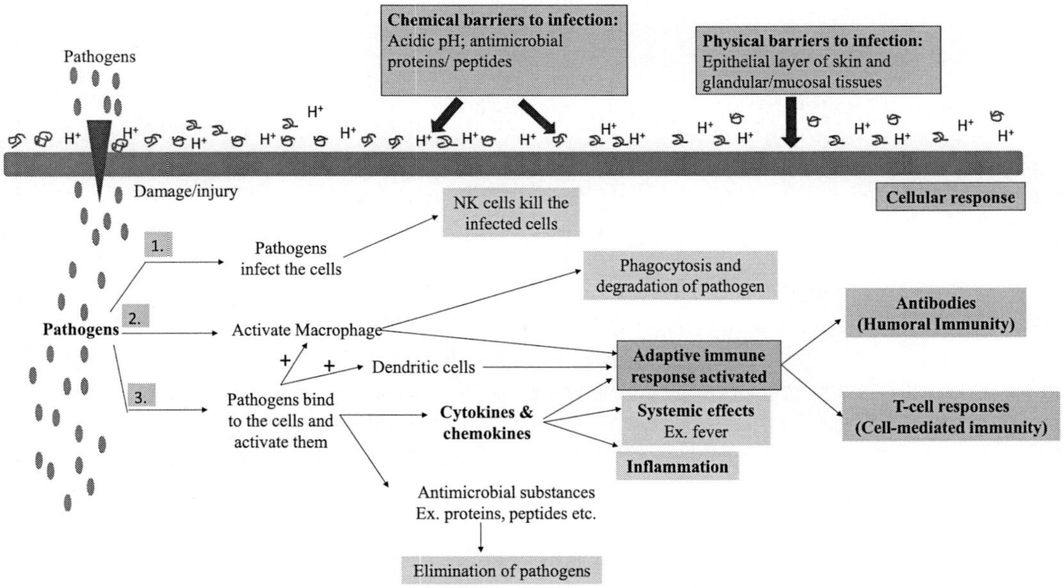

FIGURE 2.1 Various events in an innate immune response. Physical barriers (skin and mucous membrane) prevent the entry of pathogen; in case of breach of physical barrier, the chemical barrier at the surface clears the pathogen; once the chemical barrier also becomes ineffective, the cellular response takes care of the pathogens- 1. the cells infected with the pathogens are destroyed by the NK cells; 2. binding of the pathogen to the cell surface stimulates the cells to secret various antimicrobial substances that clear the pathogen or secrete cytokines and chemokines that can lead to inflammation or systemic effects; 3. pathogens may also activate the adaptive immune response leading to humoral and cell-mediated responses to clear the pathogen from the body.

BOX 2.2 PATHOGEN ASSOCIATED IMMUNOSTIMULANTS

The process of translation in prokaryotes and eukaryotes is different as in prokaryotes the translation starts with formylated methionine unlike methionine in eukaryotes. Thus, any peptide that contains formylmethionine at the N-terminus is of bacterial origin. These peptides act as very effective chemoattractants for neutrophils, which quickly reach the source of these peptides and phagocytose the bacteria.

Other molecules present on the surface of bacteria that act as immunostimulants include peptidoglycan cell wall, flagella, lipopolysaccharides (LPS) on Gram-negative bacteria and teichoic acids on Gram-positive bacteria. In fungi, the molecules on the cell wall that act as immunostimulants include zymosan, glucan and chitin while glycosylphophatidylinositol in *Plasmodium* has the same function.

In bacterial DNA, short sequences like the CpG motif consisting of unmethylated CpG flanked by two 5′ purine residues and two 3′ pyrimidine residues, can activate macrophages, activate inflammatory response and increase antibody production by B cells. The CpG motif is twenty times more common in bacterial DNA than in vertebrate DNA.

Source: Bruce A, Johnson A, Lewis J, Raff M, Roberts K, Walter P. (2002). *Molecular Biology of the Cell.* 4th Ed. Garland Science, New York.

The innate immune response gives input for activation of adaptive immune response. The various events that constitute an innate immune response, and how it activates adaptive immune response, are summarized in Figure 2.1 (Box 2.2).

2.3 CELLULAR INNATE IMMUNE RESPONSE

Once the physical and chemical barriers are broken by the pathogen, the cellular innate immune response comes into action within minutes of invasion. Cellular innate immune response refers to innate immune response due to interaction between cells of innate immunity and pathogens. The actual interaction occurs between cell surface or intracellular receptors of cells of the innate immune system and conserved molecular components on the cell surface of the pathogen. The primary cells involved in the innate immune response are macrophages and neutrophils, which phagocytose the pathogens. Natural killer cells may also be recruited which destroy the virus-infected or other altered cells. Cellular innate immune response is different from the cell-mediated immune response of adaptive immunity, as the former is non-specific, is mediated by several types of cells, especially phagocytic cells, and does not create memory cells while the latter is specific, is mediated by only T–lymphocytes and generates memory cells for every immunogen.

2.4 CELLS INVOLVED IN INNATE IMMUNITY

The innate immunity tries to protect the body from all kinds of foreign invasions. The innate immunity comprises various anatomical and physiological barriers, chemical mediators such as antimicrobial peptides and proteins, that have already been discussed in the previous sections. In addition to these, innate immune responses also involve several cells.

2.4.1 Lymphoid Cell-Natural Killer Cell

Three types of cells differentiate from a common lymphoid progenitor. The T- and B-lymphocytes are the major lymphoid cells and they are involved in the adaptive immune response while natural killer cells (NK cells), also originating from the lymphoid progenitor, play role in innate immunity. NK cells recognize molecules associated with pathogens, damaged cells or stress and initiate a quick response. Unlike B- and T-cells which have diverse receptors, NK cells carry a limited set of receptors, which activate them in case of an infection, cancer or damage. Further, NK cells are preprogrammed to respond instantly to suitable stimuli and release effector proteins from their preformed secretory granules that directly kill the altered cells by inducing apoptosis. The NK cell-mediated apoptosis is further enhanced by interferon-α (IFN-α) which are produced during early viral infections. This is an example of *positive feedback regulation in innate immunity.*

Further, the activated NK cells also secrete cytokines, primarily proinflammatory cytokines-interleukin-6 (IL-6) and tumor necrosis factor (TNF-α), Type I and Type II interferon as well as IFN-γ. NK cells express various types of receptors on their cell surfaces which are called as NK receptors. Some of these receptors are called as *activating receptors* which have an affinity for the proteins whose expression is upregulated in infected, damaged, malignant, or stressed cells. They act as signals that are received by NK cells to kill them. In addition, NK cells also express Toll-like receptors (TLRs) and binding of pathogen-associated molecular pattern molecules (PAMPs) and damage-associated molecular pattern molecules (DAMPs) on pathogens and dead/damaged cells, respectively, amplifies the activation signals.

In addition to receptors that activate the NK cells to kill the altered cells, NK cells also express *inhibitory receptors* that recognize the moieties expressed on the surface of healthy cells (like MHC proteins) and inhibit their apoptosis. In case of a virus-infected cell, MHC expression ceases and thus, there will be no inhibitory signal. In such a scenario, where the NK cell receives more activating signals as compared to inhibiting signals, the NK cell recognizes the target cell as abnormal cell and is activated to kill the abnormal cell. Thus, NK cells are a part of the innate immune response where they act to provide immediate protection, even if it is self-cells that have become harmful. The initial activation of NK cells increases the population of activated NK cells which would generate a greater response if the same ligand is encountered again. This seems like a memory and is the best example of immunological memory in the vertebrate innate immune system. The innate immune response takes care of the infection for a few days till the adaptive response takes over, in case of heavy infection.

The innate immune responses, including inflammation, are necessary to remove or destroy the pathogens, dead or damaged cells and thereby promote healing. When the levels of antimicrobial substances and phagocytic cells increase, they remove the pathogen while dendritic cells present the pathogens to lymphocytes (T- and B-lymphocytes) thereby activating the adaptive immune response. However, sometimes the innate responses may become harmful, causing tissue damage and occasionally death. Thus, there are regulatory mechanisms for innate immune responses. But nonetheless, adaptive immunity complements the innate immunity in vertebrates and there is growing evidence that adaptive and innate immunity in vertebrates have evolved together which has resulted in high degree of interaction and interdependence between the two.

2.4.2 Non-Lymphoid Cells

The cells playing major roles in the innate immune responses are mostly non-lymphoid cells and all of them are non-specific in their action. The major cells and the role they play during the innate immune response are shown in Figure 2.2 and summarized in Table 2.4.

2.4.2.1 Macrophages

Monocytes can migrate into tissues as a result of infection and may get differentiated into tissue-specific *macrophages*. Macrophages may have several roles to play like some macrophages are long-term residents in tissues where they are involved in their repair and regeneration. Many macrophages are also involved in innate immune responses wherein they may undergo several significant changes after encountering pathogens. Such macrophages are called as inflammatory macrophages.

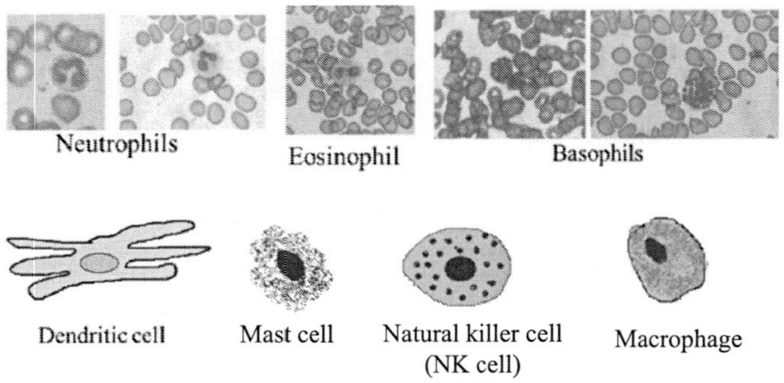

FIGURE 2.2 Different types of cells involved in innate immunity. (Size not to scale.)

Table 2.4
Major Cells of the Innate Immunity and their Functions

Major Cells of Innate Immunity	Function
Neutrophils	Phagocytosis; Release of antimicrobial peptides; Production of reaction oxygen species (ROS) and reactive nitrogen species (RNS)
Macrophages	Phagocytosis; Antigen presentation; Generation of ROS and RNS Release of cytokines; Release of inflammatory mediators; Activation of complement proteins
Dendritic cells	Phagocytosis, processing and presentation of antigen to lymphocytes; Production of ROS; Release of interferons; Release of cytokines; Costimulatory signals;
NK cells	Killing of virus-infected cells; Macrophage activation; Release of interferons;
Mast cells	Involved in inflammation; produce antibacterial peptides such as cathelidcidins, psidins and defensins; show antiviral activity;
Eosinophils	Are phagocytic but their major role is protection against parasitic organisms such as worms; they release inflammatory mediators, from granules, that attack worms
Basophils	Are non-phagocytic but contain granules that are loaded with proteins including histamine which increases permeability of blood vessels; regulates inflammation

They clear the pathogens by phagocytosis and also assist antigen-presenting cells (APCs) during the adaptive immune response.

Macrophages have receptors called FcγRs and FcαRs which trigger a signaling cascade that induces the internalization of the pathogen by forming a phagosome. The contents of the phagosome can be destroyed by a variety of mechanisms like oxidative damage, enzyme digestion, and disruption of the membrane by antibacterial peptides. The macrophages are known by different names in different tissues where they act as phagocytes. For example, in bone, they are known as osteoclasts, in the central nervous system as microglial cells and in lungs as alveolar macrophages.

2.4.2.2 Mast Cells

The precursors of mast cells are released into the blood from bone marrow and they mature only when they leave the blood and enter peripheral tissues. Mast cells are found in various tissues like connective tissues, mucosal epithelium of respiratory, urogenital and digestive tracts as well as skin. Mast cells contain several granules in the cytoplasm which contain histamine and many pharmacologically active substances. Mast cells contain TLRs and receptors for complement proteins. When the pathogen binds to the receptors, it activates mast cells to release inflammatory mediators which help to remove the pathogen. The mechanism of removal depends on the PAMP that is recognized and involves chemical mediators such as IL-4, IL-1, IL-6 and TNF-α. The inflammatory mediators released by mast cells increase vascular permeability, fluid accumulation and recruit cells such as neutrophils, eosinophils and NK cells. They also produce antibacterial peptides such as cathelidcidins, psidins and defensins and are involved in antiviral responses and recruit $CD8^+$ T-cells which release interferon-α (IFN-α) and IFN-β. Mast cells also play important role during allergic responses.

2.4.2.3 Dendritic Cells

Dendritic cells (DC) originate in the bone marrow and were discovered by Ralph Steinman and Zan Cohn in 1973 for which they received Nobel Prize in 2011. Their name comes from their resemblance to the dendrites of nerve cells as they are covered with long membranous extensions. DC can also extend and retract energetically which increases their surface area. DC consist of a diverse population of cells which may arise from two lineages *i.e.* myeloid and lymphoid lineages of hematopoietic cells (See Figure 3.3).

Dendritic cells can perform two functions *i.e.* they can engulf antigens as well as present the antigens to T-cells. During the adaptive response, the immature dendritic cells move out of the lymph nodes, and capture the intruding pathogen, process the antigen, move back to lymph nodes and present the antigens to naïve T-cells. However, when the immature dendritic cells are available in the peripheral tissues, they either engulf the pathogen by phagocytosis or internalize it by receptor-mediated endocytosis and/or by pinocytosis. However, as they mature they become more specialized in presenting the antigen to T-cells. During this transition, they lose the capacity to phagocytose and pinocytose but they gain the ability of antigen presentation and also express costimulatory molecules that are required for the activation of naïve T-cells. Once they are activated, dendritic cells migrate from peripheral tissues to lymphoid tissues to present antigens to naïve T-cells via blood or lymphatic circulation.

2.4.2.4 Eosinophils

Eosinophils constitute 5% of circulating leukocytes and are motile phagocytes that migrate from blood into the tissue spaces. Although they are phagocytes, their role in phagocytosis is insignificant but they play important role in defending the body against multicellular parasitic organisms, like worms. Whenever there is infection of worms, they cluster around the worms and release proteins from eosinophilic granules which damage the membranes of the worms.

Eosinophils express IgE (FcεR) receptors. When the receptors are activated, they get cross-linked and release inflammatory mediators, histamine and proteases from granules, that attack roundworms. Like neutrophils and basophils, eosinophils also secrete cytokines that play role in adaptive immunity. Other than defending against parasitic worms, eosinophils are involved in tissue damages accompanying asthma and allergy symptoms.

2.4.2.5 Neutrophils

The first major cells to respond to infection are neutrophils, which constitute 50%–70% of circulating leukocytes. During an inflammatory response, neutrophils are directed to the site of infection by chemokines secreted by other cells that interact with pathogen. Neutrophils engage and kill pathogens through various mechanisms like phagocytosis, chemotaxis, formation of neutrophil extracellular traps (NET) or NETosis, the release of reactive oxygen species (ROS), cytokines and antimicrobial proteins from the granules, and are thus involved in tissue remodeling. NETosis refers to the ejection of nuclear material along with antimicrobial peptides and enzymes such as defensins, cathelicidins, neutrophil elastase (NE) and myeloperoxidase (MPO). The NET kills or suppresses bacterial and fungal proliferations and thus, infections. Neutrophils release L-selectin during migration which activates them.

Neutrophils express more than 30 different receptors that modulate their migration, function and behavior during inflammatory responses. These include G-protein coupled receptors (GPCRs), Fc receptors, cytokine receptors, PRRs and adhesion receptors. Neutrophils are the main cellular components of the pus as they accumulate at the site as they have a short life span.

2.4.2.6 Basophils

Basophils are granulocytes but they are non-phagocytic. They contain large granules which are filled with basophilic proteins. Although, they are present in very small numbers in blood (1%–2% of circulating leukocytes) but when circulating antibodies bind to basophils, they release the contents of their granules by a process called degranulation. The major component of basophilic granules is histamine which increases the permeability of blood vessels and the activity of smooth muscles. Basophils, together with eosinophils, are the major cells that respond to parasitic infections, particularly helminths. Basophils can also secrete cytokines that play role in the adaptive immune response.

The composition of granules in various granular leucocytes, involved in innate immunity, differ from each other and account for their characteristic functions. The composition of granules in different cells involved in innate immunity and their functions are listed in Table 2.5.

2.5 INFLAMMATORY RESPONSE

Apart from phagocytosis, the activation of pattern recognition receptors (PRRs) such as Toll-like receptors and NOD-like receptors on cells, produces proteins and other substances that may have several functions such as antimicrobial activity, and attract the cells (chemotaxis), fluid and molecules to the site of infection. Due to this, the infected site becomes swollen and shows other signs which are together referred to as *inflammation*.

When the physical barriers of innate immunity are broken, the body responds to infection by inducing a complex cascade of events which are together known as *inflammatory response*, as shown in Figure 2.3. The inflammatory response can be of two types- acute inflammatory response (short term, followed by healing; for example, local tissue damage) and chronic inflammatory response (long term, with no complete healing). Chronic inflammatory response may lead to diseases like arthritis, cardiovascular disease, inflammatory bowel disease and Type 2 diabetes.

The characteristics of a local inflammatory response were first described by Celsus, a Roman Physician in the 1st century AD. He described the symptoms of the inflammatory response as

Table 2.5
Composition of the Granules in Neutrophils, Eosinophils, Basophils and Mast Cells

Cells of Innate Immunity	Composition of Granules	Examples	Functions
Eosinophil	Cytokines	ECP, EDN	Antiviral activity
	Chemokines	IL-4,10,13; TNF-α	Regulate adaptive immune response
	Cationic proteins	RANTES, MIP-1α	Attract leucocytes
	Ribonucleases	EPO	Stimulates synthesis of ROS
		MBP	Vasodilation and degranulation of basophils
Neutrophil	Antimicrobial proteins	Lysozyme, defensins	Kill the pathogen
	Proteases	Collagenase, Elastase	Tissue remodeling
	Protease inhibitors	α1-anti trypsin	Regulate proteases
	Histamine; Olfactomedin 4(OLFM4)		Inflammation, vasodilation
	CD177		
Basophil/Mast cell	Lipid mediators	Leukotrienes	Regulate inflammation;
	Cytokines	IL-4, IL-13	Regulate adaptive immune response;
	Histamines		Vasodilation and activation of smooth muscles

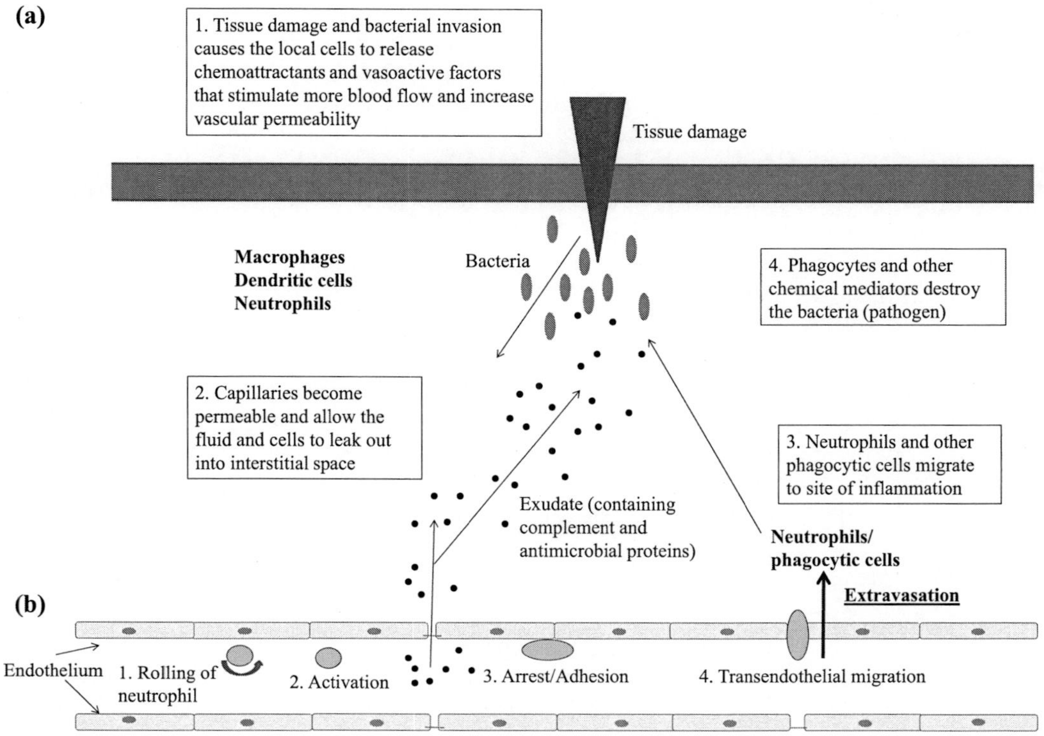

FIGURE 2.3 (a) Events in an inflammatory response. The inflammatory response involves four hallmark events – (i) redness due to a higher flow of blood at the affected site as a result of vasodilation; (ii) swelling due to leakage of exudate from vasculature; (iii) heat due to the presence of the higher amount of blood at the site and (iv) pain due to injury to nerve endings at the site; (b) Extravasation – includes various steps viz. rolling, activation, adhesion and transendothelial migration, that lead to the passage of neutrophils from capillaries to the extracellular spaces at the site of infection. (Adapted from *Kuby Immunology*, 7th Ed.)

'*rubor et tumor cum calore et dolore*' which translates to: redness and swelling with heat and pain. One more characteristic was added by the physician Galen in 2nd century AD *i.e.* loss of function (*functio laesa*). Thus, inflammation is characterized by:

i. **Vasodilation:** An increase in diameter of blood vessels, which results in increase in blood volume in that area.
ii. **Heat and redness:** Due to higher volume of blood, the tissue heats up and causes redness.
iii. **Edema:** The increase in vascular permeability leads to leakage of fluid from the vasculature, causing accumulation of fluid in that area and as a result the tissue swells (edema).

Within a few hours, leucocytes also leak from the blood vessels and enter the affected tissue. As a result, the resident tissue cells like macrophages, dendritic cells and mast cells get activated by *pathogen-associated molecular patterns* (PAMPs) and *damage-associated molecular patterns* (DAMPs) to release cytokines, chemokines and other mediators at the site or in the vicinity of infection. Due to chemokines, leucocytes are recruited at the damaged site and are activated to phagocytose the bacteria and other debris. The response is amplified due to these mediators. Thus, to summarize, the acute inflammatory response involves: the clearance of invading pathogens, damaged tissue, dead cells, activation of systemic acute phase response, initiation of wound healing and adaptive immune responses. In case the infection or tissue damage is not healed, it may become chronic inflammatory condition resulting in more local tissue damage which may have systemic consequences.

2.5.1 Inflammation and Innate Immunity

In case of local infection or tissue damage, special cells like mast cells, macrophages and dendritic cells get activated by PAMPs and DAMPs respectively, which cause phagocytosis of the invading pathogen. These cells further release several chemical mediators which play role during inflammation.

The early event to occur in inflammation is the recruitment of leukocytes at the site of infection. PRR signaling causes the resident mast cells, macrophages and dendritic cells to release early substances of cellular innate immunity like proinflammatory cytokines (TNF-α, IL-1, IL-6), chemokines, prostaglandins (after activation of COX2) and histamine and other mediators released by mast cells. These substances act on the endothelial cells of local blood vessels thereby increasing their permeability. They also increase the expression of *cell adhesion molecules* (CAM) and chemokines like IL-8 in blood vessels. Under the influence of chemoattractants and CAMs, the cells that pass through the local capillaries adhere to vascular endothelial cells in the region showing vasodilation and cross the walls of capillaries to reach the tissue spaces. This process is called as *extravasation* and has been depicted in Figure 2.3.

Figure 2.4 shows the interaction of cell adhesion molecules with neutrophils and endothelial cells. The rolling of cells is the first step in initiating the process of extravasation. Rolling is mediated by the binding of selectin molecules to sialylated carbohydrate moities on mucin-like cell adhesion molecules (CAMs). After this, chemokines bind to G-protein-linked receptors on the surface of neutrophils, thus activating the neutrophils. The activating signal, along with the forces of the blood flow, induces the integrin molecules on the neutrophils to undergo a conformational change which then binds to the *immunoglobulin-superfamily cell adhesion molecules* (ICAM) on the endothelial cells.

The interaction between integrins and ICAMs causes the migration of neutrophils into the underlying tissue *i.e.* transmigration. Neutrophils are the first cells that are recruited at the site of infection where they enhance the innate response, followed by monocytes that differentiate into macrophages which clear the pathogens and help in initiating the healing of the wound. Other than these, naïve lymphocytes also show extravasation and migrate across high-endothelial venules to enter lymph nodes (Box 2.3).

2.5.2 Acute Phase Response and Innate Immunity

As early as the 1920s and 1930s, not much was known about antibiotics. At that time pneumococcal pneumonia was prevalent and researchers noted the changes in the serum proteins during the acute phase of the disease *i.e.* the phase preceding recovery or death. These changes in the serum proteins were collectively known as the *acute phase response* (APR). The proteins whose concentrations varied during acute phase response were called as *acute phase response proteins* (APR proteins). Some of these proteins are involved in the innate immune response and are listed in Table 2.6.

The acute phase response is caused due to signals that are carried by the blood from the site of injury or infection. The major signals for APR are proinflammatory signals in the form of chemical mediators *i.e.* TNF-α, IL-1 and IL-6, which stimulate the liver cells to release APR proteins in large quantities. The APR proteins include some proteins that act as opsonins, which enhance phagocytosis, and components of the complement system, which are involved in both innate and adaptive immune responses. Other APR proteins include *membrane binding lectin* (MBL) (a protein which when binds to the microbial surface activates the complement system) and *C-reactive protein* (CRP) (a type of short Pentraxins). Pentraxins are a family of pentameric pattern-recognition proteins, which bind to ligands in a calcium-dependent reaction and are highly conserved in evolution. The CRP recognizes polysaccharides present in the pneumococcal bacteria and phosphorylcholine present in several bacterial species. CRP is an opsonin and also

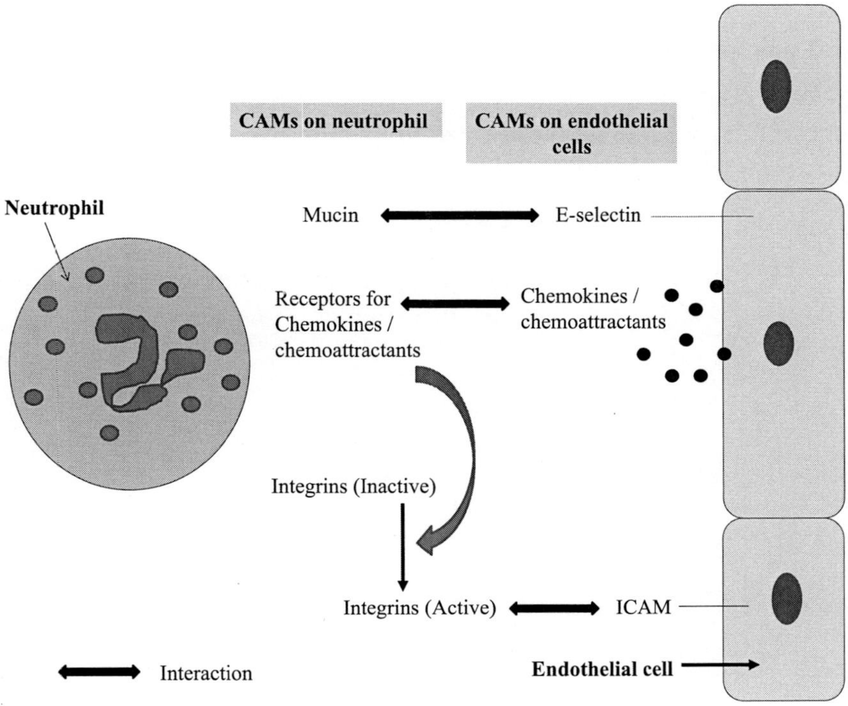

FIGURE 2.4 Initiation of extravasation. The initiation of extravasation is controlled by interactions of cell-adhesion molecules (CAMs) like mucin, chemokines and integrin with their receptors on the neutrophils and endothelial cells respectively. First, the rolling is mediated by mucin-selectin interaction followed by binding of chemokines to their receptors inducing changes in the structure of integrins. The integrins become activated and bind to ICAM (Immunoglobulin-superfamily Cell Adhesion Molecules).

BOX 2.3 EXTRAVASATION OR DIAPEDESIS OF NEUTROPHILS

The cascade of neutrophil extravasation includes rolling, adhesion, crawling and firm adhesion to the endothelial surface of the blood vessels. After the neutrophils reach a proper site for extravasation, the neutrophils undergo transendothelial migration and reach the interstitial spaces. After this, they start migrating to the site of action aided by the cytokines and chemoattractants from the dead and damaged cells at the site of infection.

During the process of extravasation, neutrophil undergoes several events like intrusion, perivascular embedment and crawling, protrusion followed by elongation of uropod and formation of microparticle. Microparticles once formed remain embedded between the endothelial cells and pericytes. Once the neutrophils have crossed the endothelial lining by extravasation, they start interstitial migration. The leucocyte extravasation is less commonly known as diapedesis. Monocytes also use the process of extravasation during their development to macrophages in the tissues.

Source: Park SA and Hyun YM (2016). Neutrophil extravasation cascade: What can we learn from two-photon intravital imaging? *Immune Netw.*, 16(6):317–321.

Table 2.6
Some of the Common Acute Phase Proteins and Their Functions

Acute Phase Protein	Function
C-reactive protein (CRP)	• CRP binds to bacteria, fungi and parasites and activates classical pathway of complement activation • Initiates agglutination but it is non-specific • Acts like an opsonin • Amplifies chemotaxis and phagocytosis by macrophages and neutrophils • Degrades nuclear components of damaged cells by binding to their chromatin
Coglutinin (CGT) A plasma molecule	• Binds to C3b of the complement system • Contributes to agglutination of C3b covered particles, their phagocytosis and elimination
Serum amyloid A (SAA)	• Decreases levels of IL-1 and TNF-α • Inhibits oxidative reaction in neutrophils
α1-Acid glycoprotein (α1AGP)	• Hampers antibody synthesis • Suppresses lymphocytes blastogenesis
β2-Microglobulin (β2-M)	• Protease inhibitor • Neutralizes neutrophils and macrophages lysosomal hydrolases
Ceruloplasmin (CPL)	Clears neutrophil superoxides
Fibrinogen (FNG)	• Coagulation and wound healing • Increases erythrocyte sedimentation rate
Haptoglobulin (HGB)	Inhibits bacterial growth by binding to iron

Source: Tirziu E. (2009).

acts via complement activation. If the CRPs are found circulating in the blood, they indicate a condition of inflammation.

Apart from CRP, other pentraxins that act as opsonins and activate the complement system during innate immune response include serum *amyloid P-component* (short pentraxin, produced in liver), *PTX3* (long pentraxin), surfactant proteins *SP-A* and *SP-D,* and proteins involved in the process of blood coagulation like *fibrinogen.* All the APR proteins are generally present in the blood at low levels at all times but their levels increase during acute phase responses which provide enhanced protection during infections.

2.6 PHAGOCYTOSIS: A MAJOR PROCESS INVOLVED IN INNATE IMMUNITY

Sometimes pathogens may cross the physical and chemical barriers as a result of some evolved mechanisms or due to disruption of the epithelia as a result of wounds, abrasions or sometimes insect bites. Once this happens, pathogens reach into the tissue spaces of the body where several cellular membrane receptors and soluble proteins recognize the microbial cell surface components, and thus detect the pathogens, and trigger a suitable defense against the pathogen i.e. phagocytosis. In tissues, neutrophils, macrophages and dendritic cells act as the phagocytic cells while in blood, monocytes act as the phagocytic cells. The main function of phagocytic cells is to engulf the particulate material like bacteria and eliminate them from the body.

Elie Metchnikoff described the process of phagocytosis in the 1880s using starfish cells and described phagocytosis to be a process that had an important role in providing immunity to the body. Most tissues in the body contain a resident population of phagocytic cells. The cell surface receptors present on the phagocytic cells recognize the molecular components on the surface of

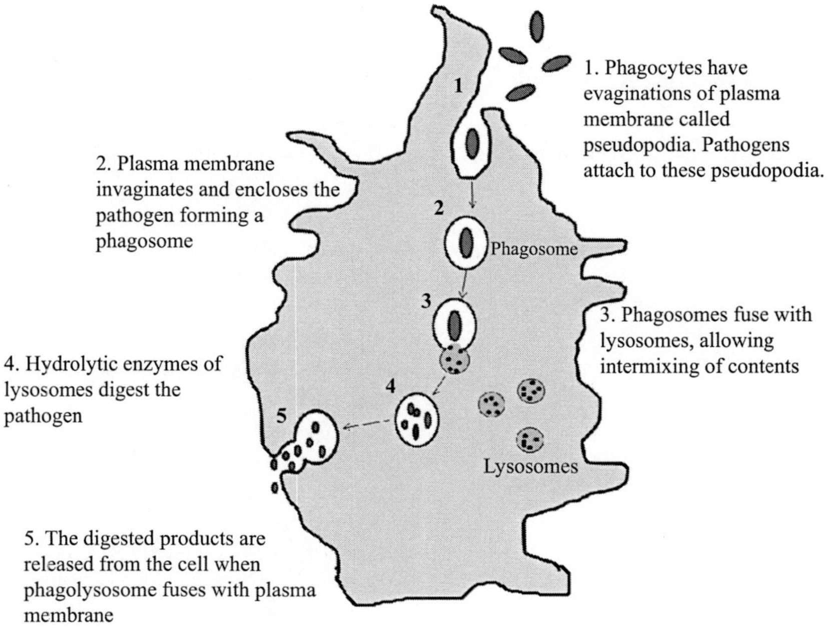

FIGURE 2.5 Steps in phagocytosis of a pathogen. The phagocytes engulf the pathogens by forming pseudopodia, which later pinch off as phagosomes. The phagosomes fuse with the lysosomes, forming the phaglysosomes, where the hydrolytic enzymes digest the pathogen. The digested products are released outside the cell when the phagolysosome fuses with the plasma membrane. Thus, the plasma membrane utilized in the formation of phagosome is restored after the release of the digested products. (Adapted from *Kuby Immunology*, 7th Ed.)

the pathogens, extend their plasma membrane to surround the pathogens and internalize them in the *phagosomes i.e.* endosomes formed during phagocytosis. After this, the phagosomes fuse with the lysozomes which release substances that kill and destroy pathogens, as shown in Figure 2.5.

2.6.1 Recognition of Microorganisms by Phagocytic Cells

There are a variety of receptors on the surface of phagocytic cells. Some of these receptors directly recognize conserved molecular components on the cell surface of microorganisms. These conserved molecular components are usually present in several copies on the surface of pathogens like bacteria, fungi, parasite or virus, and are known as *pathogen-associated molecular patterns* (PAMPs). These patterns are present in pathogenic as well as non-pathogenic microbes. PAMPs may also be called as *microbe-associated molecular patterns* (MAMPs) as these patterns of molecules are present in all the microbes.

PAMPs are recognized by the receptors known as *pattern recognition receptors* (PRRs). Some common receptors used by PAMPs have been listed in Table 2.7. The surface components present on the dead and damaged cells, known as *damage-associated molecular patterns* (DAMPs), are also recognized by PRRs.

Phagocytosis can occur either by direct recognition of PAMPs on the surface of microbes or indirectly, by recognizing soluble proteins bound to the surface of microbes, a process known as *opsonization. Opsonins* are phagocytosis enhancing proteins that bind to various conserved and repeating molecules on the microbe surface like carbohydrates, lipopolysaccharides or viral proteins. The opsonins bound to the microbe surface are recognized by opsonin receptors on the phagocytes. There are several proteins that may act as opsonins, for example, surfactant protein-

Table 2.7
Receptors Utilized by Phagocytic Cells in Humans

Type of Receptor	Examples	Molecular Components Present on Surface of Microorganisms (PAMPs)	
Pattern Recognition Receptors (PRRs)		Recognizes; Present on	
C-type lectin receptors (CLRs)	Mannose receptor Dectin 1 DC-SIGN	Mannans; fungi, bacteria and parasites β-glucans; fungi and some bacteria Mannans; bacteria, fungi and parasites	
Scavenger receptors	SR-A SR-B	Lipopolysaccharides (LPS) and lipoteichoic acid (LTA); bacteria LTA, lipopeptides, diacylglycerides; bacteria β-glucans; fungi	
Opsonin receptors		Soluble microbe-binding opsonins	Found in
Collagen-domain receptor	CD91/calreticulin	Collectins SP-A, SP-D; Mannose-Binding Lectin (MBL) L-ficolin (binds to acetylated sugars on streptococcal bacteria) C1q (Binds to lipopolysaccharides on bacterial cell wall and viral proteins)	Blood, mucosal secretions Blood and respiratory fluid Blood Blood
Complement receptors	CR1, CR3, CR4, CRIg, C1qRp	Components of complement system	Macrophages
Immunoglobulin Fc receptors	FcαR FcγRs	Specific IgA antibody-antigen complex; Specific IgG antibody-antigen complex; C-reactive protein (CRP)	All cells of immune system Serum, human saliva

A (SP-A) and SP-D found in blood and mucosal secretions. These are identified by CD91 opsonin receptor on phagocytic cells. The SP-A and SP-D opsonins play important roles in clearing out the fungal respiratory pathogen *Pneumocystis carinii*, which causes pneumonia in persons with AIDS.

Opsonins such as membrane binding lectin (MBL) (and other collectins), L-ficolin, and complement component C1q share some common structural features, as shown in Figure 2.6 and hence all of them are recognized by the same receptor *i.e. CD91 opsonin receptor.* C-reactive protein (CRP) is another opsonin which recognizes phosphorylcholine and carbohydrates present on the surface of bacteria, fungi and parasites and are recognized by the Fc receptors of IgG which are found on most phagocytes.

Complement proteins are among the most effective opsonins which are present in vertebrates as well as invertebrates and are a link between adaptive and innate immunity. The complement system of vertebrates consists of more than 30 binding proteins and enzymes. The clearance of pathogen by the complement system can be achieved by various innate soluble pattern-recognition proteins like first complement molecule C1q, MBL, ficolins, CRP and properdin etc. and microbe-bound antibodies produced as a result of the adaptive immune response. MBL is a potent opsonizing agent and it also activates complement system. It has been seen that if people have MBL deficiency they develop severe respiratory tract infections, especially pneumococcal pneumonia. However, MBL deficiency is protective against tuberculosis too, as it acts as opsonin to enhance the phagocytosis of *Mycobacterium tuberculosis* which infects macrophages leading to tuberculosis.

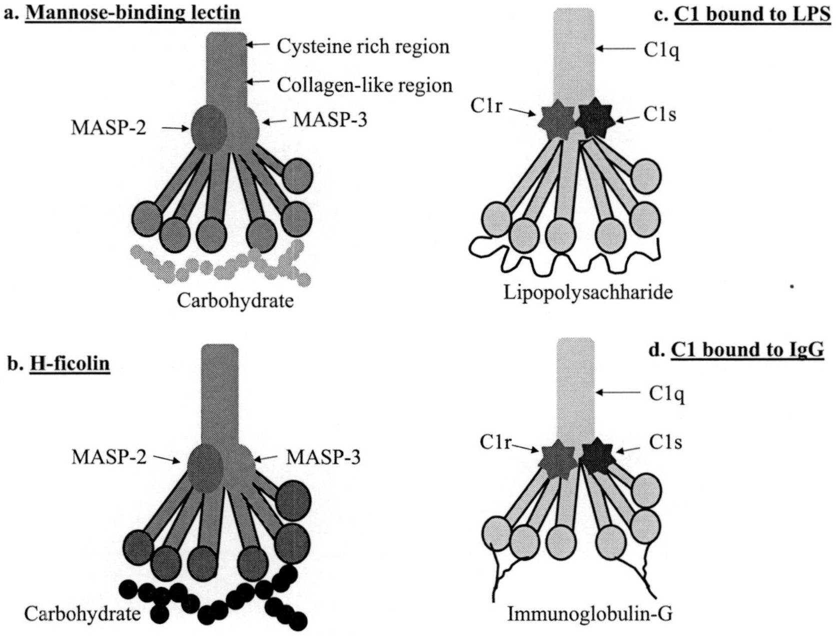

FIGURE 2.6 Structure of opsonins. (a) Mannose-binding lectin, (b) H-ficolin are polymers of many polypeptide chains whose N-terminal is rich in cysteine, followed by a collagen-rich region, an α-helical neck region and a recognition domain. MBL and H-ficolin, both have associated MBL-associated serine proteases (MASPs) which get activated once the ligand binds *i.e.* carbohydrate residues on the cell surface of pathogen which in turn activate the complement system. C1 has multimeric structure which is similar to MBL where C1q binds to LPS on the bacterial cell surface (c) or antibodies bound to antigens (d). After binding of ligands, C1r and C1s act as MASPs and further activate complement system. (Adapted from *Immunology* by Kuby, 7th Ed.)

2.6.2 Mechanisms to Kill Phagocytosed Microbes

Microbes can bind directly to PRRs or they can bind to opsonins, which are then recognized by opsonin receptors. Both these interactions activate the downstream signaling pathways. The signaling pathways cause polymerization of actin to form membrane extensions around the microbes and internalize them in a phagosome. The phagosomes fuse with the lysosomes and within neutrophils, they fuse with the primary and secondary granules forming phagolysosomes. The phagolysosomes contain several antimicrobial components that kill and destroy the internalized microbes. The antimicrobial components include proteins, peptides like defensins and cathelicidins, acid-activated hydrolytic enzymes like proteases and lysozymes, reactive oxygen species (ROS) and low pH.

Figure 2.7 depicts the chemical reactions that occur during oxidative damage to the internalized microbes. The damage occurs in macrophages, neutrophils and dendritic cells via two highly toxic species: *reactive oxygen species* (ROS), and *reactive nitrogen species* (RNS), both of which damage the intracellular components. The reactive oxygen species are generated within the phagocytes by utilizing the NADPH oxidase enzyme complex (or phagosome oxidase). This enzyme is activated when the microbes bind to the phagocytic receptors. The phagocytes need oxygen to produce ROS by NADPH oxidase by a process known as respiratory burst, during which the intake of oxygen by cell increases manyfold. NADPH oxidase converts oxygen to superoxide ion ($^{*}O_2$), hydrogen peroxide (H_2O_2) and hypochlorus acid (HClO).

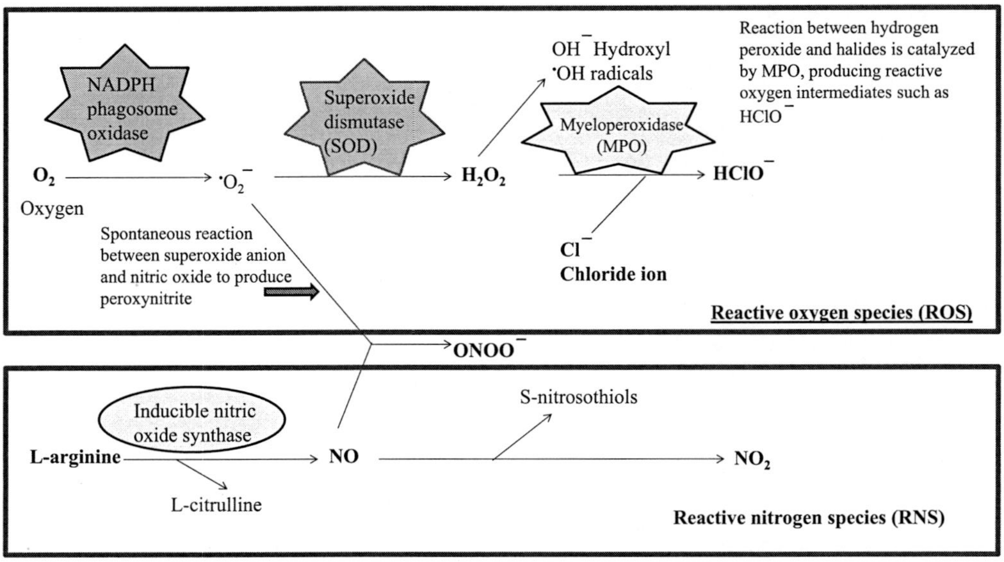

FIGURE 2.7 Generation of reactive oxygen species (ROS) [H_2O_2, Hydrogen peroxide; HClO, Hypochlorous acid; $O2^-$, Superoxide anion; OH•, Hrdroxyl radical] and reactive nitrogen species (RNS) [NO, Nitric oxide; NO_2, Nitrogen dioxide; $ONOO^-$, Peroxynitrite] during pathogen infection.

RNS are produced after activation of the gene for the enzyme *inducible nitric oxide synthase* (iNOS or NOS_2) and it is different from the related NO synthases in other tissues. The enzyme is activated after the binding of microbial components to various PRRs. The iNOS oxidizes L-arginine to L-citrulline and nitric oxide (NO), the latter being a potent antimicrobial agent. NO also produces a reactive oxygen species, peroxynitrite ($ONOO^-$) and S-nitrosothiols, which are toxic to cells/pathogens. ROS and RNS together are highly toxic to the internalized microbes and exert their effects through oxidation, hydroxylation, chlorination, nitration and S-nitrosylation of microbial molecules. In addition, they also cause the production of sulfonic acids and the destruction of iron-sulfur clusters in proteins. The defect in the process of oxidative damage to internalized microbes followed by their degradation leads to a condition known as chronic granulomatous disease (CGD). People with CGD are more prone to some bacterial and fungal infections as they have defects in subunits of NADPH oxidase rendering it ineffective in producing ROS (Box 2.4).

BOX 2.4 EFFECTOR MECHANISMS OF INNATE IMMUNE RESPONSE

The innate immune response involves fighting the infection by adopting various methods for eliminating the pathogen. The pathogens contain certain molecules on their cell surface which are known as Pathogen-Associated Molecular Patterns (PAMPs). PAMPs are recognized by pattern recognition receptors (PRRs) present on the cell surface of phagocytic cells like neutrophils, dendritic cells and macrophages. The pathogen is internalized by the phagocyte and killed and degraded with the help of lysosomes. Also, when the phagocytes are activated, they produce antimicrobial substances and inflammation-promoting cytokines and chemokines.

The surface molecules of the pathogens are recognized by C-reactive proteins (CRP), mannose-binding lectin or surfactant proteins *i.e.* SP-A or SP-D. The binding of these molecules marks the pathogens for recognition by phagocytic cells. Some bacteria and fungi can activate the complement system directly or via CRP or MBL leading to direct lysis of the pathogen. Further, inflammatory signals cause the phagocytes like neutrophils and monocytes to extravasate and reach the site of infection where they phagocytose and kill the infecting microbe. The binding of microbes to the phagocytes activates phagocytosis and releases more antimicrobial and proinflammatory molecules which intensifies the response by recruiting more phagocytes and other soluble mediators like MBL, CRP and complement from blood to the site of infection. The dendritic cells phagocytose and degrade the pathogens after which they enter the lymphatic vessels and reach the nearby lymph nodes and present the antigen-derived peptides with the help of MHC molecules to T-cells. The T-cells in turn become activated and start adaptive immune responses against pathogen. Additionally, cytokines produced during innate immune responses are also play role in the adaptive immune responses against the pathogen.

Source: Owen JA, Punt J and Stranford SA (2013). *Kuby Immunology.* 7th Ed. W.H. Freeman and Company, New York

2.7 PATTERN RECOGNITION RECEPTORS (PRRs)

The macrophages have the receptors that recognize and clear the cellular debris which may include cells that have died due to damage or toxic stimuli, from apoptosis and aging red blood cells. All these cells are recognized by the PRRs of phagocytic cells due to the presence of DAMPs. The apoptotic cells attract the phagocytic cells by releasing lysophosphatidic acid, a lipid mediator that functions as a chemoattractant. The cells destined for apoptosis express some molecules on their surface that are not expressed on the healthy cells like phospholipids (phosphatidyl serine and lysophosphatidyl choline), proteins (annexin I), and altered carbohydrates. DAMPs are recognized by receptors like phosphatidyl serine receptor and scavenger receptor SR-A1.

PRRs bind to PAMPs as well as DAMPs and stimulate signaling pathways that activate genes which play important role in innate immunity. These genes encode various proteins including antimicrobial proteins and peptides, interferons, chemokines and cytokines which attract and activate other cells and enzymes like iNOS that produce antimicrobial molecules and pro-inflammatory mediators. The PRRs may be present on the plasma membrane or in endosomes, lysosomes or in cytosol. This ensures that the intracellular as well as extracellular pathogens are recognized by the phagocytes.

2.7.1 Toll-Like Receptors (TLRs)

Toll-like receptors (TLRs) ('Toll' means 'weird' in German slang) were the first family of PRRs that were discovered and the best-characterized receptors as well. Toll membrane protein is present in *Drosophila* and a mutation in it may cause problems in the establishment of proper dorsal-ventral axis. For their characterization of *toll* and related homeobox genes playing role during embryonic development, Christiane NussleinVolhard, Eric Wieschaus, and Edward B. Lewis were awarded the Nobel Prize in Physiology and Medicine in 1995. Further, they and other scientists created several mutations in the *toll* gene and in 1996 Jules Hoffman and Bruno Lemaitre observed that mutations in the *toll* gene made flies susceptible to fungal infections (*Aspergillus fumigatus*) which were lethal while wild-type flies were immune. Several other studies followed and it was discovered that Toll and related proteins play role during innate immune response in invertebrates. Jules Hoffman was a co-winner of the 2011 Nobel Prize in Physiology and Medicine for his contribution to the role of Toll protein.

Further study of Toll protein revealed that the cytoplasmic domain of Toll was homologous to receptor for cytokine IL-1 (IL-1R) in vertebrates. In 1997, Charles Janeway and Ruslan Medzhitov discovered that there was a gene in humans for a protein similar to Toll, which was involved in the activation of genes involved in innate immunity in human cells. So, this gene and similar vertebrate genes were named as Toll-like Receptors (TLRs). In 1998, Bruce Beutler was working on mutant mice where he discovered that TLRs contribute to the normal innate immune responses in mammals. Mice which were homozygous for the mutant form of the gene *lps* were immune to toxic effects induced by lipopolysaccharide (LPS), an endotoxin and a major component of the cell wall in Gram-negative bacteria. In humans, if this endotoxin builds up due to severe bacterial infection, it can lead to septic shock wherein vital organs like the brain, heart, kidney and liver may fail, which may prove life-threatening. So, it was unexpected to observe that mutant mice were resistant to fatal doses of LPS. Beutler found that there existed a defective form of *lps* gene which coded a mutant form of TLR *i.e.* TLR4, which was different from the normal TLR by one amino acid and was not activated by LPS. This provided a confirmation that TLR4 is a cellular innate pattern recognition receptor that recognizes LPS. Beutler shared the Noble Prize in 2011 for the same.

Till 2011, in mice and human 13 TLRs were identified that function as PRRs. TLRs 1–10 are conserved in mice and human. TLR10 is not functional in mice while TLR 11–13 are expressed only in mice and not in humans. In humans, TLRs can detect a variety of PAMPs from viruses, bacteria, fungi and protozoan parasites as well as DAMPs from the various injured cells and tissues. TLRs and their target microorganisms are listed in Table 2.8.

TLRs are membrane-spanning proteins. As shown in Figure 2.8, each TLR polypeptide is made up of three segments: a *ligand-binding exterior domain* that contains many leucine-rich repeats (LRRs) which are repeating segments of 24–29 amino acids having the sequence LxxLxLxx where L is leucine and x is any other amino acid, a *transmembrane domain* and a *TIR (Toll-Interleukin receptor) domain,* which interacts with TIR domains of other members of TLR signal-transduction pathways. There are several LRRs which constitute the extracellular ligand-binding domain of the TLR polypeptide chain to look like a horse-shoe.

When the extracellular domains of the TLRs bind to their ligands *i.e.* PAMPs or DAMPs, the TLRs dimerize spontaneously. Usually, TLRs dimerize with themselves forming *homodimers* but occasionally *heterodimers* are also formed like when TLR2 dimerizes with TLR6. TLRs are found mostly on the plasma membrane but they are also found on the membranes of endosomes and lysosomes and they recognize and respond to the microbial ligands. Thus, the TLRs that recognize PAMPs on the surface of extracellular microbes are located on the plasma membrane while TLRs that recognize internal microbial ligands, for which the microbes must be dismantled, are found in

Table 2.8
TLRs and their Target Microorganisms

Toll-like Receptors	Target Microorganisms
Cell surface TLRs (recognize cell surface molecules like lipids, lipoproteins and proteins)	
TLR1	Gram-negative bacteria and mycobacteria
TLR2	Gram-positive bacteria, mycobacteria, mycobacteria, yeasts, fungi, trypanosomes, schistosomes
TLR4	Gram-negative bacteria, respiratory syncytial virus (RSV), fungi
TLR5	Bacteria
TLR6	Gram-negative bacteria, mycobacteria, yeasts and other fungi
TLR10	Not well understood
Intracellular TLRs (tecognize nucleic acids of bacteria and viruses and self- cells as in diseased conditions)	
TLR3	Viruses
TLR7	Viruses
TLR8	Viruses
TLR9	Bacterial DNA, some herpes viruses, malarial parasite heme product
TLR11 (Murine)	Toxoplasma, pathogenic bacteria of urinary system
TLR12 (Murine)	Not well understood
TLR13 (Murine)	Vesicular stomatitis virus

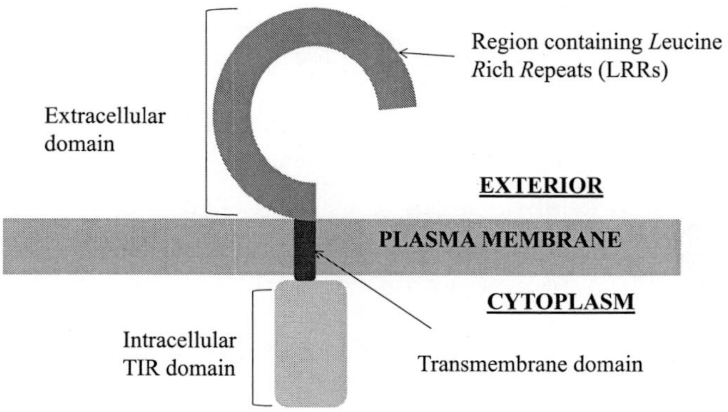

FIGURE 2.8 Structure of a Toll-like receptor. TLRs are transmembrane receptors which have an N-terminal extracellular ligand recognition/binding domain, a transmembrane domain and a C-terminal cytoplasmic signaling domain or TIR domain. The signaling domains are also called Toll IL-1 receptor (TIR) domains as they are structurally homologous to IL-1R family members. Some plant proteins also carry homologs of TIR domains which provide resistance to plants against pathogens. The transmembrane domain consists of approximately 20 uncharged residues which are mostly hydrophobic. The N-terminal extracellular domains (ECDs) of TLR consists of glycoproteins with about 550–800 amino acid residues and contain tandem copies of leucine-rich repeats (LRRs) which are 22–29 amino acid residues long and contain hydrophobic residues placed at intervals. The ECDs of all the TLRs assume horse-shoe shape but they cannot be superimposed as they vary in their curvature.

endosomes and lysosomes. However, TLR4 can move from the plasma membrane to endosomes/lysosomes after binding with the LPS or PAMPs.

In addition to PAMPs, TLR can also recognize DAMPs which are the ligands present on the surface of dead and/or damaged cells. Some examples of DAMPs are heat shock proteins, chromatin proteins, parts of extracellular matrix components such as fibronectin and hyaluronin, amyloid-β and oxidized low-density lipoprotein. The TLR signaling includes several pathways that have some shared components as shown in Figure 2.9. One such shared component is transcription factor *NF-κB* (nuclear factor kappa-light-chain-enhancer of activated B cells) which is an inducer molecule for several genes playing role during innate and inflammatory responses. NF-κB can induce several genes like genes for iNOS, chemokines, cytokines such as TNF-α, IL-1, IL-6 etc. produced by activated macrophages and dendritic cells. It can also stimulate *interferon regulatory factors* (IRFs) which are required for activation of IFN-α and IFN-β genes. After PAMP binding, the TLRs dimerize and initiate signal transduction pathways which are determined by the type of TLR involved and the initial adaptor protein. The two important adaptor molecules are *MyD88* (Myeloid differentiation factor 88) and *TRIF* (TIR domain-containing adaptor-inducing IFN-β factor). Most of the TLRs bind to MyD88 and follow the MyD88-dependent signaling pathways. However, TLR3 binds to the adaptor protein TRIF resulting in the activation of TRIF-dependent signaling pathways. TLR4 can bind to both the adaptors *i.e.* MyD88 (in plasma membrane) and TRIF (in endosomes). There are two additional adaptor molecules associated with most TLRs *i.e. TIRAP* (Toll/interleukin-1 receptor (TIR) domain-containing adaptor protein) and *TRAM* (TIRF related adaptor molecule).

2.7.2 C-Type Lectin Receptor (CLRs)

C-type lectin receptors (CLR) also belong to the family of PRRs that activate innate and inflammatory responses. They are present on the plasma membrane of several cells like macrophages, monocytes, neutrophils, B-cells, T-cell subsets and dendritic cells. CLRs commonly recognize carbohydrate moieties of the ligands on the surface of mycobacteria, viruses, parasites, fungi and allergens such as peanut and dust mite proteins. Humans have at least 15 CLRs that act as PRRs and majorly recognize sugar moieties such as mannose, fucose and glucans.

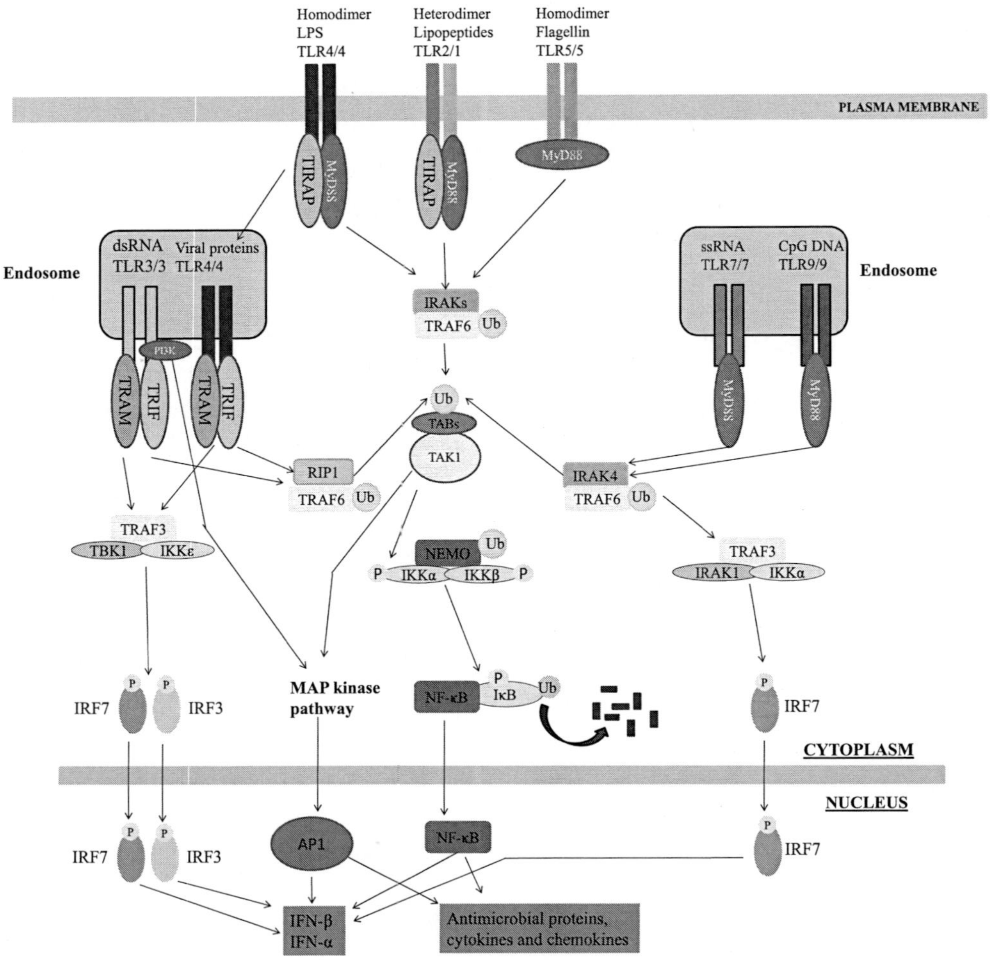

FIGURE 2.9 An overview of TLR signaling. The figure shows three cell surface signaling pathways by TLRs (induced by TLR2/1, TLR4 and TLR5 here); pathways for endosome or lysosome TLRs (TLR7, TLR9, TLR8, TLR3 and TLR4); MyD88-dependent TLR signaling pathways are activated by TIRAP and MyD88 receptor-binding adaptors while TRIF-dependent signaling pathways are activated by TRAM and TRIF receptor-binding adaptors. AP1, Activator protein-1; IFN, Interferon; IKK, IκB kinase; IRAK, Interleukin-1 receptor-associated kinase; IRF, Interferon regulatory factors; MyD88, Myeloid differentiation factor 88; NF-κB, Nuclear factor κB; TAB, TAK-1 binding protein; TAK, Transforming growth factor β (TGF-β)-activated kinase; TBK, TANK binding kinase; TIRAP, Toll/interleukin-1 receptor (TIR) domain-containing adaptor protein; TRAF, Tumour necrosis factor (TNF) associated factor; TRAM, TIRF related adaptor molecule; TRIF, TIR-domain containing adaptor-inducing IFN-β factor; Ub, Ubiquitnation. Note: Only representative signaling pathways are shown here. (Adapted from *Kuby Immunology*, 7th Ed.)

Unlike TLRs which do not promote phagocytosis, some CLRs promote phagocytosis by acting as phagocytic receptors. All the CLRs activate signaling pathways that activate transcription factors involved in expression of genes (Figure 2.10). Signals through different PRRs can affect each other, thereby enhancing or hampering the expression of genes.

2.7.3 The Retinoic Acid-Inducible Gene-I-Like Receptors (RLRs)

The retinoic acid-inducible gene-I-like receptors (RLRs) are soluble PRRs which are found in cytoplasm of cells and play a critical role in cells infected with viruses. There are three known

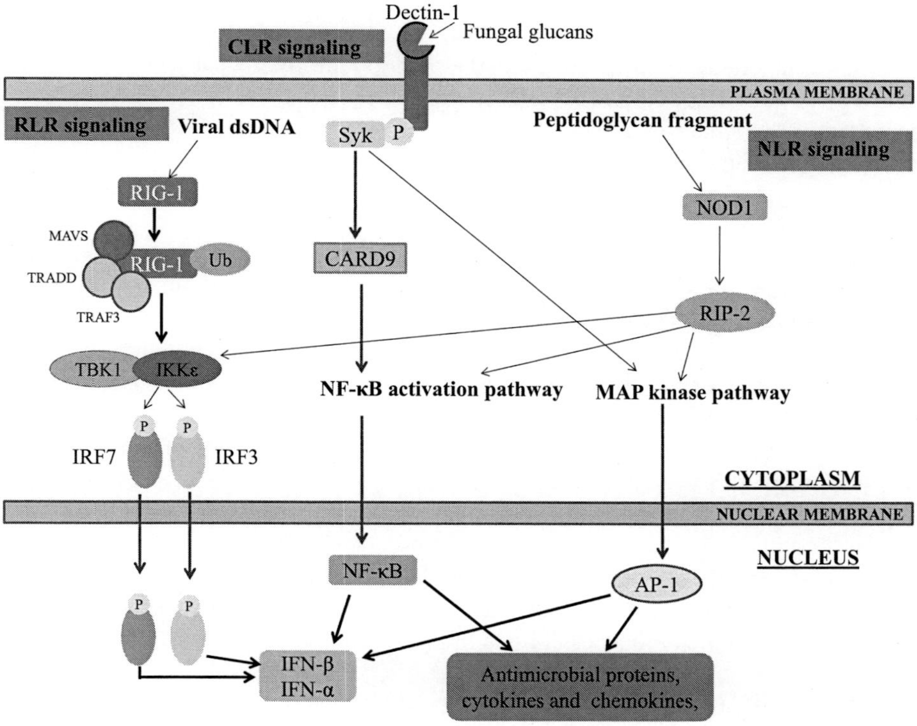

FIGURE 2.10 CLR, RLR and NLR signaling pathways. AP1, Activator protein 1; CARD, Caspase Recruitment Domains; CLR, *C*-type Lectin Receptor; IFN, Interferon; IKK, IκB kinase (an enzyme complex); IRF, Interferon regulatory factor; MAVS, Mitochondrial antiviral signaling protein; NF-κB, Nuclear factor-κB activation protein complex; NLR, NOD-like receptor; RIG, Retinoic-acid inducible gene; RIP, Receptor-interacting protein kinase family; RLR, Retinoic acid-inducible gene-I like Receptors; Syk, Syk family of tyrosine kinases; TBK, TANK binding kinase-1; TRADD, Tumor necrosis factor receptor type-1 associated DEATH domain protein; TRAF, TNF receptor-associated factor. (Adapted from *Kuby Immunology*, 7th Ed.)

RLRs - *Retinoic acid Inducible Gene-1 protein* (RIG-I), *Melanoma Differentiation-Associated protein 5* (MDA5) and *The Laboratory of Genetics and Physiology 2* (LGP2). All the RLRs are RNA helicases containing *Caspase Recruitment Domains* (CARD) which recognize viral RNAs. They can distinguish between viral and self RNA molecules based on some unique features like viral RNAs have double-stranded regions, virus-specific sequence motifs and in RIG-1, the presence of a 5′ triphosphate modification incorporated during viral RNA synthesis and processing. The schematic CLR signaling has been presented in Figure 2.10.

2.7.4 NOD like Receptors (NLRs)

Nod-like receptors or NLRs are a large family of proteins found in the cytoplasm. Intracellular PAMPs, DAMPs or other harmful substances present in cytoplasm activate NLRs. They activate innate as well as inflammatory responses. Sometimes, the trigger for inflammation may lead to tissue damage and diseases. In humans, there are approximately 23 NLR genes that have been identified while in mouse there are 34 NLR genes. NOD1 and NOD2 are the best studied NLRs. The products generated during the synthesis or breakdown of cell wall peptidoglycans of extracellular or intracellular bacteria are recognized by NLRs. These resulting peptides activate the NODs when they enter the cells. In mice, NOD1 is activated by *Trypanosoma cruzi*, a protozoan parasite that causes Chagas disease while NOD2 is activated during some viral infections like influenza. NLR signaling is represented in Figure 2.10.

2.7.5 AIM2-Like Receptors (ALRs)

The Absent in Melanoma 2 (AIM2)-like receptors (ALRs) are the recently characterized PRRs. They are involved in the recognition of cytosolic and nuclear pathogen DNA. ALRs consist of two family members *viz.* the interferon-inducible protein 16 (IFI16) and AIM2. These receptors have been associated with autoimmune diseases like systemic lupus erythematosus (SLE).

2.7.6 Relationship between CLR, RLR and NLR Signaling Pathways

The RLR signaling involves the binding via CARD-CARD interactions to the mitochondria-associated anti-viral signaling (MAVS) protein. Several MAVS proteins aggregate and involve other proteins in downstream signaling like TRADD, TRAF3 and RIP1 which activate NEMO/IKKs followed by activation of IRFs-3 and -7 and NF-κB pathway which further leads to expression of IFNs-α and -β, antimicrobial proteins, chemokines and proinflammatory proteins.

The NLR signaling is initiated by binding of PAMPs to LRR regions of the NODs followed by activation of serine-threonine kinase RIP2 through their CARDs. RIP2 further activates, IKK which translocates NF-κB from cytoplasm to nucleus. Additionally, RIP2 activates MAPK pathway and then activation of AP-1. The activated factors, NF-κB and AP-1, induce the initiation of transcription of several genes like genes of inflammatory cytokines (TNF-α, IL-1, IL-6), antimicrobial proteins and other mediators. RIP2 also stimulates TRAF-3/TBK1/IKKε which cause phosphorylation of IRFs-3 and -7 and then produce IFN-α and -β.

2.8 ANTIMICROBIAL PEPTIDES/PROTEINS INDUCED BY PRRs

Antimicrobial proteins/peptides form a part of the physical barrier to infection and are present on the epithelial lining of the body openings and the skin. These are secreted continuously, without any stimulation, by some cells and tissues. In humans, intestinal Paneth epithelial cells continuously express α-defensins and some β-defensins. Similarly, some defensins and cathelicidin LL-37 are continuously synthesized and packaged in the granules of neutrophils, such that they are ready to kill any phagocytosed bacteria, fungi, viruses and parasites. But in cells like mucosal and glandular epithelial cells, skin keratinocytes, and NK cells, the expression of antimicrobial proteins/peptides is dependent on the stimulation by PRRs. Further, the expression of antimicrobial peptides in macrophages follows a different path. When macrophage TLRs bind to the microbial ligands, it stimulates more expression of receptors for vitamin D. Binding of vitamin D to its receptor stimulates the macrophages to produce cathelicidin, which helps the macrophages to kill the bacteria.

Apart from the above-mentioned antimicrobial components, some other types of chemical mediators are also secreted which are discussed below.

2.8.1 Type I Interferons

Type I interferons (IFN-α, -β) are transcriptionally induced chemical mediators and their production is directly induced by PRRs. The stimulation is provided by the receptors which recognize viral nucleic acids and other components. Type I interferons have a primary role in controlling viral infections. Apart from this, they also activate NK cells and regulate the activities of macrophages and T-cells.

2.8.2 Cytokines

Cytokines are low molecular weight proteins or glycoproteins that are secreted by leucocytes and a variety of other cells. Cytokines are transcriptionally induced by PRRs and are not directly antimicrobial but they activate and regulate several cell types and tissues involved in the innate response, inflammation as well as adaptive response. The most important and major proinflammatory cytokines are IL-1, IL-6 and TNF-α. Table 2.9 summarizes the functions of these cytokines and other

Table 2.9
Some Proteins and Peptides Involved in Signaling through PRRs

Types	Examples	Produced by	Act on	Effects
Antimicrobial proteins	Defensins/ Cathelicidins	Epithelia (respiratory, intestinal, urogenital tracts; Kidney, skin keratinocytes); NK cells	Pathogens Monocytes, immature dendritic cells, T cells Mast cells	Inhibit, kill Chemoattractant; stimulates cytokine production Activate degranulation
	IFN-α, IFN-β	Virus-infected cells, macrophages, dendritic cells, NK cells	Virus infected cells NK cells Macrophages, T cells	Hampers/inhibits virus replication Activates Regulates activity
Cytokines	IL-1	Monocytes, macrophages, dendritic cells, keratinocytes, epithelial cells, vascular endothelial cells	Lymphocytes Bone marrow Vascular endothelium Liver Hypothalamus	Increases activity Promotes production of neutrophil Activates and increases vascular permeability Stimulates acute phase response Induces fever
	IL-6	Monocytes, macrophages, dendritic cells, NK cells, epithelial cells, vascular endothelial cells	Lymphocytes Bone marrow Vascular endothelium Liver Hypothalamus	Regulation Promotes hematopoiesis –neutrophil Activation; enhances vascular permeability Stimulates acute phase response Induces fever
	TNF-α	Monocytes, macrophages, dendritic cells, NK cells, epithelial cells, mast cells	Macrophages Vascular endothelium Liver Hypothalamus Tumors	Activation Activation; enhances vascular permeability, fluid loss, local blood clotting Stimulates acute phase response Induces fever Cytotoxic for many tumor cells
	GM-CSF	Macrophages, vascular endothelial cells	Bone marrow	Induces hematopoiesis-myeloid cells
	IL-12, IL-18	Macrophages, monocytes, dendritic cells	Naïve CD4 T cells Naïve CD8 T cells, NK cells	Stimulates T_H1 phenotype and IF-γ production Activation
	IL-10	Macrophages, dendritic and mast cells; NK cells, T cells and B cells	Macrophages and dendritic cells	Opposes inflammatory response, including production of IL-12 and T_H1 cells
Chemokines	IL-8 (CXCL8) chemokine	Macrophages, dendritic cells, vascular endothelial cells	Neutrophils, basophils, immature dendritic cells, T cells	Attracts the cells to site of injury or infection

important cytokines playing role in innate immunity. The cytokine IFN-γ, leads to the production of T_H1 cells which induce cell-mediated immunity and activates macrophages.

2.8.3 Chemokines

Chemokines can also be called as chemoattractants *i.e.* substances that induce cells to move toward higher levels of these substances and helps in recruiting cells to the site of injury or infection. Chemokines produced in response to PRRs have important roles during early innate response and inflammatory response. They help in clearing the debris at the site of tissue damage or infection and also help in amplifying the response. The first chemokine that was cloned was interleukin-8 (IL-8), also known as Chemokine (C-X-C motif) ligand 8 (CXCL8), and is produced as a result of activation by PAMPs, DAMPs and some cytokines. IL-8 attracts the neutrophils to the site of infection. Other than neutrophils, B-cells are recruited to the lamina propria, a lymphocyte-rich tissue under the intestinal epithelium, by two chemokines *viz.* Chemokine (C-C motif) ligand 28 (CCL28) and Chemokine (C-C motif) ligand 20 (CCL20) produced by intestinal epithelial cells when PAMPs bind to their TLRs. These epithelial cells also stimulate B-cells to produce IgA, which is most effective in preventing mucosal infections.

2.9 ENZYMES ACTIVATED BY PRRs: iNOS AND COX2

As shown in Figure 2.7, PRRs activate the genes for two enzymes *viz.* inducible nitric oxide synthase (iNOS) and cyclooxygenase 2 (COX2), which are antimicrobial and proinflammatory. iNOS catalyzes the reaction that leads to the formation of nitric oxide which lyses the internalized microbes. The synthesis of COX2 is induced by PRRs in monocytes, macrophages, neutrophils and mast cells and converts arachidonic acid to prostaglandins which are strong proinflammatory mediators.

2.10 REGULATION OF INNATE AND INFLAMMATORY RESPONSES

The innate and inflammatory responses are beneficial to the body if they happen at optimal levels but they may cause serious damage if they are not adequately controlled. So, there exist regulatory mechanisms that either enhance or inhibit the innate and inflammatory responses. These regulatory mechanisms control when the responses have to be stimulated or inhibited, the type of response and the duration of responses such that the pathogens are eliminated from the body without causing any damage to the organs or tissues. The regulatory mechanisms are of two types: positive feedback mechanism and negative feedback mechanism.

2.10.1 Positive Feedback Mechanism

There are several mediators that enhance the innate and inflammatory responses to protect the body from the harm caused by foreign pathogens. To amplify the initial response, PAMPs and DAMPs bind to PRRs, which release two important cytokines – TNF-α and IL-1, and activate the pathways that are similar to those activated by TLRs, which, in turn, cause the release of more TNF-α and IL-1. This amplifies the response and is an example of positive feedback regulation of innate and inflammatory response. Another example includes cells that produce Type 1 interferons produce more IFN by stimulating Interferon Alpha Receptor (IFNAR). Other positively regulated mechanisms are PRR-induced activation of genes for some TLRs, for subunits of NF-κB itself, NLRP3 and caspase-1. Enhanced activation of all these genes causes more secretion of IL-1.

2.10.2 Negative Feedback Mechanism

An uncontrolled activation of innate and inflammatory responses can lead to serious health conditions, hence there exist mechanisms that control the overstimulation of these responses and are called as negative feedback mechanisms. The proteins which are activated following the binding of ligands to PRRs play an important role in inhibiting the downstream signaling following PRR activation. One such protein is a short form of the protein MyD88. The activation of intracellular negative feedback mechanism causes the cells to become less responsive and thus limits the innate immune response. In one study it was found that when macrophages were continuously exposed to LPS, which binds to TLR4, initially they produced antimicrobial and proinflammatory mediators; but later they induced inhibitors like IκB and short form of MyD88, which inhibited activation of macrophages by LPS. Hence, this state when macrophages do not respond to LPS is called LPS tolerance or endotoxin tolerance in general which ensures that there are less chances of sepsis due to continued exposure to LPS from bacterial infection.

There are several other feedback mechanisms, which also inhibit the inflammatory effects of TNF-α and IL-1. In each of these pathways, soluble versions of the receptors or receptor-like proteins are formed that bind to the circulating cytokines and thus, prevent them from acting on and affecting other cells. Cytokine IL-10 that is produced by macrophages late during the response to PAMPs inhibits the actions of inflammatory cytokines and enhances wound healing. TGF-β is another anti-inflammatory cytokine and is produced by macrophages, dendritic and other cells after activation of PRRs by apoptotic cells. One more example of negative feedback inhibition is the circulating levels of TNF-α and IL-1 that can induce harmful inflammatory responses at high levels and act by stimulating the hypothalamus which in turn induces the adrenal medulla to secrete glucocorticoid hormones which have anti-inflammatory effects and inhibit the release of TNF-α and IL-1.

2.11 HARM CAUSED DUE TO INNATE AND INFLAMMATORY RESPONSES

The innate and inflammatory responses are beneficial to the body if they are able to clear the pathogens and other harmful substances rapidly and efficiently without causing any tissue damage or interfering with the normal functions of the body. Several mediators or key molecules are involved in the innate immune response that in clearing the pathogen as well as dead and damaged cells from the body. Thus, they help in keeping our body healthy. However, sometimes the innate and inflammatory responses can also be harmful if the pathways that are followed are altered or the molecules involved are not produced as required i.e. either they are overproduced or are underproduced. Such conditions may lead to illnesses or even death. Hence, it is important that the innate and inflammatory responses are carefully regulated so that they are beneficial for maintaining a healthy body.

The most dangerous condition that may occur as a result of inappropriate innate response is *sepsis*, which is a systemic response to infection with symptoms like fever, increased heartbeat and respiratory rate, low blood pressure and inappropriate functioning of organs due to defects in the circulatory system. Sepsis can lead to *septic shock*, which means the collapse of the circulatory and respiratory system with a 90% mortality rate. Infections in the blood, caused due to Gram-negative bacteria and collectively called as *septicemia*, can also cause sepsis. The cell wall component of Gram-negative bacteria is lipopolysaccharide which is an endotoxin and is the main cause of sepsis. In addition, LPS is a potent immunogen and activates mediators of innate immunity like TNF-α, IL-1, IL-6, chemokines and antimicrobial components. When there is systemic infection, various blood cells like monocytes, neutrophils, resident macrophages, vascular endothelial cells and some cells in spleen, liver, etc. release these soluble mediators. These mediators in turn activate the vascular endothelial cells to release chemokines, cytokines, cell-adhesion molecules (CAM) and clotting factors which amplify the inflammatory response. However, the enzymes and reactive oxygen species released from activated neutrophils or other cells damage the blood vessels. This damage together with TNF-α-induced vasodilation and increased vascular permeability causes the fluid to enter the tissues which leads to a drop in blood pressure. In addition, TNF also stimulates the clotting factors to be released from the vascular endothelial cells which cause the clotting of blood in the capillaries. Because of such influences on the blood vessels, the organs that are highly vascularized are damaged to varying extents. Higher levels of TNF-α and IL-1 in the blood adversely affect the heart. Thus, a systemic inflammatory response triggered due to septicemia can cause respiratory and circulatory failure, which may lead to a state of shock and death.

Like acute inflammatory response, the chronic inflammatory response can also have adverse effects on our health, although it may not be as severe as in the case of acute inflammatory response. The chronic inflammatory response is due to continuous activation of the innate immune response. One example is the bacteria *Helicobactor pylori* which releases a toxin that damages the stomach by disrupting the junctions between the gastric epithelial cells and induces chronic inflammation which may cause peptic ulcers and stomach cancers. Additionally, DAMPs, cholesterol (insoluble aggregates or crystals which are noninfectious) and amyloid β are associated with diseases like atherosclerosis *i.e.* hardening of arteries and Alzheimer's disease.

2.12 EVASION OF INNATE AND INFLAMMATORY RESPONSES

The pathogens need to evolve mechanisms that will help them to escape the responses that will eliminate them from the body. The replication rate of several bacteria, fungi and viruses is very high and by way of mutations, they may develop components that will help them in avoiding recognition by innate immune effector mechanisms. There are other complex mechanisms too that help other pathogens to evade innate immune effector mechanisms. Viruses work in a systematic way to evade innate effector mechanisms and evade immune responses triggered by interferons and other cytokines. They acquire genes from their hosts that have evolved and inhibit the innate and inflammatory immune response. The mechanisms by which pathogens generally avoid detection by PRRs and activation of innate and inflammatory responses have been summarized in Table 2.10.

Table 2.10
Mechanisms by Which Pathogens Evade Detection During Innate and Inflammatory Responses

Mechanisms of Evasion	Examples
Evasion of detection by PRRs	1. Mutation in *Proteobacteria flagellin* – evade recognition by TLRs 2. Altered LPS in *Helicobacter*, *Coxiella*, and *Legionella* bacteria- evade recognition by TLR4 3. HTLV-1 virus p30 protein – inhibits expression of TLR4 gene 4. Proteins from several viruses like Ebola, Influenza and Vaccinia bind to viral dsRNA in cytosol and prevents the activation of RLR
Blocking of PRR signaling pathway	1. TIR domains in Vaccinia virus protein A46R and several bacterial proteins- block MyD88 and TRIF binding to TLRs 2. Blocking of TBK1/IKK-activation of IRF3 and IRF7, required for IFN production by viruses 3. NS1 protein in West Nile virus – inhibits transport of NF-B and IRF to nucleus 4. Yop proteins from *Yersinia* bacteria – prevent inflammasome activity and transcription of IL-1 gene
Inhibition of replication or killing	1. *Salmonella* and *Listeria* bacteria – rupture the phagosome membrane and enter the cytosol 2. *Mycobacteria tuberculosis*- prevents fusion of phagosome with lysosomes 3. A protein from Vaccinia virus –binds to Type I IFNs, thus preventing their binding to IFN receptors. 4. Hepatitis C virus protein NS3–4A and Vaccinia virus protein E3L – bind protein kinase R and block IFN-mediated inhibition of protein synthesis.

2.13 KEY CONCEPTS

- Immunity is a mechanism by which the body protects itself from the harmful effects of pathogens and/or antigens.
- Our body has two types of defense mechanisms- innate immunity and adaptive immunity. Innate immune response is considered as 'first line of defense' as it is activated within hours of the entry of the pathogen.
- The innate and inflammatory mechanisms are carefully regulated by positive and negative feedback, if not, they result in pathological conditions.
- Adaptive immune response is delayed and may take days to respond but it is more persistent and effective.
- Adaptive immunity is of two types- humoral (mediated by B lymphocytes involving antibodies) and cell-mediated (mediated by T lymphocytes through cell-cell interaction).
- Immunity can also be passive, when preformed antibodies are given to someone or active when the body of a person is sensitized to produce its own antibodies.
- Sometimes body's own immune system may not respond suitably to the pathogen/antigen or may attack its own cells, such conditions lead to several immune dysfunctions including autoimmunity.
- Innate and adaptive immune responses are not separate but in fact interact with each other to provide necessary protection to the body.

2.14 SUMMARY

- Immunity is the ability of the body to protect itself against the invading pathogens which may harm the body. Innate immunity is the primitive form of protection and is spontaneous or in built in our body.
- The components of the innate immune system act very fast, may even start acting within seconds of the entry of pathogen but these components are not very specific
- The innate immunity comprises of anatomical, physical, chemical, and cellular barriers.
- The physical barriers act as the first line of defense and include: epithelial layer of the skin and epithelial layer of mucosal and glandular tissue surfaces, connected to openings in the body.
- The chemical barriers include proteins and peptides which protect us against the pathogens such as lysozyme, lactoferrin, calprotectin, defensins, cathelicidin and psoriasin.
- Cellular innate response is triggered due to interaction between cell surface or intracellular receptors with conserved molecular components on the cell surface of the pathogen. The cells involved in it are – macrophages and neutrophils, which phagocytose the pathogens. Natural killer cells may also be recruited which destroy the virus infected or other altered cells.
- Inflammation is a kind of innate response when the body reacts to the injury or the invading pathogens by leakage of blood and lymph from the endothelial cells to the surrounding tissues. Inflammation is characterized by four signs: redness, swelling, heat and pain.
- Acute phase response is the changes in the levels of the serum proteins during the acute phase of the disease. There are several acute phase response (APR) proteins like membrane binding lectin (MBL) and C-reactive protein (CRP) that are involved in innate immune responses.
- Phagocytosis is the major phenomenon during innate immune response. The pathogens are recognized by phagocytic cells due to the presence of pattern-associated molecular patterns (PAMPs) on their cell surface. The receptors that recognize PAMPs are called pattern recognition receptors (PRRs). PRRs also recognize components on the dead and damaged cells, which are known as damage-associated molecular patterns (DAMPs).
- Opsonins are the soluble molecules present on the surface of pathogens which enhance phagocytosis. For example, L-ficolin, MBL and CRP.

- Once the microbes are internalized in the phagocytic cells, they are damaged due to activation of pathways involving formation of reactive oxygen species (ROS) and reactive nitrogen species (RNS).
- Some of the common pattern recognition receptors are Toll-like receptors (TLRs), C-type lectin receptor (CLR), the Retinoic acid-inducible gene-I-Like Receptors (RLRs) and NOD like receptors (NLRs).
- PRRs induce the secretion of several antimicrobial proteins and antimicrobial peptides such as defensins, cathelicidins, Type I interferons, cytokines and chemokines. They also induce activity of nitro oxide synthase and cyclooxygenase 2 enzymes.
- The major cells that involved in innate immune response are natural killer cells, neutrophils, mast cells, macrophages, dendritic cells, eosinophils and basophils.
- Sometimes, the prolonged innate immune response may result in conditions such as sepsis. Sepsis is a systemic response to infection with symptoms like fever, increased heart beat and respiratory rate, low blood pressure and inappropriate functioning of organs due to defects in circulatory system. Sepsis can lead to septic shock, collapse of circulatory and respiratory system with 90% mortality rate.
- The innate immune responses are regulated by either positive feedback mechanism or negative feedback mechanism.
- Positive feedback response: To amplify the initial response, PAMPs and DAMPs bind to PRRs which release two important cytokines – TNF-α and IL-1- activate the pathways that are similar to those activated by TLRs, which cause more release of TNF-α and IL-1. This amplifies the response and is an example of positive feedback regulation.
- Negative feedback response: The activation of intracellular negative feedback mechanisms causes the cells to become less responsive and thus limits the innate immune response. For example, TLRs initially produce antimicrobial and proinflammatory mediators but later they produce inhibitors like IκB and the short form of MyD88 which inhibit the activation of macrophages by LPS. This is an example of a negative feedback mechanism.

QUESTIONS

1. Short answer questions:
   a. Name five antimicrobial proteins
   b. What is the significance of PRRs?
   c. Name the enzymes activated by PRRs
   d. Name the cells involved in innate immunity.
2. Give full forms of:
   a. PAMPs
   b. DAMPS
   c. CAM
   d. APR
   e. CRP
   f. TLRs
   g. LRRs
3. Give contribution of the following scientists:
   a. Elie Metchnikoff
   b. Christiane NussleinVolhard, Eric Wieschaus, and Edward B. Lewis
4. Define:
   a. Inflammation
   b. Sepsis
   c. Innate immunity

BIBLIOGRAPHY

1. Arango, D.G., and Descoteaux, A. 2014. Macrophage cytokines: involvement in immunity and infectious diseases. *Frontiers in Immunology* 5:491. https://doi.org/10.3389/fimmu.2014.00491
2. Aristizábal, B., and González, A. 2013. Innate immune system. In *Autoimmunity- From Bench to Bedside*, eds. Anaya, J.-M., Shoenfeld, Y., Rojas-Villarraga, A., Levy, R.A. and Cervera, R., Bogota (Columbia): El Rosario University Press. Retrieved from: https://www.ncbi.nlm.nih.gov/books/NBK459455/
3. Battersby, A.J., Khara, J., Wright, V.J., et al. 2016. Antimicrobial proteins and peptides in early life: ontogeny and translational opportunities. *Frontiers in Immunology* 7:309. https://doi.org/10.3389/fimmu.2016.00309
4. Botos, I., Segal, D.M., and Davies, D.R. 2011. The structural biology of toll-like receptors. *Structure* 19, no. 4:447–459. https://www.cell.com/structure/comments/S0969-2126(11)00072-4
5. Bruce, A., Johnson, A., Lewis, J., et al. 2002. *Molecular Biology of the Cell*. 4th Ed. New York: Garland Science
6. Caneparo, V., Landolfo, S., Gariglio, M., and De Andrea, M. 2018. The absent in melanoma 2-like receptor IFN-inducible protein 16 as an inflammasome regulator in systemic lupus erythematosus: the dark side of sensing microbes. *Frontiers in Immunology* 9:1180. https://doi.org/10.3389/fimmu.2018.01180
7. Cray, C., Zaias, J., and Altman, N.H. 2009. Acute phase response in animals: a review. *Comparative Medicine* 59, no. 6:517–526.
8. Jaitley, S., and Saraswathi, T.R. 2012. Pathophysiology of Langerhans cells. *Journal of Oral Maxillofacial Pathology* 16, no. 2:239–244.
9. Janeway, C.A., Travers, P., Walport, M., and Shlomchik, M. 2001. *Immunobiology: The Immune System in Health and Disease*. 5th Ed. New York: Garland Publishing.
10. Kaufmann S. (2019). Immunology's coming of age. *Frontiers in immunology* 10:684. https://dx.doi.org/10.3389%2Ffimmu.2019.00684
11. Kawasaki, T., and Kawai, T. 2014. Toll-like receptor signaling pathways. *Frontiers in Immunology* 5:461. https://dx.doi.org/10.3389/fimmu.2014.00461
12. Krystel-Whittemore, M., Dileepan, K.N., and Wood, J.G. 2016. Mast cell: a multi-functional master cell. *Frontiers in Immunology* 6:620. https://doi.org/10.3389/fimmu.2015.00620
13. Kuroda, K., Okumura, K., Isogai, H., and Isogai, E. 2015. The human cathelicidin antimicrobial peptide LL-37 and mimics are potential anticancer drugs. *Frontiers in Oncology* 5:144. https://doi.org/10.3389/fonc.2015.00144
14. Mantovani, A., Garlanda, C., Doni, A., and Bottazzi, B. 2008. Petraxins in innate immunity: from C-reactive protein to the long pentraxin PTX3. *Journal of Clinical Immunology* 28, no. 1:1–13.
15. Miller, E.A., Beasley, D.E., Dunn, R.R., and Archie, E.A. 2016. Lactobacilli Dominance and Vaginal pH: Why Is the Human Vaginal Microbiome Unique? *Frontiers in Microbiology* 7:1936. https://doi.org/10.3389/fmicb.2016.01936
16. Mortaz, E., Alipoor, S.D., Adcock, I.M., Mumby, S. and Koenderman, L. 2018. Update on Neutrophil Function in Severe Inflammation. *Frontiers in Immunology* 9:2171. https://doi.org/10.3389/fimmu.2018.02171
17. Ouellette, A.J. 2015. Defensins in enteric mucosal immunity. In *Mucosal Immunology*. 4th Ed., eds. Mestecky, J., Russell, M.W., Cheroutre, H., Strober, W., Kelsall, B.L., and Lambrecht, B.N., Vol. 1, pp. 271–285. https://doi.org/10.1016/B978-0-12-415847-4.00016-1
18. Owen, J.A., Punt, J., and Stranford, S.A. 2013. *Kuby Immunology*. 7th Ed. New York: W.H. Freeman and Company.
19. Park, S.A, and Hyun, Y.M. 2016. Neutrophil extravasation cascade: what can we learn from two-photon intravital imaging? *Immune Network* 16, no. 6:317–321.
20. Sánchez, M.L., Martínez, M.M.B., and Maffia, P.C. 2013. Natural antimicrobial peptides: pleiotropic molecules in host defense. *CellBio* 2:200–210.
21. Tirziu, E. 2009. Acute-phase proteins in immune response. *Lucrări Ştiinţifice Medicină Veterinară* 42, no. 1:329–339.
22. Topham, N.J., and Hewitt, E.W. 2009. Natural killer cell cytotoxicity: how do they pull the trigger? *Immunology* 128, no. 1:7–15.
23. Tsai, C.-Y., Yu, J.F., Wang, Y.W., et al. 2014. An alternative staining method for counting red-eared slider turtle (*Trachemys scripta*) blood cells using crystal violet in cells diluted with 0.45% sodium chloride. *Journal of Veterinary Diagnostic Investigation* 26, no. 5:610–615.

3 Adaptive Immunity

Meena Yadav
University of Delhi

CONTENTS

DOI: 10.1201/9781003286424-3

3.1 INTRODUCTION

Innate immunity has been discussed in detail in Chapter 2, this chapter focuses on the more advanced form of protection i.e. adaptive immune response or adaptive immunity. Adaptive immunity is naturally not present in the body but is generated, depending upon the antigen encountered, hence it is known as the 'acquired immunity'. Since it takes days to get activated, it is also known as 'delayed immunity' and as it comes into action, only if innate immunity is breached or is not able to tackle the pathogen load, it may also be called 'second line of defense'. It is present only in vertebrates as it is more specialized and was probably acquired during evolution. This chapter discusses the concept of adaptive immune response, cells involved in adaptive immune response, cell-mediated immunity, humoral immunity, active immunity, passive immunity, immune dysfunctions, tolerance, autoimmunity etc.

3.2 UNDERSTANDING ADAPTIVE IMMUNE RESPONSE

In case of failure of innate immune responses, vertebrates possess another mechanism to protect the body *i.e.* adaptive immune response. Adaptive immune response is mediated by two major cell types *i.e.* lymphocytes (B lymphocytes or B-cells and T lymphocytes or T-cells) and antigen-presenting cells (APCs). The lymphocytes either produce antibodies against pathogen or they become effector T-cells, which destroy the pathogens through cell-cell interaction. Adaptive immunity takes a longer time but it is more antigen-specific as compared to innate immunity. It takes about 5–6 days for an adaptive immune response to start against a pathogen and only when the innate immunity mechanism is penetrated. The slow response of the adaptive immunity is due to two factors: *one*, fewer cells that possess specific receptors for antigens and *second*, prior encounter and categorizing of antigens after innate immune response.

After antigens interact with B-cells and T-cells, both cells undergo selection and proliferation. However, once these cells become activated they can deal with the infection in an effective manner and destroy the pathogen, thus providing protection against the pathogen.

3.3 CELLS INVOLVED IN ADAPTIVE IMMUNE RESPONSE

As discussed in the previous section, the major types of cells that play a role during adaptive immune response are T-lymphocytes, B-lymphocytes and antigen-presenting cells (APCs).

3.3.1 Lymphocytes

Lymphocytes are the major cells playing role in adaptive immunity. In the circulating blood they constitute 20%–40% of total cells, while in lymph, 99% of the cells are lymphocytes. The lymphocytes are of two types i.e. B-lymphocytes or B-cells and T-lymphocytes or T-cells. In circulating blood, approximately 15%–25% are B-cells and 40%–75% are T-cells. T-cells and B-cells can't be distinguished under a microscope, however, the presence of surface proteins differentiates T- lymphocytes from B-lymphocytes. There are some unique proteins expressed on the surface of

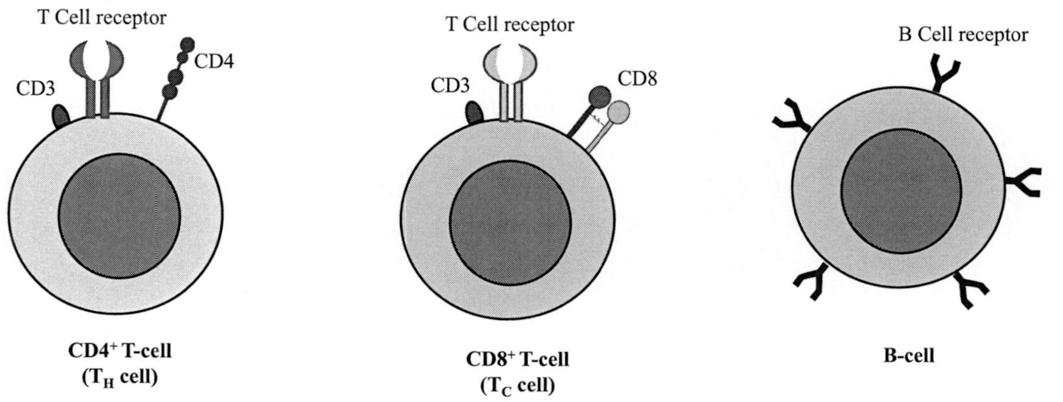

FIGURE 3.1 T-lymphocytes and B-lymphocytes. T- and B-lymphocytes are the major cells involved in providing adaptive immunity. T-cells are of two types: T helper cells (T_H cells) and T cytotoxic cells (T_C cells). T_H and T_C cells are characterized by the presence of T-cell receptors (TCRs) and CD3 in both. Also, T_H cells contain CD4 and T_C cells containing CD8 glycoprotein molecules exclusively on their cell surfaces respectively. CD4 and CD8 act as coreceptors for the T-cells along with TCR. B-cells contain only B-cell receptors on their cell surface.

lymphocytes and they are called as *cluster differentiation* (CD or *cluster of designation*). While T-lymphocytes contain CD4, CD8 etc., B-lymphocytes contain CD19, CD45, CD22, CD32 etc. Of all the CD proteins, CD4 and CD8 hold special importance as they are key factors that differentiate T_H cells from T_C cells, the two subtypes of T-cells. These were discovered by an international group of researchers in 1982 while studying various types of antibodies produced. Other than CD proteins, each B and T-cell expresses an antigen-specific receptor on its cell surface *i.e. B-cell receptor* (BCR) on B-cells and *T-cell receptor* (TCR) on T-cells. Unlike B-cells which can identify both soluble as well as particulate antigens, T-cells can recognize the antigens only when they are processed and presented with the help of *major histocompatibility complex* (MHC) molecules.

There are two types of T-cells *viz.* T-helper cells (T_H) and T-cytotoxic cells (T_C). The two types of T-cells can be differentiated by the presence of CD4 or CD8 glycoproteins on their cell surface, which also serve as co-receptors. The T-cells display CD4 on their cell surface and function as T_H cells ($CD4^+$ cells). They recognize antigens presented with class II MHC molecules, while T-cells, which display CD8, function as T_C cells ($CD8^+$ cells), and recognize antigens presented by class I MHC molecules. In human peripheral blood, the ratio of $CD4^+$–$CD8^+$ cells is 2:1 and any deviation in this ratio is an indicator of immunodeficiency diseases, autoimmune disorders etc. The basic differences in the structure of B-cell and T-cells are shown in Figure 3.1.

The CD4 protein is a single-chain molecule consisting of four domains as in immunoglobulins. The first two domains *i.e.* D1 and D2 are packed together tightly and are joined to the similar structure formed by the other two domains *i.e.* D3 and D4, by a flexible hinge. The CD8 molecule consists of two subunits *i.e.* α and β chains, which are covalently linked to each other by a disulfide bond and thus, is a heterodimer. The structure of CD4 and CD8 glycoproteins is represented in Figure 3.2.

The *naïve CD8+ cells* recognize the class I MHC -peptide complex on the surface of the APCs and get activated, divide and differentiate to produce effector cells called as *cytotoxic T lymphocyte* (CTL). The CTL performs an important function of eliminating the cells which express class I MHC-peptide complex such as virus-infected cells, tumor cells or foreign graft cells. However, to perform this function naive T_C cells also need the support of T_H cells.

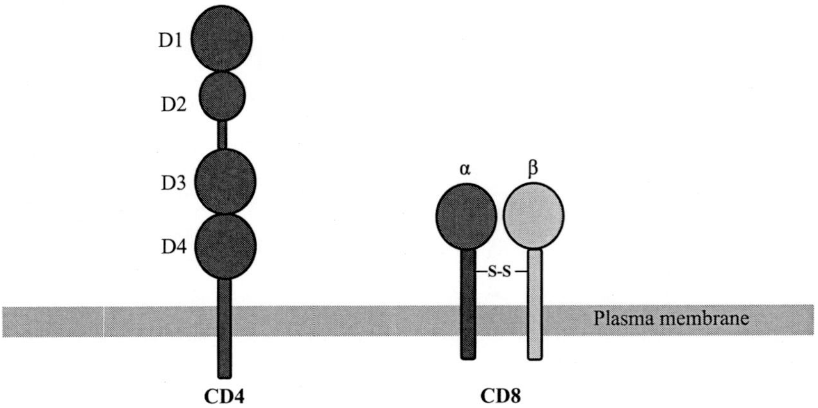

FIGURE 3.2 Structure of CD4 and CD8 glycoproteins. CD4 and CD8 are transmembrane glycoproteins present on T-cell surfaces. CD4 is expressed on T_H1, T_H2 and T_H17 helper cells as well as T regulatory cells (T_{REG}) while CD8 is expressed on cytotoxic T lymphocytes (CTLs) and T_{REG} cells. CD4 contains a single chain consisting of four domains D1, D2, D3 and D4; D1 and D2 are joined to D3 and D4 via a flexible hinge. CD8 contains two polypeptide chains joined to each other by disulfide bonds and consisting of one domain each *i.e.* α and β. CD4 and CD8 increase the T-cell signaling by binding with MHC class II or MHC class I molecules on the APCs respectively.

The *naïve CD4⁺ cells* browse the surface of antigen-presenting cells and recognize class II MHC-peptide complex, which activates them leading to their proliferation and differentiation. The $CD4^+$ cells differentiate to form a variety of effector T_H-cell subtypes:

i. **T helper type I cells (T_H1):** respond to intracellular pathogens; produce interferon-γ (IFN-γ), interleukin-2 (IL-2) and tumor necrosis factor-β (TNF-β). They activate macrophages and are responsible for cell-mediated immunity.
ii. **T helper type 2 cells (T_H2):** respond to many extracellular pathogens such as parasites, bacteria etc.; they activate and maintain humoral immunity.
iii. **T helper type 17 cells (T_H17):** secrete IL-17 and play role in cell-mediated immunity; particularly at the mucosal and epithelial layers; may be in defense against fungi.
iv. **T follicular helper cell (T_{FHC}):** These are subsets of $CD4^+$ T cells; participate in humoral immunity and B-cell development in germinal centers.
v. **One more type of $CD4^+$ cell is T regulatory cell (T_{REG}):** These cells suppress an immune response and maintain homeostasis and self-tolerance. These cells can be produced during maturation of autoreactive cells (natural T_{REG} or nT_{REG}) in thymus but they can also be produced at the site of infection in an antigen-dependent manner (induced T_{REG} or iT_{REG}). These cells are characterized by the presence of CD4 and CD25 on their cell surface.

3.3.2 Antigen Presenting Cells (APCs)

The cellular immune response involves cells that recognize antigens, process them and present them to cells such as naïve T-cells for further action. The classical antigen-presenting cells are B-cells, dendritic cells, and macrophages.

Dendritic cells: The dendritic cells are named so as they appear 'tree-like' or have 'dendritic shape'. DCs are leukocytes derived from bone marrow and have been discussed in Chapter 2. The DC precursors leave the bone marrow and reside in every non-lymphoid

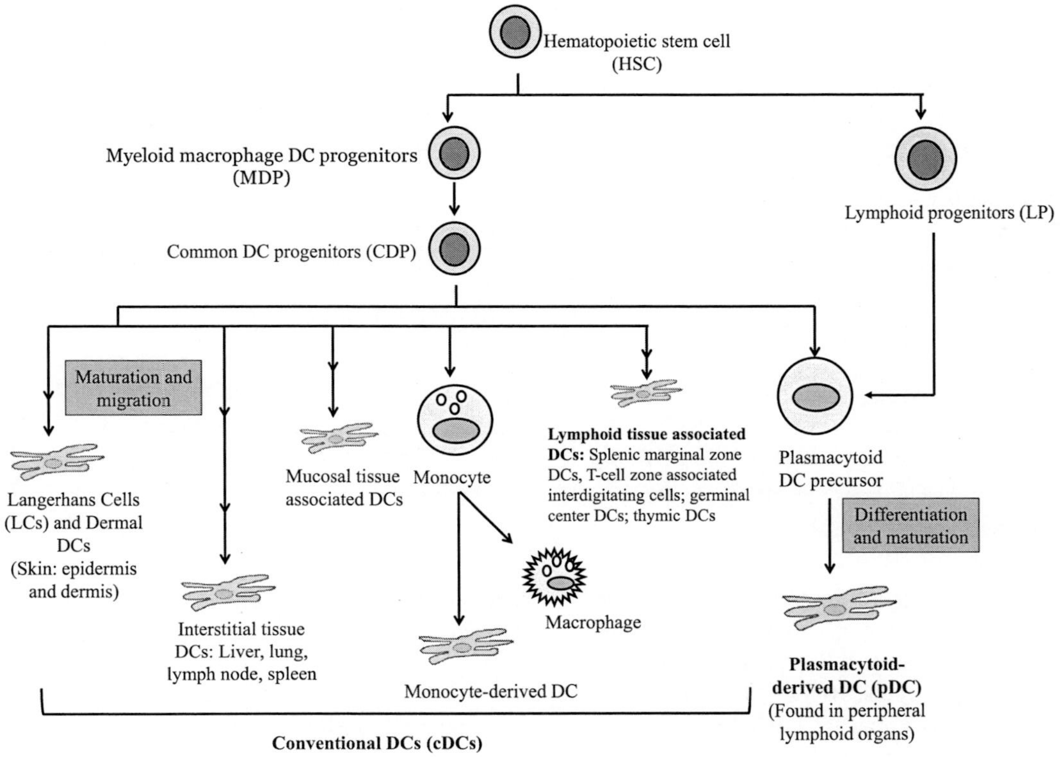

FIGURE 3.3 Origin of dendritic cells (DCs). All types of dendritic cells are derived from hematopoietic stem cells and are descendants of common myeloid precursor cell. In humans, all DCs express high levels of class II MHC (HLA-DR) molecules but lack other marker molecules like CD3 (as present in T-cells), CD19/20 (as present in B-cells) and CD56 (as present in NK cells). Plasmacytoid-derived DCs (pDCs) express CD123, CD303 and CD304; monocyte-derived DC or $CD14^+$ DCs are found in tissues and lymph nodes; Langerhans cells (LCs) are present in the squamous epithelial layer of oral mucosa and epidermis of skin while interstitial DCs are found in interstitial spaces of non-lymphoid organs.

tissue as immature DCs. When pathogens enter the tissues, immature DCs recognize them through pattern recognition receptors (PRRs) and leave the non-lymphoid tissue and migrate to lymph nodes. During this migration, they mature and express costimulatory molecules such as CD80/CD86. The origin of dendritic cells is depicted in Figure 3.3. DCs are the most effective APCs as they capture, process the proteins breaking them into small peptides and present them with class II MHC molecules on their cell surface. They can activate naïve T-cells through MHC-peptide complex and costimulatory molecules.

Among the DCs, langerhans cells (LCs) are skin-resident immature and specialized dendritic cells that ingest antigen by phagocytosis or endocytosis. They are found in all layers of epithelium but are more prominent in the stratum spinosum. Langerhans cells were first described by Paul Langerhans in 1868 as cells similar in appearance to dendritic cells. Langerhans cells reside in the skin. They migrate from epithelium to regional lymph nodes where they undergo maturation and transform into interdigitating cells (IDCs). In lymph nodes they act as potent activators of naïve T-cells by presenting the processed antigens. Along with macrophages, Langerhans cells activate Type IV hypersensitivity response.

Table 3.1
Various Types of Antigen Presenting Cells (APCs)

Professional APCs	**Target Cells of Professional APCs**
1. Dendritic cells	CD4+ & CD8+ cells
2. Macrophages	CD4+
3. B-cells	Activated CD4+
Nonprofessional APCs	
1. Thymic epithelial cells	8. Hepatocytes
2. Vascular endothelial cells	9. Hepatic stellate cells
3. Fibroblasts (skin)	10. Thyroid epithelial cells
4. Liver sinusoidal endothelial cells	11. Cholangiocytes
5. Glial Cells (brain)	12. Lymphatic endothelial cells
6. Pancreatic beta cells	13. Mast cells

Macrophages: Macrophages also act as APCs and are activated after phagocytosis of particulate antigens, after which they present the antigen in a class II MHC-peptide complex. They possess costimulatory molecules such as B7.

B-lymphocytes: B-cells, in addition to being the primary cells involved in adaptive immune response, also act as APCs and constitutively express class II MHC molecules on their cell surface but they must be activated before they can express costimulatory molecules. They activate naïve T-cells.

Thus, B-cells, DCs, and macrophages are known as *professional antigen-presenting cells* as they constitutively express class II MHC molecules, provide a costimulatory signal to T-cells and also secrete cytokines/chemokines.

However, other than these, there are some more cells that can process and present the antigen, and they are called as *nonprofessional antigen-presenting cells.* Table 3.1 lists professional and some non-professional APCs. The nonprofessional APCs lack class II MHC and other costimulatory molecules but they can be induced to express class II MHC molecules or a costimulatory signal. The non-professional APCs function only for a short duration such as during a sustained inflammatory response. Recent studies suggest that granulocytes i.e. neutrophil, eosinophil and basophil, may be considered as APCs as they express class II MHC molecule and costimulatory signals for T-cells under certain conditions.

3.4 CELL-MEDIATED IMMUNITY

During the early years of research in the field of immunology, there were controversies regarding the concept of humoral immunity and Metchnikoff's concept of cell-mediated immunity. However, later it was proved that both the concepts are correct and both complement each other.

During an experiment, Merrill Chase, at The Rockefeller Institute, succeeded in imparting immunity against tuberculosis by simply transferring white blood cells from one guinea pig to another. This was tried and was possible because no effective vaccine or antibody-based therapy against tuberculosis could be produced before it. Due to the success of Chase's experiment, hopes for cell-based immunity were strengthened.

In the 1950s, due to improved cell culture and transfer techniques, it was proved that lymphocyte was the cell involved in both cellular and humoral immunity. Further, experiments by Bruce Glick on chickens at Mississippi State University demonstrated that there exist two types of lymphocytes *viz.* T-lymphocytes derived from the thymus and B-lymphocytes derived from bursa of Fabricius which is an outgrowth of the cloaca in birds. The organ equivalent to bursa of Fabricius in mammals is bone marrow, the place where B-cells mature. We now know that B-cells are the main mediators of humoral immunity while T-cells are mediators of cellular immunity or cell-mediated immunity and both types of immunities are interdependent (Discussed in detail in Section 3.10).

The cell-mediated immune response starts when a naïve T-cell recognizes the antigen presented by an APC. Before the interaction of naïve T-cells and APCs, the latter recognize the pathogens via their PRRs. APCs get activated when antigens bind to PRRs and then they also express costimulatory ligands (for example, CD80 and CD86) and produce cytokines that activate T-cells. CD28 is the most common costimulatory receptor. The pathogens may be extracellular (opsonized intracellular pathogen) or cells infected by intracellular pathogens. In both cases, the pathogens get processed and their peptides are presented with the help of class I MHC or class II MHC molecules on the surface of the APCs. Such APCs then reach either the lymph nodes or spleen. These APCs are then scanned by naïve $CD8^+$ or $CD4^+$ T-cells which recognize class I MHC-peptide and class II MHC-peptide complexes respectively. Each T-cell expresses a unique receptor, for each antigen, which is formed due to gene rearrangement during T-cell development in the thymus. During development, T-cells undergo a selection process in the thymus, such that they are tolerant to self-antigens but recognize the rest as non-self-antigens or 'foreign antigens'. Some of the naïve T-cells become $CD8^+$ cytotoxic T-cells and some become $CD4^+$ helper T-cells. When a naïve $CD8^+$ or $CD4^+$ T-cell binds to MHC-peptide complexes on APCs, it becomes activated.

3.4.1 What Happens When T-Cells Get Activated?

When $CD4^+$ and $CD8^+$ T-cells complete their development in the thymus, they leave the thymus and enter the blood circulation where they exist as cells in the G_0 stage (condensed chromatin, little transcriptional activity and little cytoplasm) and they remain naïve until they encounter an antigen. These cells keep circulating in the blood, lymph, and secondary lymphoid organs, searching for the antigens. Each naïve T-cell, once fully developed, circulates between blood and lymph nodes every 12–24 hours and 1 in every 10^5 naïve T-cells has a chance to encounter a specific antigen. During this circulation, if a naïve T-cell does not encounter its specific antigen, it exits the circulation and enters efferent lymphatics and finally drains into the thoracic duct and rejoins the blood. However, if a naïve T-cell encounters a specific MHC-peptide complex on the surface of an APC, it gets activated, proliferates and produces clones of cells by clonal expansion, which eliminates the pathogens. Figure 3.4 shows the process of T-cell activation, clonal expansion and differentiation. The resulting effector $CD4^+$ T-cell and effector $CD8^+$ T-cells activate B-cells, macrophages etc., and kill infected target cells respectively. The memory $CD4^+$ T-cell and $CD8^+$ T-cell remain in the body to quickly recognize the same antigen upon any further invasion.

In 1987, Helen Quill and Ron Schwartz demonstrated that if there is no functional APC, the high-affinity TCR-MHC interactions lead to T-cell non-responsiveness. This phenomenon was called *T-cell-anergy* and suggested that two signals are essential for complete activation of T-cells i.e. *Signal I*: Binding of antigen to TCR (enhanced by coreceptors and adhesion molecules) and *Signal II*: interaction between costimulatory receptors on T-cells and costimulatory ligands on APCs.

Both the signals together activate T-cells which produce cytokines, which in turn stimulate T-cells to enter cell cycle and proliferate.

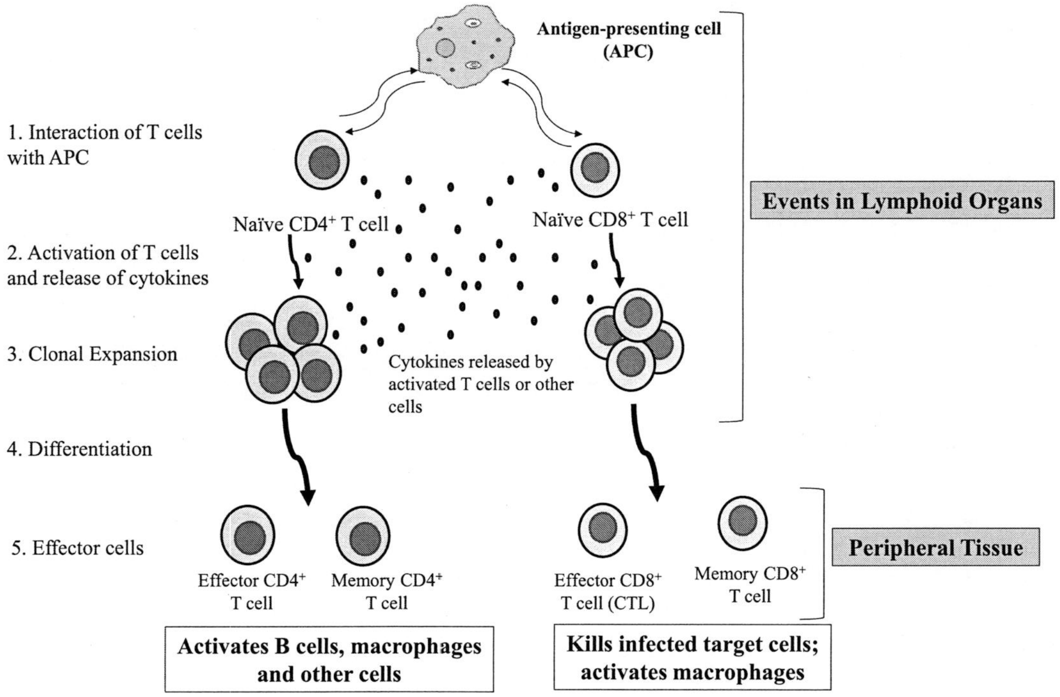

FIGURE 3.4 T-cell activation and differentiation. The APCs carrying antigen bound to MHC molecules are recognized by the naïve CD4+ and CD8+ T-cells. The CD4+ T-cells once activated start to proliferate and differentiate into effector CD4+ T-cells and memory CD4+ T-cells. The effector CD4+ T-cells start secreting cytokines which activate B-cells, macrophages etc. The effector CD4+ T-cells also stimulate CD8+ T-cells to proliferate by clonal expansion and differentiate into two populations of cells *i.e.* effector CD8+ T-cells (CTLs) and memory CD8+ T-cells. The CTLs kill the infected cells and also activate macrophages.

The costimulatory receptors on T-cells are of two types *i.e.* negative costimulatory receptors and positive costimulatory receptors. The negative costimulatory receptors inhibit TCR signaling and are involved in maintaining peripheral T-cell tolerance. Activation of these receptors also reduces inflammatory response after natural infection or during chronic infection. The naïve T-cells do not possess negative costimulatory receptors and are thus activated in the secondary lymphoid tissues when an immune response is initiated, whereas effector T-cells upregulate the expression of negative costimulatory receptors at the culmination of an immune response. CD28 is a 44 kDa glycoprotein and was the first costimulatory molecule to be discovered that acted as positive costimulatory receptor. Another positive costimulatory receptor is ICOS (*i*nducible *co*stimulator). An example of a negative costimulatory receptor is CTLA-4 (cytotoxic T-lymphocyte associated protein-4). Some of the major costimulatory molecules have been summarized in Table 3.2.

Apart from the MHC-peptide complex and costimulatory signal, the T-cell activation is also influenced by soluble cytokines produced by APCs and T-cells. This signal can be called as *signal III.* The cytokines bind to cytokine receptors and activate T-cells which results in a cascade of

Table 3.2
Some of the Major Costimulatory Molecules on T-lymphocytes and Their Ligands

Costimulatory Receptor on T-Cells	Costimulatory Ligand	Function
	Positive Costimulation	
CD28 (B7/CD28 family)	CD80 (B7-1) or CD86 (B7-2) (Expressed by professional APCs and medullary thymic epithelium)	Activates naïve T-cells
ICOS (*i*nducible T-cell *co*stimulator) (B7/CD28 family)	ICOS-L (ICOS-Ligand) (expressed by B-cells, some APCs and T-cells)	Maintains activity of differentiated T-cells; characteristic of T-cell and B-cell interaction
CD40 (TNFR/TNF family)	CD154	Important for functions of B-cells and DCs
OX40 (TNFR superfamily)	OX40L	Regulates functions of effector T-cells and Treg responses
CD27 (TNFR/TNF superfamily)	CD70	Stimulates proliferation and survival of activated T-cells
LIGHT or lymphotoxin α (LT α)	Herpes Virus Entry Mediator (HVEM)	Activates T-cells
	Negative Costimulation	
CTLA-4 (B7/CD28 family)	CD80 (B7-1) or CD86 (B7-2) (Expressed by professional APCs and medullary thymic epithelium)	Negative regulation of immune response
Programmed cell death-1 (PD-1) (B7/CD28 family)	PD-L1 or PD-L2 (Expressed by professional APCs, few T and B-cells, and tumour cells)	Negative regulation of immune response; regulation of differentiation of T_{REG}
BTLA (*B* and *T Lymphocyte Attenuator*) (B7/CD28 family) or CD160	HVEM (*h*erpes *v*irus-*e*ntry *m*ediator) (Expressed by few APCs, T- and B-cells)	Inhibits T-cells
CD244	CD48 (Ligand is Expressed on hematopoietic cells)	Inhibit CD8+ T cells
Lymphocyte activation gene 3 (LAG-3)	class II MHC	Inhibits CD4+ and CD8+ T-cells

intracellular signals that promote proliferation and survival of T-cells. The most potent cytokine is IL-2. Thus, complete activation of naïve T-cells requires three signals *viz.* signal I, signal II and signal III as shown in Figure 3.5.

Now the question arises, which cells provide signal I and signal II to naïve T-cells? Almost all cells in our body express class I MHC molecules while only professional cells like dendritic cells, activated macrophages or activated B-cells express class II MHC molecules. MHC-peptide complex forms signal I. Signal II refers to costimulatory molecules on APCs and their receptors on T-cells. The costimulatory molecules are present only in professional APCs and thymic epithelial cells. The properties of professional APCs have been summarized in Table 3.3. Further, the resting and activated APCs display different types of ligands and they are activated by different signals as depicted in Table 3.4.

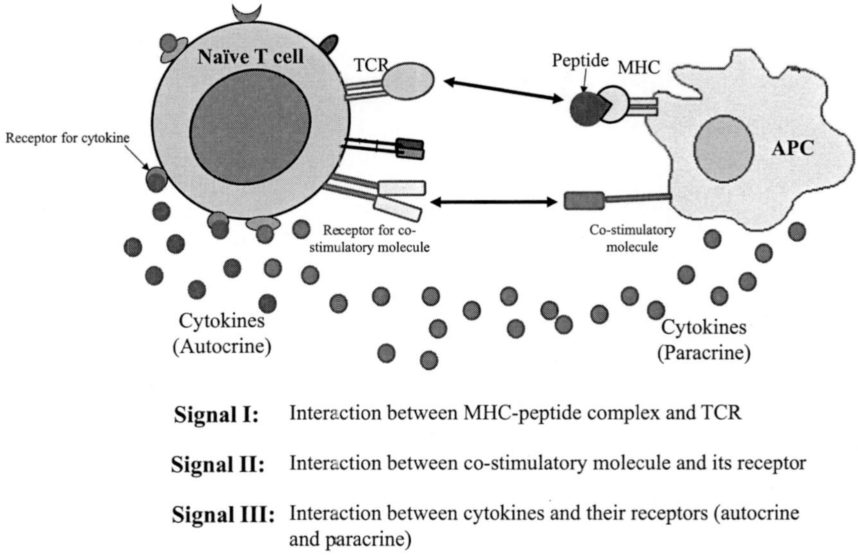

FIGURE 3.5 Signals for activation of naïve T-cell. Signal I to a naïve T-cell is provided by TCR-MHC-peptide complex, CD4 and CD8 receptors as well as adhesion molecules. Signal II is provided by other costimulatory molecules like CD28, ICOS etc. Signal I and signal II together act on naïve T-cell which initiates a cascade of signal transduction leading to the production and activation of cytokines and transcription factors (which form signal III) that cause T-cell proliferation and differentiation. T-cell proliferation is supported by the action of cytokines like IL-2 which can act in an autocrine manner (*i.e.* acting on the self cells) or in a paracrine manner (*i.e.* produced by one cell and acting on other cells).

Table 3.3
Characteristics of Antigen Presenting Cells (APCs)

	Types of APCs		
Characteristics	**Dendritic Cell**	**Macrophage**	**B-cell**
Uptake of antigen	Endocytosis Phagocytosis	Phagocytosis	Receptor mediated endocytosis
Activated by	PRRs	PRRs and T-cells	Antigens
Expression of class II MHC	Increases with activation (however, constitutive expression at very low levels)	Increases with activation	Increases with activation
Costimulatory signals	CD80/86	CD80/86	CD80/86
Activation of T-cells	Naïve T-cells; Effector T-cells; Memory T-cells	Effector T-cells; Memory T-cells	Effector T-cells; Memory T-cells
Found in	When resting: circulation and peripheral tissues When activated: secondary lymphoid tissues (SLOs) (T-cell zones) and tertiary tissues	When resting: found in circulation and peripheral tissues When activated: in secondary lymphoid tissues (subcapsular cortex of lymph node and marginal zones of spleen) and peripheral tissues	When resting: in circulation and secondary lymphoid tissues (follicles) When activated: in secondary lymphoid tissues (B-cell/T-cell zone interface, germinal centers and marginal zones)

Source: Adapted from *Immunology by Kuby*, 7th Ed.

Table 3.4
Ligands on Resting and Activated APCs

Types of APCs	Ligands Expressed on Resting APCs	Ligands Expressed on Activated APCs	Activated by
Dendritic cell	Class I MHC	Class I MHC Class II MHC CD80/86	PAMPs and cytokines
Macrophage	Class I MHC	Class I MHC Class II MHC CD80/86	PAMPs and T-cell help (IFN-γ)
B lymphocyte	BCR, class I MHC class II MHC	Class I MHC Class II MHC CD80/86	Antigens

3.5 HUMORAL IMMUNITY

For several decades there was debate over which of the two-the cells or the soluble mediators- contribute to immunity. During the early 1900s, several researchers were engaged in finding out the protective agent of immunity. They named the soluble component, based on its activity, as antitoxin, agglutinin, precipitin etc. Thus, they thought that several molecules contributed to immunity. However, after the work of Elvin Kabat, a fraction of serum known as gamma globulin, now known as immunoglobulin (Ig) or antibodies (Ab), was thought to confer immunity. Since these antibodies are found in body fluids (known as *humors* at that time), the events that they mediated were called as *humoral immunity.* The humoral immunity can be of two types: passive and active humoral immunity and has been discussed in detail in Sections 3.5.2 and 3.5.3.

3.5.1 Primary and Secondary Immune Response

Immunologic memory is a feature of the adaptive immunity and it allows the immune system to respond much more swiftly and more efficiently during the second encounter with the same antigen. The vertebrate immune system not only adapts to the encounters with the antigens but also develops a memory for the same for future use.

The B-cells and T-cells express a wide variety of receptors which are specific to each and every antigen encountered, but if a specific lymphocyte divides, the resulting daughter cells will have identical receptors for the same antigen. Thus, the population of T-cells carrying identical receptors for an antigen produced from the same parent T-cell is called a clone. The B- and T-cells which are mature but have not encountered any antigen are called naïve cells. When the naïve B-cells encounter an antigen, they get activated and differentiate to produce two populations of cells: effector B-cells (plasma cells) and memory B-cells. Plasma cells are highly specialized cells that produce identical antibodies to tackle the pathogen. A single plasma cell can secrete a few hundred to more than a thousand antibody molecules per second. The plasma cells do not divide further and they usually have a short life span of 1 or 2 weeks. The memory cells can remain in the body for decades, after the primary response, depending on the antigen encountered. Memory cells are so specific and rapid in their action that when the same antigen invades again, the invading organism is cleared from the body without showing any symptoms. This immunological memory possessed by residual B- and T-cells is the basis of vaccination. The kind of response produced by the body when the antigen enters the body for the first time is called the primary immune response while the response produced when the same antigen enters the

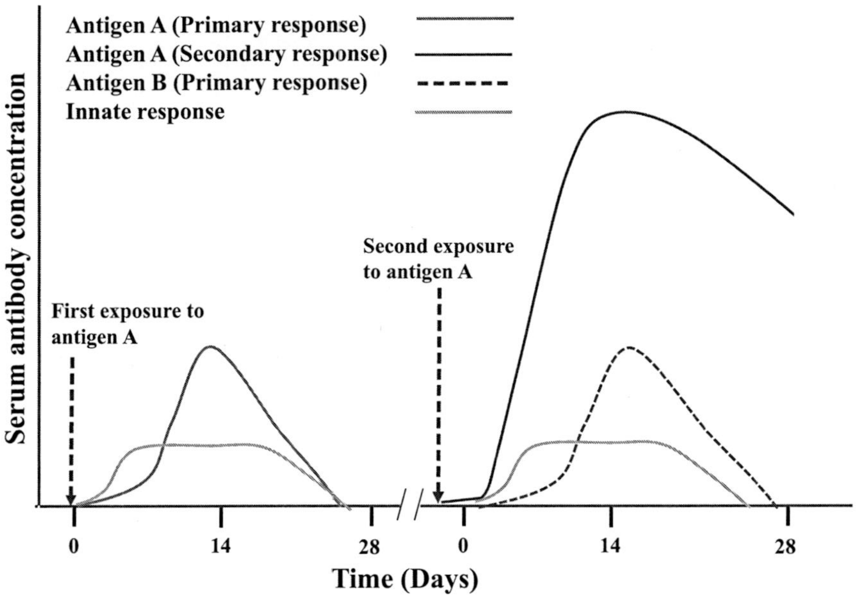

FIGURE 3.6 Primary and secondary immune response: When an animal is exposed to, say antigen A, for the first time it elicits a primary response characterized by low magnitude and short duration which peaks at about 10–20 days after exposure. However, the second exposure of antigen A causes a secondary response which is greater in magnitude and peaks in less time *i.e.* about 1–4 days after exposure and is more antigen-specific than the primary response of antigen A. In the figure, it can be seen that antigen B when injected for the first time also develops primary response in the same way as was produced in case of first exposure of antigen A. The innate immune response occurs at the same magnitude irrespective of how many times the same antigen is entering the body as it doesn't have any memory and is non-specific.

body subsequently is called the secondary response. Figure 3.6 shows a comparative account of primary and secondary immune response and also compares them with innate immune response.

3.5.2 Passive Immunity

The transfer of immunity by administering 'preformed antibodies' to individuals is called passive immunity. This is so because the individual who is receiving it did not make the antibodies against the pathogen. The passive immunity can be natural *i.e.* natural transfer of preformed antibodies between individuals or artificial *i.e.* transfer of preformed antibodies between individuals by injecting or administering the antibodies.

3.5.2.1 Natural Passive Immunity

An example of natural passive immunity is the migration of maternal IgG antibodies from the placenta into the developing fetus, especially in the third trimester. If the mother has had chickenpox, she will have antibodies against the condition and the same may be passed to the baby. Similarly, if the mother has antibodies against any other condition, the same may be passed on to the baby. Mothers secrete thick yellowish milk called colostrum during the first few days before they start to produce the 'true' breast milk which is rich in antibodies (IgA) and protects the baby. However, immunity in newborn babies is short-lived and decreases after a few weeks or months.

Maternal antibodies provide protection to the developing fetus against several diseases like diphtheria, streptococci, rubeola, tetanus, rubella, mumps and polio virus.

3.5.2.2 Artificial Passive Immunity

Passive immunity was the concept given by Emil von Behring and Shibasaburo Kitasato, which was put into clinical practice. They showed that immunity elicited in one animal can be transferred to a healthy animal by injecting the serum from the immunized animal. At that time, doctors prepared antiserum (serum containing antibodies, obtained from the pathogen-exposed individual, especially horses in those days) and gave it to patients suffering from diphtheria and tetanus. Even today, there are therapies in which antibodies are given to susceptible individuals. We can inject a recipient with preformed antibodies i.e. antiserum, from immune individuals. Some of the conditions in which passive immunization can be done are:

i. Cases of immune deficiency, like congenital of B-cell defects
ii. Cases of exposure to any toxin or venom which has an immediate threat to life
iii. Cases of exposure to pathogens that can be fatal before any effective immune response can develop

Children who are born with congenital immune deficiency, and children who have the problem of acute respiratory failure, caused by respiratory syncytial virus (RSV), can be given passive immunization. Passive immunization can be used for unvaccinated people in case of exposure to a pathogen or it can be used by travelers and health care workers who may expect to get exposed to pathogens for which they do not have immunity. Some of the common diseases that can be managed by passive immunization are listed in Table 3.5. Since passive immunization does not activate host's immune response, it does not generate any memory and the protection is short lived.

Table 3.5
Passive Immunization Available for Some Conditions

Diseases	Agents for Passive Immunization
Rubella	Human immunoglobulin (Ig)
Spider bite	Horse antivenin
Scorpion bite	Horse Fab fragments
Diphtheria	Horse antitoxin
Hepatitis A and B	Pooled human Ig
Erythroblastosis fetalis	Rh Ig
Respiratory disease	Monoclonal anti-RSV
Snake bite	Horse antivenin;
Viper and pit viper bites including rattle snakes	Crotalidae, polyvalent
Tetanus	Pooled human Ig or horse antitoxin
Varicella zoster virus	Human polyclonal antibody
Botulism	Horse antitoxin
Measles	Pooled human Ig
Rabies	Human or horse polyclonal antibody
Cytomegalovirus	Human Ig (IV=intravenous)
Vaccinia	Vaccinia Ig
Hypogammaglobulinemia	Human Ig (IV/subcutaneous)
Botulism	Heptavalent A, B, C, D, E, F, G antitoxin for adults; horse Ig for infants
Idiopathic thrombocytopenic purpura	Human Ig (IV)

Although passive immunity is an effective method for protection against some diseases but it should be used carefully as it carries certain risks. For example, usually antibodies are produced in some other animal, like horse, hence in some cases, the recipient may develop a strong response against the isotypic determinant present in a foreign antibody or certain unique parts of horse antibody. Some recipients may develop IgE antibodies against horse-specific determinants. High levels of such IgE antibodies can lead to pervasive degranulation of mast-cell resulting in systemic anaphylaxis. In some recipients, IgG or IgM antibodies may be produced against the foreign antibody which may form complement-activating immune complexes. Deposition of these complexes in the tissues can lead to type III hypersensitivity reactions. Moreover, if we use human antiserum or gamma globulin, the recipients may generate anti-allotype antibodies against foreign antibodies. The intensity of anti-allotype antibodies is less compared to anti-isotypic antibodies.

Passive immunity can also be advised as a preventive measure to strengthen the immune potential of people with the compromised immune system or those who may be exposed in future to certain microbes. Passive immunity is a quick solution to the pathogen attack but it is limited and short-lived as the cells that produced these antibodies are not being transferred (Box 3.1).

BOX 3.1 RACE OF MERCY – WINTER OF 1925

Nome is a small remote town located on the coast of Alaska, on the shores of Berring sea. In 1925, there were only 1,400 inhabitants and in the surrounding areas, there were approximately 10,000 people. There were no roads or railways or air transport at that time. Ships and dog sleds were the only means of transport in 1925. However, during winter ships could not be used as the water froze. In November 1924, the last ship with the necessary supplies to the town had left for Nome.

At that time, Dr. Curtis Welch examined a sick girl in Nome and advised her rest as she appeared to have a sore throat. The girl could not survive the next day. Three more children died with similar symptoms between December 24th, 1924 and January 21st, 1925. The doctor then realized that these were cases of diphtheria which had become epidemic. To his horror, he realized that the last ship carrying the necessities did not carry antitoxin for diphtheria and he was left with only a few vials of antitoxin which had expired. A team of doctor and four nurses went door to door treating the patients as much as they could with the available resources. On 22nd January 1925, Dr. Welch sent an emergent telegram to the authorities, requesting them to send 1 million units of antitoxin to Nome. Due to the telegram, the matter was taken seriously by the authorities and the newspapers came up with the story of the distressed town for a few days.

The antitoxin could not be sent by air transport, ship or road transport due to the unkind weather. The only option was to send it by using dog sled (8–10 dogs) and musher (the driver). The Governor of the state ordered a relay team consisting of 20 mushers and 150 dogs to transport the serum quickly. The relay race began braving temperatures of −40°C which dropped further to −70°C with wind chill.

Leonhard Seppala, a Norwegian, was the best musher with his lead dog Togo, who started from Nome to reach Shaktoolik (270 km far away), collected the serum and returned to Golovin. The last stretch of the relay, 85 km, was covered by the musher Gunnar Kaasen with his lead dog named Balto. Kaasen arrived in Nome with the serum at 5 am on 2nd February 1925 and handed over the serum to Dr. Welch. The administration of the serum saved all the lives except seven who all had died before the serum had arrived.

The country was grateful for the efforts of mushers and dogs. To honor the brave dogs, they erected a statue of Balto in Central Park, New York in December 1925.

Source: Vinodh, NS (2017). The race of mercy- Winter of 1925. Retrieved from: https://medium.com/@nsvinodh/the-race-of-mercy-winter-of-1925-2db6c1736760.

3.5.3 Active Immunity

Unlike passive immunity, the aim of active immunization is to trigger adaptive immune response and generate immunologic memory. If the active immunization is successful, a subsequent exposure to the same pathogen elicits a secondary immune response that removes the pathogen or any of its products that may result in disease. In active immunization, the activated T- and B-cells proliferate and provide immunity as well as form immunologic memory.

Active immunity can be of two types: Natural active immunity, which can be attained by natural infection with a pathogen or artificial active immunity which can be attained by administration of a vaccine.

For achieving artificial active immunity, Mission Indradhanush was launched by the Government of India under the National Health Mission on December 25, 2014, to immunize all children below 2 years to provide active immunization against diseases as mentioned in Table 3.6. In 2016, some more vaccines were added which include Rubella, Japanese Encephalitis, Injectable Polio vaccine Bivalent and Rotavirus.

In the United States, in addition to the vaccines recommended in India, they also recommend Hib vaccine, Meningococcal vaccine (MCV4) against *Neisseria meningitides*, Measles-Mumps-Rubella (MMR) combined vaccine, Pneumococcal conjugate vaccine (PCV) and Influenza virus vaccine for children. Additionally, in 2006, they also recommended a vaccine for young females, age 11 years onwards against sexually transmitted human papilloma virus (HPV) which causes cervical cancer in women. Further, in 2012, they also recommended a vaccine for HPV for males of 11–12 years of age.

Children usually need booster doses (repeated inoculations) at appropriate time intervals for maximum effectiveness. During the initial months of life, the reason for the booster is due to circulating maternal antibodies. For example, let us take the case of the DTP vaccine. This vaccine is given when the infant is 2 months old. At this time, maternal antibodies which are passively acquired bind to the epitopes on the DTP vaccine and thus block the ensuing activation of adaptive immune response by DTP. Thus, there is a need to give this vaccine many times after the maternal antibodies have been cleared from the infant circulation so that the adaptive immune system of the infant can be activated for protection and immunological memory. The same principle is applicable to the MMR vaccine. International travelers are routinely immunized, based on the country they are visiting, against endemic diseases like yellow fever, plague, cholera, hepatitis, typhoid, meningitis, polio and typhus.

Table 3.6
Active Immunization under Mission Indradhanush[a]

Time of Active Immunization	Vaccines
Just after birth	BCG[a], HEP B, OPV
At 6 weeks	DPT, HEP B, OPV
At 10 weeks	DPT, HEP B, OPV
At 14 weeks	IPV, DPT, HEP B, OPV
At 9–12 months	Measles, JE, Vitamin A
At 16–24 months	Measles, JE, Vitamin A, OPV booster, DPT booster
In 5–6 years	DPT booster
In 10–16 years	TT

Source: Arora (2017).

BCG, Bacillus Calmette-Guérin; DPT, Diptheria-Tetanus-Pertussis; IPV, Inactivated Polio Vaccine [DPT, HEP B and Hib Vaccine is not given if pentavalent vaccine is given]; JE, Japanese Encephalitis; OPV, Oral Poliovirus Vaccine; Pentavalent vaccine, Diptheria, HEP B, Pertussis, Tetanus and Haemophilus influenza type b; TT, Tetanus-Toxoid.

3.6 COMPARISON OF HUMORAL AND CELL-MEDIATED IMMUNITY

The humoral and cell-mediated immunity involve the activity of various effector molecules, all of which ultimately discourage the propagation and further infection by the invading pathogen. For example, viral infection can be handled by the humoral as well as cell-mediated immune response. Figure 3.7 shows the interaction between humoral and cell-mediated immune responses. The various mechanisms used by humoral and cell-mediated immune responses in case of viral infection have been summarized in Table 3.7 (Box 3.2).

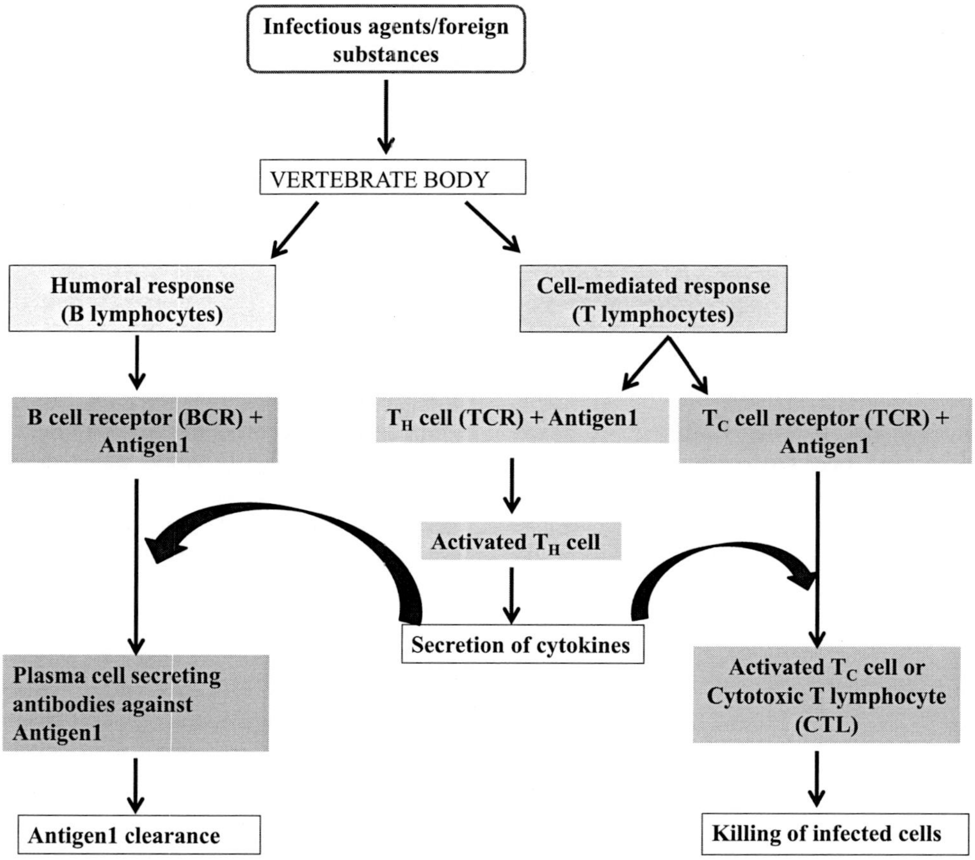

FIGURE 3.7 Interaction between humoral and cell-mediated immune responses. The entry of a pathogen or an antigen triggers the vertebrate body to produce two kinds of responses i.e. humoral response and cell-mediated response. *Humoral response*: The binding of antigen to BCR induces B-cells to mature into plasma cells which secrete volumes of antibodies which bind the antigens and clear them from body. However, the complete activation of B-cell also requires chemical signals in the form of cytokines from the T_H cell. *Cell-mediated response:* Similarly, T_H cell and T_C cell get activated by binding with the antigen: the activated T_H cell secretes cytokines, which activate B-cell and T_C cell, and T_C cell differentiates and matures into cytotoxic T lymphocyte (CTL), which kills the infected or altered cells.

Table 3.7
Comparison of Mechanisms of Humoral and Cell-Mediated Immunity Against Viruses

Humoral Response		Cell-Mediated Response	
Effector Cell/ Molecule	**Function**	**Effector Cell/ Molecule**	**Function**
Antibody (secretory IgA)	Inhibits viruses to bind to host-cells and thus prevents infection	IFN-λ secreted by T_H or T_C cells	Antiviral activity
IgM, IgG and IgA	Inhibits fusion of viral envelope and host-cell plasma membrane	Cytotoxic T lymphocytes	Kill self cells infected with virus
IgG and IgM	Increases phagocytosis of viral particles (opsonization)	NK cells and macrophages	Kill self cells infected with virus by antibody-dependent T-cell-mediated cytotoxicity (ADCC)
IgM	Agglutination of viral particles		
Complement activated by IgG or IgM	Stimulates opsonization by C3b; lyses the enveloped viral particles by membrane-attack complex (MAC)		

BOX 3.2 ANTITOXIN AND SERUM THERAPY

Shibasaburo Kitasato (1852–1931) and Emil von Behring (1854–1917) worked on immunizing guinea pigs using diphtheria toxin treated with heat. They showed that the blood products [*sera* (plural); *serum* singular)] extracted from the guinea pigs carried a substance that played a vital role in preventing the harmful effects of *Corynebacterium diphtheriae* and its toxin when the guinea pigs were again exposed to lethal doses of the bacteria and its toxin. Further, they also showed that it was possible to cure diphtheria in an animal by injecting the serum obtained from an immunized animal. Thus, they said that the substance in serum that prevented or cured diptheria are called as *antitoxin* and this treatment as *serum therapy.*

Behring started with the human trial of the vaccine in mid-January 1892 against diphtheria but it was not very successful. He then tried it in larger animals, like sheep, horses, to standardize it before the vaccine was ready for clinical trials. In 1984, according to a report 220 children who had contracted diphtheria were vaccinated with 77% success. Thus, they could successfully use horse and sheep to produce enough antitoxins that could be used for treating humans.

In 1901, Emil Von Behring won the first Nobel Prize for medicine for his work for successful therapy for diphtheria and tetanus.

Source: The College of Physicians of Philadelphia. The history of vaccines. Retrieved from: https://www.historyofvaccines.org/content/articles/passive-immunization.
Kaufmann, S.H.E. (2017). Remembering Emil von Behring: From tetanus treatment to antibody cooperation with phagocytes. *mBio*, doi: 10.1128/mBio.00117-17.

3.7 BEHAVIOR OF CELLS DURING ADAPTIVE IMMUNE RESPONSE

The first and crucial step in the adaptive immune response is the activation of naïve T-cells by antigen-presenting cells. This happens in lymphoid tissues and organs. $CD4^+$ T-cells are required for the initiation of both humoral and cell-mediated adaptive immune responses. The first experimental proof about the live $CD4^+$ T-cells interacting with antigens presented by dendritic cells came from the Cahalan laboratory, where they showed live that naïve T-cells traveled faster through paracortex (10 μm/min) before engaging with the antigen. After binding with the MHC-peptide complex, their speed slowed down to 4–6 μm/min. Thus, these "arrested" T-cells were actually engaged with the dendritic cells.

3.7.1 Behavior of T-Cells during Adaptive Immune Response

The antigen-presenting cells express co-stimulatory signals (cell surface receptors) that are needed for the complete activation of T-cells. The two best characterized co-stimulatory molecules are B7.1 and B7.2. The naïve T-cells recognize the antigens only when APCs present processed antigen to TCR and B7 (co-stimulatory molecule) to the CD28 receptor on the T-cell. Naïve $CD8^+$ T-cells that leave the thymus are predetermined to become cytotoxic cells but they do not show any of the functions of the differentiated effector cells *i.e.* T_C cells. However, $CD4^+$ cells behave differently and are more complex in their behavior. The naïve $CD4^+$ cells when activated get differentiated into effector cells, which can be either T_H1 cells or T_H2 cells. Both the effector $CD4^+$ cells secrete different kinds of cytokines and thus, differ in their functions. This crucial decision of the fate of the $CD4^+$ cells is fixed during the clonal expansion of the $CD4^+$ cells after their first encounter with the antigen as shown in Figure 3.4.

The factors that determine whether $CD4^+$ cell will form T_H1 cell or T_H2 cell are not clearly understood. However, some of the factors include cytokines like interferons, IL-12 and IL-4, costimulators acting during the response and the nature of the MHC-peptide complex. Figure 3.8 shows the process of differentiation of $CD8^+$ and $CD4^+$ cells. When pathogens interact with PRRs present on dendritic cells or other APCs, the APCs produce polarizing cytokines which determine differentiation of which T_H subset will be favored. The polarizing cytokines interact with the cytokine receptors and signal the transcription of unique master gene regulators. The master regulator genes in turn regulate the expression of various genes, including those for effector cytokines, and thus define the functions of each subset of T-helper cell as shown in Figure 3.9. Once the selection of T_H1 or T_H2 cell has been done, the effects are intense and specific as the selective production of T_H1 cells initiates cell-mediated immunity while the predominant production of T_H2 cells initiates humoral immunity as shown in Figure 3.8.

Let us take one example to understand the differential functions of T-cells. The causative organism for leprosy is *Mycobacterium leprae.* After infection, *M. laprae* enters and develops in macrophage vesicles. For the immune response to be effective, there should be activation of macrophages by T_H1 cells. In patients suffering from tuberculoid leprosy, T_H1 cells are preferentially activated resulting in few live bacteria and less antibody production. The activated macrophages cause damage to the skin and peripheral nerves due to inflammatory responses, the disease persists and progresses slowly and the patient generally survives. However, sometimes in tuberculoid leprosy patients, T_H2 cells are preferentially induced and as a result, the body reacts by producing humoral response. The antibodies thus produced during humoral response do not reach the intracellular bacteria inside the macrophages and the patients develop another condition known as lepromatous leprosy which is characterized by an abundant growth of *M. leprae* in macrophages. This condition causes vast tissue destruction which can ultimately be fatal.

Naïve $CD8^+$ T-cells differentiate to form cytotoxic T-cells (T_C cells) and require more costimulatory signals to become effector cells as compared to naïve $CD4^+$ cells. The first stimulus for

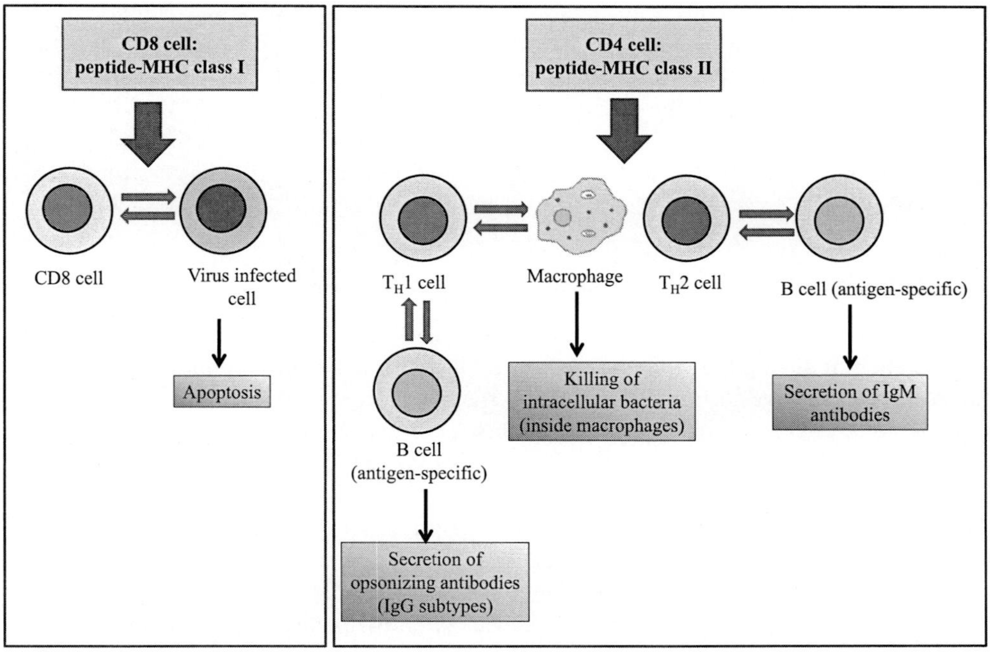

FIGURE 3.8 Differentiation of $CD8^+$ and $CD4^+$ cells. The $CD8^+$ cells get activated by class I MHC-peptide complex presented by APCs and differentiate to form cytotoxic T lymphocytes (CTLs) that eliminate the virus-infected cells. The $CD4^+$ cells are activated by class II MHC-peptide complex presented by APCs and get differentiated into T_H1 cells and T_H2 cells. The T_H1 cells in turn activate 2 types of cells: the macrophages and B-cells, the former kills the microbes by phagocytosing it and the latter secrete opsonizing antibodies (IgG). The T_H2 cells activate B-cells exclusively and induce B-cells to secrete IgM antibodies.

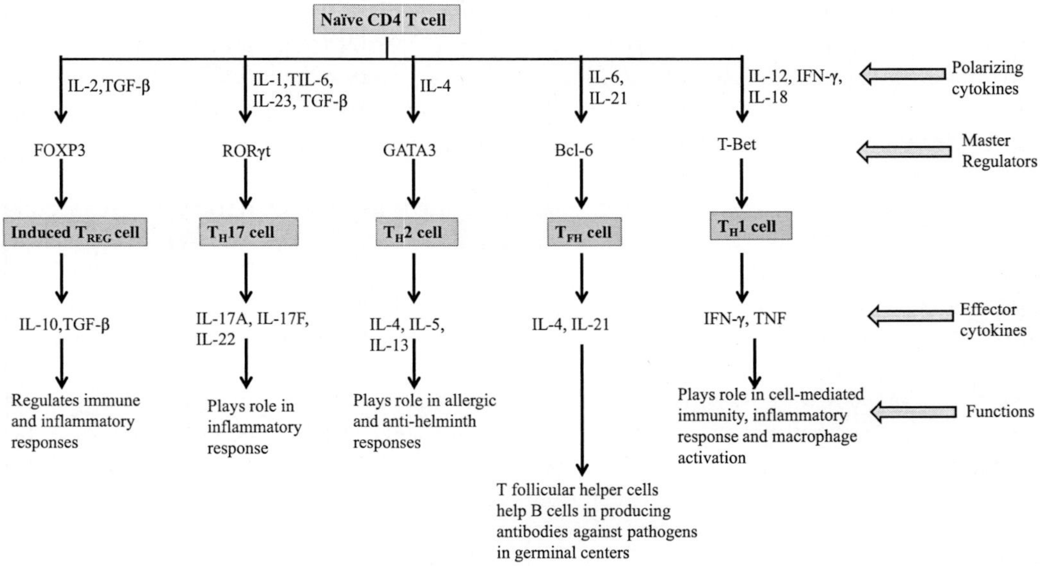

FIGURE 3.9 Differentiation of naïve T_H cell to produce different subsets. The naïve $CD4^+$ T-cells, when activated by APCs, start to proliferate. However, the differentiation of the newly formed $CD4^+$ T-cells depends on the availability of cytokines in the immediate vicinity and results in the formation of different types of T helper cells such as T_{REG} cells, T_H17 cells, and T_H2 cells, T_{FH} cells and T_H1 cells. All these differentiated $CD4^+$ cells have specific functions.

activation of $CD8^+$ T-cells comes from dendritic cells which possess intrinsic costimulatory activity. Interaction with dendritic cells stimulates $CD8^+$ cells to secrete IL-2 which is autocrine in function and stimulates $CD8^+$ T-cells to proliferate and differentiate into T_C cells. During cellular responses against some viruses and tissue grafts, T_C cells require the availability of $CD4^+$ T-cells also. The binding of CD40 ligand on $CD4^+$ T-cell to the CD40 receptor on the surface of APC stimulates the expression of B7 on the cell surface and thus helps APC to co-stimulate $CD8^+$ T-cell directly.

3.7.2 Behavior of B-Cells during Adaptive Immune Response

Like T-cells, the activation of B-cells also requires two signals. When the B-cells recognize and bind to the antigen during their travel through the B-cell follicle in lymph nodes, they become partially activated. However, for full activation they need help from the activated T-cells. The activated T-cells release cytokines that stimulate differentiation of B-cells and also provide ligands that stimulate CD40 signaling. Various effector helper T-cells – T_H1, T_H2, T_{FH} provide costimulatory signals that help B-cells to get activated, differentiate and become effector B-cells that produce copious amounts of highly specific antibodies and memory B-cells. The activated B-cells receive help from T-cells at the junction between the follicle and the paracortex of the lymph node. The necessary evidence for this was provided by the UCSF-Cyster lab, San Fransisco, California showing dynamic images of antigen-specific B-cells interacting with the T-cells.

The interaction between T-cells and B-cells is intimate and stable. The molecules involved in this interaction include TCR, CD28 and adhesion molecules on T-cells with class II MHC-peptide complexes, CD80/86 molecules and other adhesion molecules on B-cells. During this interaction, T-cells deliver cytokines like IL-4 that stimulate B-cell to undergo differentiation. Another interaction is observed between CD40L on T-cell and CD40 on B-cell which further assists in stimulating B-cell to proliferate and differentiate.

After the B-cells have been stimulated by T-cells, the activated B-cells travel to the outer edges of the follicle where B-cells continue to proliferate while some still continue to get T-cell help. Some of the activated B-cells differentiate into effector cells or plasma cells and leave the lymph node while others return to the follicle and form a *germinal center* where they carry out more rounds of proliferation and somatic hypermutation, a process in which point mutations accumulate in the variable (V) region of the heavy and light chains with the aim to produce high-affinity antibodies.

There are two major zones in the germinal center *i.e. the dark zone* (rich in proliferating B-cells and is the main site of somatic hypermutation) and *the light zone* (rich in follicular dendritic cells and their reticular network and provide the scaffold for further B-cell/T-cell interactions). The light zone is the primary site where B-cells differentiate to become plasma cells and memory cells. Another important event that occurs here is the positive selection of B-cells specific for every antigen.

Based on the model of positive selection, researchers assumed that B-cells shuffle between the light and dark zones but dynamic experiments by Cyster and other labs have shown that germinal center B-cells prefer to travel ***within*** and ***not between*** the light and dark zones. However, a small percentage of cells travelled between the zones. The B-cell proliferation was not confined to the dark zone but was also observed in the light zone. Experiments also showed that B-cells interacted briefly with T_H cells in the light zone which further challenged another assumption that high-affinity B-cell clones make stable connections with T-cells. The behavior of B-cells and T-cells after they have interacted with the antigen in the lymph node is represented briefly in Figure 3.10.

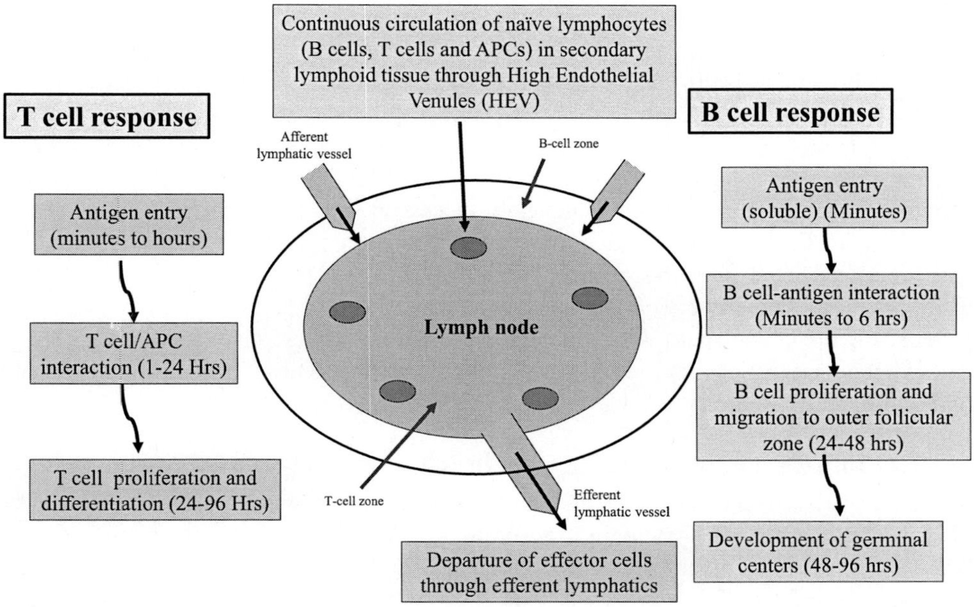

FIGURE 3.10 Overview of the behaviour of B-cells and T-cells in lymph node (after entry of antigen).

3.8 IMMUNE DYSFUNCTIONS

Immune dysfunction refers to a condition when the immune system of a person is compromised such that it results in failures and subsequent diseases. Immune dysfunction can be categorized into three groups:

a. **Immune deficiency:** The immune system becomes insufficient to protect the body against invading pathogens.
b. **Autoimmunity:** Immune cells target the self-cells and tissues.
c. **Hypersensitivity:** It refers to a profound attack on the common but benign antigens

Immune dysfunction results when there is no proper regulation and the system attacks something that it shouldn't and does not attack when it should. During an autoimmune disease, the immune system attacks an improper target. Some of the common autoimmune disorders and their etiology is discussed in the following sections.

3.8.1 Tolerance

Autoimmune disorders are caused when the immune system fails to differentiate between self-cells and non-self cells. Autoimmune disorders can cause serious harm to the cells and organs and sometimes may even be fatal. Although there are some autoimmune disorders but at large the body protects itself and there are mechanisms that check anti-self immune attack. The mechanism in the body that controls the self-immune attack is called *self-tolerance or tolerance.* The mechanism of self-tolerance is complicated and if the process is working correctly, host T-cells and tissues remain undisturbed and only the non-self cells or invaders are attacked. The mechanism of self-tolerance also causes rejection in the case of transplanted tissue or cells that may carry genetically non-related proteins.

Tolerance is a term that is used for the ability of the immune system to not identify self cells and tissues for an immune attack. This means that the immune system 'knows' the cells that are not to be attacked. This memory is the result of several processes and the first step of the process is called *central tolerance.* In central tolerance, clones of T- and B-lymphocytes that recognize self-antigens are deleted before maturation. This deletion occurs in *primary lymphoid organs* which implies that such B-lymphocyte clones are destroyed in bone marrow while T-lymphocyte clones are destroyed in the thymus. Central tolerance is not completely successful as many lymphocytes may enter the secondary lymphoid organs, hence some more mechanisms are needed to ensure that such lymphocytes are also deleted from the system. These additional mechanisms together constitute the *peripheral tolerance.* The peripheral tolerance, thus, inactivates the self-reactive lymphocytes and also generates lymphocytes that inhibit immune response against self cells and tissues. Also, the action of lymphocytes that are against self-cells is further limited as they are destroyed by apoptosis due to their limited life span.

3.8.1.1 Central Tolerance and Development of Self-Reactive T- and B-Lymphocytes

Central tolerance effectively works by deleting the self-reactive lymphocytes during their development. Let us consider one phenomenon *i.e.* generation of diversity in T- and B-lymphocyte receptors. The rearrangement of genes involving V region, D and J-region gives rise to a multitude of T-cell receptors including TCRs against self-cells. If this were to happen, the mature T-lymphocytes would lead to autoimmune disorders.

The exact mechanisms that underlie the central tolerance are not fully understood but during development, the lymphocytes undergo a process called *negative selection.* Negative selection refers to the death of some cells carrying self-reactive TCRs or immunoglobulin receptors (Ig receptors).

Tiegs and his group showed that during development B-cells can undergo *receptor editing.* In receptor editing, the V region of the TCR is edited *i.e.* the V(D)J gene rearrangement is changed and as a result, the edited TCR is less reactive and has a low affinity for self-antigens. Thus, negative selection *i.e.* clonal deletion or apoptosis, and receptor editing, together, are able to eliminate several self-reactive T- and B-cell clones in thymus and bone marrow respectively.

3.8.1.2 Peripheral Tolerance and Circulating Self-Reactive Lymphocytes

We have discussed in the above section that central tolerance is not sufficient to delete all the self-reactive lymphocytes. Two factors play important role during central tolerance and in the entry of self-reactive lymphocytes into the peripheral system: one, all self-reactive antigens are not expressed in the primary lymphoid organs where central tolerance works using negative selection, and second, self-antigens have an affinity toward lymphocytes before clonal deletion is stimulated which allows the weak self-reactive clones to survive the weeding-out process. As we know that the activation of T-cells is dependent on two factors – one, binding of TCR to antigen presented by self-MHC (*M*ajor *H*istocompatibility *C*omplex) molecule and second, a costimulatory signal. The costimulatory signal is provided by the interaction between CD28 on T- cell and CD80/86 on APC. There are other molecules on the T-cells that can also bind to CD80/86 and one such molecule is CTLA-4 which inhibits T-cell activation rather than activating it. Thus, there are molecules that can either stimulate or inhibit T-cells and are known as immunomodulatory molecules. CTLA-4 expression happens only after T-cell has been activated, thus regulating the immune response. It has been observed in experiments that mice which lack CTLA-4, show excessive proliferation of lymphocytes and severe autoimmune diseases suggesting the possible role of CTLA-4 in peripheral tolerance.

On similar lines, B-cells also participate in peripheral tolerance. The experiments show that when B-cells encounter soluble antigens and there is no costimulatory signal from T-cells, they become anergic and never migrate to germinal centers. Thus, the B-cells become tolerant to the same antigen as the T-cells.

3.8.2 Mechanisms involved in Tolerance

3.8.2.1 T-Cell Anergy

The mechanisms that cause peripheral tolerance in the body may vary. For example, when an antigen binds to a lymphocyte, the immune system gets activated. However, if the same antigen is presented in an altered form, time or location, the encounter may lead to tolerance. Thus, some antigens induce tolerance and are known as *tolerogens*. The same antigen may act both as immunogen and tolerogen depending on - in which form it is presented and where it is presented. For example, if an antigen is presented to a T-lymphocyte, then without co-stimulation the interaction results in the form of tolerance called *anergy* and the same antigen when presented with costimulatory signals can become a strong immunogen. Some antigens if ingested can be tolerated while if the same antigens are given as intradermal or subcutaneous injection, they can be immunogenic. The tolerogens are antigen-specific and hence, if an immune response is inactivated it does not result in general immune suppression.

In an adult, if an antigen enters the body, the body initiates an immune response against it with the aim to eradicate the antigen. However, in the fetus when it is growing, the immune system is not developed and any exposure to antigen is frequently tolerated and results in tolerance, thereby 'identifying' the antigen as 'self'. There are other factors also that contribute to the development of tolerance which include high doses of antigen, availability of antigen in the host for long term, oral or intravenous entry of antigen, absence of adjuvants (compounds that increase the immunogenicity), less costimulation and antigen presentation by immature or not suitably activated antigen-presenting cells (APCs).

3.8.2.2 Apoptosis

During the 1960s, it was believed that the self-reactive lymphocytes were eliminated from the thymus and bone marrow during development. As per this understanding, if the body fails to eliminate self-reactive lymphocytes, it may lead to autoimmune disorders. However, lately, it has been seen that a healthy person may also possess mature self-reactive lymphocytes but they may not have any autoimmune disorder. This suggests that there must be mechanisms that regulate the activity of these self-reactive lymphocytes. One such mechanism to induce self-tolerance is natural cell death (*i.e.* apoptosis) or cell anergy in lymphocytes. Apoptosis is an important phenomenon that is central to central and peripheral tolerance.

Other than central and peripheral tolerance, the other mechanism that ensures that self-reactive lymphocytes are deleted is sequestration or compartmentation of antigens. For example, the anterior chamber and lens of the eye are one such site where antigens may be sequestered and due to lack of lymphatic drainage, the sequestered antigens may never encounter lymphocytes. Thus, when the antigen is not exposed to immune cells, there will not be any reactivity. But because of the sequestration, the antigen never encounters developing lymphocytes and no active tolerance is developed for it. In certain cases, when the barriers are broken and such sequestered antigen encounters the lymphocytes, it will be seen as foreign as it was not previously exposed. So, if there is trauma in one eye, the immune cells may react against the antigen and cause inflammation followed by tissue destruction and impairment in vision. The other eye may also become inflamed due to the sudden entry of clones of the recently activated lymphocytes. There are some privileged tissues like the eyes and central nervous system (CNS) which lie in a locally immunosuppressive environment and bias the immune response toward tolerance.

3.8.2.3 T-Regulatory Cells

In addition to anergy and apoptosis, T-regulatory cells also play a significant role in maintaining tolerance. T_{REG} cells are present in active form in secondary lymphoid tissues and at the sites of inflammatory response where they recognize self-antigens and sometimes foreign antigens also via TCR interactions. However, when T_{REG} cells interact with these antigens in periphery, they down-regulate the immune response. There are two ways by which T_{REG} are generated: Naturally, in thymus (nT_{REG} cells) and after stimulation by antigens in periphery (iT_{REG} cells).

It is believed that nT_{REG} cells are involved in responses against self-antigens, thus preventing autoimmune diseases while iT_{REG} cells act against foreign antigens at the mucosal surfaces which are in constant touch with the outside world.

3.9 SOME COMMON AUTOIMMUNE DISEASES

When the immune system breaks down and it starts attacking the self cells, the condition results in autoimmunity and can result in several diseases. The symptoms of autoimmunity vary depending on the type of tissues involved. For example, in multiple sclerosis, there is an autoimmune attack on myelin sheaths in the brain and spinal cord, a response which originates as the production of auto-reactive T-lymphocytes in peripheral immune system, and results in neuromuscular dysfunction. In Crohn's disease, there is an attack on intestinal tissues that destroys the epithelial layer in the gut resulting in poor absorption of food. Rheumatoid arthritis is a common example of an autoimmune disorder wherein the immune system attacks the joints of hands, arms, feet and legs. Autoimmune diseases can be caused due to genetic as well as environmental factors.

Immunodeficiency is a condition when some component of the innate or adaptive immunity is either defective or absent. There are two reasons for the development of immunodeficiency – one, genetic factors, and such condition is known as *primary immunodeficiency* and two, damage due to chemical, physical or biological agents, a condition known as *secondary immunodeficiency* (Box 3.3).

BOX 3.3 PRIMARY AND SECONDARY IMMUNODEFICIENCY DISEASES

There exist more than 400 medical conditions that can be regarded as the primary immunodeficiency diseases (PIDs) as they are caused due to genetic factors. Some of the common ones are given below:

Primary Immunodeficiency Disease	Characteristics
Autoimmune polyglandular syndrome type I (APS-1)	Symptoms like autoimmunity against various organs; heightened susceptibility to infection by *Candida* yeast.
Autoimmune Lymphoproliferative Syndrome (ALPS)	A rare autoimmune disorder; causes lower RBC count, platelets and WBCs. There is higher risk of infection and hemorrhage.
Caspase Eight Deficiency state (CEDS)	A rare genetic disorder caused due to mutation in the gene *CASP8*; enlargement of spleen and lymph nodes; frequent infections in sinus and lung; frequent viral infections and compromised immune system
BENTA disease	A rare genetic disorder caused by mutation in the gene *CARD11*; Increase in the number of some cells of immune system during infancy; immunodeficiency; enlargement of spleen and lymph nodes; an increased risk of lymphoma

(*Continued*)

BOX 3.3 (*Continued*) PRIMARY AND SECONDARY IMMUNODEFICIENCY DISEASES

Primary Immunodeficiency Disease	Characteristics
Chronic Granulomatous Disease (CGD)	Inability of phagocytes to kill invading bacteria nad fungi; mutation in one of any five genes can lead to this disease
Common Variable Immunodefieiency (CVID)	Caused due to several genetic abnormalities leading to insufficient production of immune cells and thus antibodies
Severe Combined Immunodeficiency (SCID)	It is a group of rare disorders caused due to mutations in genes playing role during development and maturation of B and T lymphocytes; it can be fatal.

In addition to the above mentioned primary immunodeficiency diseases, there are many other diseases belonging to this group which include: *CARD9* deficiency and other syndromes leading to susceptibility to cardiasis; congenital neutropenia syndromes; *CTLA4* deficiency; *DOCK8* deficiency; *GATA2* deficiency; Glycosylation disorders accompanied by immunodeficiency; hyper-immunoglobulin E syndrome; hype-immunoglobulin M syndrome; IF-γ, IL-12 and IL-23 deficiencies; leukocyte adhesion deficiency (LAD); *PLCG2*-associated antibody deficiency and immune disregulation (PLAID); *LRBA* deficiency; hyperactivity of PI3 kinase resulting in PI3 kinase disease; *STAT3* gain-of-function disease; Wiskott-Aldrich syndrome (WAS); Warts, hypogammaglobulinemia, infections and Myelokathexis (WHIMS); X-linked lymphoproliferative disease (XLP) and X-linked immunodeficiency with magnesium defect; X-linked agammaglobulinemia (XLA); Epstein barr-virus (EBV) infection and neoplasia (XMEN disease) to name a few.

The secondary immune deficiency (SID) diseases are more common than PID diseases. The secondary immune deficiency diseases are caused due to several factors like severe malnutrition; newborn period and advanced age; surgery and trauma; some chronic diseases like diabetes and uremia; infectious diseases like infection with HIV; immunosuppressive medication like radiation therapy, cytotoxic chemotherapy, surgical removal of bone marrow prior to transplantation, radiation therapy; certain types of cancers like leukemia as well as the absence of spleen.

Some of the examples of the secondary immune deficiency diseases are Acquired Immunodeficiency Syndrome (AIDS) caused due to HIV infection; cancers of immune system such as leukemia; diseases due to immune-complexes such as multiple myeloma and viral hepatitis.

Source: Types of primary immunodeficiency diseases. Retrieved from: https://www.niaid.nih.gov/diseases-conditions/types-pidds

Immune deficiency disorders- Types. Retrieved from https://www.nationaljewish.org/conditions/immune-deficiency-disorders/types.

3.9.1 Acquired Immune Deficiency Syndrome (AIDS)

The extreme immunodeficiency can result into a condition called as severe combined immunodeficiency (SCID) which affects both B and T-cells and severely affects the adaptive immunity. If it is left untreated, it can result in death during early stages. The most common form of SCID is Acquired Immune Deficiency Syndrome (AIDS) which is caused due to infection with Human Immunodeficiency Virus-I (HIV-I). HIV exists in two forms *viz.* HIV-I and HIV-II which are closely related to each other. HIV-II is restricted to West Africa but is now spreading to India. HIV-I is more virulent form of HIV and is the cause of AIDS worldwide. HIV was first identified as an opportunistic infection in the people of the United States in 1981. This group of patients had unusual infections like fungal pathogen *Pneumocystis carinii* which causes *P. carinii* pneumonia (PCP) in people with immunodeficiency. After evaluation, it was confirmed that the patients had a severe deficiency in cell-mediated immune responses and a marked reduction in T_H cells. After its discovery in the 1980s, AIDS has spread all over the world. As in December 2011, nearly 34 million people were affected with HIV-I all over the world. HIV-I spreads through intimate contact of body fluids.

HIV persists and is carried in various cells like CD4 T-cells, dendritic cells and macrophages. The virus also exists in free form and is carried in the blood, semen, vaginal fluid or milk. The most common routes of HIV infection are sexual intercourse, use of contaminated needles and use of blood or blood products for therapy. During an HIV infection, the immune system does not recognize this virus appropriately and hence is not able to remove it effectively. Thus, there remains persistent infection as HIV remains hidden inside the genomes of T_H cells, target cells and cells involved in adaptive immunity. With time, more and more T_H cells are lost and as the disease progresses further, a significant number of T_H cells are lost or become dysfunctional which ultimately results in the collapse of the immune system. Most patients having HIV infection eventually develop AIDS after an incubation period known as clinical latency or quiescence. During this period, the virus replicates continuously and there is a gradual decline in the function as well as numbers of CD4 T-cells.

There exist certain anti-retroviral treatments which can prolong the life of AIDS patients almost indefinitely but still, there is no vaccine or cure for the disease. The major effects HIV-I has on the immune system have been described in Table 3.8.

3.9.2 Other Autoimmune Disorders

Talking further of autoimmunity, some autoimmune disorders are caused due to failure of immune tolerance. Immune tolerance can be generated in the body by repeated injection or ingestion of low doses of antigens. This happens because the antigens stimulate T_{REG} cells to suppress the immune response to the allergen. Autoimmune diseases are caused due to destruction of self-proteins, cells and organs by self-reactive lymphocytes or auto-antibodies. For example, in Hashimoto's thyroiditis auto-antibodies are the major players which cause severe damage to the thyroid tissue. Another example is rheumatoid arthritis (RA) in which tissues are destroyed directly by T-cells. The self-reactive T-cells attack the tissues in joints which causes inflammation and tissue destruction. Some of the common autoimmune disorders mediated by T-cell response are listed in Table 3.9 and have been discussed in the below section.

3.9.2.1 Rheumatoid Arthritis

Rheumatoid arthritis is an autoimmune disorder which is primarily observed in women between 40 and 60 years of age. Rheumatoid arthritis is an example of secondary immunodeficiency. It is caused by self-reactive antibodies called rheumatoid factors, which cause chronic inflammation of the joints of hands, feet, arms and legs. Other than the joints, rheumatoid arthritis affects the hematologic, cardiovascular and respiratory systems. The rheumatoid factors are determinants for the Fc region of IgG. The most common rheumatoid factor is the IgM antibody that binds to the circulating IgG antibody. The IgM-IgG complexes are deposited in the joints. These immune complexes can activate the complement system which causes type III hypersensitivity reactions leading

Table 3.8
Common Abnormalities Observed in the Working of Immune System during HIV Infection

Immunologic Abnormalities	Early Stage of Infection	Late Stage of Infection
Lymph node structure	Dendritic cells are destroyed; some destruction of lymphoid tissue in gastrointestinal tract	Severe damage to tissues and tissue necrosis; loss of germinal centers and follicular dendritic cells; inefficient antigen trapping and inefficient activation of T and B-cells
T_H cells	Decrease in the CD4+ T-cells, especially in gut; loss of proliferative response to specific antigen *in vitro*	Further reduction in TH cells
Antibody production	Increased non-specific IgA and IgG production but decreased IgM synthesis	HIV-I specific B-cells not proliferating; no detectable anti-HIV-I antibodies; presence of B-cells with low expression of CD21
Cytokine production	Augmented levels of some cytokines	Shift in production of cytokines, initially from T_H1 and later to T_H2 cells
Delayed-type hypersensitivity	Reduction in proliferation of T_H1 cells	No DTH response;
T_C cells	Normal reactivity	Reduction in CTL activity as T_C cells not converted to CTLs properly.

Table 3.9
Some Common Autoimmune Disorders

Autoimmune Disease	Affected Organs/Cells	Effector Molecules
Hemolytic anemia	RBC membrane proteins	Auto-antibodies
Grave's disease	TSH receptor	Auto-antibodies (stimulating)
Type 1 diabetes mellitus	Pancreatic beta cells	Auto-antibodies, T_H1 cells
Myasthenia gravis	Acetycholine receptor	Auto-antibodies (blocking)
Pernicious anemia	Gastric parietal cells	Auto-antibodies
Addison's disease	Adrenal gland	Auto-antibodies
Myocardial infarction	Heart	Auto-antibodies
Hashimoto's thyroiditis	Thyroid proteins and cells	Auto-antibodies, T_H1 cells
Spontaneous infertility	Sperm	Auto-antibodies
	Systemic Autoimmune Diseases	
Multiple sclerosis	Brain or white matter	Auto-antibodies; T_H1 cells, Tc cells
Ankylosing spondylitis	Vertebrae	Immune complexes
Systemic lupus erythematosus (SLE)	DNA, nuclear protein, RBC and platelet membranes	Auto-antibodies; immune complexes
Goodpasture's syndrome	Type IV collagen in renal glomeruli and pulmonary alveoli	Auto-antibodies
Rheumatoid arthritis	Connective tissue; IgG	Auto-antibodies; immune complexes
Scleroderma	Heart, lungs, GI tract, kidney; nuclei	Auto-antibodies
Autoimmune polyendocrinopathy-candidiasis-ectodermal dystrophy (APECED)	Multiorgan, inactivity of *aire* gene	Defective central tolerance
Immune dysregulation polyendocrinopathy enteropathy X-linked (IPEX)	Multiorgan, inactivity of *FoxP3* gene	Missing T_{REG} cells

to chronic inflammation of the joints. Rheumatoid arthritis can be treated by reducing the inflammation of joints by using non-specific drugs like nonsteroidal anti-inflammatory drugs (NSAIDs) and corticosteroids.

The common symptoms of secondary immunodeficiency include heightened susceptibility to common pathogens and many a times opportunistic infections. The effect of the infection depends on the degree of immune deficiency. In most of such cases, the withdrawal of external factors causing the immunodeficiency can result in the restoration of immune function.

3.9.2.2 Systemic Lupus Erythematosus (SLE)

Systemic lupus erythematosus (SLE) is more common in women than in men, like other autoimmune disorders. The symptoms usually occur between 20 and 40 years of age and are common in women of African, American and Hispanic origin than in Caucasians. In identical twins, if one person is suffering from SLE, the other twin has a 60% chance of suffering from SLE. The close relatives of a person having SLE have 25% chances of getting SLE but only 2% of these individuals develop SLE.

Individuals having SLE produce auto-antibodies (IgG) for several antigens like DNA, histones, RBCs, platelets, leucocytes, clotting factors and nucleoprotein particles like nucleosomes and spliceosomes. They also produce auto-antibodies against a small cytoplasmic ribonucleoprotein complex containing two proteins *viz.* Ro and La (These proteins were named after the first two alphabets of the surnames of the two patients from whom autoantibodies against these proteins were isolated). The symptoms of SLE include fever, arthritis, weakness, skin rashes and kidney dysfunction. If auto-antibodies are produced against RBCs and platelets, they can cause complement-mediated lysis of RBCs and platelets which may lead to hemolytic anemia and thrombocytopenia respectively. The complex of auto-antibodies with different nuclear antigens may sometimes get deposited along the walls of the small blood vessels resulting in type III hypersensitivity reaction. These complexes activate complement system which lead to formation of membrane-attack complex (MAC). MAC and other activated complement fragments attack the walls of the blood vessels and result in conditions such as vasculitis and glomerulonephritis. In severe conditions, the serum levels of complement fragments increase excessively which leads to aggregation of neutrophils and their attachment to the vascular endothelium. Over a period of time, neutrophils decrease in number (the condition is called neutropenia) and vasculitis (*i.e.* occlusion of blood vessels) occurs which can cause large scale tissue damage. The clinical investigation of SLE includes detection of antinuclear antibodies against dsDNA or ssDNA, histones, nucleoprotein and nucleolar RNA.

3.9.2.3 Myasthenia Gravis

In myasthenia gravis, autoantibodies are produced against the α chain of nicotinic acetylcholine receptor. The acetylcholine receptor is present at the neuromuscular junctions and its blockage by the autoantibodies can block the neuromuscular transmission. The antibodies cause the acetylcholine receptors to get internalized and are then degraded. People suffering from myasthenia gravis become progressively weaker which may prove to be fatal.

In normal conditions, acetylcholine released from motor neurons at the neuromuscular junction binds to the acetylcholine receptors (AChRs) on the skeletal muscles and triggers muscle contraction. Due to autoantibodies in people with myasthenia gravis, the number of receptors for acetylcholine decreases and the skeletal muscles no longer respond to acetylcholine resulting in loss of function of muscles.

3.9.2.4 Multiple Sclerosis

In western countries, multiple sclerosis (MS) is the most common cause of neurologic disability. It occurs 2–3 times more in women than in men and occurs during 20–40 years of age. The disease is characterized by the production of self-reactive T-cells that form inflammatory lesions in the myelin sheath of the nerve cells in the brain and spinal cord. Myelin sheath insulates the nerve cell and any break in the sheath can cause numerous neurologic dysfunctions including loss of vision, paralysis or numbness in limbs.

The exact cause of developing multiple sclerosis is not well understood. Infection with Epstein-Barr virus (EBV) may make a person susceptible to multiple sclerosis. Infection with some other viruses can cause diseases in which myelin sheath is destroyed, and may be linked to MS but there is no definitive proof for the same.

3.9.2.5 Goodpasture's Syndrome

Goodpasture's syndrome is a type of type II hypersensitivity reaction wherein autoantibodies are produced against α3 chain of type IV collagen which is generally found in the basement membrane of renal glomeruli and in some cases in basement membranes of pulmonary alveoli. When the autoantibodies are produced against type IV collagen they bind its α3 chain and thus ligate Fcγ receptors and activate neutrophils, monocytes, mast cells and tissue basophils. This process attracts more neutrophils in the glomeruli which increases the damage to the tissues. The damage caused by these autoantibodies may prove to be fatal. The autoantibodies can also activate the complement system locally which may amplify the tissue injury.

3.9.2.6 Wegener's Granulomatosis

Wegener's Granulomatosis is also an autoimmune disease and is characterized by severe necrotizing vasculitis. The autoantibodies are produced against a granule proteinase of neutrophils and are known as anti-neutrophil cytoplasm antibodies (ANCA). The antigen in this case is a protein called proteinase-3 which is present in abundance as serine proteinase in neutrophil granules. The presence of ANCA is associated with the onset of disease, but generally, people having high levels of ANCA may not develop any symptoms. However, if such people develop an infection, the chances of showing the symptoms of Wegener's granulomatosis increase. This can be explained as follows:

The resting or naïve neutrophils do not express proteinase-3 on their cell surface. When the neutrophils get activated due to some infection, proteinase-3 is expressed on their cell surface. The autoantibodies against proteinase-3 can now bind to them and cause tissue injury.

3.9.3 Factors Affecting Autoimmunity

Unregulated immune activation can lead to conditions like autoimmunity. Let us now look at the various factors that contribute to autoimmunity. It has previously been postulated, by experiments on germ-free mouse models, data collected from identical twins and epidemiologic studies of geographic associations that environmental factors as well as genes contribute to the development of autoimmunity.

3.9.3.1 Environmental Factors

It has been observed that certain autoimmune syndromes are more common in some geographical locations or in a specific climate. Thus, there might be a link between environmental factors or lifestyle with the development of autoimmunity. For example, interaction between gut microflora and the systemic immune system helps in regulating peripheral tolerance which may play important role in developing autoimmune diseases. Infections can also contribute to autoimmunity. For example, an infection resulting in tissue pathology may release the sequestered self-antigens which are presented and this results in immune activation instead of tolerance. Similarly, certain microbes may have chemical features that resemble the self components which may activate the immune system with the capability to cross-react.

Table 3.10
Genes Involved in Autoimmune Disorders

Genes	Autoimmune Disorder	Effector Cells/Organs
Aire	Mutant form of *aire* gene causes autoimmune polyendocrinopathy –candidiasis-ectodermal-dystrophy (APECED), a systemic disease	Defective deletion or inactivation of autoreactive T-cells in thymus
Human leukocyte antigen (HLA)	Rheumatoid arthritis (RA)	Anticitrullinated Abs
FoxP3	Mutant form of *FoxP3* gene causes immune dysregulation polyendocrinopathy and X-linked syndrome (IPEX)	Inability to produce T_{REG} to maintain pheripheral tolerance
PTPN22, TRAF1-C5, PADI4, STAT4	Make individuals susceptible to RA	-
HLA and non-HLA genes- TRAF3IP2, IFIHI, PTPN22, STAT4	Increase SLE susceptibility	-

3.9.3.2 Genetic Factors

There are certain alleles within the MHC gene that contribute to autoimmune diseases. For example, there is a strong association of the HLA allele with ankylosing spondylitis, an inflammatory disease of vertebral joints. People having one HLA allele *i.e.* HLA-B27 are 90 times more prone to developing ankylosing spondylitis than people possessing other HLA alleles at the same locus. However, the presence of the HLA-B27 allele does not mean that the person will develop ankylosing spondylitis which implies that there are other factors also that contribute to the development of ankylosing spondylitis. Ankylosing spondylitis is seen in 90% of males. Some of the other genes involved in developing autoimmune disorders are listed in Table 3.10.

3.10 INTERACTION BETWEEN INNATE AND ADAPTIVE IMMUNITY

The innate immunity provides us with several layers of protection as has been discussed in Chapter 2. However, innate immunity is not sufficient to provide us complete protection against the invading microbes or other antigens. This is because there are many microbes which can easily cross the layers of protection of innate immunity and result in pathological conditions. To counter this, our body has powerful mechanism in the form of adaptive immune system which utilizes the antibodies or cell-mediated mechanisms to resolve the infections. Thus, it may be said that our innate immune system initiates and also regulates the adaptive immune system to take care of the infection.

Whenever pathogens cross the innate barriers, the innate system immediately starts processes to clear the infection but at the same time it also sends signals to the adaptive immune system. At the site of infection, the innate cells *i.e.* dendritic cells, mast cells, resident macrophages and newly recruited neutrophils recognize the pathogens by their PRRs and start to produce antimicrobial and proinflammatory molecules that slow down the infection and simultaneously they also initiate the processes for recognition of pathogens by lymphocytes. The lymphocytes, thus, activate the adaptive immune response whose effects are visible days after the infection but the response is strong and mediated by the antibodies as well as cells.

Table 3.11
Differential Signaling through Dendritic Cell PRRs

Stimulus for the Activation of Dendritic Cell	Target	Interaction between Groups	Result of Interaction
Fungi → TLR2/6 on dendritic cell → Release of IL-10, RA, TGF-β	Naïve T_H cell	class II MHC-antigen complex/ CD80/86 on dendritic cell and TCR	Activation of naïve T_H cell to form T_{REG} cell and secretion of IL-10 and TGF-β
Fungi → Dectin-1 receptor on dendritic cell → release of IL-6, IL-23	Naïve T_H cell	class II MHC-antigen complex/ CD80/86 on dendritic cell and TCR	Activation of naïve T_H cell to form T_H17 cell and secretion of IL-17
Helminths → TLR2/1 on dendritic cell → release of IL-10	Naïve T_H cell	class II MHC-antigen complex/ CD80/86 on dendritic cell and TCR	Activation of naïve T_H cell to form T_H2 cell and secretion of IL-4, IL-5 and IL-13
Bacteria → TLR4 or TLR5 and viral infection → (viruses bind to TLR 3, 7 and 9) on dendritic cell → Release of IL-12	Naïve T_H cell	class II MHC-antigen complex/ CD80/86 on dendritic cell and TCR	Activation of naïve T_H cell to form T_H1 cell and secretion of IFN-γ

The adaptive immune response starts with the delivery of pathogens to the lymphoid tissues where T- and B-cells recognize them and respond accordingly. This can be explained as follows: dendritic cells bind the microbes through PRRs and carry the microbes via lymphatic vessels to adjacent secondary lymphoid tissues such as draining lymph nodes where dendritic cells can present the microbes or microbial components to other cells. Dendritic cells internalize the microbes and present the antigens with the help of class II MHC molecules on their cell surface. Along with class II MHC molecules, the dendritic cells also express costimulatory membrane proteins like CD80 or CD86 which are recognized by T_H cells. The type of stimulation of dendritic cells by PAMPs of the microbes and other downstream signaling pathways decides the kind of cytokines that will be secreted by the naïve T_H ($CD4^+$) cells after activation. The different types of T_H cells produced trigger varied types of adaptive immune responses which are effective in eliminating the pathogen.

The dendritic cells also activate T-cytotoxic cells which are needed to eliminate the cells from our body that have become harmful due to infection with viruses, intracellular pathogens or by malignant tumor formation. The naïve T_C cells are activated by the mature dendritic cells *i.e.* the dendritic cells whose PRRs have interacted with the microbial PAMPs. The mature dendritic cells also internalize and process the pathogens and present the antigens on the cell surface with the help of class I MHC molecules. The antigen- class I MHC complex together with the costimulatory molecules CD80/CD86 activate the naïve T_C cells which get converted into cytotoxic T lymphocytes (CTLs). Certain cytokines produced by innate cells like dendritic cells, macrophages, NK cells etc. further stimulate the activation and differentiation of T_C cells. Table 3.11 presents different signaling mechanisms used by dendritic cell to activate T_H cells.

The innate and adaptive immune responses interact with each other to provide protection to the body from the invading pathogens. Both the immune responses involve the cells as well as several molecules and respond in different ways. Table 3.12 compares innate and adaptive immune responses.

Table 3.12
Comparison of Innate and Adaptive Immunity

Characteristics	Innate Immunity	Adaptive Immunity
Existence	Already present	It is generated after exposure to a pathogen
Activation	Rapid; within minutes or hours	Slow; takes days
Specificity	Non-specific; recognizes molecules or molecular patterns on the cell surface of pathogens or dead/damaged cells	Highly specific; discriminates even minor differences between the molecules of microbial and non-microbial molecules
Diversity	Number of receptors are limited, which are usually conserved and germ-line coded	Receptors are highly diverse- arise due to genetic recombinations of the receptor genes
Memory responses	Few (seen in invertebrates and NK cells)	Persistent memory, with subsequent responses with greater vigour than the first response
Potential to fight a pathogen	Low and limited	High
Inheritance	It is inherited from the parents as it is encoded in genes	Not inherited; it is acquired during lifetime of an individual
Discrimination of self/ non-self cells	Strong	Very good; occasional failure may result in autoimmune diseases
Soluble components	Several antimicrobial peptides, proteins and other mediators	Antibodies and cytokines
Major cells involved	Phagocytes (macrophages, monocytes, neutrophils), NK cells, other leucocytes, epithelial and endothelial cells	B lymphocytes, T lymphocytes and antigen presenting cells (APCs)
Allergic reaction	None	May show hypersensitivity reactions

3.11 KEY CONCEPTS

- Immunity is a mechanism by which body protects itself from the harmful effects of the pathogens and/or antigens.
- Our body has two types of defense mechanisms- innate immunity and adaptive immunity. Innate immune response is considered as 'first line of defense' as it is activated within hours of the entry of the pathogen.
- The innate and inflammatory mechanisms are carefully regulated by positive and negative feedback, if not, they result in pathological conditions.
- Adaptive immune response is delayed and may take days to respond but it is more persistent and effective.
- Adaptive immunity is of two types- humoral (mediated by B lymphocytes involving antibodies) and cell-mediated (mediated by T lymphocytes through cell-cell interaction).
- Immunity can also be passive, when preformed antibodies are given to someone or active when the body of a person is sensitized to produce its own antibodies.
- Sometimes body's own immune system may not respond suitably to the pathogen/antigen or may attack its own cells, such conditions lead to several immune dysfunctions including autoimmunity.
- Innate and adaptive immune responses are not separate but in fact interact with each other to provide necessary protection to the body.

3.12 SUMMARY

- Adaptive immunity is also known as acquired immunity or delayed immunity. It takes larger time but is more antigen specific as compared to innate immunity.
- Two major types of cells play role in adaptive immunity *viz.* lymphocytes and antigen presenting cells (APCs). Lymphocytes are of two types: T-lymphocytes and B-lymphocytes while there are several APCs.
- Some unique proteins are expressed on the surface of lymphocytes, known as cluster differentiation proteins (CD or cluster of designation). CDs are used to distinguish between the T-lymphocytes and B- lymphocytes.
- Other than CD, each B- and T- cell expresses an antigen specific receptor on its cell surface *i.e.* B-cell receptor (BCR) and T-cell receptor (TCR) respectively.
- There are two major types of T-cells *viz.* T-helper cells (T_H) or $CD4^+$ T-cells and T-cytotoxic cells (T_C) or $CD8^+$ T-cells.
- The naïve $CD8^+$ T-cells recognize class I MHC-peptide complex on the surface of antigen presenting cells and they get activated, divide and differentiate to produce effector cells called as cytotoxic T-lymphocytes (CTLs).
- $CD4^+$ T-cells differentiate to form a variety of effector T-cell subtypes: *T-helper type I cells* (T_H1) which respond to intracellular pathogens; *T-helper type 2 cells* (T_H2) which respond to many extracellular pathogens; *T-helper type 17 cells* (T_H17) which secrete IL-17 and play role in cell-mediated immunity and may be in defense against fungi, and *T-follicular helper cell* (T_{FHC}) which participate in humoral immunity and B-cell development in germinal centers. One more type of $CD4^+$ cell is *T-regulatory cell* (T_{REG}) and it can inhibit an immune response.
- The APCs help the T-cells to recognize the antigen. The major APCs are dendritic cells, macrophages and B-lymphocytes, which are also known as professional APCs.
- There are two types of adaptive immunity: Cell-mediated immunity, also known as T-cell mediated immunity, and humoral immunity or B-cell mediated immunity.
- Cell-mediated immunity: The APCs carrying antigen bound to MHC molecules are recognized by the naïve $CD4^+$ and $CD8^+$ T-cells. The $CD4^+$ T-cells proliferate and differentiate into effector $CD4^+$ T-cells, which secrete cytokines that activate B-cells as well as macrophages, and memory $CD4^+$ T-cells. The activated $CD8^+$ T-cells differentiate into two populations of cells *i.e.* effector $CD8^+$ T-cells (CTLs) and memory $CD8^+$ T-cells. The CTLs kill the infected cells and also activate macrophages.
- Humoral immunity concerns with the production of antibodies against a pathogen. The antibodies bind to the pathogens and help in their destruction and removal from the system. The B-cells when activated get differentiated into effector B-cells or plasma cells and memory B-cells. The primary function of the plasma cells is to secrete antibodies.
- When the humoral response is for the first time for an antigen it is called as primary immune response and when the same antigen is encountered subsequent times, the response is heightened with huge production of antibodies, then it is called secondary immune response.
- The transfer of immunity by administering preformed antibodies to individuals is called passive immunity. When the natural preformed antibodies are transferred, it is called natural passive immunity and when the artificially produced antibodies are transferred it is called artificial passive immunity.
- Active immunity is obtained when the body itself produces antibodies against an antigen. It can also be of two types: natural active immunity when the antigen enters the body on

its own and artificial active immunity when we introduce attenuated form of the pathogen in the form of vaccines to sensitize body to produce antibodies.

- Immune dysfunction refers to a condition when the immune system of a person is compromised such that it results in failures and subsequent diseases. It can be of three types: immune deficiency, autoimmune diseases and hypersensitivity.
- The mechanism in the body that controls the self-immune attack is called self-tolerance or tolerance. It can be of two types: central tolerance and peripheral tolerance.
- In central tolerance, clones of T- and B-cells that recognize self-antigens are deleted before maturation and it occurs in primary lymphoid organs. Central tolerance is not completely successful as many lymphocytes may enter the secondary lymphoid organs hence some more mechanisms are needed to ensure that such lymphocytes are also deleted from the system. These additional mechanisms together constitute the peripheral tolerance. The peripheral tolerance inactivates the self-reactive lymphocytes and also generates lymphocytes that inhibit immune reaction against self cells and tissues.
- The antigens that induce tolerance and are known as tolerogens. if an antigen is presented to a T-cell, then without co-stimulation the interaction results in the form of tolerance called anergy and the same antigen when presented with costimulatory signals can become a strong immunogen.
- Immunodeficiency can be of two types: primary (due to genetic factors) and secondary (due to chemical, physical or biological agents).
- The extreme immunodeficiency can result into a condition called as severe combined immunodeficiency (SCID) which affects both B- and T-cells and severely affects the adaptive immunity. If it is left untreated, it can result in death during early stages. The most common form of SCID is Acquired Immune Deficiency Syndrome (AIDS) which is caused due to infection with Human Immunodeficiency Virus-I (HIV-I).
- When the immune system breaks down and it starts attacking the self cells, the condition results in autoimmunity, and can result in several autoimmune diseases.
- Multiple sclerosis (autoimmune attack on a protein in nerve sheaths in brain and central nervous system that results in neuromuscular dysfunction), Crohn's disease (attack on intestinal tissues), Hashimoto's thyroiditis (auto-antibodies in the thyroid), rheumatoid arthritis (RA) in which tissue destruction is mediated directly by T-cells, systemic lupus erythematosus (the individuals having SLE produce auto-antibodies (IgG) for several antigens like DNA, histones, RBCs, platelets, leucocytes, clotting factors and nucleoprotein particles), myasthenia gravis (autoantibodies are formed against the α chain of nicotinic acetylcholine receptor) and multiple sclerosis (self-reactive T-cells that form inflammatory lesions in the myelin sheath of the nerve cells in brain and spinal cord).
- The innate and adaptive immune responses are vital components of immune system and are interconnected to ensure our body remains free from diseases.

QUESTIONS

1. Short answer questions:
   a. Name types of T-cells
   b. Which type of complex is recognized by naïve $CD8^+$ cells?
   c. How many signals are needed for activating a T-cell?
   d. Name the cells involved in adaptive immunity.
2. Give full forms of:
   a. APCs
   b. TLR
   c. MMR
   d. CTLA

   e. CD
   f. ICOS
   g. MHC
   h. ADCC
3. Give the contribution of the following scientists:
   a. Elie Metchnikoff
   b. Merrill Chase
   c. Bruce Glick
   d. Helen Quill and Ron Schwartz
   e. Emil von Behring and Shibasaburo Kitasato
   f. Christiane NussleinVolhard, Eric Wieschaus, and Edward B. Lewis
4. Define:
   a. Adaptive immunity
   b. B-cell activation
   c. Passive immunity
   d. Autoimmunity

BIBLIOGRAPHY

1. Anaya, J.-M., Shoenfeld, Y., Rojas-Villarraga, A., Levy, R.A., and Cervera, R. (2013). *Autoimmunity: From Bench to Bedside*. El Rosario University Press, Bogota (Colombia).
2. Arora, S. (2017). National immunization schedule India: A review. *Research and Reviews: A Journal of Immunology* 7(3):6–10.
3. Bruce, A., Johnson, A., Lewis, J., Raff, M., Roberts, K., and Walter, P. (2002). *Molecular Biology of the Cell*. 4th Ed. Garland Science, New York.
4. Ceccarelli, F., Agmon-Levin, N., and Perricone, C. (2017). Genetic factors of autoimmune diseases 2017. *Journal of Immunology Research* 2017:2789242. https://dx.doi.org/10.1155%2F2017%2F2789242
5. Chinen, J., and Shearer, W.T. (2010). Secondary immunodeficiencies, including HIV. *The Journal of Allergy and Clinical Immunology* 125(2):S195–S203.
6. Collin, M., McGovern, N., and Haniffa, M. (2013). Human dendritic cell subset. *Immunology* 140(1):22–30.
7. Cray, C., Zaias, J., and Altman, N.H. (2009). Acute phase response in animals: A review. *Comparative Medicine* 59(6):517–526.
8. Eiz-vesper, B., and Schmetzer, H.M. (2020). Antigen-presenting cells: Potential of proven und new players in immune therapies. *Transfusion Medicine and Hemotherapy* 47: 429–431. https://doi.org/10.1159/000512729
9. Ferri, F.F. (2022). Clinical practice guidelines – immunizations and chemoprophylaxis. In *Ferri's Clinical Advisor 2022*. pp. 2039–2101. Elsevier.
10. Jaitley, S., and Saraswathi, T.R. (2012). Pathophysiology of Langerhans cells. *Journal of Oral and Maxillofacial Pathology* 16(2): 239–244
11. Janeway, C.A., Travers, P., Walport, M., and Shlomchik, M. (2001). *Immunobiology: The Immune System in Health and Disease*. 5th Ed. Garland Publishing, New York.
12. Kaufmann, S.H.E. (2017). Remembering Emil von Behring: From tetanus treatment to antibody cooperation with phagocytes. *mBio*. https://doi.org/10.1128/mBio.00117-17
13. Keane, R.W., Dietrich, W.D., and Vaccari, J.P.R. (2018). Inflammasome proteins as biomarkers of multiple sclerosis. *Frontiers in Neurology* 9:135.
14. Li, Y., Wang, W., Yang, F., Xu, Y., Feng, C., and Zhao, Y. (2019). The regulatory roles of neutrophils in adaptive immunity. *Cell Communication and Signaling* 17:147. https://doi.org/10.1186/s12964-019-0471-y
15. Li, Y., Yin, Y., and Mariuzza, R.A. (2013). Structural and biophysical insights into the role of CD4 and CD8 in T-cell activation. *Frontiers in Immunology* 4:206. https://doi.org/10.3389/fimmu.2013.00206
16. McNerney, M.E., Lee, K.M., and Kumar, V. (2005). 2B4 (CD244) is a non-MHC binding receptor with multiple functions on natural killer cells and CD8+ T cells. *Molecular Immunology* 42(4):489–494. https://doi.org/10.1016/j.molimm.2004.07.032
17. Mehrfeld, C., Zenner, S., Kornek, M., and Lukacs-Kornek, V. (2018). The contribution of non-professional antigen-presenting cells to immunity and tolerance in the liver. *Frontiers in Immunology* 9:635. https://doi.org/10.3389/fimmu.2018.00635

18. Musumeci A, Lutz K, Winheim E and Krug AB (2019) What makes a pDC: Recent advances in understanding plasmacytoid DC development and heterogeneity. *Frontiers in Immunology* 10:1222. https://doi.org/10.3389/fimmu.2019.01222
19. National Institute of Allergy and Infectious Diseases. Types of primary immunodeficiency diseases. Retrieved from: https://www.niaid.nih.gov/diseases-conditions/types-pidds
20. National Jewish Health. Immune deficiency disorders: Types. Retrieved from: https://www.nationaljewish.org/conditions/immune-deficiency-disorders/types
21. Owen, J.A., Punt, J., and Stranford, S.A. (2013). *Kuby Immunology*. 7th Ed. W.H. Freeman and Company, New York.
22. Rosales, C. (2017). Fcγ receptor heterogeneity in leukocyte functional responses. *Frontiers in Immunology* 8:280. https://dx.doi.org/10.3389%2Ffimmu.2017.00280
23. Sanjabi, S., Oh, S.A., and Li, M.O. (2017). Regulation of the immune response by TGF-β: From conception to autoimmunity and infection. *Cold Spring Harbor Perspectives in Biology* 9(6):a022236. https://dx.doi.org/10.1101%2Fcshperspect.a022236
24. Sharpe, A.H. (2009). Mechanisms of costimulation. *Immunological Reviews* 229(1):5–11. https://dx.doi.org/10.1111%2Fj.1600-065X.2009.00784.x
25. The College of Physicians of Philadelphia. The history of vaccines. Retrieved from: https://www.historyofvaccines.org/content/articles/passive-immunization
26. Tiegs, S.L., Russell, D.M., and Nemazee, D. (2011). Receptor editing in self-reactive bone marrow B-cells. *The Journal of Immunology* 186(3):1313–1324.
27. Vinodh, N.S. (2017). The race of mercy- Winter of 1925. Retrieved from: https://medium.com/@nsvinodh/the-race-of-mercy-winter-of-1925-2db6c1736760

4 Antigens

Priya Singh
University of Delhi

CONTENTS

4.1 INTRODUCTION

Antigen is the foreign molecule recognized by the immune cell receptors specifically the B-cell receptors (BCR) and/or T-cell receptors (TCR) bringing about an interaction of the two. By verbatim antigen is "any molecule that induces the generation of antibodies in response to its interaction with the immune cells" and the extension of the term includes the generation of immune reactions to prevent sickness in entirety.

These antigens are specific chemical patterns/molecules on the surface of the invading infectious agents (viruses, bacteria or protozoa, *etc.*) or any other foreign substance (pollens, dust, transplants/grafts *etc.*) (Figure 4.1). The antigens can be broadly classified into two types: foreign antigens and self-antigens also termed as heteroantigens and autoantigens respectively. As the terms suggest heteroantigens are the ones that originate outside the body, *e.g.* microbes, poison, venom, food proteins, graft tissue, RBCs in donor's blood *etc.* and autoantigens are from one's own body. Immune system otherwise has a very good capacity of segregating self and non-self-molecules but in few people, this system is revoked, where they lack the tolerance to certain self-molecules, recognizing them as non-self, thus inducing the immune response resulting in the auto-immune diseases.

DOI: 10.1201/9781003286424-4

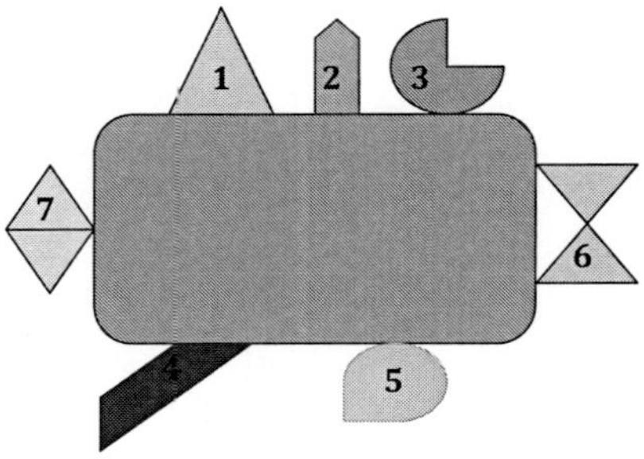

FIGURE 4.1 Schematic representation of the antigen. The surface of the antigen has several antigenic determinants (epitopes) marked as 1–7. When such antigen is encountered by immune system, antibody specific for each antigenic determinant is produced. The different antibodies that act on the antigen together are termed polyclonal antibodies while antibody specific for a single epitope is the monoclonal antibody.

The immune response depends on the interaction of the antigens with the lymphocytes or their products. Many factors that govern this interaction and kind of immune response are discussed in detail later. In this chapter, a detailed account of antigens, immunogens, antigenicity, immunogenicity, epitopes, haptens, and adjuvants will be stated.

4.2 ANTIGENICITY AND IMMUNOGENICITY

Immunocompetent lymphocytes have the responsibility to recognize the foreign particles and mount an immune response specifically against foreign particles in order to inactivate them or their source. This recognition is done using the specific receptors present either on the surface of lymphocytes or secreted receptor molecules circulating in the blood. The binding of these particles to the receptors may or may not induce the immune response and this property is used to differentiate antigenicity from immunogenicity. While antigenicity is the ability of a foreign particle to bind to the products of the immune response, immunogenicity is the ability to induce the immune system to respond to the binding of certain foreign particles. Antigenicity refers to the simple recognition and binding of the foreign molecules to the receptors present in the immune system (i.e. secreted antibodies and/or surface receptors on T-cells). This interaction will eventually neutralize the antigens and their effect. On the other hand, immunogenicity is evoking the immune response on binding of certain foreign molecules with the receptors and this eventually unfolds the cascade of immune reactions to that foreign molecule leading to its neutralization. So, immunogenicity is the ability of foreign molecules to induce humoral and/or cell-mediated immune reactions (Figure 4.2):

B-cells + Immunogen → Effector B-cells + Memory B-cells
↓
Plasma B-cell → Secretes antibody
+
Antigen

T-cells + Immunogen → Effector T-cells + Memory T-cells
↓
CTLs, T_H *etc.* → Secretes cytokines & cytotoxic factors
+
Antigen

FIGURE 4.2 Depiction of immunogenic aspects of the immunogen on its encounter with the B- or T-cells. There is production of effector and memory cells from the B- and T-cells encountering immunogen and the effector cells take the responsibility to eradicate the immunogen and its cause. The memory cells have the responsibility to get rid of the reinfection with the same immunogen in future as soon as the immune system encounters it.

4.3 IMMUNOGENS

The foreign particles may or may not induce immune responses. The molecules that simply bind to the immune receptors are termed antigens and this term is very broad to accommodate a wide range of molecules. Many of them might simply exist there and never cause harm or illicit any immune response. In contrast to antigens, we have immunogens, a generic term used for the foreign molecules that stimulate one or more immune effector responses by the immune cells on their binding to the receptors. This can be simply put as antigens can bind to the products of immune response while immunogens binding results in induction of specific immune effector mechanisms. So, all molecules recognized as non-self are antigens, all the molecules that can bind final products (Ab, BCR or TCR) of the immune responses are termed as antigens and the substance/molecules that induce a specific immune response are called immunogens. Neither all foreign molecules are antigens nor are all antigens foreign (*e.g.* self-antigens). All the antigens are not immunogens but all immunogens are antigens.

4.4 HAPTENS

As stated above, many antigenic molecules simply lack the capacity of immunogenicity. Haptens are such small molecules that are antigenic, but they themselves are not capable of inducing a specific immune response as they lack immunogenicity. Examples include dinitrophenol (DNP), aminobenzene, *o*- aminobenzoic acid, *p*-aminobenzoic acid, *m*-aminobenzoic acid, some peptide and steroid hormones, few drugs like penicillin, *etc.* The word 'hapten' is derived from '*haptein*', a Greek word for '*to fasten*'. Haptens are made immunogenic by tightly fastening them with carrier molecules, which are bigger and mostly immunogenic proteins, through chemical coupling. This provides an immunogenic hapten-carrier conjugate and makes haptens important research and valuable diagnostic tools. If an organism immunized using the hapten-carrier conjugate then it produces antibodies against three antigenic determinants: hapten, carrier and hapten-carrier combination. Hapten then reacts with the antibodies specific to it generated as a result of immune response. When present alone hapten lack the ability to induce immune function but when conjugated with carrier molecule in multiple numbers, hapten comes in the range of immune cells recognition system and eventually set in immune reactions due to its acquired immunogenic property owing to carrier molecule (Figure 4.3).

As it showcases a wide range of antigenic sites/determinants on its surface, the hapten-carrier conjugate provides immunologists and researchers with an effective tool to study the effects of minor changes in the chemical structure of immunogen on the immune system specificity. In the early years of the 20th century, Karl Landsteiner took up elaborate study of antigenic properties of synthetic hapten molecules and his work deciphered the highly specific nature of antibody binding. He used small organic molecules as haptens that otherwise are non-immunogenic but act so in combination with a carrier molecule. His experiments included immunization of rabbits with hapten-carrier conjugates, isolating the serum containing anti-hapten antibodies and then test the reaction of sera with that hapten, injected hapten-carrier conjugate and few other different hapten-carrier conjugates. This helped Landsteiner pin-pointing the reaction of anti-hapten antibodies and not the antibodies for original carrier molecule epitopes. He also tested the antibodies for their reaction with other closely related hapten molecules, if they bind then it was a cross-reaction. Landsteiner used aminobenzene (aniline) and its carboxyl derivatives, *o*-aminobenzene, *p*-aminobenzene, or *m*-aminobenzene, as hapten coupled to carrier molecule to study the various degrees of cross-reactivity. Here antisera for a particular molecule reacted with that specifically without any cross-reactions. Such simple experiments were then used to check which hapten molecules/their modifications prevented or allowed cross-reactions, thus, giving insights into minor details of antigen-antibody reactions and their specificity. His study opened the new avenue of immunological research and diagnostic tools.

Since then many test kits have been developed using the peptide hormones, steroid hormones and certain drugs as haptens in conjugation with large protein carrier molecules. These conjugates are used to raise antibodies specific for the hapten molecule and these antibodies are eventually used to detect absence/presence of various substances in samples. For example, a type of home pregnancy test kit uses the antihapten antibodies raised against human chorionic gonadotropin (HCG) hormone, the kit precisely detects HCG in women's urine which is sign of pregnancy. Similarly, precise quantity of certain drugs in the body can be traced using anti-drug antibodies as is done in case of asthma patients where high doses of the drug (leukotriene) can be dangerous and regular monitoring of circulating drug is required to prescribe dose (Table 4.1).

The hapten-carrier conjugate experiments clearly delineate the concepts of antigen, immunogen, antigenicity and immunogenicity. But in practice, there are much more to this as multiple factors are involved in determining the property of eliciting an immune response. Immunogenicity depends on the intrinsic properties of antigens to an extent, but at the same time, it is largely affected by the manner of immunogen presentation to immune system and a number of properties of the biological system involved in antigen encounter. So, the next section describes the factors that play a vital role in immunogenicity.

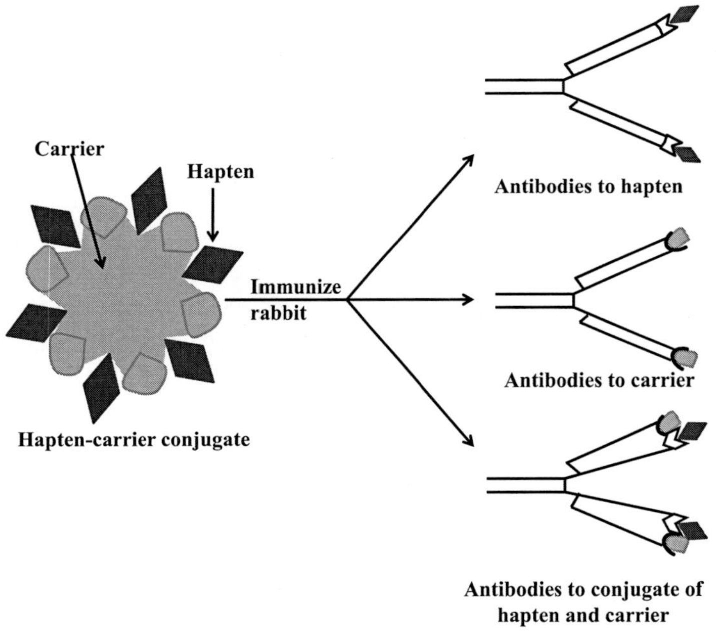

Injection with:	Antibodies formed:
Hapten (DNP)	None
Protein carrier (BSA)	Anti-BSA
Hapten-carrier conjugate (DNA-BSA)	Anti-DNAP (major) Anti-BSA (minor) Anti-DNP-BSA (minor)

FIGURE 4.3 This is the illustration of hapten-carrier conjugate which is an immunogen, while hapten is an antigen that cannot be immunogenic by itself. Multiple copies of the hapten (*e.g.* dinitrophenol or DNP; $C_6H_4N_2O_5$) are carried on the surface of carrier (*e.g.* bovine serum albumin or BSA). Immunization with only hapten, DNP, does not result in any immune response while with carrier shows three types of Ab in serum *viz.* anti-DNP, anti-BSA, anti-DNP-BSA and anti-DNP are majorly produced. This indicates that here DNP-hapten is the immunodominant epitope.

TABLE 4.1
Examples of the Test Kits Using Anti-Hapten (Small Molecules) Antibodies

Small Molecule (Hapten) Used	Example of the Kit Available (https://www.creative-diagnostics.com/Hapten.htm)
Growth promoters	DAGA-177K (Thyroxine KLH Conjugated)
Immunosuppressive drugs	DAG-WZ018A (Sirolimus or rapamycin, BSA-conjugated)
Other drugs	DAG1207 (Ivermectin, HRP-conjugated)
Toxic heavy metal	DAGA-011B (Aflatoxin BSA Conjugated)
Pesticides	1OVA (For detecting Parathion-methyl)
Allergen	DAGA-3086 (Dog allergen, can f5. [His], Recombinant protein from *Pichia pastoris* for EIA)
Drugs of abuse	DAGA-091K (Malachite Green KLH Conjugated)

4.5 FACTORS INFLUENCING IMMUNOGENICITY

There are properties that are shared by most of the immunogens that affect immunogenicity in similar ways. Along with these properties of immunogens, certain contributions come from the biological system that determines the extent of immunogenicity. So, multiple factors play an individual role in the immunogenicity of any substance. The immunogen foreignness, chemical composition and complexity, molecular size and ability to get processed and presented by MHC molecule determine its immunogenicity.

4.5.1 Foreignness or Phylogenetic Unrelatedness

An antigen should be primarily recognized as non-self by the receptors of lymphocytes and only then does it trigger the immune system. The capacity of the immune system to recognize non-self is termed tolerance of the self-antigens, a property that makes the immune system unresponsive to the self-antigens. This capacity is built up during the primary stages of lymphocyte development where they encounter self-antigens early in life and those which react with self are eliminated, releasing only the survivors that recognize non-self-antigens. Some self-antigens that are not presented at this stage are, if encountered later, recognized as non-self and immune response is mounted. When an organism is encountered with an antigen, its immunogenicity depends on its foreignness. In general, the greater the phylogenetic distance between two organisms, the greater the structural differences between their constituent molecules. So their molecules will be more antigenic for each other's immune system. For example, the common protein used in experiments, bovine serum albumin (BSA) is not immunogenic in cattle but is highly immunogenic when it is injected into a rabbit and much more immunogenic in chicken than in a sheep. There are some exceptions to this as there exist conserved molecules in distantly related organisms as well, for example, collagen and cytochrome *c*. Such molecules are not very much immunogenic across diverging species owing to their structure conservation throughout evolution. In contrast to this, some self-molecules, for example corneal cells, sperm, some brain cells, that are effectively sequestered from immune system, if encountered with self-immune system can induce a highly immunogenic response in the host.

4.5.2 Chemical Composition and Complexity

The heterogeneity and complex chemical composition of substances play important role in their immunogenicity. For example, homopolymer compounds (composed of repeating units of single sugar or amino acid) do not elicit immune responses regardless of their large size. While heteropolymers (composed of two or more than two types of constituent molecules) are highly immunogenic in comparison with homopolymers, proving that chemical complexity contributes to immunogenicity to a larger extent. In the case of proteins, the chemical complexity can be attributed to all four levels of structural organization- primary, secondary, tertiary and quaternary and thence affect its immunogenicity greatly. For example, protein with 3° structure will elicit stronger immune response as compared to a primary structure 1° protein (Figure 4.4).

Otherwise non-immunogenic lipid or other molecules can be conjugated with carriers like BSA or keyhole limpet hemocyanin (KLH) to obtain antibodies against such antigens (that can act as haptens). These antibodies are highly specific and are of practical importance to carry out clinical assays of medically important lipids as discussed above in topic of haptens.

4.5.3 Molecular Size

It has been observed that the size of the immunogen plays a major role in its immunogenicity, and most active ones have a molecular mass ≥100 kDa (*e.g.*, γ-globulin –160 kDa). Large molecules remain insoluble and can easily be processed and presented by macrophages upon ingestion. In general, molecules with molecular mass less than 3–10 kDa (*e.g.,* Insulin –6 kDa) are poor immunogens with some exceptions as substances of even 1 kDa molecular mass elicited high immune response.

4.5.4 Predisposition to Processing and Presentation by MHC

Antigen processing and presentation are important for the cell-mediated immune response by the T-lymphocytes which require processed antigen to be presented on the cell surface by self MHC molecules. T-lymphocytes then recognize the MHC-antigen complex using T-cell receptors and other co-stimulatory signals. Before recognition by TCR proteins must be degraded to peptides and presented. Antigens arising outside cells (*e.g.* bacteria) are phagocytosed, processed by exogenous (endocytic) pathway and then presented by Class-II MHCs present on the surface of antigen presenting cells (APCs). While those arising inside cells cytoplasm (*e.g.* viral particles) are processed by endogenous (cytosolic) pathway and presented by the Class-I MHC molecules present on the surface of almost

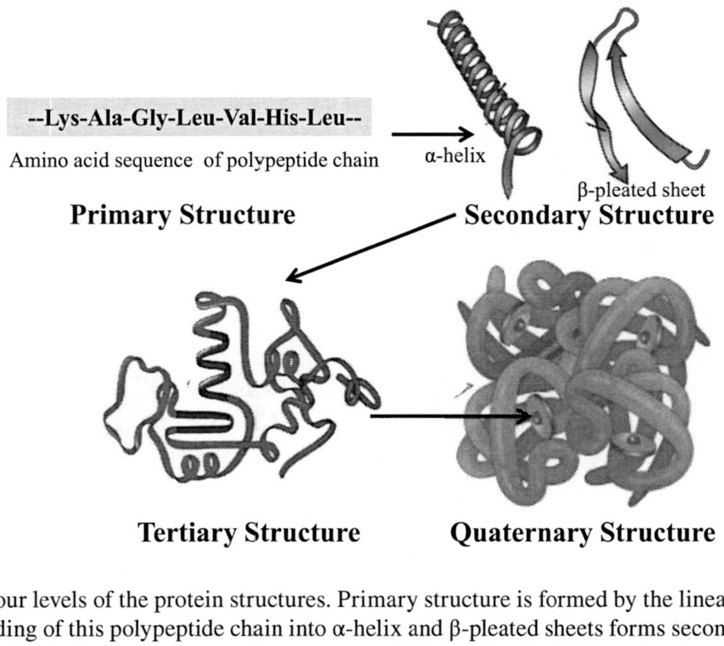

FIGURE 4.4 Four levels of the protein structures. Primary structure is formed by the linear arrangement of amino acids. Folding of this polypeptide chain into α-helix and β-pleated sheets forms secondary structure of proteins. The next level of organization, tertiary structure, is formed by the folding of regions between secondary structures giving the shape to molecule and functional attributes. Quaternary structure is formed when two or more polypeptide chains combines forming polymer. (OpenStax College, CC by 3.0.)

all nucleated cells. Due to the requirement of antigen processing and presentation for T-lymphocyte immune function large, insoluble or aggregated macromolecules are generally more immunogenic than small, soluble ones as the former can be phagocytosed and processed readily. Macromolecules that cannot be degraded and presented with MHC molecules are poor immunogens. For example, polymers of D-amino acids that are stereoisomers of L-amino acids (naturally occurring in biological systems) are not degraded by enzymes within APCs that can only degrade proteins containing L-amino acids. So polymers of D-amino acids cannot be processed and thus are poor immunogens.

4.5.5 Other Factors

The above stated four properties of antigen affect its immunogenicity greatly and along with these, the biological system it encounters also contributes to immunogenicity. The factors that contribute to this are the Host's genetic make-up, the manner in which the substance is introduced into host and lastly the use of agents (adjuvants) to enhance immunogenicity.

4.5.5.1 Genotype of the Host

A major factor determining immune responsiveness is genotype of the host. The genetic constitution influences immune response by the recipient animal as the genetics of an individual always govern the overall biological functions and so is immune function. That is the reason for the difference in immune response and predisposition toward the same immunogen is different in any two individuals. As observed by Hugh McDevitt in two different inbred mice that posed a very different immune response for a single synthetic polypeptide immunogen. The response of an individual is also influenced by the genes encoding B-cell, T-cell, and other immune receptors along with those encoding regulatory proteins that function in the immune response.

4.5.5.2 Dose and Route of Administration of Immunogen

wwImmunogens exhibit a particular dose-response curve. An insufficient dose will not stimulate an immune response either because it fails to activate enough lymphocytes or because its low dose induced immunogenic unresponsiveness or tolerance. Conversely, an excessively high dose can also induce tolerance. A single dose of most of the experimental immunogens will not induce a strong response rather repeated administration or booster doses can increase clonal proliferation of immunogen-specific B- or T-cells and thus increase the lymphocyte population specific for the immunogen. For example, 0.5 mg of Pneumococcal capsular polysaccharide did not elicit immune response in mice but 5×10^{-4}mg of repeated administration of booster doses resulted in sufficient immunogenicity.

Experimental immunogens are administered variously depending on the target immune response. Usually parenterally (around gut), sometimes orally and other common routes of administration are: Intravenous (*iv;* into a vein), intradermal (*id;* into skin), subcutaneous (*sc;* beneath skin), intramuscular (*im;* into mucle) and intraperitoneal (*ip;* into the peritoneal cavity). These routes affect immune response greatly as *iv* injection takes antigen to spleen while *sc* moves antigen to local lymph nodes first and for obvious reasons, immune response varies.

4.5.5.3 Presence or Absence of Adjuvants

Adjuvants are the helpers of antigen that make them more and more immunogenic. These are discussed in detail in the next sub-topic.

4.6 ADJUVANTS

The word adjuvant has been derived from the Latin word *adjuvare* meaning *"to help"*. These are the substances that when mixed with antigen and injected with it, enhance immunogenicity of it. So, an adjuvant is a pharmacological or immunological agent that modifies the effect of other agents. If we immunize using antigens directly then it will lead to poor immune response and will be removed from the body rapidly. To evade the chances of poor immunogenicity and rapid removal from system, the antigen is first combined with an adjuvant, which is the substance that helps in stimulating and enhancing a sustained immune response against that antigen through the creation of *depot effect.* Adjuvants are very useful when the antigen is in small quantity only. Also, they help variously in antigen presentation to the immune system. Although it is not very clear how adjuvants augment the immune response, some of the adjuvants, for example, polyribonucleotide and bacterial Lipopolysaccharide (LPS) are known to act as ligands for toll-like receptors (TLRs) on dendritic cells (DCs) and macrophages thus they stimulate immune system via innate immune response activation primarily.

Immunological functions or effects of adjuvants:

- Provides physical protection to antigen.
- Increases the persistence of antigen in the system and its interaction with immune cells is prolonged.
- Enhances the co-stimulatory signals that are vital for cell-mediated immune function.
- Increases local inflammation by inducing the release of danger signal in greater amount (increases chemokines secretion by releasing cells).
- Stimulates non-specific proliferation of lymphocytes (through chemokines signals).
- Increases the innate immune response to antigen by interacting with pattern recognition receptors (PRRs) on or within accessory cells.

Some of the examples of adjuvants include:

- **Aluminium potassium sulphate (ALUM)** *is approved for use in humans. For use as adjuvant antigen is mixed and precipitated with Alum salt and is injected. Alum results in a slower release of antigen and thus increases exposure time from days to weeks. It also increases the size of antigen enhancing the chances of phagocytosis.*
- **Water-in-oil adjuvants** *prolong antigen persistence.*
    - *In* ***Freund's incomplete adjuvant (FIA),*** *the antigen is in an aqueous solution mixed with mineral oil and an emulsifying agent (e.g. mannide monooleate). The emulsifying agent disperses the oil into small droplets around antigen. This is injected into animal where it is released slowly from the site of injection.*
    - *The* ***Freund's complete adjuvant (FCA)*** *is the first of its kind deliberately formulated adjuvant developed by Julus Freund and is highly effective. It contains heat-killed Mycobacterium. The muramyl dipeptide of the mycobacterial cell wall activates DCs and macrophages making FCA more effective. The activated DCs and macrophages are more phagocytic than non-activated ones and express high levels of molecules that trigger co-stimulation and enhance the T-cell immune response. Thus, antigen presentation and requisite costimulatory signals are increased in the presence of FCA.*

The Alum, FIA and FCA adjuvants also stimulate local chronic inflammation attracting phagocytes and lymphocytes at the site of injection that often results in dense macrophage-rich mass of cells called as granuloma.

4.7 EPITOPES

Immune cells do not recognize the complete immunogen molecule rather they interact with discrete sites on the macromolecule called epitopes or antigenic determinants. Usually, these are low molecular weight hydrophilic sites. In an antigen there may be several different epitopes, making it multivalent antigen. But one epitope will certainly be dominant in its immunogenicity and the immune response is majorly directed towards this dominant epitope. Eventually, with the evolution of immune response, it covers other epitopes as well the mechanism is termed as epitope spreading. It has been found using the small antigens that B- and T-lymphocytes recognize very different epitopes on the same antigen surface. For example, glucagon injection in mice showed antibodies specific for amino-terminal while T-lymphocytes responded to the carboxyl-terminal portion.

Immune cells may interact with a complex antigen on several structural levels like 1°, 2°, 3°, 4° structures of proteins and branch points in branched of polysaccharides. B-lymphocytes interact with the soluble antigens through the binding of antibodies with highly exposed surface epitopes. T-lymphocytes recognise the epitopes on derived peptides, from enzymatic digestion of antigenic proteins, presented by MHC molecules.

There are different types of epitopes. Sperm whale myoglobin contains five different epitopes that are linear and sequential in the arrangement of 6–8 amino acids and recognized in native fragments only as it remains unrecognizable in bends of α-helix conformation (Figure 4.5). Such epitopes are termed sequential epitopes. The globular proteins show the presence of conformational epitopes recognized by B-lymphocytes after they come together by protein folding that brings together the sequential as well as non-sequential epitopes near to each other. For example, hen egg white lysozyme (HEL) non-sequential/conformational epitopes. The residues that constitute conformational epitopes are far apart in the primary amino acid sequence but they are brought together by spatially juxtaposing in the tertiary structure of the molecule, these epitopes depend on the native protein conformation for their topographical structure that is recognized by the immune system. New epitopes may arise by the chemical reactions leading to changes in structure and composition of otherwise non-immunogenic substances; these are termed as neoantigenic epitopes. Neoantigenic epitopes are newly formed antigens that have not been previously recognized by the immune system. Proteins can form new epitopes when they undergo modifications such as glycosylation, proteolysis, phosphorylation *etc.* that alters their structure (Figure 4.6).

4.8 B- AND T-LYMPHOCYTES EPITOPES

B- and T-lymphocytes recognize different epitopes of the same antigen as explained above. TCR detects mostly linear and sequential epitopes that are present in processed antigen in complex with MHC molecules. Mostly these epitopes are internal amino acid sequences that are exposed only after proteolysis of native protein by the APCs and then presented on MHC molecules on the cell surface. Epitopes recognized by TCR are amphipathic as they comprise of both hydrophilic and hydrophobic amino acids. Hydrophilic amino acids interact with TCR while the hydrophobic portion of it called *agretope* interacts with class I and II cell surface markers.

Epitopes recognized by BCR in globular proteins are located in exposed hydrophilic regions and they can be linear or conformational epitopes. Linear epitopes can have 6–7 amino acids while globular protein epitopes have more amino acids (15–22). Table 4.2 below enlists the properties of B- and T-lymphocytes epitopes.

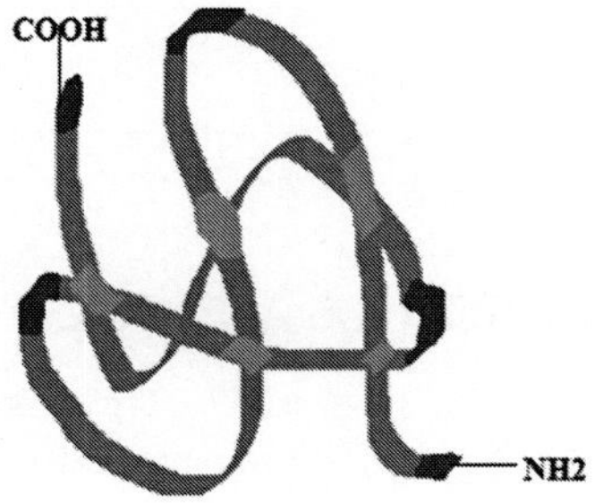

FIGURE 4.5 Protein antigens normally contain both sequential and non-sequential B-cell epitopes. A hypothetical illustration of a protein showing locations of five sequential B-cell epitopes (red) as the constituent amino acids are arranged in sequence one after another and they are detected as epitope due to their sequential arrangement. There are five non-sequential epitopes (green) as amino acids therein come closer and detected as epitope due to the protein folding otherwise they are located far from each other.

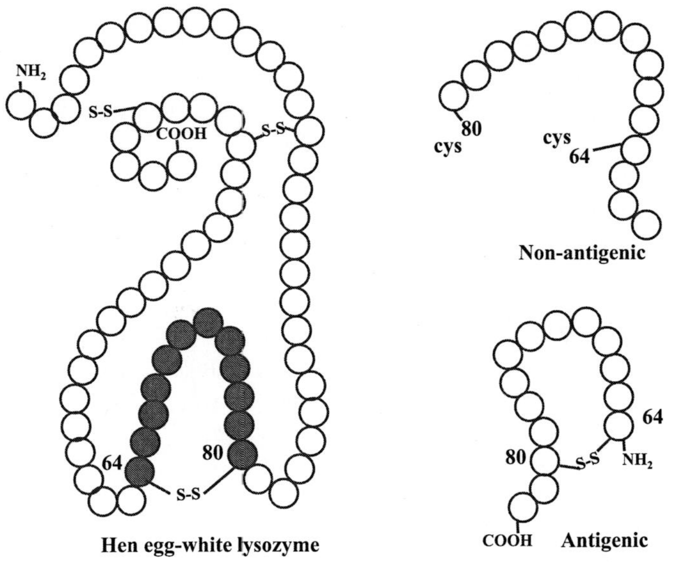

FIGURE 4.6 Tertiary protein structure of the hen egg-white lysozyme (HEL) showing that the conformational determinants to which antibody binds are maintained by the inter-chain disulfide bonds. The cysteine residue at positions 64 and 80 constitutes one of the conformational determinants in HEL, and if disturbed, (disulfide bonds) it results in loss of antigenicity of the molecule. (Adapted from D. Benjamin et al., 1984, The Antigenic Structure of Proteins: A Reappraisal. *Annual Review of Immunology* 2:67; Redrawn by Author.)

TABLE 4.2
Properties of B- and T-Lymphocytes

B-Lymphocyte Epitope	T-Lymphocyte Epitope
Have a complementary shape to that of antigen binding site on an antibody for Ag-Ab interaction.	Processed and presented to the T-cells via MHC I and MHC II complex.
Size of the epitope should not be larger than the receptor.	T-cell receptor cannot bind to the free peptides or antigens.
Comprises of hydrophilic amino acids. Hydrophobic amino acids are buried inside and thus cannot be an epitope for B-cell receptor or antibody.	Unlike B-cell epitopes, the T-cell epitopes can be internal epitopes and may have a hydrophobic amino acid sequence.
Contain sequential or non-sequential amino acids.	–
Tend to show site mobility and are present in the flexible region	The antigenic peptide which is processed by the APC or nucleated cells interact with T-cell receptor and a trimolecular complex can be observed- Class II MHC+peptide+T-cell receptor.

4.9 EXOGENOUS AND ENDOGENOUS ANTIGENS

There are two different mechanisms or pathways used by the immune system to eliminate extracellular and intracellular antigens. The antigens that are derived from outside host cells, the bacteria, viruses and fungi, are the exogenous antigens and they are taken up by the endocytosis. They are processed in the endocytic pathway occurring in APCs and presented on the membrane with class II MHC molecules. The endogenous antigens are those antigens that are generated within the cell, for example, mycobacteria and *Leishmania* that are intracellular parasites, and cytosol-borne viruses. They are processed in the cytosolic pathway in all nucleated cells and presented on the membrane with class I MHC molecules. Immunogenic substances are generated in the cell presented on the cell surface and there they interact with immune effector cells. In some diseases, immunogens are released from infected cells. For example, in systemic lupus erythematosus, immunogenic DNA is released from infected cells.

Additionally, there are a number of isoantigen/allogenic antigens, another type of endogenous antigens that are not immunogenic in host, yet they elicit a high immune response in some but not all individuals of the same species on their administration. Allogenic antigens are genetically controlled by antigenic determinants that distinguish one individual species from another. Examples of alloantigens include serum proteins, HLA (MHC) antigens, certain blood cell surface antigens *etc.* Then there are super-antigens, class of antigens (viral or bacterial proteins) that bind simultaneously to TCR and class II MHC molecules, and it results in excessive activation of the immune system (Non-specific activation of T-cells resulting in polyclonal T-lymphocyte activation and massive cytokine release. Super-antigens stimulate whole families of lymphocytic receptors, and bind to all members of a given TCR family).

4.10 SUMMARY

- Antigens have always been defined as the molecules that interact with the immune effector cells through BCR or TCR when complexed with MHC but the more accurate terminology segregates antigens and immunogens based on their property of immunogenicity and antigenicity respectively.
- All immunogens are antigens but vice-versa is not true. Haptens are small molecules that are otherwise antigens and become immunogens when conjugated with carrier molecules. Hapten-conjugate carrier has resulted in the development of many diagnostic kits and research on antigen-antibody interactions.
- The immunogenicity of a substance is governed by certain intrinsic properties of it like molecular mass, foreignness, chemical heterogeneity and susceptibility to antigen processing and presentation.
- In addition to these, the biological system encountered also governs immunogenicity by way of the recipient's genetic composition, dose, route of administration, and presence/absence of helper agents (adjuvants).
- Immune system does take account of complete immunogen rather it searches for and interacts with specific antigenic determinants or epitopes.
- They vary greatly in their structure, composition, conformation, effector functions elicited and specificity.
- Thus, there are a number of types of epitopes, sequential or non-conformational, non-sequential or conformational, neoantigenic, and agretope. Depending on the source, antigens may be exogenous or endogenous, and allogenic.

QUESTIONS

1. Define:
   a. Superantigens
   b. Immunogenicity
   c. Haptens
   d. Adjuvant
   e. Epitope
   f. Isoantigen
2. Distinguish between the following:
   a. Neoantigenic epitopes and conformational epitopes
   b. Immunogenicity and antigenicity
   c. Autologous antigens and allogenic antigens

d. Exogenous antigens and endogenous antigens
e. B-cell epitope and T-cell epitope
f. Sequential epitope and non-sequential epitope

3. Expand the following:
   a. TCR
   b. BCR
   c. MHC
   d. DNP
   e. APC
   f. HEL
   g. FIA
   h. FCA
   i. TLR
   j. hCG
   k. DC
   l. PRR
4. Give definition and properties of immunogen (factor affecting immunogenicity).
5. Define adjuvants in detail with examples.

BIBLIOGRAPHY

Abbas, Abul K., Andrew H. Lichtman, and Jordan S. Pober (1994). *Cellular and Molecular Immunology*. W.B. Saunders, Philadelphia, United States.

Benjamin, David C., Jay A. Berzofsky, Iain J. East, Frank RN Gurd, Charles Hannum, Sydney J. Leach, Emanuel Margoliash et al. The Antigenic Structure of Proteins: A Reappraisal. *Annual Review of Immunology* 2, no. 1 (1984): 67–101.

Flaherty, Dennis K. (2002). Immunogenicity and Antigenicity. In *Immunotoxicology and Risk Assessment*. Springer-Verlag, New York Inc.

Kindt, Thomas J., Richard A. Goldsby, Barbara A. Osborne, and Janis Kuby. (2007) *Kuby Immunology*. W.H. Freeman, New York.

5 Antibodies

Jyoti Singh
Maitreyi College, University of Delhi

CONTENTS

5.1 INTRODUCTION

Antibodies are the best-understood proteins, which are secreted by B-lymphocytes and are involved in immune response reactions. Antibodies occupy a central place in the immune system as these are the molecules that recognize the foreign entity or antigen. These are heterogeneous molecules that

DOI: 10.1201/9781003286424-5

are formed in response to a variety of antigens. Antigens have multiple epitopes which trigger the antibody production by plasma B cells carrying the same specificity of antibody molecule as carried by secreting cells. Thus, the membrane-bound antibody confers specificity on B-cell, whereas secreted antibodies carry out various effector functions of neutralizing the pathogen, complement activation, transcytosis, etc., in blood. A remarkable feature of all the antibodies is that they share a common structure and based on certain subtle differences, antibodies can be classified into different classes and subclasses. These are the main components of humoral-mediated immunity in the body.

5.2 ANTIBODY AND ITS BASIC STRUCTURE

The first distinct feature of any effective immune response is to recognize the antigen which is done by two molecules in the body viz. immunoglobulins (antibody attached to the surface of B-cell membrane and TCR (T-cell receptor on T-lymphocytes). Before the discovery of an immunoglobulin molecule, scientists were clueless about the element or the entity responsible for humoral immunity. The evidence of an "antibody" like the agent was first reported in the year 1890 by Emil Von Behring and Kitasato, in the blood which could neutralize diphtheria toxin. Since then, the science of antibodies has seen many advances and new findings emerged. Its discovery was a big step toward understanding the mechanism behind the humoral response.

5.2.1 Discovery of Antibody Molecule

It was the experiment of A. Tiselius & P. Pederson and M. Heidelberger & E.A. Kabat (1939) led to the initial findings that the majority of the antibodies lie in the globulin fraction of blood serum [that is why the term Immunoglobulin]. This experiment paved the path for future work which helped in the elucidation of the structure of a typical Immunoglobulin (Ig) molecule. The fact that antibodies can precipitate multivalent protein antigens which are separable was the basis of their analysis. In their classic experiment, the scientists immunized rabbits with ovalbumin (protein of egg white). The serum was collected and pooled, and two aliquots were made. One of the aliquots was treated with ovalbumin against which the antibodies must have formed. On incubating the aliquot with its antigen (in this case ovalbumin) the precipitate was formed which was removed. The other aliquot was left untreated. Both the aliquots (treated/ untreated) were then electrophoresed.

Albumins are considered as fast migrating proteins *per se*. Their results showed that the untreated sample had four peaks corresponding to Albumin (major peak) and Globulins (slightly less than albumin), which were further divided into alpha (α), beta (β), and gamma (γ) [Figure 5.1]. The electrophoretic profile of the treated aliquot exhibited a significant reduction in gamma globulin peak indicating that the antibody indeed resides in this fraction of globulin. To differentiate it from the other proteins present in globulin fraction. The name, Immunoglobulin, was given (the fraction of gamma globulin that confers immunity). Interestingly, today we know that a small amount of immunoglobulins do reside in other fractions (α, β) of globulin.

5.2.2 Structure Elucidation of Immunoglobulin

The hard work and astute observations of three investigators, Rodney Porter, G. Edelman, and A. Nisonoff gave us the present-day structure of a typical Immunoglobulin molecule. They worked separately but on very similar lines which relied on ultra-centrifugation and column chromatography techniques. When the gamma globulin fraction was ultracentrifuged, it was sedimented at 19S, a high molecular weight fraction and 7S, a low molecular weight fraction.

Porter used this low molecular fraction for his further studies. He subjected it to enzyme papain digestion (limited cleavage of the molecule) and obtained three fragments two of which had the antigen-binding ability (Figure 5.2). Therefore, the fragment was named Fab (Fragment having antigen-binding ability) and another fragment that got crystallized during cold storage was termed

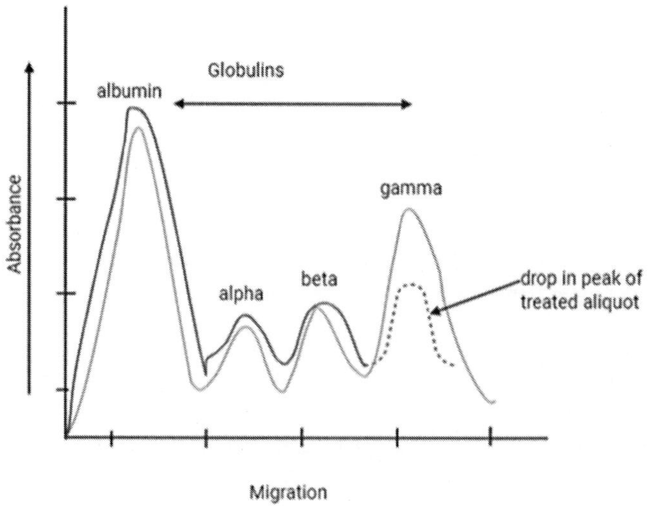

FIGURE 5.1 Electrophoretic profile showing two major peaks in serum; Albumin and Globulins. (Adapted from *Immunology by Kuby*, 1997.)

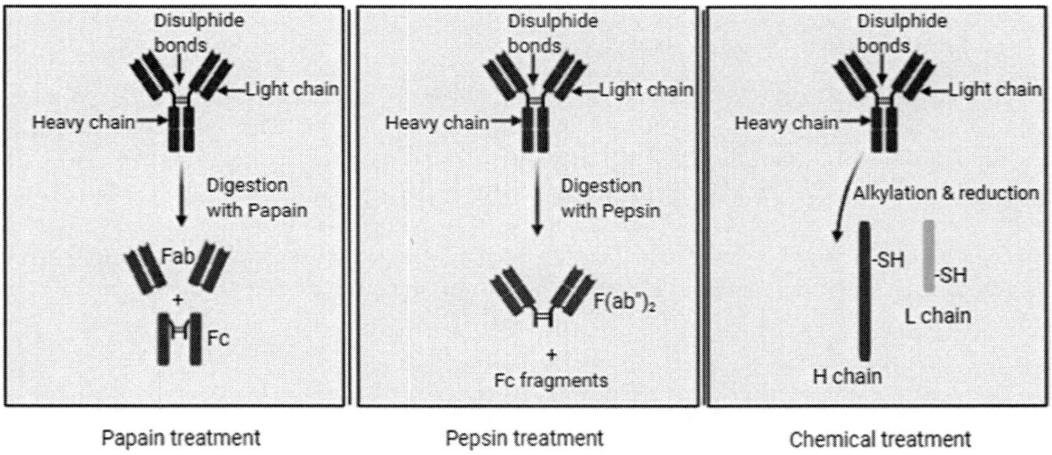

FIGURE 5.2 Various treatments of Immunoglobulin G to decipher its structure. (Author)

Fc (Fragment crystallizable). These fragments were resolved by gel electrophoresis. The molecular weights were found to be 45,000 and 50,000 D, respectively. This experiment helped the scientists to conclude that the antibody is bivalent in nature, i.e., two antigen-binding sites are available on an antibody molecule.

A similar approach was undertaken by Nisonoff, using a different enzyme, pepsin. The use of Pepsin resulted in the formation of a single fragment of molecular weight 100,000 D. This fragment still could bind with antigen and to differentiate it with Fab; it was designated as F(ab')$_2$. Fc couldn't be retrieved as it was digested into multiple fragments. Edelman, on the other hand, in order to dissociate polypeptide chains, used alkylation and mercaptoethanol reduction. These reagents resulted in the reduction of disulfide bonds. The investigators performed gel filtration chromatography to isolate and measure the size of protein fragments generated by reduction. It was shown that each Ig G contained two heavy chains of 50 KD and two light chains of 22 KD.

Further problem was to fit these polypeptide chains into a model that could answer all the attributes of present-day antibody structure and function. To solve this, another amazing l experiment

TABLE 5.1
Summary of Various Treatment Procedures

Treatment/Fragment	Fab	F(ab′)2	Fc	Enzyme Treatment (Above/Below Hinge)
Pepsin action on Ig G	–	+	Multiple fragments of Fc	Below
Papain action on Ig G	+		++ (intact fragment)	Above
Chemical treatment of Ig G	Individual H & L chains of IgG			

was carried out. Two separate goats were immunized with purified Fc and Fab fragments of rabbit Ig G. The antiserum was collected and allowed to react with the heavy and light chains obtained from the reduction experiment. Anti-Fab could bind with both heavy and light chains whereas Fc could bind with heavy chains. Thus, the model for antibody could be deduced as having light (L) and heavy (H) chains in dimers which were disulfide bonded to each other resulting in four chained structures or more appropriately as a dimer of the dimer. Later on, with the advancement of protein chemistry techniques, it was evident that the amino-terminal end of both types of chain resided in Fab fragment and carboxy-terminal was toward Fc fragment (Table 5.1). Rodney Porter and G. Edelman shared Nobel Prize for this breakthrough work in 1972.

5.2.3 Fine Structure of Immunoglobulin

The antibody is a glycoprotein molecule where carbohydrate moiety is attached to the carboxy-terminal end. It belongs to the immunoglobulin family of proteins. The antibody is roughly a *Y-shaped* molecule where arms of Y bear antigen-binding sites at their tips. Typically, it consists of two *light* (L, 25 kD) chains toward the exterior and two *heavy* (H, 50 kD) chains on the interior of the molecule. Light chains are joined to the heavy chains by disulfide bonds (inter-chain) and heavy chains are also disulfide bonded (hinge region) to each other along with multiple non-covalent interactions (Figure 5.3). The amino acid sequencing of both heavy and

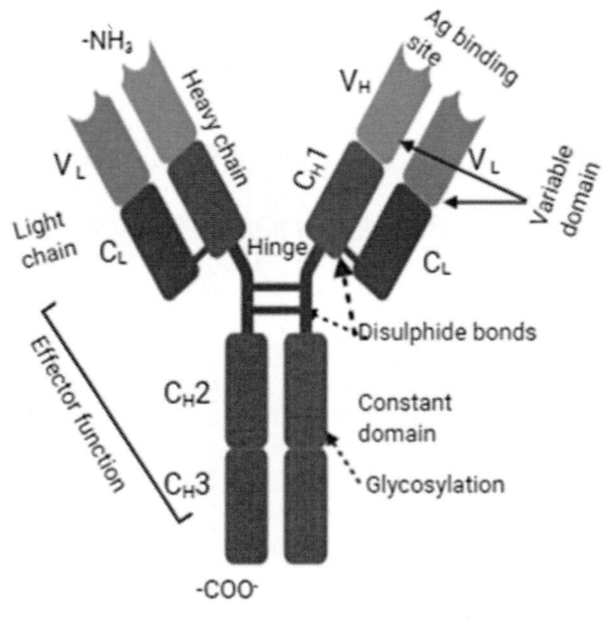

FIGURE 5.3 Schematic representation of a typical antibody (IgG) molecule. (Author)

light chains revealed that at A-terminal (amino-terminal end), a region of approximately 110 amino acids, is presently showing a high degree of variability. These regions are known as V_H and V_L in heavy and light chains respectively (variable regions of heavy and light chains). The rest of the regions which is more or less conserved in terms of amino acid sequences are known as constant region (C_H & C_L in heavy and light chain, respectively). This was confirmed by the work of Hilschman and Craig.

MULTIPLE MYELOMA

Plasma cells secrete antibodies in considerable amount, once an effective immune response is invoked. However, the secretion is heterogenous in nature and for amino acid sequencing sufficient amounts of homogenous protein are required. Despite antibodies having similar structure and chemical properties, their antigen-binding specificities vary implying that their amino acids are different.

Sequencing studies became possible with the accidental discovery of cancerous lymphoid cells (plasma cells) that secrete a large amount of antibodies without any antigenic trigger, a condition known as multiple myeloma and the cells are known as myeloma cells. Plasma cells have a life span of very few days during which it secretes antibodies of a particular specificity and eventually dies. But in patients with multiple myeloma, the cells don't die and continue secreting homogenous antibodies in large amounts using its intact protein machinery. Their homogeneity was established by G. Edelman and M. Poulik by starch gel electrophoresis in 1961. Some myeloma patients also secrete these proteins in urine as described first by Henry Bence Jones (thus named Bence Jones proteins) but their nature remained obscure until 1961. These proteins were interesting, for they were low molecular weight, abundantly secreted, and homogenous in nature. It was further suggested by Edelman & Poulik that Bence Jones proteins represented one of the chains (L) of immunoglobulin molecule but were not incorporated into myeloma protein and thus, were excreted in urine. Multiple myeloma is inducible and can be done by injecting mineral oil into the peritoneal cavity of mouse. The colonies of cancerous cells developed in such a way are called as plasmacytomas (MOPCs, mineral oil-induced plasmacytoma cells). These tools prove helpful in immunological research.

The antibody belongs to the immunoglobulin family of proteins. Sequencing studies (see box) of IgG further suggested that there are two major variations of the constant region of light chain (subtypes) i.e., kappa chain (κ) and lambda chain (λ) and lambda can further be subdivided into λ1, λ2, λ3 & λ4 with subtle amino acid changes. There is no functional difference attributed to antibodies having kappa or lambda chains, *per se*. In humans, 60% of light chains are kappa and the rest are lambda. In mice, on other hand, κ type is predominantly found (95%) and the reverse is true for cattle. At any point of time, the immunoglobulin molecule can have either κ or λ but never both (Table 5.2).

As is the case with the light chain, the heavy chain also showed great variation in sequences of amino acids (V_H). The remaining constant region also exhibited five basic patterns that are responsible for determining classes of immunoglobulins. These five sequences have been named and designated as: Gamma (γ), Alpha (α), Mu (μ), Epsilon (ε), and Delta (δ). Each of these heavy chain variants is known as isotype, present in different classes of immunoglobulin. Thus, the presence of a certain isotype in immunoglobulin determines its class. On basis of this, there are five classes of immunoglobulin which are immunoglobulin G (Ig G), immunoglobulin

TABLE 5.2
Immunoglobulin Classes and their Features

Class	Ig G	Ig A	Ig M	Ig E	Ig D
Subclass	Υ1, Υ2, Υ3, Υ4	α1, α2	None	None	None
Heavy chain isotype	Υ	α	μ	ε	δ
Light chain	K or λ	K or λ	K or λ	K or λ	K or λ
No. of C_H domain	3	3	4	4	3
Hinge	+	+	–	–	+
J-chain	Absent	Present	Present	None	None

Source: Author.

A (Ig A), immunoglobulin M (Ig M), immunoglobulin E (Ig E), and immunoglobulin D (Ig D) [Table 5.2]. Similarly, minor differences in the amino acid sequences of these isotypes can lead to sub-classification into sub-isotypes as found for γ heavy chain (γ1, γ2, γ3, γ4) and for α (α1, α2). Thus, antibody classes can further be categorized into subclasses, Ig G1, Ig G2, Ig G3, Ig G4 and Ig A1, Ig A2 (Table 5.2). The structural details discussed in this chapter are in the reference to IgG molecule as this is the most abundantly found antibody. The general structural features of all isotypes are the same but the subtle modifications in sub-isotypes confer certain functional differences.

The immunoglobulin G or Ig G is the most abundant class of antibody found in plasma. Because of this fact, this is the best understood class of antibody. The subclasses – particularly in humans – have very similar primary sequences, the greatest differences being observed in the hinge region. The existence of subclasses is an important feature as they show marked differences in their ability to trigger effector functions.

The amino acid sequencing of the light chain revealed that the N-terminal half of the chain has a length of 100–110 amino acids. A similar finding (of 100–110 amino acids at the N-terminal end) was obtained when the heavy chain was sequenced. Both these regions are V_L and V_H (variable domains of light and heavy chains, respectively), which correspond to the Hypervariable region. The length of the constant region varies among isotypes, for example, it is 330 amino acids for α, δ, γ, and 440 amino acids for ε and μ. Thus, a light chain appears to have one V_L and C_L whereas, a heavy chain consists of V_H and C_H1, C_H2, C_H3 (constant domains) and sometimes C_H4 (depending on class). There is a hinge region as well in an immunoglobulin molecule which imparts flexibility to the entire structure.

To summarize, an antibody or an immunoglobulin has two light and two heavy chains, each of them has one variable domain in a variable region and the rest of the domains (constant domains) are constituted by the constant region. A hinge region gives flexibility to the molecule. The hypervariable region at the N-terminal end has CDR which interacts with the epitopes present on an antigen. The C-terminal half of the molecule has constant region domains, which play an active role in biological effector functions.

5.2.4 Immunoglobulin Ultrastructure: Multiple Domains with High Degree Organization

The antibody molecule displays all the structural organizations that are typical of a large protein, i.e., primary, secondary, tertiary, and quaternary structures. The molecule belongs to the

Immunoglobulin superfamily. It is characterized by the presence of one or more immunoglobulin domains. A series of repeating units are found in both the light and heavy chains, which are homologous in nature. These homologous units have a chain length of 110 amino acid residues that can fold into a globular motif called as Immunoglobulin domain. As revealed from various studies, the antibody contains multiple domains in the form of variable and constant domains, in both the light (V_L and C_L) and heavy (V_H, C_H1, C_H2, C_H3 & C_H4) chains (Figure 5.3).

The key to understanding the relationship between the structure and function of an antibody molecule is the basic structure of the domain. X-crystallography studies have shown that the immunoglobulin domains are further folded onto themselves to make a compact structure known as the immunoglobulin fold. The structure composes of a *sandwich* of two layers of the β-pleated sheet. Each layer of the sheet displays three to five strands of polypeptide chain arranged in an antiparallel fashion (Figure 5.4). The two layers of sheets are joined to each other by a disulfide bridge and together form a barrel-shaped structure roughly, which is known as β-barrel. While the β-barrel represents the tertiary level organization of the antibody molecule, the secondary level is constituted by the β-sheets. The β-strands have hydrophilic and hydrophobic amino acid residues arranged in an alternate pattern. The amino acids are stacked in such a manner that internal volume, β-pleated sheets are tightly packed with hydrophobic ones. The overall arrangement is stabilized by an internal disulfide bond joining two sheets in the central position and through hydrophobic interactions. The structure in an immunoglobulin fold can be visualized as a *sandwich* of two β-sheets (representing *bread*) pinned by a disulfide bond (a *toothpick*) and the hydrophobic interactions in between (as *butter*).

The structures of two types of domains, i.e., constant and variable domains, are similar with minor differences. In a constant domain, one sheet has three strands, whereas another one has four β-strands. The loops or bends joining the strands do not show or show little secondary structure. The amino acid residues residing in the strands tend to be conserved while the amino acids present in the loop show diversity. An extra loop is present in the variable domain and β-strands in the domain appear to be more distorted.

All the domains except C_H2 are involved in domain pairing. Thus, the domains are in lateral association with each other in heavy and light chains. Two sugar chains are interposed between C_H2 domains. On the same polypeptide chain, there are weaker cis- interactions with adjacent domains. This non-covalent interaction between corresponding domains on either of the polypeptide chain

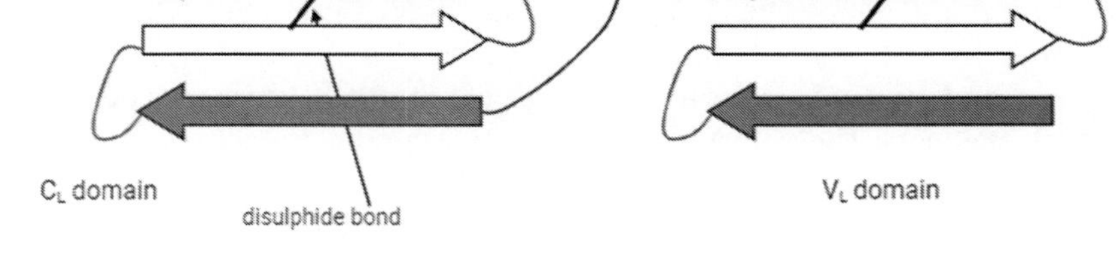

FIGURE 5.4 Diagram of immunoglobulin Light chain depicting immunoglobulin fold. (Author)

constitutes the quaternary structure of an antibody molecule. Consequently, the Fab fragment exhibits paring between V_H and V_L domains, C_H1 and C_L domains. The paring between V_H and V_L is possible due to contact between the respective three strands β-sheet layers. On the contrary, C_H1 and C_L are paired through two four-stranded layers. The overall arrangement is further stabilized by a disulfide bond juxtaposed between C_H1 and C_L. The cis- interactions between "lengthwise" domains, i.e., V_H and C_H1 and VL and CL domains are very limited and weak.

Various crystallographic studies combined with antibody sequencing have revealed that antibodies can identify a diverse repertoire of antigens. The most intriguing fact is that the basic structure of an antibody molecule does not change much. Therefore, the recognition of an enormous range of antigens by the antibody is plausible when certain region shows variability and the structural studies have confirmed this fact. Within variable domains of both heavy and light chains lies a region that shows more variability in terms of amino acid residues. This region of short sequences is designated as the Hypervariable region. Experimental studies have indicated the presence of three such *hotspots* on light and heavy chains. At the structural level, these correspond to three projecting loops, connecting strands of β-sheet that make up the variable domain. These regions consist of ten amino acids and are supported by the rest of the region of the variable domain called as frameshift region (FR) which is less variable relatively. Topologically, all the hypervariable regions of light and heavy chains are clustered at one place and interact with an epitope of the antigen. The structure of this region together is complementary to the structure of the epitope. For this reason, the hypervariable regions are also known as Complementarity determining regions (CDRs). CDRs are denoted as CDR1, CDR2, and CDR3, starting from the N-terminus of heavy and light chain polypeptides. Among these CDRs, CDR3 of both V_H and V_L show far greater variability. The specificity of individual antibodies is attributed to these CDRs, especially CDR3. The frameshift regions and constant domains are thus responsible for maintaining similar structures across different antibody types, despite having different antigenic specificity (see box).

Among all the Complementarity determining regions (CDRs), CDR3 shows far greater variability in terms of amino acid residues. This was shown by the experimental studies done by Elvin Kabat and Tai Te Wu. When a plot was drawn between the residues of CDR1, CDR2, CDR3 of light chain and variability among all antibodies, interesting results were found. Wu and Kabat were intrigued by the possibility that an analysis of the diversity present in Ab polypeptides would reflect the nature of the mechanism used to generate the diversity. They started working on Bence-Jones proteins (myeloma light chains) isolated from urine of myeloma patients or by inducing tumors in mice. For the first time, both the scientists assembled all of the existing amino acid sequences of Ig chains, analyzed the variability of 77 light chains and a few heavy chains by alignment of their similar sequence features, and then analyzed three aspects of this variability and proposed a specific somatic theory of Ab diversity. They noticed the role of conserved amino acids played in the structural framework of Ab. It was with their work that ensued, multigene families could indeed evolve in a residue-conserved, species-specific manner through gene conversion, apart from the functional selection. They created a metric of variability at each amino acid residue position (variability = number of different amino acids at a given position/frequency of the most common amino acid at that position), and this variability was plotted across the 107 residues of the light chain variable region. Immunologists all over the world later used the pioneering concept of CDRs and it established the function of CDRs in determining diversity shown by antibodies.

5.2.5 Hinge Region

This is the region present in the heavy chain which is rich in specific amino acid residues. It shows a considerable difference between classes and subclasses and shows no homology with other regions, *per se*. This region which is rich in cysteine and proline imparts flexibility to the molecule and is designated as a hinge (interdomain). As shown in Figure 5.7, the position and numbers of the hinge also vary within the subclass of IgG. The position of the hinge differs in terms of amino acid residues (12–15) involved. Two disulfide bonds link the heavy chains together in human Ig G1 whereas human IgG2 contains four disulfides and a shorter hinge. This results in lower flexibility of human IgG (h IgG2) as compared to hIgG1. A pronounced effect is also found in the properties of antibody molecules by a minor change in the hinge region. The absence of one of the proline residues in the hinge of IgG4 coupled with substitution in the CH3 domain allows IgG4 to form half-antibodies. The hinge is the region that is susceptible to the proteolytic attack as also seen in the case of enzyme treatment (by pepsin or papain). δ, α, and γ show extended peptide region (corresponding to hinge), whereas ε and μ do not show such region. Thus, IgG, IgA, and IgD have hinge regions and IgE and IgM have not.

5.2.6 Glycosylation

Immunoglobulins are glycoproteins where glycan moiety is associated with the Fc region of the molecule. The effector functions of the molecule are found to be influenced by the extent of glycosylation. For example, when an HIV antibody which otherwise fails to neutralize the infection, acquires the same property after post-translational modification, in cell lines. Experimental evidence suggests that there is an increase in the content of sialic acid, N-acetylglucosamine and fucose. The extent of glycosylation also varies among isotypes, but it occurs invariably in the C_H2 domain of the constant region of a heavy chain. This domain appears to play an important role in the binding multitude of molecules like complement Clq and Fc receptors (FcR) and also FcRn (Fc receptors in neonates).

5.3 SECRETED AND MEMBRANE-BOUND ANTIBODY

As is clear from the above discussion, the antibody can exist in two forms, secreted (in blood and extracellular fluid, sIg) and membrane-bound antibody (present as B lymphocyte receptor, mIg). The two forms differ in the carboxy-terminal region of their heavy chains (Figure 5.5). The membrane-bound form has two additional structures present at the C-terminal end, an α- helical transmembrane anchor region comprising hydrophobic amino acid residues and next to it, an intracellular juxta membrane positively charged region. This region helps in the anchorage of molecules to the membrane. The cytoplasmic portions in mIgM and mIgD are relatively shorter than that of mIgA and mIgE. In secretory immunoglobulin, the carboxy-terminal portion is hydrophilic in nature.

5.4 ANTIGENIC DETERMINANTS OF IMMUNOGLOBULINS

Immunoglobulins being glycoproteins in nature, are themselves capable of inciting an immune response when injected into an individual or species. The antigenic analysis by means of heterologous antisera was performed on human heavy chain and light chain domains and it was deduced that the group of determinants or antigenic determinants can be broadly categorized into three types, isotypic, allotypic and idiotypic.

5.4.1 Isotypic Determinants

The term "*iso*" means "same", implying that these epitopic determinants are the same in all the members of the *same* species. The considerable number of evidence suggest that these are constituted by the epitopes present in the constant region of the heavy chain. Thus, in a manner, these

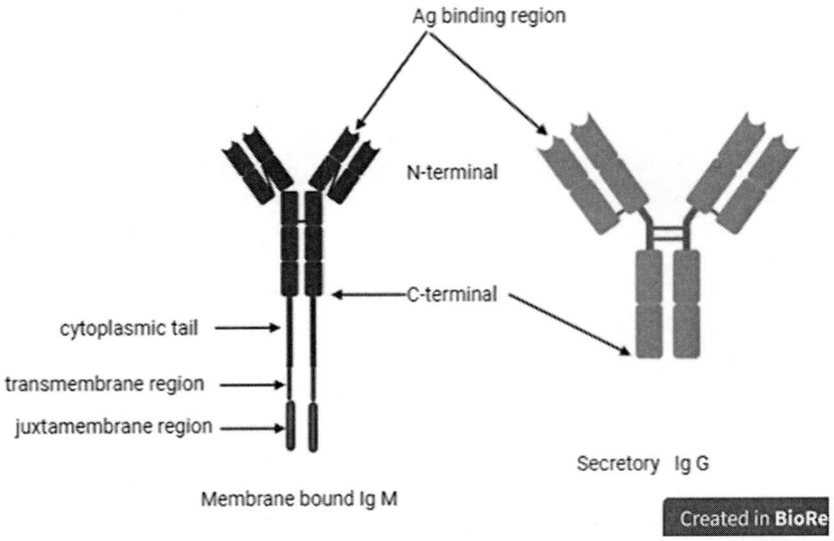

FIGURE 5.5 Diagram comparing structures of Secretory and Membrane-bound Antibody. (Author)

reflect the class and subclass (heavy chain) and light chain type and subtype, within a species. These are characteristic features of antibodies in a particular species but are not the same in different species. As a result, anti-antibody is generated on inoculation with an antibody of one species to another species. Each isotype, in a species, is encoded by separate constant region genes and this region is inherited in all members of the same species. The anti-isotype antibodies find their use in determining the class of immunoglobulin which is bound to the membrane or during an immune response, the first antibodies to be formed.

5.4.2 Allotypic Determinants

The word "allo" denotes "different," i.e. different in the same species. Despite the inheritance of the same set of genes for isotypes in a species, multiple alleles exist. These alleles encode few amino acid differences across the different members of the same species. Thus, some members of the same species may have polypeptide regions with substituted amino acids. These are called allotypic determinants and the sum of all the allotypic determinants present on an antibody is known as Allotype. In other words, it can be said that multiple alleles of the same isotypic genes encode allotype determinants and in the same way, are also shown to be present in constant region domains. The allotypes are found for all the four IgG subclasses and for one Ig A (Ig A2) subclass and κ- light chain, in humans. There are 25 allotypes for IgG, also known as Gm markers. The κ-light chain markers are designated as κm1, κm2, and κm3.

The allotypic antibodies can be produced by injecting antibodies from one individual to another individual of the same species. The classic example of allotypic antibody production is that of blood transfusion or secretion of maternal antibodies during pregnancy in response to fetal immunoglobulin.

5.4.3 Idiotypic Determinants

The constant region has a pivotal role to play in determining antigenic determinants of the antibody. But the variable region also plays its role. The variable regions of both heavy and light polypeptide chains are not only involved in antigen binding but also act as an immunogen. As a result, the determinants present in this region are represented as idiotypic determinants or idiotope and the sum total of all individual determinants is called Idiotype. The region may comprise the actual binding site of antigen or may lie beyond the binding site. Based on this, there are three known idiotopes, which are α, β and γ. The α-Idioptope resides outside the antigen-binding site whereas γ-idiotope is formed by the amino acids of the antigen-binding site or CDRs. The β-idiotope lies very close to the antigen-binding site.

5.5 ANTIBODY HALF-LIFE

The half-lives of various antibody classes differ considerably. The shortest life span is that of IgE (2 days) but Ig E bound to Fc receptors on mast cell can stay for long in the cell. Circulating Ig M has a half-life of 4 days approximately while Ig A has 3 days. The Ig molecule having maximum half-life is, Ig G ranging from 21 to 28 days. The longest half-life of IgG has a therapeutic advantage as well. The complex of a molecule involved in immune response and IgG can be used in the treatment of certain autoimmune diseases like rheumatoid arthritis or psoriasis.

5.6 PARATOPE VS EPITOPE VS AGRETOPE

Paratope is that portion of the antibody that binds with the antigen or the part which physically interacts with an antigen and epitope is the part of the antigen, which binds with immunoglobulin. The antigens have multiple epitopes on their surface if it is a B-cell epitope whereas internal peptides can also act as an epitope in the case of T-cell epitope.

T-cell epitopes interact with T-lymphocyte as a ternary complex. It comprises of antigen-presenting cell (APC), attached to it is an MHC class I/II molecule, interacting with processed antigen, which eventually binds with a TCR on T-lymphocyte. An Agretope is that portion of processed antigen which interacts with the MHC molecule.

5.7 BIOLOGICAL FUNCTIONS AND STRUCTURE OF IMMUNOGLOBULIN CLASSES

The antibody molecules, either in the secreted form or bound form, interact with the antigen for its elimination. It triggers a variety of effector mechanisms for its further elimination or neutralization. The biological functions of an immunoglobulin molecule involve complement activation, antibody-dependent cell-mediated cytotoxicity (ADCC), opsonization, hypersensitive reactions, exosecretions and transcytosis (Figure 5.6). The variable region of the antibody binds the antigen in CDR regions, the biological effector functions are mediated by the constant regions of a heavy chain. The functional properties of all the subclasses differ considerably which is essentially correlated to their structures (Table 5.3). The independent functions of various subclasses will be discussed individually but general functions are briefed here.

5.7.1 Functions of Immunoglobulin

5.7.1.1 Neutralization

This effect is seen when the body is attacked by some microbes or infectious chemicals produced by these microbes. The microbes attack the host cell and gain entry by interacting with the surface molecules of the host, through pili or viral envelope proteins. The antibodies prevent the entry of pathogens by binding to these structures and the infection is thus stopped. The forces involved could be as simple as the steric repulsions. The toxins produced by microbes, sometimes, are also implicated in pathological infections. A very good example is *Clostridium tetanii*, when it enters and produces a toxin, the toxin attaches to the receptors in neuromuscular junctions. This inhibits muscular movement leading to paralysis. Such toxins are neutralized by the antibodies through steric hindrances between toxins and the host cells.

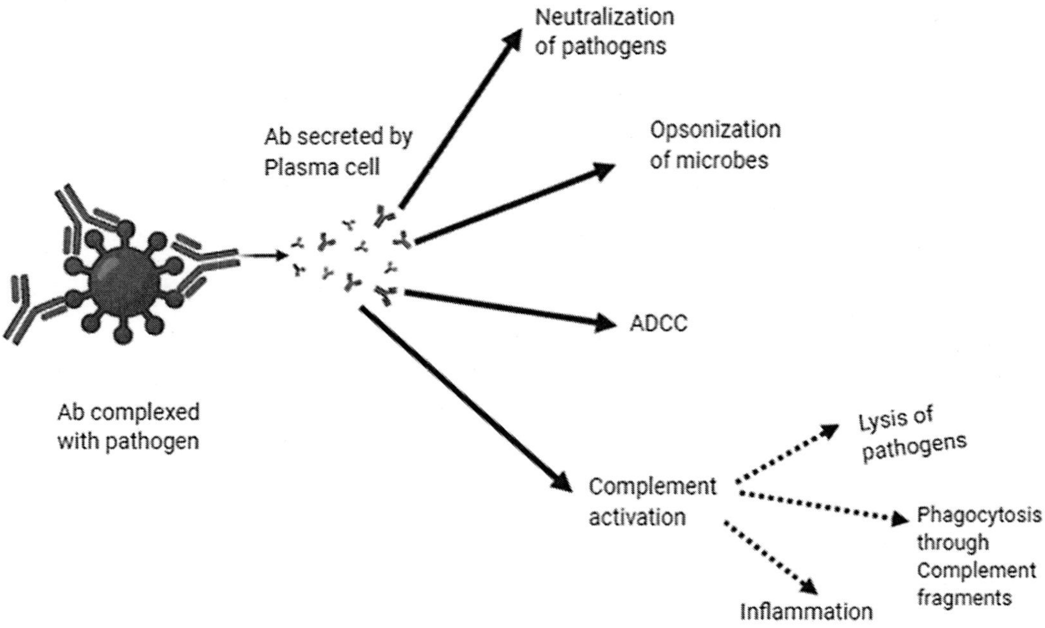

FIGURE 5.6 Origin and various functions of the antibody. (Author)

TABLE 5.3
Properties and Structures of Different Subclasses of Immunoglobulin

Class	Subtype (heavy chain)	Subclass	Serum concentration mg/ml	Hinge	Secreted form	Structure
Ig G	Υ	Ig G1, Ig G2, Ig G3, Ig G4	13.5 on an average	+	Monomer	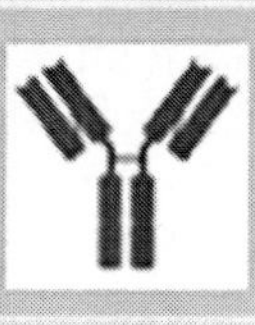
Ig M	μ	None	1.5	-	Monomer, Pentamer	
Ig A	α	Ig A1, Ig A2	3.5	+	Monomer, Dimer, Trimer	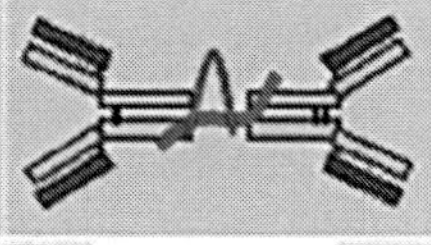
Ig E	ε	None	0.05	-	Monomer	
Ig D	δ	None	Very low	+	Monomer	

This function can be mediated by any of the classes of immunoglobulin as it merely involves interaction between pathogen/toxin and antigen-binding regions. Generally, it is shown by Ig G and Ig A.

5.7.1.2 Opsonization

The process of rendering the pathogen or cell susceptible to phagocytosis by macrophages/neutrophils is referred to as opsonization. This is possible because of the presence of Fc receptors on the surface of phagocytes (a general term for cells exhibiting phagocytosis). These receptors can bind to the Fc portion of IgG which can specifically bind to their specific antigenic epitopes. The process of opsonization can also be mediated through complement proteins like C3b, and the molecules which mediate this process are called opsonins. Thus, opsonins could be either antibody or complement activator proteins.

5.7.1.3 Antibody-dependent Cell-mediated Cytotoxicity (ADCC)

The process of ADCC is much like that of opsonization, except for phagocytosis. The leukocytes and especially the natural killer cells (NK cells) exhibit this process in which, the antibody-coated particles get bound to and are eventually destroyed by the NK cells. The process was first shown to be associated with NK cells that have the Fc receptor, FcγRIIIA on their surface. It is a CD16 molecule, having a low affinity for clustered IgG (it does not bind monomeric IgG circulating in the blood). Therefore, ADCC is possible only when the target cell is coated with antibodies. NK cells reflect a similar kind of killing mechanism as that of CTLs (Cytotoxic T- Lymphocyte). These have perforins and granzymes enclosed in their cellular compartments. When the NK cells bind to an antibody-coated target cell, vesicular content is released and eventually, proteolytic events are initiated leading to activation of Caspase. Caspase is a combination of Cysteine, Aspartate and Protease. Caspases are believed to be involved in the process of apoptosis.

5.7.1.4 Activation of Complement System

The complement system is one of the most important mediums of humoral immunity. It consists of all those molecules, in the serum and cell surface proteins that interact with each other and a cascade of reactions takes place, resulting in the formation of a cytocidal membrane attack complex (MAC). The complement system can be mediated through three pathways, all culminating at the common step of Convertase formation and consequently eliminating the microbes by changing the membrane integrity of the cell. The complement system gets activated by microbial entry or by the antibody-coated microbes. The proteolytic reactions then occur in a sequential manner to bring out inflammation, phagocytosis, and cell lysis. The process is described later in the unit.

5.7.1.5 Transcytosis

It is a specialized process of transfer of immunoglobulin molecule across the mucosal surface, in gastrointestinal, urogenital, secretion in breast milk or movement through the placenta. Transcytosis is a function of the Fc region and the process has much importance in imparting immunity in neonates which have a lesser ability to invoke an appropriate immune response. In neonates, passive immunity is the major immune defense mechanism. Maternal antibody (IgG) can cross the placenta and Ig A is secreted in breast milk. Consequently, a newly born baby contains the same type of IgG as that of the mother. FcRn (neonatal receptor) is involved in the transportation of IgG across the placenta and intestinal epithelium of the baby. The movement of IgG occurs in humans and mice. In humans, it takes place in the third trimester of the gestation period. The transepithelial transport of Ig A essentially depends on the presence of poly Ig receptors.

5.8 IMMUNOGLOBULIN CLASSES

As discussed earlier in this chapter, antibodies can be classified into five classes based on differences in the heavy chain.

5.8.1 Immunoglobulin G

Immunoglobulin G or IgG is the most predominant of all the classes of antibodies, as shown by its serum concentration. Its average serum concentration constitutes about 80% of the total immunoglobulins. The half-life of IgG is also relatively long and all structural elucidations of antibody *in toto* are also being carried out using this class. IgG is invariably present as a monomer in blood. It consists of two gamma heavy chains and based on the differences found in the structure and function of constant regions, particularly in C_H1 & C_H4, four IgG subclasses are reported in serum, viz. IgG1, IgG2, IgG3 and IgG4. The numbers 1, 2, 3, and 4 are assigned in the order of their decreasing serum levels. The difference in the size of hinge and number, the position of interchain disulfide bonds varies among four subclasses (Figure 5.7). The cooperative interactions with multivalent antigens which are the functions of the C_H domain get affected due to the subtle differences resulting in altered antibody flexibility and functional affinity. The differences in constant domains are also reflected in certain functions shown by IgG subclasses. For example, activation of the complement system is a marked function of IgG and IgM but IgG4 fails to bind to C1q and thus cannot activate the complement cascade. Among the rest of the IgG subclasses, the most efficient activator is IgG3 followed by IgG1 and IgG2 is the least efficient of all. IgG2 cannot cross the placenta and thus, does not show the process of transcytosis which is very well exhibited by IgG1, IgG3 and IgG4. The affinity of IgG subclasses toward Fc receptors also varies considerably. IgG1 and IgG3 can strongly bind to all three classes of FcR, i.e. I, II, and III. IgG2 binds only to FcRII while IgG4 binds weakly to FcRII and FcRIII. Interestingly, in secondary immune response, there is a skewing of antibody subclass. In general, IgG1 and IgG3 are secreted in response to protein antigen whereas IgG2 and IgG4 are found to be associated with polysaccharide antigen. This potential can be fully utilized in designing vaccines. Neutralization of viruses and toxins is also mediated by IgG but the differences are again shown in the interaction. For example, in patients suffering from HIV, the IgG3 subclass is much more effective in neutralization as compared to IgG1. This is presumably attributed to an increase in flexibility, improving the antibody access to the virus and bringing conformational change in the oligomeric structure of the virus.

5.8.2 Immunoglobulin A

Immunoglobulin A or IgA is the most predominant class found in the external secretions and mucosal surface, though the serum concentrations comprise only 10%–15% of total immunoglobulins. It also surpasses the serum IgM levels, which is also secretory immunoglobulin in nature. It is prominently present in all the external secretions like tears, saliva, breast milk, colostrum and also extensively present in the mucus of broncho, urinogenital and digestive tracts. IgA is primarily a monomer, however at the mucosa, the secretory form is a dimer (sometimes trimer or a tetramer), where two monomer units are joined together by a J-chain or Joining chain and a secretory

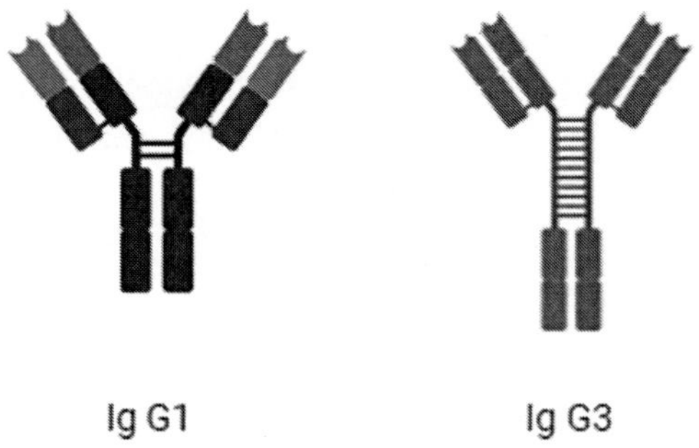

FIGURE 5.7 Structure of subclasses of IgG and positions of Hinge region. (Author)

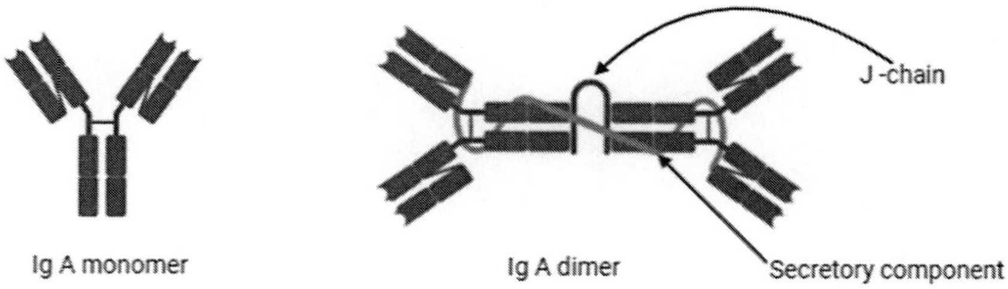

FIGURE 5.8 Structure of dimeric IgA showing secretory component. (Author)

component (Figure 5.8). As the name signifies, the J-chain facilitates the joining of two monomer units. In C_H3 domains of IgA, there are short tailpieces to which the J-chain binds through disulfide bonds. The dimeric IgA is transported across the epithelial layer through poly IgR(pIgR) receptors, the portion of which gets cleaved resulting in the formation of a secretory component. The secretory component is a polypeptide of molecular weight 70KD and is produced by the mucous membranes of the epithelial cells. It comprises five domains that bind to the Fc region domains of IgA dimer. The structure is stabilized by a disulfide bond between the fifth domain of the secretory component and the C_H2 domain of the dimer. There are two subclasses of IgA viz. IgA1 and IgA2 that differ in their hinge regions. The length of the IgA1 hinge region is more having a duplicate stretch of amino acid residues. This extra length makes IgA1 more sensitive toward protease digestion. The absence of duplicated stretch also explains the predominant presence of IgA2 in the mucosal secretions which is less susceptible to proteolytic attack. At the same time, 90% of total serum IgA is of IgA1 type. Immunoglobulin A exhibits well-defined phenomena of neutralization and ADCC of viruses and bacteria. The secretory IgA mounts an effective defense against bacteria like *Neisseria gonorrhoeae* and polio, influenza viruses. Since it is a prominent component in breast milk and colostrum (first milk), it is advisable for newborn babies to have first milk soon after birth. The healthcare providers across the world advocate the practice of breastfeeding at least for 6 months as the baby's immune system is very weak and the passive immunity provided by the mothers to their young ones proves to be beneficial.

The IgA is produced in an astonishing amount! An average human being can produce 5 —15 μg of secretory IgA, every day. In the jejunum portion alone, plasma cells responsible for its secretion are lined up in the mucosal epithelium in the range of 2.5×10^{10}. This number of cells exceeds the total number of plasma cells present in the bone marrow, spleen, and lymph nodes.

As discussed earlier, the transport of IgA is mediated through poly IgR. These tend to express on the basolateral surface of mucosal and glandular epithelium of the glands. Once the poly IgA get binds to these pIgRs, the entire complex is then transported into the lumen across the epithelial layer. This involves the mechanism of receptor-mediated endocytosis via coated pits. The endocytic vesicles fuse with the plasma membrane and pIgR is then cleaved, which eventually becomes a secretory component of secretory IgA. The position of this secretory component in the dimeric IgA is such that it hides the sites that are susceptible to the proteolytic attack in the hinge region. Thus, the secretory component confers a protective element to IgA.

5.8.3 Immunoglobulin M

Immunoglobulin M or IgM is the most versatile of all the classes of antibodies. The serum concentration accounts for 5%–10% of the total antibodies. It is the first immunoglobulin type that is expressed on the B-cell surface during its course of development. The naïve B- cells co-express CD79a and CD79b in addition to IgM, which is present in its monomeric form (Figure 5.9). Once the B cells leave the bone marrow, their life span is very few days or weeks. The B-cells are rescued from apoptosis by interaction with cognate antigen, as proposed by Burnet in his clonal selection theory. During the process of its maturation, the B-lymphocyte upon activation co-express IgD

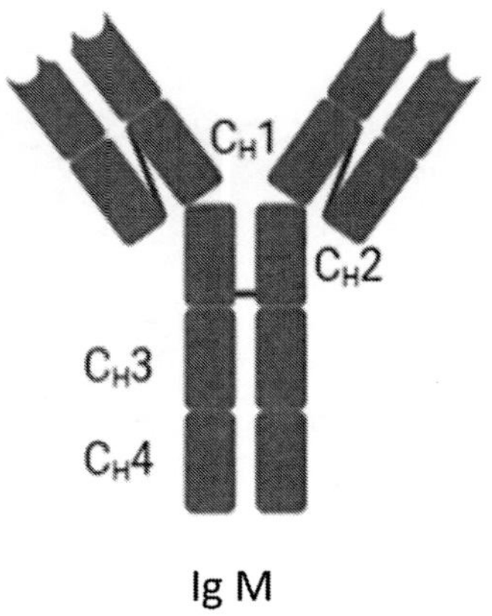

FIGURE 5.9 Monomeric IgM (note extra constant heavy chain domain). (Author)

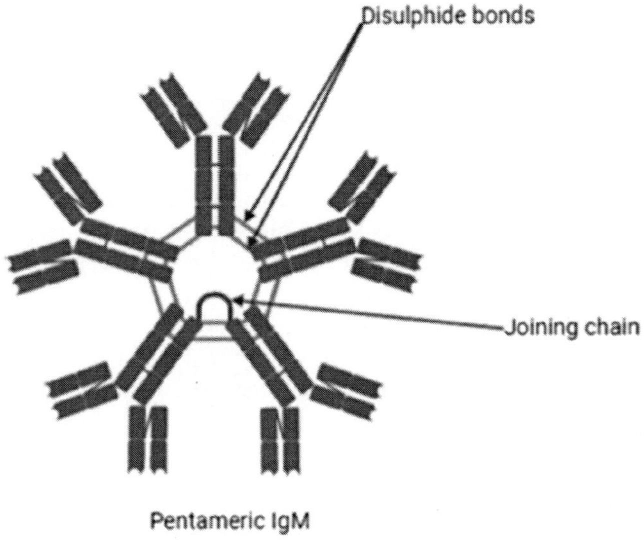

FIGURE 5.10 Structure of pentameric IgM joined by J-chain in center. (Author)

molecule. Interestingly, both IgM and IgD have the same antigenic specificity, which again is a unique feature shown by the vertebrate immune system. The secretory IgM in the blood or other secretions exists in pentameric form (Figure 5.10). In this form, five monomeric units are arranged, with their C4 domains oriented toward the center of this star pattern and joined together by J-chain which is bound to two of the monomers through disulfide bond (Figure 5.10). The J-chain facilitates the secretion at the mucosal surface as in the case of IgA. It is also considered to be transported through the mucosal surface with the help of polymeric Ig receptors. Thus, IgM plays an accessory role as a secretory antibody although a major type is IgA. In patients, having a deficiency of IgA, IgM therapy remarkably improves IgA substitution.

IgM exhibits low affinity but this can be compensated by its high avidity, as the antibody exists primarily in its pentameric form in secretions. This also helps in rendering a pathogen to

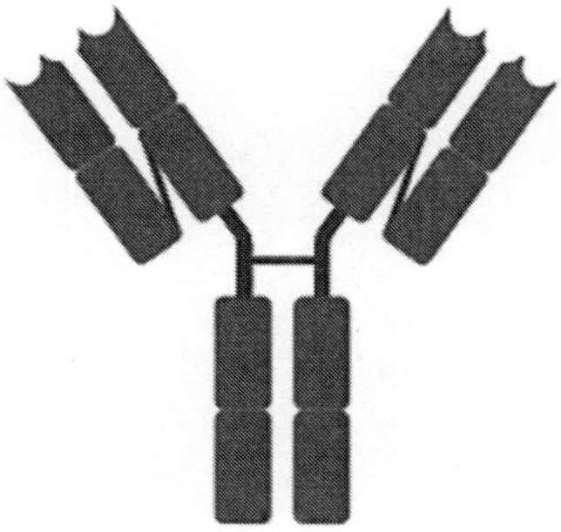

FIGURE 5.11 Diagram showing the structure of Ig D. (Author)

opsonization. Polymeric IgM is known to play an efficient role in complement activation. IgM is the first type of antibody class formed in response to an infection.

5.8.4 Immunoglobulin D

IgD is probably the most enigmatic class of antibodies as no specific function is attributed to it *per se*. The serum concentration of IgD is very low and it constitutes approximately 0.2% of the total immunoglobulins in serum. The half-life of IgD is also short, attributed to the sensitivity of the hinge toward proteolysis. It is present on all the naïve B-cells as a membrane receptor and like IgM requires co-expression of CD79a and CD79b. During the process of maturation most of these naïve B-cells, leaving the bone marrow, tend to lose IgD and acquire IgM, by the process of class switching. As mentioned earlier, the hinge region is more susceptible to proteolytic attack pertaining to the difference in its structure as compared to the rest of the antibody classes (Figure 5.11). This sensitivity is also extended to the C3 domain in the C-terminus region. The other two domains, i.e. C_H1 and C_H2 are similar to that of other isotypes but C_H3 lacks proline residues prominently, which plays a significant role in loop formation in β-strands. Additionally, the region also contains N-linked carbohydrates at Asn_{316} and Asn_{347}, unlike any of the isotypes.

5.8.5 Immunoglobulin E

IgE is the least abundant class found in serum with a concentration of only 0.3 μg/mL in a normal individual, however, at the same time, an allergic person can have serum concentrations increasing several folds (Figure 5.12). The molecule was discovered by Carl Prausnitz and Heinz Küstner in 1921. The results of their experiment included wheal and flare reaction which is also known as P-K reaction in their respect. The discovery was the result of astute observation of both the scientists (see box). It has the shortest serum half-life. Despite its shortest half-life, the IgE remains in the body for several weeks and months owing to its high affinity toward FceR1 receptors. Thus, it tends to be present in bound form to its receptors present on basophils and mast cells. Once they are activated through the antigen or rather an allergen, pharmacological mediators such as histamines are released after degranulation affecting inflammatory response. Thus, the hypersensitivity reactions are mediated by the IgE class of antibodies. These hypersensitivity reactions can even result in anaphylactic shock which can be fatal. The persons suffering from atopy always have high levels of IgE in their blood.

Prausnitz was head of the Hydrophobia Department of the Hygienic Institute at the University of Breslau, Germany. It was a simple and clever experiment reported in 1921, by Carl Prausnitz and Heinz Küstner that led to the discovery of the Ig E molecule. Küstner was highly allergic to fish but only in the cooked form. He came to Praustniz with this observation and asked for his help to analyze this situation. Praustniz decided to carry out experiment by inoculating himself intradermally with Küstner's serum. His arm served as a recipient of this experimentation. The following day he again injected himself with extract of fish, at the same place. To his astonishment, he developed a wheal and flare reaction. The reaction was temporary but fast and only occurred if the extract was preboiled. The reaction is also known as P-K reaction. The experiment also reinforces the fact that allergy can be transferred to a healthy person through the serum of an allergic individual.

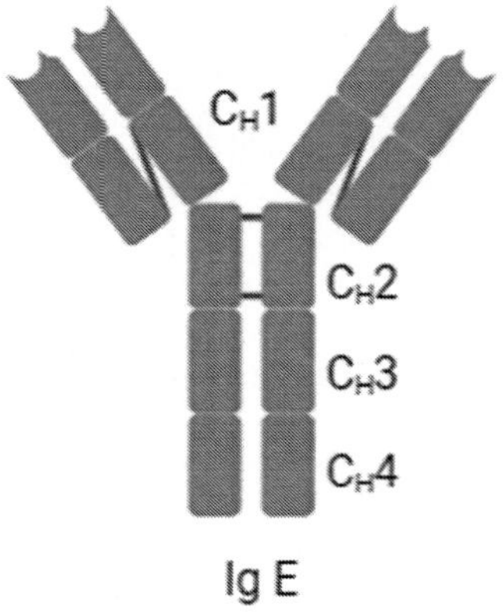

FIGURE 5.12 Diagram showing structure of Ig E. (Author)

5.9 MONOCLONAL ANTIBODIES

The antibodies that arise from a variety of clones of B-cells are designated as Polyclonal antibodies as found *in vivo*. The sera containing these are called Polyclonal sera. This type of sera is specific to different epitopes which are produced in response to the pathogen entering the human body. These multitudes of antibodies are required by a healthy immune system. In contrast to Polyclonal antibodies, Monoclonal antibodies are the ones that are generated from a single clone of B-cells having predefined specificity for a single type of epitope. Monoclonal antibodies or MAb are produced *in vitro*, by Hybridoma technology. The polyclonal antibody has limited use in research and diagnostic purposes whereas, MAb has become an indispensable tool for research and pharmaco-diagnostic methods.

5.9.1 Hybridoma Technology for MAb Production

It is almost impossible to use polyclonal antibodies in diagnostic labs; therefore, scientists felt a need to develop a technology to synthesize antibodies as per the requirement and specificity. A quantum leap occurred with the landmark work of G. Köhler and C. Milstein in the year 1975 who was working in Cambridge University. Later in 1984, both workers were awarded the Nobel Prize for their breakthrough development in the field of immunology. This paved the way for the researchers to design antibodies according to the desired specificity. Both Köhler and Milstein worked independently but later combined their work to get a single type of cell through Hybridoma technology secreting antibodies unlimitedly. They developed a hybrid cell by the fusion of myeloma cell (cancerous plasma cell) with spleen cell (splenocyte, antibody-producing B-cell) from a mouse pre-immunized with sheep red blood cell (SRBC). The resultant hybrid was an immortal cell (a feature of myeloma cell) and had the capability to secrete a large amount of antibodies specific for sheep or red blood cells (Figure 5.13).

To discuss it briefly, the fusion of two cells could be made since the early 1970s. Utilizing this approach Köhler and Milstein fused mouse myeloma and mouse spleen cells. The fusion can be obtained by using inactivated Sendai virus or polyethylene glycol (PEG). Both the fusing agents allow the fusion of the plasma membrane. Hence, a single hybrid cell is being formed carrying initially two different nuclei and two cytoplasms. The resultant cell is referred to as a heterokaryon. A heterokaryon could be bi- nucleated or multinucleated. In later stages, during the course of cell division, the nuclear membranes also disintegrate giving rise to a cell with a large nucleus containing chromosomes of both the parent cells (multinucleated cells would die eventually because of their instability). The hybrid cells or hybridomas continue dividing and only a small percentage of cells reach maturity. There could be three possibilities during the fusion process; fusion of like cells i.e. two parents if denoted as A and B then fusion could result in either AA or BB. The desired hybrid is AB and only a small percentage of cells represent this combination. Eventually, these hybrid cells are required to be separated from unfused cells or AA/BB cells.

The fused cells are grown in a selective HAT medium (Hypoxanthine, Aminopterin, Thymidine) in order to kill those cells which cannot survive in this medium such as parent-parent cells or unfused cells. If such cells are not destroyed, they have the ability to multiply relatively faster and outgrow hybridoma cells. The unfused cells have a limited lifespan and these would die naturally with time. The HAT selection works on the fact that the mammalian system can synthesize its nucleotides in two different ways, the de novo pathway and salvage pathway. If one stops functioning, the other route is taken by the cell to synthesize nucleotides. The *de novo* pathway can be blocked by aminopterin (the folic acid analogue). On blockage by aminopterin, as in HAT medium, the cell undertakes the salvage pathway. The enzymes involved in the pathway are Hypoxanthine-guanine phosphoribosyl transferase (HGPRT) and Thymidine kinase (TK). Improper functioning resulting from a mutation in any of these enzymes can lead to blocking of the salvage pathway. Thus, at the time of fusion of cells mutation is being induced in either cell with respect to HGPRT

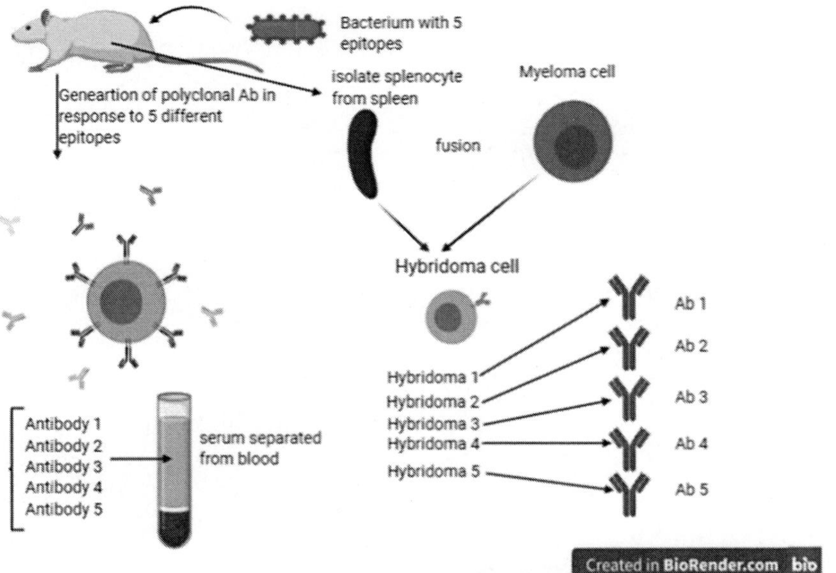

FIGURE 5.13 Diagram showing the difference in the generation of Polyclonal and Monoclonal Antibody. (Author)

and TK. Consequently, when cells are grown in a HAT medium which blocks *de novo* synthesis of DNA, the cell has to rely on the salvage pathway. Only those cells which have a full complement of both the enzymes to carry out the salvage pathway can survive on an HAT medium and such cells are Hybrid cells only. Resultantly, the rest of the unfused or fused parent cells (deficient in either of the enzymes) will not survive.

The cells continually grow in the HAT selection medium in the culture vials and after 7–10 days, most of the wells contain dead cells. Only a few wells contain a cluster of viable cells where each cluster represents a clone of hybridoma cells secreting an ample amount of antibody with pre-determined specificity. The viable cells can be assessed using an inverted phase-contrast microscope. In order to maintain monoclonality, the single cells should be transferred to new culture vials at regular intervals.

The next step is to screen the subclones for the presence of the antibody. The antibody is present in the medium in which cells are growing and this can be achieved effectively by applying RIA or ELISA. The identification of hybridoma secreting MAb is followed by recloning (Figure 5.14). This is done by limiting the culture dilutions while further transferring cells to make fresh clones. The step is to ensure that the culture retains its monoclonal nature.

The last step is the propagation of the desired hybridoma. When the subclones are transferred to the fresh culture wells, the secretion of antibody is fairly low (10–100 μg/mL). This defeats the very purpose of monoclonal antibody preparation (In fact Kohler and Milstein never got their work patented to ensure the technique to be used by as many people for the advantage of humanity). Thus, later on, several years after the technique's discovery, a method was devised to transfer the hybridoma in the peritoneal cavity of histocompatible mice at much higher concentrations (1–25 mg/mL). In mice, the antibody was secreted in ascites and it can be easily isolated and purified from ascites fluid. Several modifications of the technique have been done to increase the yield of antibody.

Thus, in all, the following steps are involved in monoclonal antibody production (Figure 5.14):

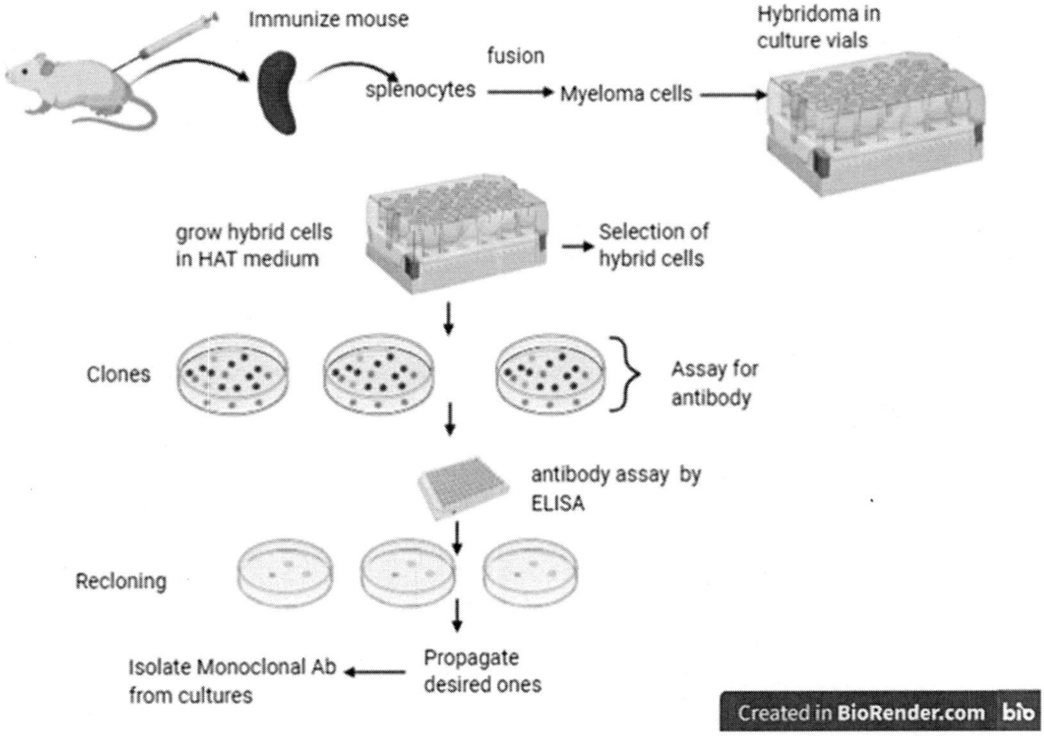

FIGURE 5.14 Standard protocol of Monoclonal antibody production. (Author)

i. Generation of hybridoma cell by fusing the two parent cells comprising of myeloma parent and antibody-producing cell
ii. Selection of hybridoma in HAT medium
iii. Screening of clones to assess antibody production of the desired specificity
iv. Subcloning to ensure monoclonality
v. Propagation of hybridoma

5.9.2 Monoclonal Antibody: Usage and Application

Since the discovery of monoclonal antibody in 1975, the usefulness of molecules has been much explored. With day-by-day novel modifications in classical technique, the application of MAbs and their derivatives continue to hold an elaborate potential. The potential of monoclonal antibodies has been even tapped in the development of anticancer drugs like Rituximab (Mabthera), Cetuximab (Erbitux), Trastuzumab (Herceptin) for chronic lymphocyte leukemia, advanced bowel cancer and breast, stomach cancer treatments respectively.

There are many positive attributes of monoclonal antibodies including:

a. Defined specificity and sensitivity.
b. Ability to enhance other diagnostic techniques (EIA, latex agglutination, fluorescence immunoassay).
c. Reduction in reaction time.
d. Consistent and reliable source of reagent.
e. Allows fine-tuning of proportions of immunoglobulins directed at a given antigen in the diagnosis and treatment of various diseases.

In the beginning, MAbs were used as diagnostic reagents primarily for the detection of pregnancy, pathogenic infections and matching histocompatibility antigens. Classical RIA being highly sensitive can also employ MAb for its successful usage. Conjugating radioisotopes with monoclonal antibodies proves beneficial in the detection of tumors or malignancy in primitive stages of cancer which otherwise is difficult to be located by the other techniques. Similarly, immunofluorescence (IFA) finds its use in monoclonal antibodies. The combination of two has immensely improved the diagnostic accuracy. Several laboratories have reported successful use of MAbs in detecting and differentiating HSV type 1 and type 2 (Herpes simplex virus). Before the availability of MAb, it was not possible to differentiate between two strains of this virus, serologically. Using an indirect IFA with MAbs, rapid detection of genital HSV detection has been possible.

Monoclonal antibodies can be synthesized against numerous tissue markers using membrane fractions or specific tissue antigens. This helps in the detection of liver and renal diseases, HIV-AIDS, cancer and transplant rejection. Monoclonal antibodies to alpha-fetoprotein and melanoma-associated antigens have been produced for the immunodiagnosis of tumors. ELISA can also become faster and more sensitive employing MAbs as reported in the case for quantification and detection of immunotoxins produced by *Shigella* and *Hemophilus influenza*.

Monoclonal antibodies can also act as enzymes and such antibodies are termed Abzymes. When hapten-carrier conjugate was fused with myeloma cells and the resultant Monoclonal antibody was incubated with ester substrate, the hydrolysis got accelerated many folds. This implies that MAbs can also be designed in such a way to behave like enzymes catalyzing the enzymatic reactions.

More recently several types of monoclonal antibodies have been synthesized like Chimeric MAbs, Humanoid MAbs and Human MAbs which play a crucial role in the treatment of incurable diseases, e.g. Mabthera, Erbitux, Herceptin are good examples in this regard.

5.10 SUMMARY

- An antibody molecule is generated in response to an antigenic exposure. It is a glycoprotein consisting of two heavy chains on the inner side and two light chains on the outer side, the dimer of a dimer.
- A fraction having the antigen-binding capacity (Fab) and the fraction that can be crystallized (Fc), were isolated using pepsin and papain treatments.
- Each light chain can have two variants in the form of kappa (κ) and lambda (λ), the heavy chain has five isotypes (α, μ, δ, γ and ε) giving rise to five classes of Immunoglobulin i.e. Ig A, Ig M, Ig D, Ig G and Ig E, respectively.
- The fine structure of the antibody reveals the presence of repeating units called domains. Each light and the heavy chain have constant and variable domains namely, V_L & C_L in the light chain and V_H & C_H in the heavy chain. The light chain has one variable and one constant domain.
- The number of domains in a heavy chain varies among classes according to the absence or presence of the hinge region. The antibody class lacking hinge (Ig E & Ig M) has an extra heavy chain domain.
- Generally, two distinct regions can be seen in an immunoglobulin molecule, CDR (Complementarity determining region) and Frameshift region (FR). The CDR or hypervariable region constitutes the antigen-binding site.
- A different class of antibody molecule has a specific effector function as well in addition to its generalized biological function. e.g. Ig A is a major secretory antibody class, Ig M is an effective complement mediator, Ig E is involved in allergic reactions.
- Immunoglobulins are not only found in secreted form (sIg) but also expressed as membrane-bound form (mIg) on the surface of B-lymphocyte. Subtle differences at the carboxy-terminal end of the immunoglobulin molecule determine the difference in sIg and mIg.
- Monoclonal antibody production through hybridoma technology was one of the major breakthrough in Immunology. Monoclonal antibodies are formed by a single clone of cells *in vitro* in contrast to Polyclonal antibodies found in sera, *in vivo* that are generated in response to antigen with multiple epitopes giving rise to multiple clones of B-cells.

QUESTIONS

1 Define antibody molecule. Are the terms antibody and immunoglobulin the same?
2 Describe the basic structure of an antibody. Which experiment proved that antibody is present in the serum fraction?
3 Discuss in detail the experiment through which scientists could determine the current structure of antibodies.
4 Are the antibodies present in serum exclusively? Justify your answer.
5 What is an immunoglobulin domain? What are the different domains present in an immunoglobulin molecule?
6 Does the entire antibody molecule interact with antigen? If not, which portion is responsible for the interaction with antigen?
7 Define isotype. How are isotypes related to various classes of antibodies?
8 Discuss various effector functions of an antibody molecule.
9 Write down basic differences between secreted and membrane-bound antibody.
10 What are different antigenic determinants present on the antibody molecule?
11 What are monoclonal antibodies? How do they differ from polyclonal antibodies?
12 Which antibody type would be effective *in vivo*, polyclonal or monoclonal?

13 Describe hybridoma technology in detail. What are the various steps involved in the production of monoclonal antibodies?
14 Discuss the different advantages of monoclonal antibodies.

BIBLIOGRAPHY

1. Tiselius, A., Kabat, E. A., 1938. An electrophoretic study of immune sera and purified antibody preparations. *Science*, 87:312.
2. Edelman, G. M., Antibody Structure and Molecular Immunology, Nobel Lecture, December 12, 1972 (https://www.nobelprize.org/uploads/2018/06/edelman-lecture.pdf, accessed on July 10, 2020).
3. Edelman, G. M., Poulik, M. D., 1961. Studies on structural units of the gamma-globulins. *J. Exp. Med.,* 113(5):861–884. doi: 10.1084/jem.
4. Edelman, G.M., Gally, J.A., 1962. The nature of Bence-Jones proteins. *J. Exp. Med.*, 116:207–227.
5. Kohler, G., Milstein, C., 1975. Continuous cultures of fused cells secreting antibody of predefined specificity. *Nature*, 256:495–497.
6. Schwartz, J., Edelman, G. M., 1963. Comparison of Bence-Jones proteins and L polypeptide chains of myeloma globulins after hydrolysis with trypsin. *J. Exp. Med.,* 118(1):41.
7. Taeye, S.W., Rispens, T., Vidarsson, G., 2019. The Ligands for Human IgG and their effector functions. *Antibodies*, 8:30. doi: 10.3390/antib8020030.

6 Antigen–Antibody Interactions

Jyoti Singh
Maitreyi College, University of Delhi

CONTENTS

6.1 INTRODUCTION

Antigen-antibody reactions are the biomolecular reactions involving antigen and its specific antibody. However, specificities are not always absolute and non-specific reactions occur in nature that may arise due to the relatedness of antigens. An analogy can be drawn between lock and key, by assuming key as antigen and lock as the antibody. Generally, the entire molecule of antigen and antibody take part in such reactions. The studies involving X-ray crystallography confirm that antigenic epitopes snuggle into the cleft formed by antigen-binding sites of antibodies. These reactions form the basis for the development of various immunoassays. The association often results from noncovalent bonds including hydrogen bonds, electrostatic bonds, Van der Waals forces and hydrophobic interactions. These forces are weak in nature however, multiple bonding between antibody and antigen ensures the tight complex between them.

6.2 STRENGTH OF ANTIGEN-ANTIBODY REACTION

The strength of the reaction can be assessed in terms of avidity and affinity. Affinity is the intensity of strength between a single antigenic epitope and a single paratope of an antibody i.e between a

DOI: 10.1201/9781003286424-6

single antigen-binding site of an antibody. In other words, it is the total of all the forces of attraction or repulsions at the binding site. It can be expressed by the following equation:

$$Ag + Ab \xrightarrow[k_{-1}]{k_1} Ag - Ab$$

Where k_1 = forward rate constant or association constant
$k-_1$ = reverse rate constant or dissociation constant

The ratio of association to dissociation constant is the measure of the affinity, denoted by K_a that in turn can be shown as:

$$K_a = k_1 / k_{-1}$$

The molar concentrations of antigen (Ag) i.e. [Ag], antibody (Ab), [Ab] and of antigen-antibody complex (Ag–Ab), [AgAb] can be used to calculate the value for K_a. Small antigens tend to have high values of k_1 whereas larger antigens have smaller values. The k_{-1} also plays a major role in determining the affinity of the antigen towards an antibody. The high-affinity complexes have very high values for K_{-1}. K_d is the dissociation constant of the above reaction and can be expressed as a reciprocal of K_a and acts as an appropriate indicator of the stability of an antigen-antibody complex. Very stable complexes have low values of K_d and vice-versa is true for less stable complexes.

Avidity is a better measure of the overall strength of a reaction between an antigen and an antibody. In nature, a majority of the antigens predominantly are multi-epitopic or have copies of the same epitope. In that case, affinity does not explain the strength of the reaction effectively. Avidity thus is a measure of the total of all the interactions between a multivalent antibody and all the epitopes involved in the reaction (Valency refers to the number of binding sites present on an antibody molecule, which is two in general and varies among different classes of immunoglobulin). Avidity is influenced by the valency of the antibody. Low affinity can be compensated with high avidity and a very good example of such a situation is IgM which is pentavalent when not in bound form. IgG has a higher affinity as compared to IgM, however, IgM has high avidity owing to ten binding sites present in the molecule, making it a better candidate to bind antigens. It also is an efficient complement activator. Reactions involving multivalent antibodies and multiple epitopes are more stable and are easier to detect.

6.3 CROSS-REACTIVITY AND SPECIFICITY

The antibody molecule exhibits a high degree of specificity which refers to the ability of the antibody or group of antibodies to bind with only one type of antigen at a given time. The antibody can easily identify an antigen based on differences in primary, secondary or tertiary structure (in the case of protein antigen) and isomeric form of an antigen if any. However, despite showing a high degree of specificity, the antibodies do show certain levels of cross-reactivity. This can result from sharing an identical or similar epitope between two unrelated antigens. Generally, this occurs in the case of glycoprotein antigens where sugar moiety is the determining factor. ABO blood group antigens are the best-understood examples of this phenomenon of cross-reactivity.

The blood group antigens are glycoproteins present on the surface of red blood cells. The minor change in the terminal sugar of the glycoprotein determines its identity as A or B antigen. These antigens share similarities with the antigens of gut microbes which are harmlessly present in the human intestine. An individual having blood group A would have antigen A on RBCs and antibodies against antigen B in its plasma. Anti-B antibodies are formed by exposure of these antigens of gut microbes and not by any other exposure. The antibodies against antigen A would have been eliminated during the developmental stage when antibodies to self epitopes are removed (antigen A is self in this case). Antibodies B are induced against Antigen B which is considered as non-self or foreign by the individual's body. The situation is vice-versa for an individual with a B blood group. This phenomenon plays a crucial role during blood transfusion reactions.

Cross-reactivity is not just shown by the blood group antigens but also by certain vaccines and infectious agents like *Streptococcus*. In the case of vaccines, *Vaccinia* virus displays cross-reacting epitopes with that of *Variola* virus, the causative agent of smallpox. Edward Jenner in his classical experiment used cowpox pustules fluid for treating smallpox patients where cowpox (caused by Vaccinia virus) is less infectious as compared to smallpox which was a deadly disease. The experiment laid the foundation stone for future work in the field of immunological studies.

6.4 AGGLUTINATION REACTIONS

Agglutination or clumping reactions are the interactions between a particulate antigen and its complementary antibody. The reactions result in visible clumps of insoluble antigens (RBCs, bacteria or some inert particle) and the antibody. The antibodies mediating these reactions are called agglutinins. The antigens are referred to as agglutinogens. The reaction is based on the formation of a large complex by cross-linking (agglomeration) polyvalent antigen and antibody, resulting in the aggregate (agglutinate) that can be seen by the naked eye. The formation of such agglutinate is directly correlated with the concentrations of respective antigens and antibodies. The large complexes are markedly found in the zone of equivalence where concentrations of both the components are approximately equal. Small complexes can still be found when either of the antigens or antibodies is in excess. However, these small complexes do not result in visible clumps and may result in false-negative results. The absence of agglutinates at high concentrations of antibodies is termed as prozone effect. The prozone effect can be caused by high concentrations of antibodies. At high concentrations, the number of binding sites available on it can exceed the antigenic determinant sites of antigens. Resultantly, most antibodies bind univalent and not multivalently. Thus, no cross-linking occurs. The prozone effect thus observed can be overcome by simply diluting the antibody solution to its optimum concentrations. The effect is also pronounced when polyclonal sera are used, which might contain incomplete antibodies. These are the antibodies that when present in high concentrations, bind the antigen but do not induce agglutination, e.g., IgG class which mostly blocks the antigenic sites but does not agglutinate the antigen. This also makes the complex less accessible to multivalent antibodies like IgM. The restricted flexibility at the hinge region can be one of the reasons for the prozone effect. Thus, agglutination is influenced by the class of antibody in addition to the amounts of antigen and antibody.

Agglutination reactions are mostly used as qualitative diagnostic tests like hemagglutination and bacterial agglutination. The reaction has a considerable amount of reproducibility, specificity and is cost-effective. Hemagglutinations are used in determining blood types in humans. The test is performed on the glass slides where RBCs are mixed with the antisera. The visible clumping is evident where an appropriate antigen is present. The test is very useful prior to blood transfusions to check the compatibility of donor and recipient.

Bacterial agglutination finds its application in sensing the presence of an infectious agent like bacteria in the sample. The serum of the patient infected with a particular bacterium is serially diluted and the bacterium is added to it. The last tube that shows visible agglutination represents the serum antibody titer. The antibody titer refers to the reciprocal value of the highest dilution of the antibody showing a positive reaction.

There is another variation of agglutination reaction viz., agglutination inhibition that can be considered as a sensitive method to detect minor quantities of antigen. The method has been very useful in the development of earlier kits meant for pregnancy detection. These kits used to have latex particles coated with human chorionic gonadotropin (HCG) and antibody to HCG i.e., anti-HCG. The addition of the urine sample of a pregnant woman carrying HCG would inhibit the agglutination of latex particles when the antibody was added. The method is also helpful in the detection of illegal drug usage by individuals. Dope tests for athletes and other sportspersons have been developed on this principle. In clinical laboratories, the method is used widely for the detection of exposure of individuals to certain viruses which may cause agglutination of RBCs. If a person has antiviral antibodies in his serum, these will bind to the virus and would interfere with hemagglutination by the virus.

6.5 IMMUNOPRECIPITATION REACTIONS

The interaction between a soluble antigen and an antibody in the fluid or semi-solid medium leads to the development of a visible precipitate. The antibodies mediating such reactions are referred to as precipitins. Immunoprecipitation in the fluid medium includes the use of serial dilution of one of the components and keeping the other one constant. The reaction occurs within a few minutes or may take time of a day or two to show visible precipitate. This visible development of precipitate is due to the formation of ionic bonds among antibody molecules within the lattice, causing it to lose its charge thus becoming insoluble. There are certain prerequisites for the precipitation reactions to occur viz., the antibody should be bivalent and antigen must be bivalent or multivalent i.e., it should have at least two copies of the same or different epitopes.

6.5.1 Immunoprecipitation in Fluids

The precipitation reactions in the fluid employ the technique of mixing of antibody and antigens in a tube or series of tubes with serially diluted antigen or antibody. The reaction initiates with increasing turbidity in the reaction medium, changing the appearance of the precipitating or flocculating aggregates. An estimate of gradation can be made by the naked eyes or more precisely by turbidimetric or nephelometric methods. An adequate measurement is done by weighing the precipitate or by chemical analyses like N-determination or isotope labelling. More the precipitate, more is the reaction. The method can be easily used to determine the closeness between the species. More related species show less precipitation and the distantly related species exhibit more precipitation. The binding is regulated by the structural complementarities eventually providing serological specificity.

Two stages of the reaction can be identified, one is the rapid binding of the antibody to the antigen within a few seconds and the second stage is the formation of an antigen-antibody complex in the form of a precipitate that lasts for hours or days. The binding in the initial stages involves physical forces like van der Waals forces, hydrophobic interactions, hydrogen bonding and electrostatic bonds.

When a graph is plotted for such antigen-antibody mixture, it is evident that the precipitation occurs in the zone of equivalence, which is a quantitatively balanced system, where the ratio of the antigen to the antibody is optimal (Figure 6.1a). In this zone, a large multimolecular lattice is formed and as the complex of antigen and antibody increases in size, it precipitates out of the solution. On either side of the zone of equivalence are the zone of antigen excess and the zone of antibody excess. The precipitation is not possible in these zones as the conditions are not favourable for the formation of an extended lattice network. In the zone of antibody excess, the unreacted antibody can be found along with the small soluble complex where a single antigen is bound by several antibody molecules. The same is true for the zone of antigen excess.

6.5.2 Immunoprecipitation in Gel

Several investigators in the 1940s modified the phenomenon of precipitation to study various types of antigen-antibody reactions and employed the technique in determining the purity of antigens. In 1946, J. Oudin who was working at Pasteur Institute, layered the antibody solution on the antigen solution so that there is no mixing of two solutions. The method was later modified much in 1948 by Orjan Ouchterlony. He was working in State Bacteriological Laboratory in Stockholm, Sweden.

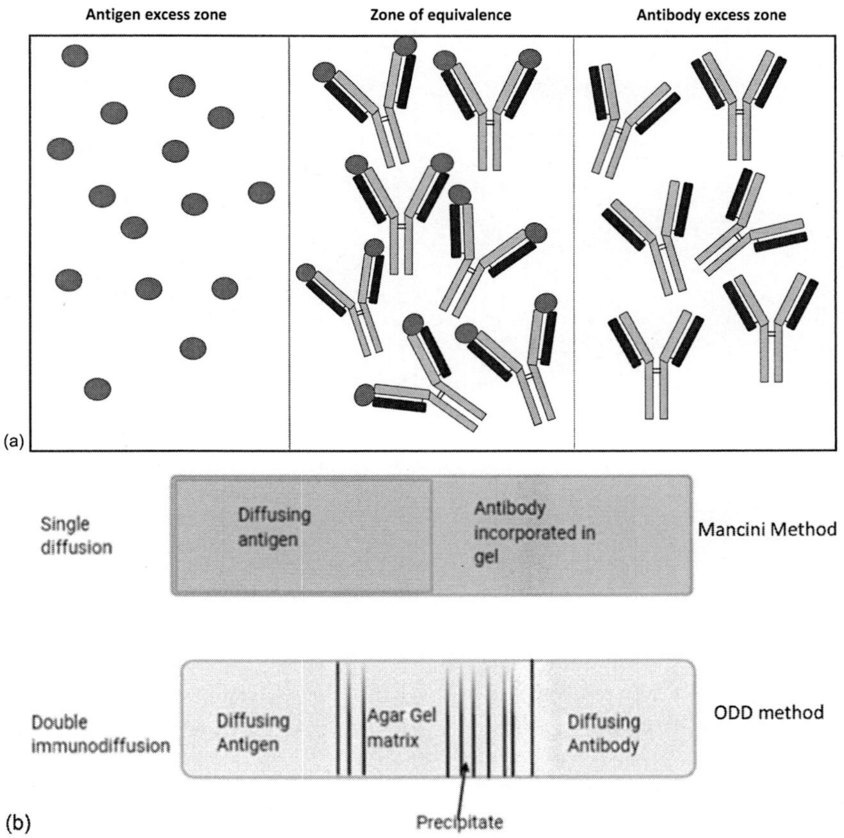

FIGURE 6.1 (a) Schematic representation of a slide showing precipitate in the zone of equivalence using the ODD technique (author). (b) Diagram showing diffusion of antigen and antibody in gel. (Author.)

Ouchterlony used agar solution in place of liquid medium (called immunodiffusion) [Figure 6.1b]. Agar, on cooling, forms a layer of gel in a petridish or a glass slide. Two wells were created while the gel was being solidified. Since he was working with diphtheria bacteria, he placed the diphtheria toxin in one well and a specific antibody to the toxin in another. A visible line of precipitation could be observed in the zone of equivalence. This is referred to as double immunodiffusion or Ouchterlony double diffusion (DID or ODD). The antigen and antibody both diffuse in gel and the antibodies precipitate out the antigens in the zone of equivalence (Figure 6.1a).

This is in contrast to Mancini's method, where there is a diffusion of a single component, generally antigen. G. Mancini who was working at the University of Ferrara, Italy, developed the method in 1965. He referred to the technique as radial immunodiffusion which was based on the migration of the antigen from well through agar gel incorporated uniformly with an antibody. The diffusing antigen establishes a concentration gradient through the gel and when the antigen and antibody are in optimal proportions (in the zone of equivalence), the precipitation occurs in the form of a white ring. The diameter of the ring is directly proportional to the concentration of antigen placed in the well (Figure 6.2). By comparing the area of the precipitin ring with the standard curve (easily obtained by measuring the area of the precipitin ring of known antigen concentration), the concentration of the antigen sample can be determined.

Mancini method was used in laboratories to quantitate serum levels of IgM, IgG and IgA by using anti-isotype of antibody in agar solution. But it is not considered to be a very sensitive method for quantification purposes as other highly sensitive methods are now available. ODD, on the other

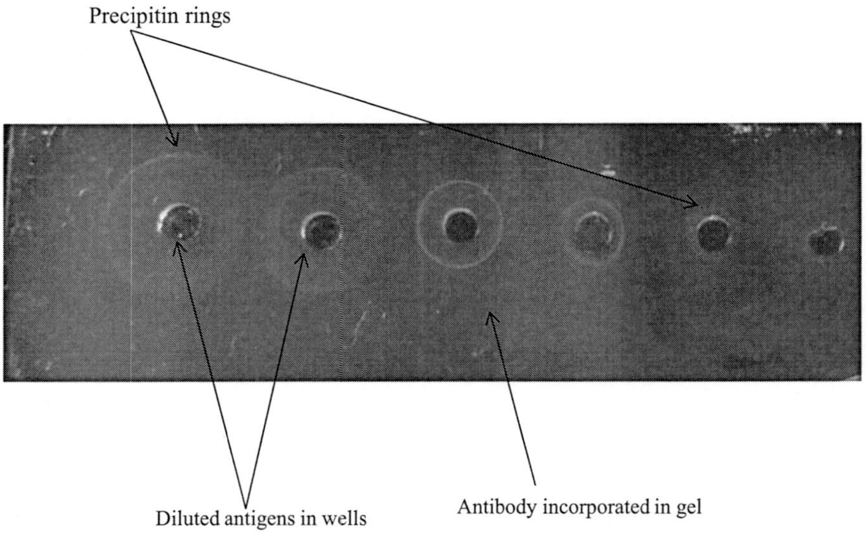

FIGURE 6.2 Slide showing the pattern of ring formation by Mancini's method. (Author.)

hand, is a purely qualitative method. ODD is employed to assess the purity of antigen or to study the relationship between different antigens. Three types of patterns can develop when Ouchterlony's method is employed (Figure 6.3):

a. **Pattern of identity:** When precipitin lines from two antigenic wells meet with each other and form a single line. The antiserum forms a single line of precipitation with antigen and two later meet each other. This occurs because two antigens share similar epitopes.
b. **Pattern of partial identity:** The antiserum forms lines of precipitation with both the antigen but with a formation of a spur in the pattern. This indicates that there is a partial similarity in the epitopes of two antigens along with possession of a unique epitope by one antigen. This unique antigenic epitope results in the formation of a spur.

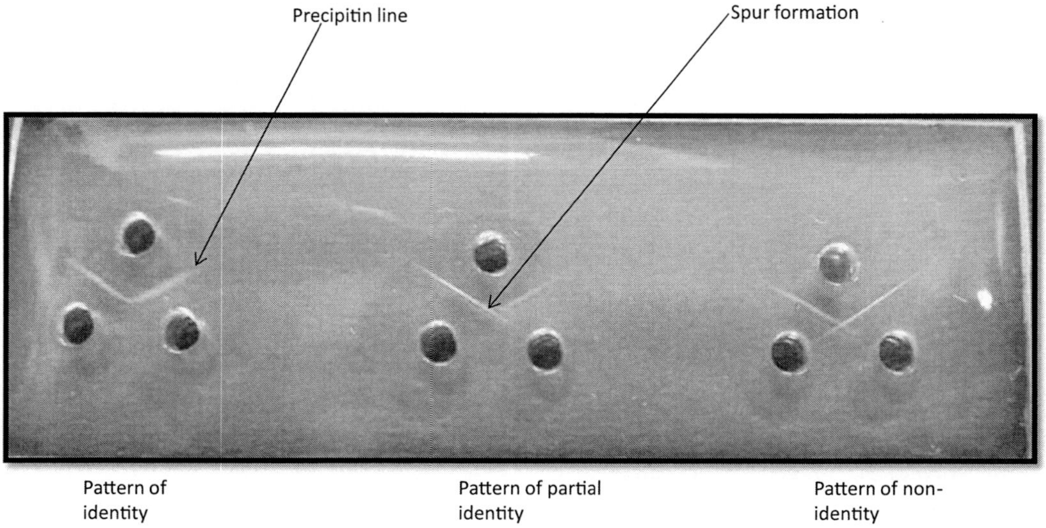

FIGURE 6.3 Slide showing various patterns of antigen-antibody reaction using the ODD technique. (Author.)

c. **Pattern of non-identity:** When there are no common or shared epitopes between two antigens, the pattern of non-identity develops. The antiserum forms two precipitin lines that cross each other.

6.6 IMMUNOELECTROPHORESIS

Immunoelectrophoresis is a combined technique of separation of antigens by electrophoresis followed by immunoprecipitation in the gel by immunodiffusion. The separation is done under the influence of electric current as in normal electrophoresis. It is a semi-quantitative technique to quantify antigen and to ensure the purity of antiserum. The technique was first described by Pierre Graber and Curtis A. Williams in 1953 using human serum proteins. Serum proteins were separated by electrophoresis in agar gel and later a trough was cut in the direction parallel to the migration of proteins. The trough was loaded with a specific antibody that diffuses in the gel to form a precipitin band with an appropriate protein antigen (Figure 6.4). The method can also be employed to assess the homogeneity of antiserum i.e., whether the antiserum placed in the trough is a mixture of antibodies specific to two or three antigens or it has only one antibody specific to one antigen, depending on the number of bands obtained after precipitation.

Several variations of this technique are available, such as counter-electrophoresis, crossed electrophoresis and most importantly rocket electrophoresis developed by Carl Laurell in 1966. In rocket electrophoresis, a negatively charged antigen is used in electrophoresis and the antibody is incorporated into the gel itself. The precipitate formed between the antigen and an antibody has the shape of a rocket, the height of which depends on the concentration of antigen in the well (Figure 6.5).

The immunoelectrophoresis-based methods are routinely used in research and clinical laboratories. The technique can be used to detect the presence or absence of specific proteins characteristically present in a particular disease such as Bence-Jones proteins in myeloma patients. The method is also useful in separating a mixture of microbial and non-microbial antigens. It helps in the detection of abnormally low levels of certain serum proteins like transferrin.

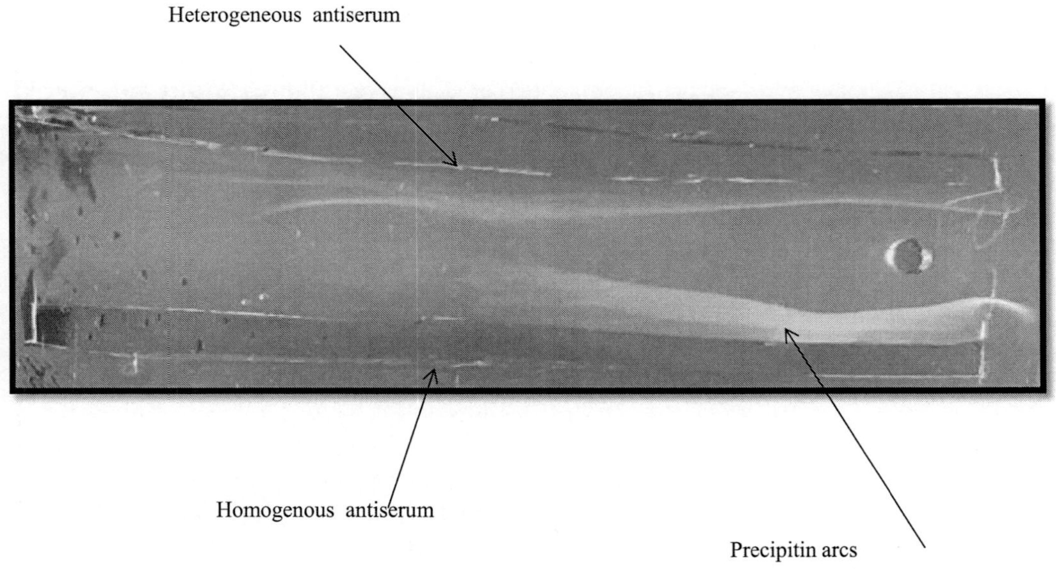

FIGURE 6.4 Picture showing precipitin arcs using two different sera. The trough on the bottom loaded with homogenous serum, showing a single precipitin arc whereas the upper trough with heterogenous sera resulting in more than one arc. (Author.)

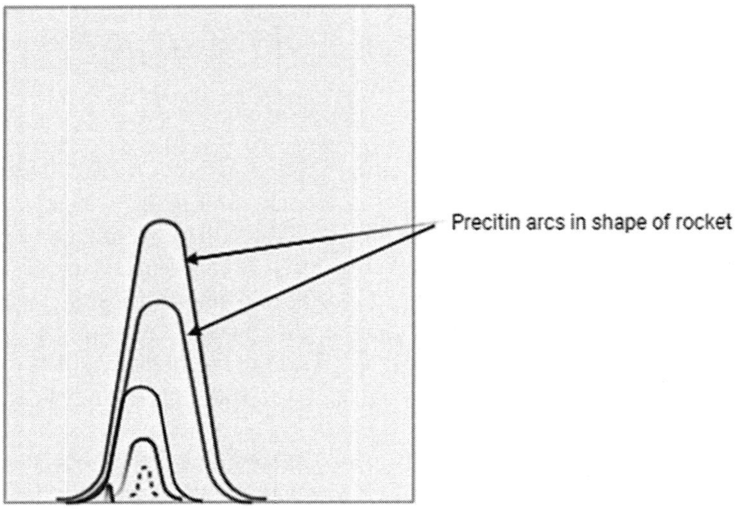

FIGURE 6.5 Fused rocket electrophoresis of human serum (schematic representation); more is the height of the rocket; more is the concentration of antigen in the sample. (Author.)

6.7 RADIOIMMUNOASSAY

Radioimmunoassay or RIA is a highly sensitive technique to measure the amount of antigen or antibody in a given sample. It is a qualitative method that, makes use of radioisotopes for the detection (thus called radioimmunoassay) of an analyte. This *in vitro* assay was first developed by S. Berson and Rosalyn Yalow, in 1960 for quantifying endogenous insulin production in diabetic patients.

This exquisite technique soon found its use in the measurement of other hormones, serum proteins, certain drugs and vitamins, which hitherto could not be measured in earlier times. The method initially was met with much resistance but this major achievement in the field of immunology was recognized by awarding Nobel Prize to R. Yalow for her contribution to the field of physiology and medicine, in 1977.

The classical method of RIA is based on the competitive binding between labelled and unlabeled antigens. The solution of an antibody of appropriate specificity is first allowed to react with labelled antigen. The mixture is later allowed to incubate with the unlabeled sample. As the increasing amount of antigen compete for available antibody and thus displaces labelled antigen, the amount of displaced radioactivity can be measured. The amount of displaced labelled antigen is directly proportional to the unlabeled antigen in the sample. The radioisotopes of iodine, I^{125} (a gamma emitter) can be used for radiolabeling. The emission can be measured with the help of an instrument called a gamma counter. The technique involves the following steps: antigen-antibody complex formation; competitive binding between labelled and unlabeled antigen for specific antibody and finally measurement of radioactivity. The technique is highly sensitive as nanograms or picograms of the analyte can be quantified.

Many variations of the initial assay have been developed by various groups of workers. The procedure essentially consists of the separation of antigen-antibody complexes from unbound reactants. One of the approaches could be precipitating antigen-antibody complex with the secondary anti-isotype antiserum e.g., if the primary antibody used for the development of antigen-antibody complex is rabbit IgG then goat anti-rabbit IgG can be used to precipitate the complex. Another method involves the use of *Staphylococcus aureus*. The protein A of *S. aureus* has the capability of binding with IgG with high affinity. Thus, a complex containing IgG can be precipitated out by using heat-killed *S. aureus*. Ammonium sulphate is another good alternative that usually precipitates IgG but not the antigens (Farr technique of RIA).

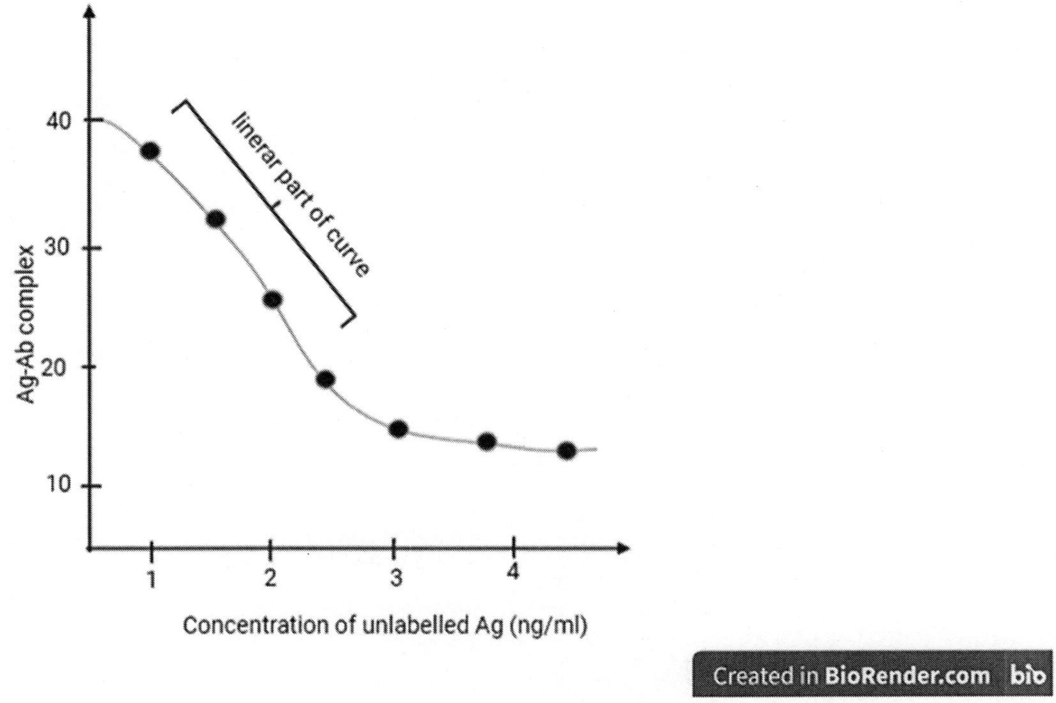

FIGURE 6.6 A standard curve between the unlabeled antigen (known concentration of Hepatitis B antigen) and antibody specific to HBs Ag, Hepatitis B virus surface antigen. (Author.)

The last step is the measurement of radioactivity emitted by radioisotopes which can be done by using a scintillation counter or gamma counter (if the gamma-emitting isotope is used). The amount of radioactivity is proportional to the unlabeled antigen in the sample. A binding curve should be plotted between bound antigens versus known concentrations of unlabeled antigen (Figure 6.6). Subsequently, unknown concentrations of unlabeled antigens can be deduced. The technique is highly advantageous in quantifying a particular antigen in low concentration, in large samples. Owing to its high sensitivity and high precision, the method is employed to screen samples for hepatitis B infection. Similarly, it is also possible to estimate the ligands in biological fluids such as circulating antibodies or binding proteins in serum/plasma, steroid receptors in neoplastic breast tissues. At the same time, RIA has several limitations and the major one is the potential hazard of handling radioactive isotopes and the disposal of radioactive waste. The cost of reagents and instruments used is also high.

6.8 ENZYME LINKED IMMUNOSORBENT ASSAY (ELISA)

Berson and Yalow made a very significant contribution in the field of science by not taking patent of their analytical technique as it led to the enormous modifications and development of immunoassays in successive years. ELISA is another quantitative technique that has almost the same sensitivity as RIA. However, it is a safe and cost-effective technique as compared to RIA. Enzyme-linked immunosorbent assay or more commonly known as ELISA is similar in principle to RIA i.e., there is a binding of antigen and antibody resulting in the generation of an antigen-antibody complex. This antigen-antibody complex later combines with the secondary antibody that is tagged to an enzyme. On adding a suitable chromogenic substrate, a coloured product is formed. The optical density of the resultant solution can be read in a specialized spectrophotometer (ELISA reader).

Thus, ELISA differs essentially from RIA, in terms of the detecting agent. While RIA employs radioisotopes, ELISA makes use of enzyme/enzymatic reaction for the final step detection.

The method was first developed by Perlmann, Engvall and a team of Weiland, independently. It involves the adsorption of the analyte (antigen/antibody) on the surface of a microtiter well plate made up of polystyrene. The analyte is allowed to react with a specific antigen or antibody depending on which component has been coated at the first step. ELISA thus can be of various types viz. Direct, Indirect, Sandwich and Competitive (Figure 6.7a–d). The first step may vary in all variants

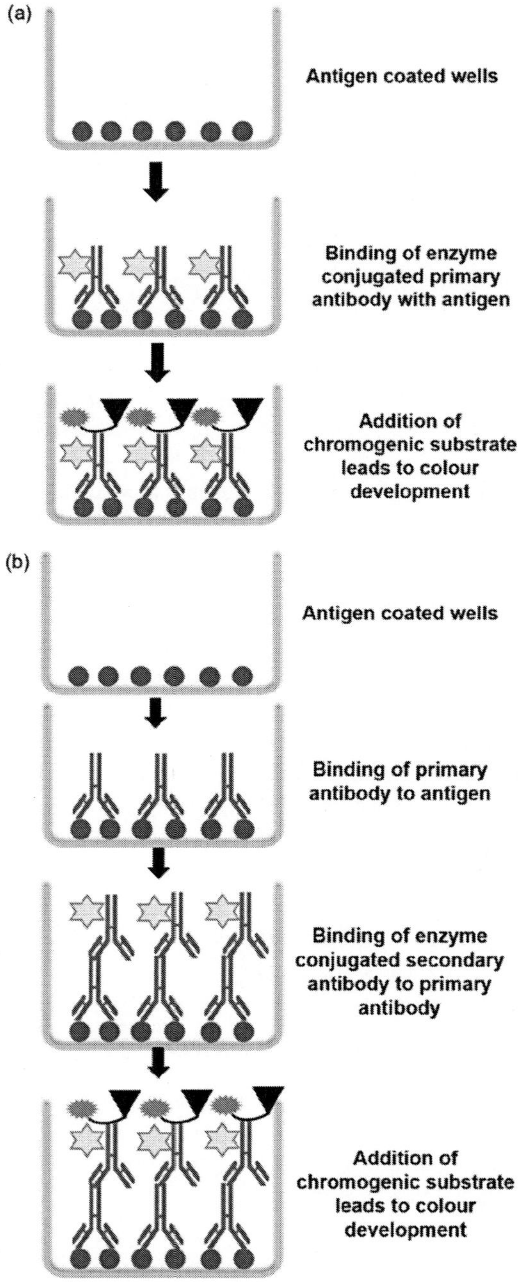

FIGURE 6.7 (a) Diagram showing various steps involved in Direct ELISA. (Courtesy: Dr. Princy Hira). (b) Diagram showing various steps involved in Indirect ELISA. (Courtesy: Dr. Princy Hira). (c) Diagram showing steps involved in Sandwich ELISA. (Author). (d) Diagram showing steps involved in Competitive ELISA. (Author.)

(Continued)

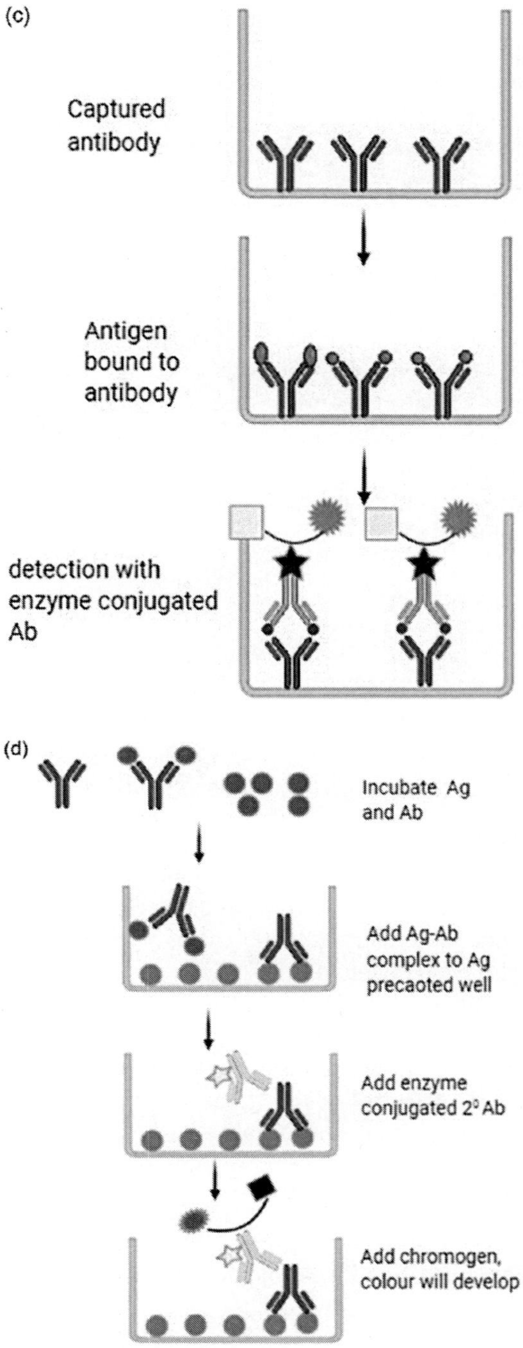

FIGURE 6.7 (*CONTINUED*) (a) Diagram showing various steps involved in Direct ELISA. (Courtesy: Dr. Princy Hira). (b) Diagram showing various steps involved in Indirect ELISA. (Courtesy: Dr. Princy Hira). (c) Diagram showing steps involved in Sandwich ELISA. (Author). (d) Diagram showing steps involved in Competitive ELISA. (Author.)

of ELISA but the rest of the steps i.e., washing of loosely bound antigen/antibody, the addition of the secondary antibody tagged with the enzyme followed again by washing and finally, the addition of an appropriate chromogenic substrate, are similar. The most commonly used enzymes for ELISA are horseradish peroxidase (HRP), alkaline phosphatase (ALP) and β-galactosidase. The assay is generally performed in a multi-well plate wherein adsorption can be done easily. Additionally, the usage of such a system allows the analysis of multiple samples at the same time.

6.8.1 Types of ELISA

The basic procedure of ELISA can be modified to get variants of ELISA without affecting its basic principle. The key step that involves adsorption of antigen whether directly or indirectly depends on the type of ELISA that is being followed. Whatsoever way the antigen is being captured (direct capturing as in Direct ELISA or through a pre-coated "captured primary antibody, as in a sandwich ELISA), it is the last step of detection that majorly determines the sensitivity of an ELISA. Typically, there are four variants, direct, indirect, sandwich & competitive ELISA. Washing after each step is essential in each variant of ELISA. Each of the methods has its own benefits, disadvantages, application and suitability (Table 6.1).

6.8.1.1 Direct ELISA

In this variant of ELISA, the antigen is immobilized on the multi-well titer plate and it is detected by an antibody that is being conjugated directly with the enzyme. Thus, there is no involvement of another antibody (secondary antibody, 2°). This is a quick method of detection as one less step is required. Since no involvement of 2° antibodies so no issue of cross-reactivity will appear (Figure 6.7a). However, the labelling of primary antibodies for each specific ELISA can be expensive.

6.8.1.2 Indirect ELISA

In the Indirect type of ELISA, the antigen is immobilized on the surface of the well and this step is followed by incubating the target protein (antigen) with the primary antibody (1°) specific to the antigen. The 1° antibody is allowed to bind with the antigen. Next, an enzyme-conjugated secondary antibody is incubated with the primary antibody that is being raised against host species. It is noteworthy here that while the primary antibody has specificity to bind with the epitopes of antigen, the secondary antibody can bind to the epitopes of 1° antibody only (Figure 6.7b). The intensity of colour produced is proportional to the amount of antibody present in the sample. It is a method of choice to detect specific antibodies in a serum sample by substituting the serum for the primary antibody. Indirect ELISA has many advantages over Direct type e.g., the maximum immunoreactivity of 1° antibody is retained and amplification of signal can be achieved.

Table 6.1
List of Various Types of Enzyme-Linked Immunosorbent Assay with their Advantages and Disadvantages

Type of ELISA	Advantages	Disadvantages
Direct	Short assay, cross-reactivity minimum	High background, less sensitive, low flexibility
Indirect	Signal gets amplified due to 2° Ab, high flexibility, same 2° Ab can be used for many 1° Ab	Lengthy protocol, cross-reactivity possible
Sandwich	High specificity, suitable for complex sample matrices	has to be designed as same antigen binds to two antibodies
Competitive	Highly suitable for small antigens, Base ELISA can be direct or indirect, high specificity	depends on Base ELISA, requires conjugated antigen

6.8.1.3 Sandwich ELISA

It is one of the most used types of ELISA. As the name indicates, one of the components gets sandwiched between two molecules of the second component. Here, the antigen will be sandwiched between two antibody molecules. Interestingly, it is necessary that two different antibodies should be specific for the same antigen but for non-overlapping epitopes (Figure 6.7c). Such antibodies are called matching antibody pairs. The method is also different from the above two methods in the sense that here the antibody is adsorbed on the surface of titer wells. This facilitates the immobilization of the antigen. Hence, firstly the antibody is captured (known as capture antibody) followed by the addition of antigen to the wells. The other antibody (called detection antibody) is then provided in the assay system. The addition of chromogenic substrate is the final step where the choice of substrate depends based on enzyme used for conjugation with detection antibody. As far as the advantages of the format are concerned, it is compatible with complex samples. Moreover, this variant has the highest specificity and sensitivity among all other types.

6.8.1.4 Competitive ELISA

This ELISA type is a preferred choice where the molecule which is to be detected is very small. Sandwich ELISA cannot be carried out for such molecules. A key difference as compared to other ELISA types is that there is a competition between the bound antigen and free antigen for the primary antibody. The optical density read in the last step is inversely proportional to the amount of the antigen present in the sample. Briefly, the primary antibody is incubated with the sample antigen resulting in the formation of the antigen-antibody complex. Prior to this, the antigen (reference antigen) is immobilized on the surface of a multi-well plate (Figure 6.7d). The complex is then added to antigen-coated wells. Any unbound antigen is then washed off. The competition is going to occur between reference antigen and sample antigen, to bind with primary antibody. The more is the antigen in the sample, the more would be binding of the sample antigen to the given antibody. The colour produced will be low after the enzyme-substrate reaction as a small amount of primary antibody will be available to bind with coated antigen.

6.9 CHEMILUMINESCENCE

Since the successful development of ELISA and its other variants, there has been a continuing trend towards the development of different types of labels in place of the chromogen. Chemiluminescence generally refers to the light produced in certain chemical reactions. This has proved to be a convenient and highly sensitive endpoint. The new assay is termed CLIA, Chemiluminescent Linked Immunoassay. The "label" or the detecting agent used in this technique is a luminescent molecule i.e., a molecule that emits visible or near-visible radiation (λ= 300–800nm) when there is an electron transition from an excited state to the ground state. The resulting potential energy is released in the form of light. The advantage over spectrophotometric measurements is that where luminescence is an absolute measurement, absorbance is relative in nature. This analytical procedure can be either competitive or non-competitive and at the same time direct or indirect based on the use of luminophores (direct), enzyme markers (indirect method), e.g., ALP with 1,2-dioxetane aryl phosphate (AMPPD), HRP with luminol and acridinium and ruthenium esters for the former method.

The method is advantageous in terms of high signal intensity, absence of interfering substances and therefore very high specificity, fast analytical signals, high stability of the conjugates, low consumption of reagents and reduced incubation time. However, several limitations are also associated with the technique, such as limited antigen detection, high cost and closed analytical system. Regardless of its current performance, CLIA is destined to develop further. A new flow-injection system (FI-CLIA) is already in place that involves the rapid introduction of micro-bubbles into the system. This ensures reduction in incubation time, more efficient reaction mixture and increasing temperature control, thus, overall improvement in reaction kinetics. CLIA is a choice of interest in a variety of autoimmune diseases detection where there might be extremely low concentrations of antibodies in the patient's serum (limit of detection-zeptomole 10^{-21} mole).

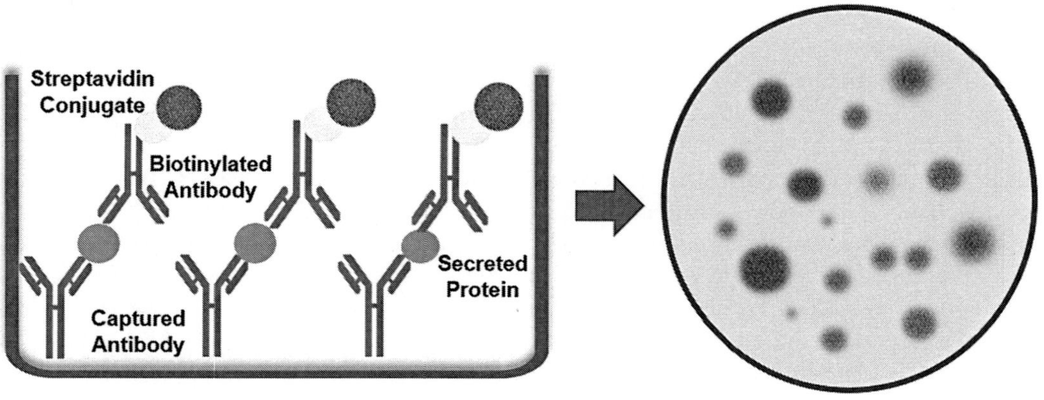

FIGURE 6.8 Steps involved in ELISPOT assay. (Courtesy: Dr. Princy Hira.)

6.10 ELISPOT ASSAY

Elispot is an immunoassay that exploits the principle of ELISA, allowing detection of secretory analyte at the single-cell level. It is also a highly sensitive technique that makes it easy to screen cells secreting low amounts of various molecules (antibodies, interferons or other cytokines). Enzyme-linked immunospot or commonly known as ELISPOT was first reported by Sedgewick and Holt and later by Czerkinsky and coworkers in the same year. In this technique, the antigen was immobilized to a solid support to bind antibodies released by cultured splenocytes. Despite using the principle of sandwich ELISA, ELISPOT differs from it in two important ways. Firstly, ELISA measures the quantity of secreted analyte, whereas ELISPOT quantifies the number of cells secreting that analyte. Secondly, ELISA is essentially done in cell-free media whereas ELISPOT is designed in such a way that live cells are directly cultured in the ELISPOT plate.

The procedure consists of the following steps: (i) Monoclonal antibodies specific to the analyte to be detected are coated on the wells; (ii) the cells are allowed to be incubated with the captured antibodies or the cultured cells that are responsible for secreting analyte are placed in ELISPOT wells; (iii) the next step follows washing of the cultured cells; (iv) the reaction medium is incubated with biotinylated secondary antibodies(in case of biotinylated antibody, enzyme-conjugated with streptavidin is used for detection); and (v) addition of appropriate chromogenic substrate (depending on the choice of enzyme) which leads to the development of spots (Figure 6.8). The number of developed spots is directly correlated to the number of cells secreting the analyte to be detected. Variations in the procedure are possible like the use of either ALP or HRP as enzyme conjugates or the use of a chromogen. Alkaline phosphatase (ALP) based BCIP/NBT (5-bromo-4-chloro-3-indolyl phosphate p-toluidine salt/nitroblue tetrazolium chloride) is the most commonly used enzyme-chromogen as it is a safe and stable complex.

ELISPOT is widely used in research laboratories and for diagnostic purposes. Researchers have reported it to be an effective and more reliable approach for the detection of mycobacterium infections in HIV-positive patients. ELISPOT is also a method of choice for the detection of allergies, such as individuals allergic to nickel e.g., it was observed that mononuclear cells in the blood of individuals allergic to Nickel secreted significantly higher quantities of interleukins. In recent times, IFN-γ-based ELISPOT is widely used for large-scale trials involved in the development of HIV vaccine or vaccines for other pathogens.

6.11 IMMUNOFLUORESCENCE

Immunofluorescence is a technique used to detect antigens in biological samples. The biological samples could be individual cells, tissue sections or cultured cells. The detecting agents are fluorescent compounds such as fluorescein, rhodamine etc., which are conjugated chemically

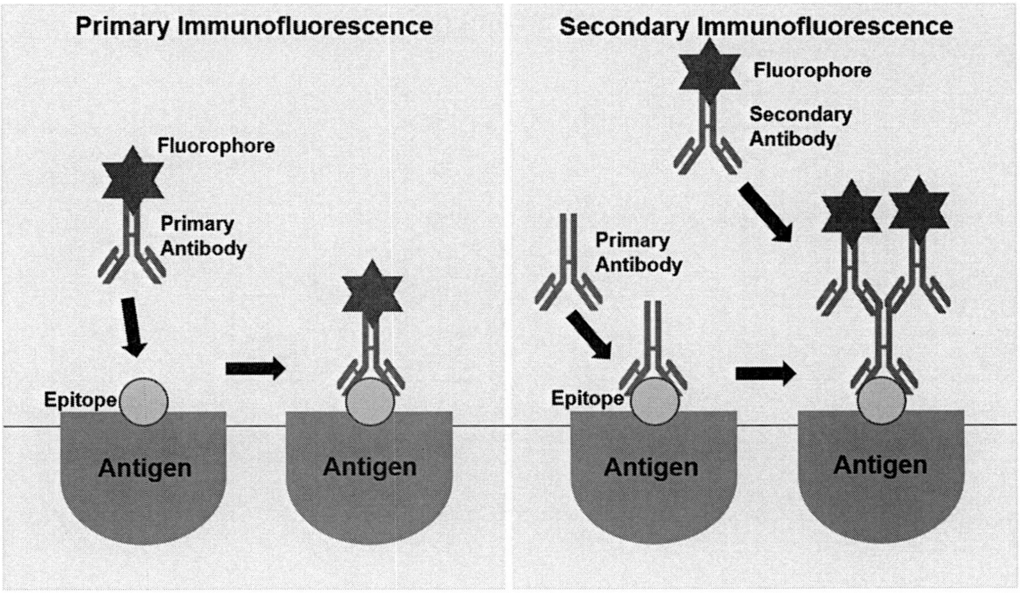

FIGURE 6.9 (a) Diagram showing steps involved in Immunofluorescence technique. (a) Primary Immunofluorescence, (b) Secondary Immunofluorescence. (Courtesy: Dr. Princy Hira).

with the antibodies. When these dyes absorb light of a shorter wavelength (high energy), they get excited and start fluorescing, emitting light of lower energy. The fluorescence can be visualized using a fluorescent microscope. The method can be indirect or direct, depending on the antibodies used. If only a single or primary antibody is directly conjugated with the fluorescent compound, the method is direct fluorescence. The method, however, is of limited use as only a limited number of molecules can be conjugated with the primary antibody. The indirect method is more popular in which an unlabeled antibody specifically binds to the antigen, which in turn binds with the secondary antibody conjugated with the fluorophore (Figure 6.9).

6.12 WESTERN BLOTTING

Western Blotting is a technique used for the detection of the proteins present in the gel. The technique essentially consists of three important parts i.e., SDS-PAGE, transferring of protein bands on to membrane and lastly treating the membrane with a specific antibody.

The given sample is first electrophoresed by SDS-PAGE to separate the mixture of proteins. The separation is done based on molecular weight. The mixture gets separated into a number of bands. The protein bands are then transferred to the supporting membrane by the technique of electroblotting. Generally, a nylon or nitrocellulose membrane is used for this purpose, however, PVDF (Polyvinylidene Difluoride) membrane can also be used. Nitrocellulose membrane has high affinity and good retention ability for the proteins but it is quite brittle, thus not very easy to handle. PVDF, on the other hand, serves as an excellent supporting medium as it is easy to handle and long-term storage is possible. An electrotransfer apparatus is used in the presence of a buffer for the process of transferring proteins from the gel to the membrane. A sandwich of the gel is created by putting sponge, layers of filter papers, gel, membrane and filter papers. The entire setup is kept in the transfer apparatus for some time under the influence of an electric current. An electric field is provided in the direction perpendicular to the surface of the gel. Such an arrangement allows the protein bands to move to the membrane. The membrane with transferred protein bands is incubated with the primary antibody followed by the secondary antibody. Washing and blocking of the membrane are the two essential steps in the Western

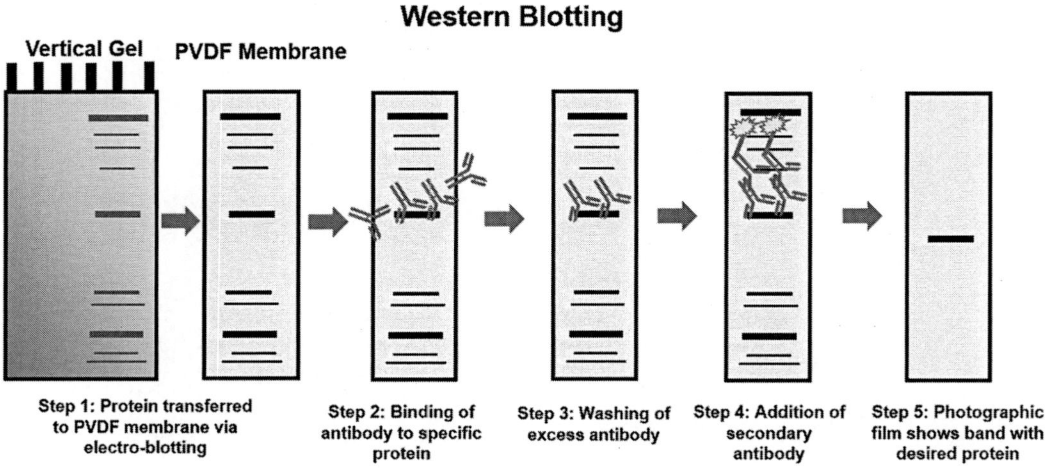

FIGURE 6.10 Schematic representation of the steps involved in Western blotting. (Courtesy: Dr. Princy Hira.)

blotting technique. These help in minimizing the background during the visualization of the results. HRP is often the enzyme of choice that is tagged to an antibody and in the presence of an appropriate substrate, the bands can be visualized easily. The method thus can find its use in the identification, quantification and determination of the size of the desired protein in a mixture. It has become a powerful tool for use in research laboratories. A schematic representation of the technique is given in Figure 6.10

6.13 SUMMARY

- Antigen–antibody interactions constitute an important basis and play a pivotal role in the development of various immunological techniques. Affinity and avidity are two integral features of an antibody molecule. Avidity is a better measure of the strength of interaction between antigen and antibody.
- Hemagglutination and bacterial agglutination reactions are meant for determining the presence of antigens to antibodies on RBCs and for antibodies specific for certain bacterial strains, respectively.
- Precipitation reactions involve the formation of a visible precipitate, mediated by antibodies. It can be carried out in liquid or gel media.
- Radioimmunoassay is a sensitive method for the measurement of antigens or antibodies in a given sample. The method is an excellent choice of interest where the amount of analyte is very less in the sample. The technique uses the radioisotope of Iodine or Hydrogen.
- ELISA is another sensitive, cost-effective and safe method for quantification of antigen and antibody in a sample. In an Enzyme-linked immunosorbent assay (ELISA), the antigen or antibody is adsorbed on a supporting medium and further detection takes place with the help of an enzyme-conjugated antibody. A chromogenic substrate is used for this reaction. CLIA on the other hand uses luminogenic substances using the same principle as ELISA.
- Biotin can also be conjugated along with streptavidin along with a particular enzyme to increase the sensitivity of an immunoassay.
- The western blotting technique is used for the detection of proteins in a given sample using antibodies. SDS-PAGE is followed by the transfer of protein bands on the membrane which is then followed by incubation with antibodies.

QUESTIONS

1. Differentiate between avidity and affinity.
2. What is a precipitation reaction? Discuss two different methods through which immunoprecipitation can be carried out in a gel medium.
3. Differentiate between RIA and ELISA.
4. Discuss various types of ELISA with the help of suitable diagrams.
5. What is chemiluminescence or CLIA? How does it differ from ELISA?
6. Through which method can one quantify the number of cells involved in the secretion of cytokines?
7. Enlist the types of ELISA responsible for the quantification of antigen or antibody.
8. Briefly discuss Radioimmunoassay. How can the free antigen be separated from antibody-antigen precipitate?
9. Describe the technique of Immunofluorescence with the help of suitable diagrams.
10. What do you understand by the term cross-reactivity? Add a note on its significance.
11. Differentiate between the terms agglutinogen, agglutinin and agglutinate.
12. How does the technique of ELISA differ from ELISPOT?

BIBLIOGRAPHY

1. Sela-Culang, I., Kunik, V., and Ofran, Y. 2013. The structural basis of antibody-antigen recognition. *Front Immunol*, doi: 10.3389/fimmu.2013.00302.
2. Edward, J.M. 2016. *A Historical Perspective on Evidence-Based Immunology*, Elsevier Publications, https://doi.org/10.1016/B978-0-12-398381-7.00039-3, accessed on November 15, 2021.
3. Cinquanta, L., Fontana, D.E., and Bizzaro, N. 2017. Chemiluminescent immunoassay technology: what does it change in autoantibody detection? *Auto Immuno Highlights*, 8(1):9. doi: 10.1007/s13317-017-0097-2.
4. Sedgwick, J.D., and Holt, P.G. 1983. A solid-phase immunoenzymatic technique for the enumeration of specific antibody-secreting cells. *J. Immunol. Methods*, 57:301–309.
5. Czerkinsky, C.C., Nilsson, L.A., Nygren, H., Ouchterlony, O., and Tarkowski, A. 1983. A solid-phase enzyme-linked immunospot (ELISPOT) assay for enumeration of specific antibody-secreting cells. *J. Immunol. Methods*, 65:109–121.
6. Kalyuzhny, A.E. 2005. Chemistry and Biology of EliSpot assay. In *Handbook of ELISPOT, Methods in Molecular Biology*, vol. 302. edited by: A. E. Kalyuzhny, Humana Press Inc., Totowa, NJ. Retrieved from https://beckassets.blob.core.windows.net/product/readingsa.
7. Kohler, G., and Milstein, C., 1975. Continuous cultures of fused cells secreting antibody of predefined specificity. *Nature*, 256:495–497.
8. Webster, D.M., Henry, A.H., and Rees, A.R. 1994. Antigen-antibody interactions. *Current Opin Structural Biol*, 231:505–510.
9. Hodget, R., Heaton, R., and Robert-Parker, J.M. 1988 Antigen-antibody interaction. *J. Biol. Chem.*, 263(24):11768–1175.

7 Major Histocompatibility Complex and Antigen Presentation

Shanuja Beri and Nidhi Garg
University of Delhi

CONTENTS

DOI: 10.1201/9781003286424-7

7.1 INTRODUCTION

Apart from the innate immunity found in mammals comprising of physical and chemical barriers present right from birth, adaptive immunity builds up in an individual only after antigen exposure and comprises both humoral immunity and cell-mediated cytotoxicity.

Humoral immunity is mediated by antibodies whereas the cell-mediated cytotoxic response is produced by the T-cells and regulated by the major histocompatibility complex also known as MHC. MHC is a cluster of genes that are present in all almost all vertebrates and encodes for glycoproteins which the cells express on their surface. These are responsible for antigen presentation and lymphocyte recognition, differentiating between self and non-self. These glycoproteins or histocompatibility molecules are also named as Major Histocompatibility Complex or MHC, and behave as antigens when introduced into another individual. The MHC belong to the immunoglobulin supergene family and encodes for three classes known as MHC Class I, MHC Class II and MHC Class III molecules. In humans these are known as human leukocyte antigens- HLA Class I, Class I and Class III. The gene cluster is called the MHC or HLA complex. It was the genes that were recognized first due to their role in the rejection of a graft. Peter A. Gorer characterized the MHC complex of mouse in 1936 naming it the H2 complex. Snell got the Nobel Prize in 1980 for discovering the role of group II genes in tissue rejection which are now known as H2. Jean Dausset described the HLA complex in humans in 1958.

7.2 ORGANIZATION OF CLASS I, CLASS II, AND CLASS III GENES

The MHC gene cluster is known as the HLA system in humans and as H2 system in mice (Figure 7.1). Each class of MHC gene is polygenic having more than one locus. The mapping of the locus was done by cloning and sequencing techniques. MHC genes are co-dominant and both the sets, one inherited from the father and another from mother express proteins which are present on the surface of all the cells. There is no recombination and the linked complex alleles found on the same chromosome is called a haplotype. This cluster of genes is found on the shorter p arm of human chromosome 6 and on chromosome 17 in mouse. The complex is made up of 128 genes and 96 pseudogenes consisting of four million base pairs. This gene cluster is highly polymorphic having many alleles at different loci. This gives rise to a large diversity of MHC within a species.

1. MHC Class I genes encode for glycoproteins which are found on all nucleated cell surfaces and which present antigens to T_C cells.
2. MHC Class II genes encode for glycoproteins which are expressed on the surface of specialized antigen-presenting cells (APCs) such as macrophages, dendritic cells, and B cells. These APCs then process the antigens for presenting to T_H cells.
3. MHC Class III genes encode many components of the complement system, soluble serum proteins and tumour necrosis factors (TNF).

7.2.1 Genetic Organization of Human Leukocyte Antigens (HLA) Complex

The HLA Class I, II and III molecules have their genetic loci placed in tandem on chromosome 6 in three distinct segments with the HLA Class I being the telomeric most of the cluster and Class II the most centromeric. Figure 7.2.

7.2.1.1 HLA Class I Molecules

The HLA Class I molecules are encoded by HLA A, B and C genes which express six types of molecules in an individual, three from each chromosome of the homologous pair. These MHC glycoprotein molecules found on all the nucleated cell surfaces present the antigens to the T_C cells and are a heterodimer consisting of a heavy chain (called α chain) and a light chain (called the β-2) polypeptides.

MOUSE H-2 COMPLEX		
MHC CLASS	REGIONS	GENE PRODUCTS
I	K & D	K Region- H-2K D Region- H-2D, H-2L
II	IA & IE	IA Region- IA αβ IE Region- IE αβ
III	S	C'Proteins, TNF-α, TNF-β

HUMAN HLA COMPLEX		
MHC CLASS	REGIONS	GENE PRODUCTS
I	A B C	A Region- HLA-A B Region- HLA-B B Region- HLA-C
II	DP DQ DR	DP Region- DP αβ DQ Region- DQ αβ DR Region- DR αβ
III	C4, C2, BF	C'Proteins, TNF-α, TNF-β

FIGURE 7.1 The above figure shows the organization of the MHC known as H-2 complex in mouse and HLA complex in human. In-spite of being two completely different genus the distribution of genes in three distinct groups indicates common origin. (Copyright Authors.)

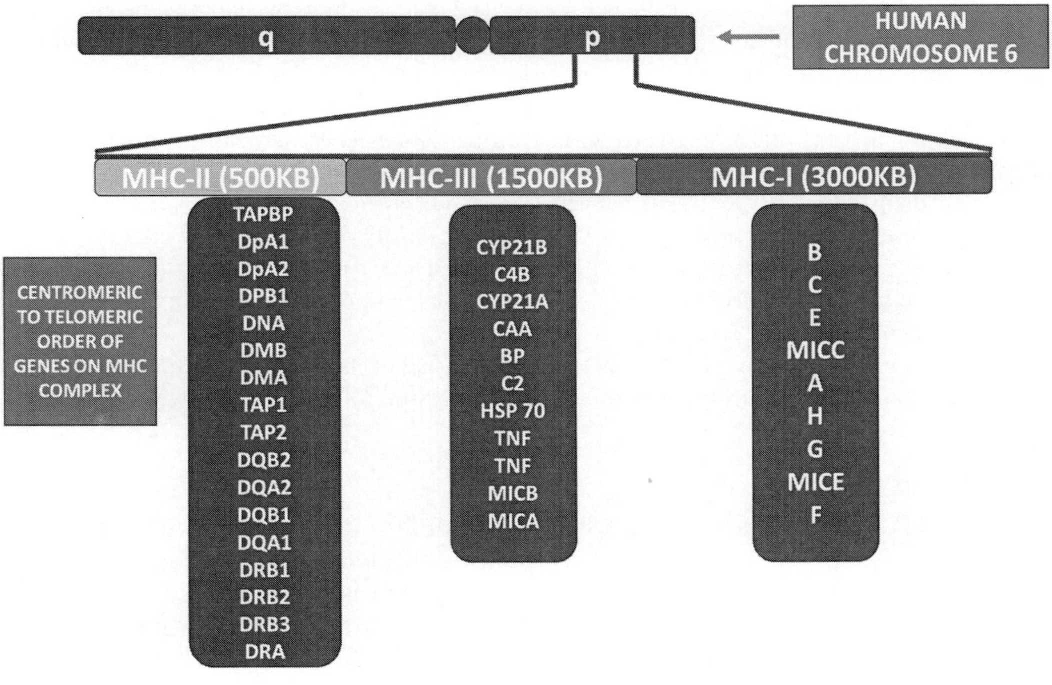

FIGURE 7.2 Diagrammatic representation of Human chromosome 6 showing the distribution of MHC class I, II and III complex and their genes. Class III genes are located in between Class 1 and Class II genes. (Copyright Authors.)

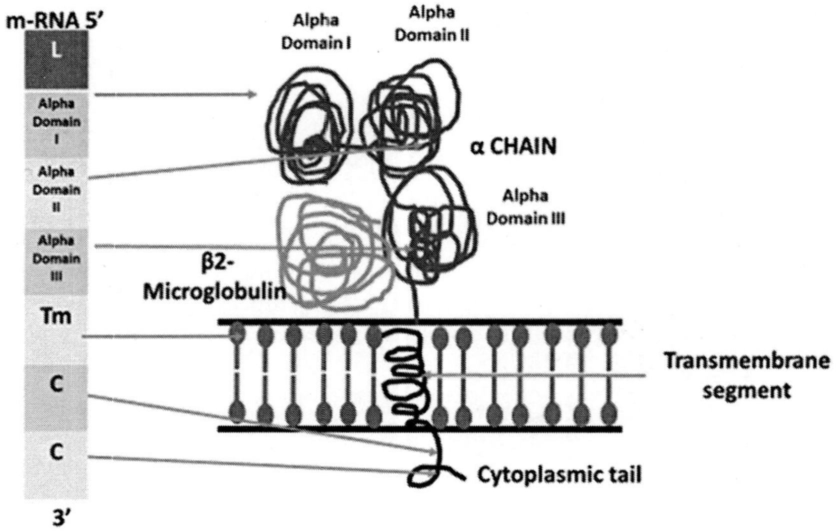

FIGURE 7.3 Diagrammatic representation of the structure of Class I MHC mRNA and the protein it encodes. Leader exon (L) is removed post transcription. (Copyright Authors.)

Each HLA gene encoding for the α chain has 7 exons out of which exons 2 and 3 code for the alpha 1 and alpha 2 domains. The alpha 3 domain is encoded by exon 4 and the leader peptide L is encoded by exon 1 and is removed post transcription whereas exon 5 encodes for the transmembrane region and exon 6 and 7 for the cytoplasmic tail. Figure 7.3. The exons 2 and 3 show maximum polymorphism resulting in a large number of alleles and encoding numerous alpha 1 and alpha 2 domain variants. These variants are important because alpha 1 and alpha 2 domains form the antigenic peptide binding cleft interacting with innumerable foreign peptides and allows an individual to recognize a large number of foreign peptides. HLA genes have over a hundred alleles and are identified by alphanumeric abbreviations such as HLA-B27 with closely related alleles classified into subtypes like HLA-B27*40. This classification is important as it helps in identifying immunodeficiency diseases as well as in determining histocompatibility between tissues for transplant.

Included along with these three classical genes are non-classical genes HLA E, G, F, H, J, K, L, cda 12 and 8 pseudo genes like MICC and MICE genes. HLA E, F, and G are the ones expressed and do not show polymorphism encoding for proteins which have a role as minor histocompatibility molecules. The peptide β molecule from HLA G associates with β-2 microglobulin and mediates the recognition of antigens by NK cells.

The β-2 microglobulin gene (β-2M gene) is found outside of the MHC locus, on chromosome 15 in humans, and has three exons. The polypeptide expressed by this gene is largely folded into β-sheets.

7.2.1.2 HLA Class II Molecules

The Class II HLA polypeptide molecule is made of two different polypeptide chains known as α chain and β chain. The Class II HLA region consists mainly of three gene families DP, DQ and DR. These three gene families each have a varying number of genes known as A and B genes which code for α and β chains respectively. The DP and DQ gene families have only one gene each for the α and β chains along with pseudogenes (DPA2, DPB2, DQA2, DQB2, DQB3) which are not expressed. The products of DQA1 and DQB1 bind together to form the DQ molecules and the DPA1 and DPB1 products form the DP molecules. There is only one DRA gene along with nine DRB genes. The invariable α chain is coded by DRA complexes with the β chains coded by DRB1, DRB3, DRB4 and DRB5 loci to form heterodimers found on the surface of APCs.

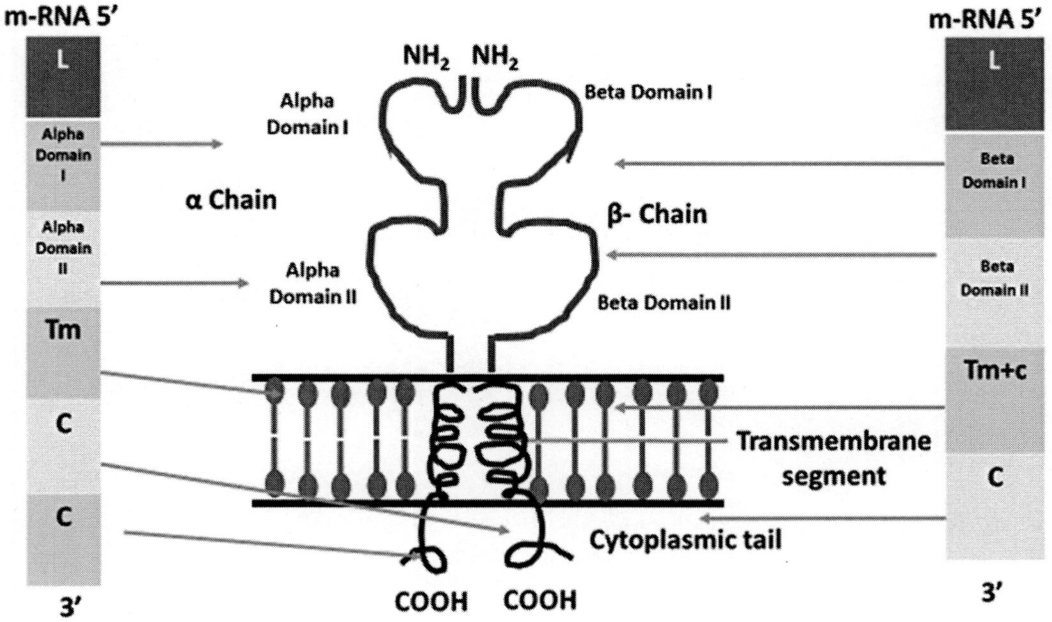

FIGURE 7.4 Diagrammatic representation of the structure of Class II MHC mRNA and the protein it encodes. Leader exon (L) is removed post transcription. (Copyright Authors.)

The MHC II loci also have other genes such as DM, DN and DO. The HLA Class II gene has 6 exons which encode for the alpha or beta domains, connecting peptides (CP), transmembrane proteins (TM) and cytoplasmic proteins (CM) (Figure 7.4). Exon 1 encodes for the Leader (L) which is removed post transcription.

Interspersed within the MHC locus are genes which transcribe many proteins involved in antigen presentation like the transporter associated with antigen processing (TAP) and proteosomes.

7.2.1.3 HLA Class III Molecules

The HLA Class III genes lie in the centre of the HLA gene cluster and have no genes coding for the membrane-bound HLA molecules (Figure 7.2). Their gene products are functionally diverse and play no role in antigen presentation. Two genes encode for the C4a and C4b subunits of C4 complement protein, one gene for factor B and one gene for the C2 complement protein, two steroid 21 hydroxylase enzymes (21-OHA, 21-OHB), inflammatory cytokines, tumour necrosis factors (TNF α and TNF β) and two heat shock proteins (HSPs). The TNF are multifunctional molecules involved in inflammation and healing by proliferation of cells, differentiation and apoptosis. The genes involved in inflammation, both local and systemic are found in the telomeric region of the HLA Class III cluster and this region has one of the highest concentrations of genes in the human genome. Genes which have this common function are now grouped under HLA Class IV and include the TNF family and HSP70 and molecules of this region could be involved in auto-immune diseases.

7.2.2 Genetic Organization of the Mouse H2 Locus

The MHC complex in mouse is called the H2 locus and consists of genes expressing the MHC Class I, MHC Class II and MHC Class III. In the 1940s Peter Gorer developed antisera against an antigen named antigen 2 which helped detect an allele in the histocompatibility locus H. The locus was then named H2. This locus is present on chromosome 17 and is homologous to the HLA

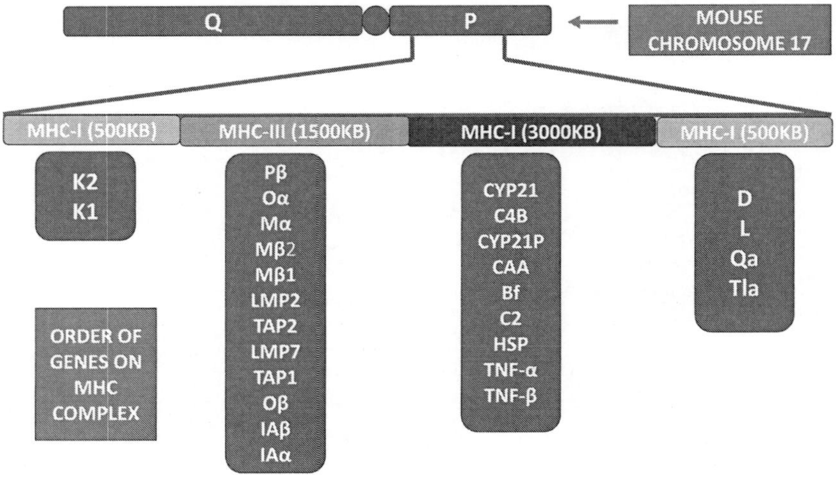

FIGURE 7.5 The above figure shows the arrangement of Class II and Class III MHC genes in between Class I MHC genes on Chromosome 17 of mouse genome. (Copyright Authors.)

complex of humans encoding for the antigens expressed on the surface of nucleated cells, APCs, as well as proteins of the complement system.

The MHC Class I molecules are encoded by the K and D segments which are separated by the Class II and Class III genes unlike the HLA Class I (Figure 7.5). It also has other genes like L, Q1, Q2, Qn and T1, T2, Tn TM and TN.

The MHC Class II genes are around 600 kb and are present in the I segment which has two gene loci IA and IE which code for the functional molecule and other genes distributed in this segment. The genes IAβ and IAα are present in the IA locus and IEβ, IEβ2 and IEα are found in the IE locus. The other genes are Pβ, Oβ, Oα, $M\beta_2$, $M\beta_1$, Mα, LMP, TAP2, TAP1, and LMP7. Out of these, IEβ2, Oβ, and Oα transcribe non-functional mRNA. Pβ is a pseudogene. The genes LMP and LMP7 encode for subunits of proteosomes and TAP1 and TAP2 encodes for the MHC 1 transporter peptides. The remaining genes $M\beta_2$, $M\beta_1$ and Mα encode for peptides which bring about the association of antigenic peptides and the MHC Class II molecules. The two mouse loci IA and IE are homologous to HLA Class II DQ and DR respectively. The gene segments have six exons expressing connecting peptides (CP), transmembrane proteins (TM) and cytoplasmic proteins (CM).

7.3 INHERITANCE

As the HLA molecule is encoded by more than one gene it is polygenic. The four loci D, B, C and A are located very close to one another on chromosome 6 and are always inherited together as linked genes. This set of alleles becomes specific to an individual who then has one such set on the paternal and one set on the maternal chromosome. The alleles co-dominantly express the antigenic determinants or specificities (Figure 7.6). The combination of specificities on one chromosome is known as a haplotype and one haplotype is thus inherited from each parent. This results in an individual having two different haplotypes with heterozygous alleles at each locus.

A maximum of six Class I specificities and similarly six specificities of Class II can be inherited and these 12 loci are involved in HLA expression in the tissue type (Figure 7.6). Moreover, as the DR α and β chains from the same parents can combine in the cis arrangement or from different parents in the trans arrangement additional specificities in the DR region arise. Sometimes crossing over may take place between the two parental chromosomes giving rise to recombinant new

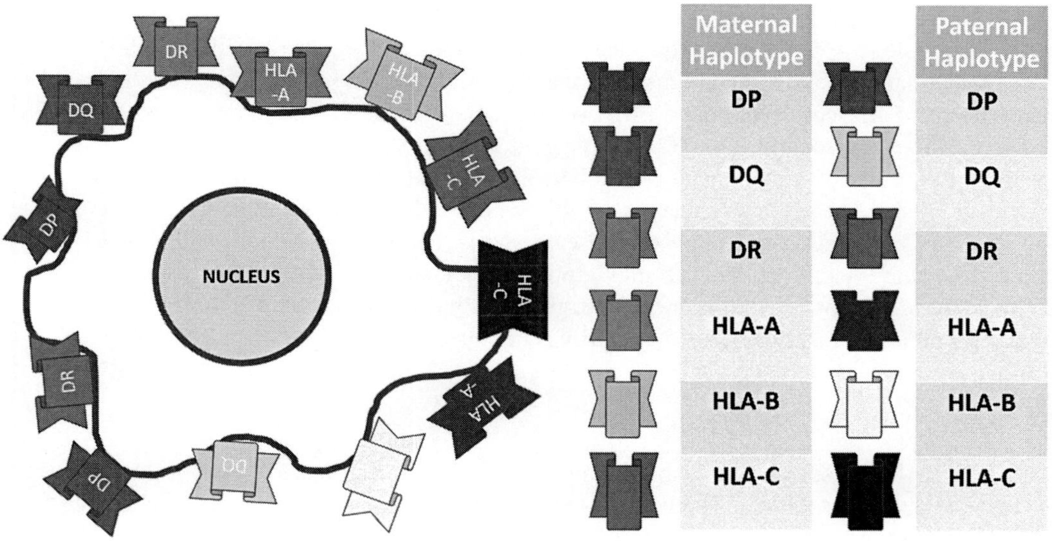

FIGURE 7.6 The above diagram shows the co-dominant expression of both maternal and paternal MHC Class I (HLA-A, B and C) and Class II (DP, DQ and DR) molecules on the surface of a mouse cell. (Copyright Authors.)

haplotypes but there is no somatic recombination. The progeny does not have the same two haplotypes as the parents but have one in four chance of similarity with their siblings.

When an epitope is located on a single antigen this is known as the HLA-private antigen example the HLA-A2 whose antigenic determinant is on only one molecule but when the same epitope is found on different antigens then these molecules are known as HLA-public antigens example the HLA-Bw4 epitope.

7.4 GENETIC POLYMORPHISM

The MHC Class I and Class II molecules can bind to many different antigenic peptides and many antigenic peptides can bind to these two classes of MHC molecules, making the possibility of binding between MHC molecules and peptides very large. Due to the polygenic locus each individual can express six different types of Class I and 12 different types of Class II molecules which is a small number compared to the immensely large number of antigenic peptides that need to be recognized and subsequently presented to the T cells. This large number of MHC molecules, capable of recognizing diverse antigenic peptides, arise due to the variation in the alleles. This variation corresponds to the antigenic peptide binding domains forming the peptide cleft of the MHC molecules and allows different peptides to be recognized. The variation is found in MHC molecules within species and within individuals, giving them a very broad specificity to antigens.

The MHC is thus polymorphic because the different alleles of the different genes are present in duplicated copies with similar structures and functions. Each cell expresses about ten thousand variant copies of each Class I MHC molecule resulting in many kinds of Class I molecules being displayed on the cell surface. The polymorphism of the alleles arises due to the difference in the DNA sequence resulting in a variation of nearly 20 amino acids. There are nearly 4,500 variants of HLA-A, 5,000 of HLA-B and 3,000 of HLA-C alleles reported. Similarly, the Class II MHC genes have thousands of variants for each of the DP, DQ and DR gene loci. As each haplotype contains genes for both classes, the possible combinations of the numerous alleles of these genes, in one haplotype are enormous (Figure 7.9). Tandem duplication takes place in the genes and each homologous chromosome can have two or three genes resulting in the two to twenty different allelic products of each loci being expressed simultaneously. They could be HLA-DRα1– HLA-DRβ1; HLA-DRα2 – HLA-DRβ2; HLA-DRα1 – HLA-DRβ2; HLA-DRα2 – HLA-DRβ1; etc.

On analyzing the observed combinations, the diversity resulting from this polymorphism is less than expected. Some combinations of alleles seem to appear in individuals more frequently together leading to linkage disequilibrium. Thus, linkage disequilibrium is that phenomenon wherein the observed frequency of a combination of alleles is different than the expected frequency of the individual alleles. The most probable reason for this disequilibrium may be that certain combination of alleles may impart resistance to a disease and is therefore selected for and appears with a greater frequency within the population. Another theory suggests that only certain areas of the chromosome show crossing over known as hotspots and may show recombinant types whereas other regions will remain linked leading to disequilibrium.

7.5 CELLULAR DISTRIBUTION

The Class I MHC molecules are displayed on the surface of all nucleated cells but the expression level is not the same. The cells which are in the front line of defense such as the lymphocytes have the highest level of expression & resulting display, of nearly 500,000 molecules per cell. They are less abundant on non-immune cells like liver cells and fibroblasts and are least in numbers on the cell surface of neural and skeletal cells.

The Class II MHC molecules are expressed by only those cells which can present antigens such as the macrophages, dendritic cells, B cells and thymic epithelial cells. These APCs show large variation in the expression levels of the Class II MHC depending on the stage of differentiation they are. The activated macrophages, monocytes and mature B cells express a much higher level of MHC molecules as compared to the inactivated or immature cells.

7.6 REGULATION OF MHC GENES

Differential gene expression of both MHC Class I and Class II takes place in the different cell types and even at different stages of cell differentiation. This indicates an active regulation at the transcriptional level which can be both positive and negative. This regulation can be carried out by transcriptional activators as well as by different cytokines and virus interference.

Certain factors such as the Class II trans activator (CIITA) and regulatory Factor X (RFX) have been identified which bind to the promoter region of the Class II MHC genes and are involved in their expression. Cells with mutations which lead to defects in these factors do not have the Class II MHC molecules expressed on their cell surface. This results in a disorder known as bare lymphocyte syndrome and the consequence is immunodeficiency as T cells cannot be activated.

Certain cytokines such as TNF and interferons α, β and γ also act as enhancers of the transcription of Class I MHC molecules. The transcription of Class I MHC α chain is up-regulated when specific transcription factors bind to the promoter region of the Class I MHC gene. The formation of these transcription factors is induced by IFN-γ which also up-regulates β-2 microglobulin, the proteasome sub-units and the transporter subunits (TAP). On the other hand, the expression of Class II MHC by B cells is down-regulated by IFN-γ and up-regulated by IL-4. The expression levels of Class II MHC are also reduced by prostaglandins and corticosteroids.

Viruses have a direct or indirect negative regulatory effect on the expression and function of MHC Class I molecules. Adenovirus 12 down-regulates the transcription of TAP genes which play an important role in the transport and loading of peptides in the endoplasmic reticulum as a result integration of MHC Class I molecule does not take place in the cell membrane. Cytomegalovirus disrupts the assembly of the MHC Class I molecule by preventing the binding of β-2 microglobulin. The virus does this as a likely strategy to evade detection by MHC molecules.

7.7 THE MHC MOLECULE: STRUCTURE AND FUNCTION

The MHC molecules of Class I and Class II are both glycoproteins found on the cell membrane and are very similar in structure and function. Both bind to peptide ligands forming stable complexes and presenting them at the cell surface. It is these peptide-MHC complexes which are recognized by T cells and mediate cell-mediated immune response. The Class III MHC molecules are not structurally or functionally related to each other or to the Class I and Class II MHC molecules and have different roles to play including in the complement system.

7.7.1 Class I MHC Molecule

The Class I MHC molecule is made up of two polypeptide chains which are very dissimilar to each other. An α-chain associates non-covalently with a small β-2 microglobulin molecule to form the functional molecule (Figure 7.3).

The α-chain polypeptide has a molecular weight of 45 kDa and is about 350 amino acids long. It is glycosylated and is anchored into the plasma membrane of all nucleated cells. The α-chain has four regions:

1. An extra-cellular hydrophilic region starting at the N-terminal and having 281 amino acids. This region has three extracellular domains known as alpha-I, alpha-II and alpha-III, each containing approximately 90 amino acids.
2. An immunoglobulin-like region resembles the C- region of an immunoglobulin.
3. A transmembrane region which is hydrophobic and has 40 amino acids.
4. An intracellular region at the C-terminal which is hydrophilic and has 30 amino acids.

The β-2 microglobulin is 12 kDa and associates with the alpha-III domain in a non-covalent manner. It is also around 90 amino acids.

The alpha-I and alpha-II domains are found on top of the MHC molecule and are arranged in a deep V-shaped peptide binding cleft measuring about 25 Å X 10 Å X 11 Å. The alpha-I and alpha-II domains have 4 β-sheets and 1 α-helix each respectively, forming a platform of eight anti-parallel β-strands traversed by two long α-helical regions. The long α-helices form the sides and the β-sheets form the bottom of the cleft. Any allelic differences in the amino acid sequences are only found in this region of the molecule. The alpha III and β-2 microglobulin domains have a great homology to the constant region domain of immunoglobulins. Both the alpha III and β-2 microglobulin domains have seven β-sheets each, with four of these being outer and three inner (Figure 7.7). The β-sheets of each domain are connected by β-peptide strands which are thin polypeptide chains and sulphide bridges that stabilize all domains. The alpha-III domain is highly conserved among the class-I MHC molecules and is recognized by CD-8 membrane receptors of T_C cells.

Function: The MHC molecules are involved in antigen presentation. The Class-I MHC molecules play a vital role in recognizing intracellular viral and bacterial antigens. When a virus infects a cell, its proteins are degraded using proteosomes. The endogenously processed viral peptide fragments bind to the peptide binding cleft formed by alpha I and alpha II domains. These antigenic peptides having 8–10 amino acids can be accommodated and are presented to the T_C CD-8 cells and appear as a bulge if they are longer than the cleft. The CD-8 cells then recognize these as non-self-peptides and can kill antigenic peptide displaying cells by cell-mediated immunity. This antigen processing and subsequent presentation of non-self-peptides is one of the main functions of Class I MHC molecules. This function of recognizing and differentiating self-peptides from non-self-peptides also leads to transplant rejection where the recipient identifies the donor MHC molecule as antigenic and proceeds to reject it as foreign.

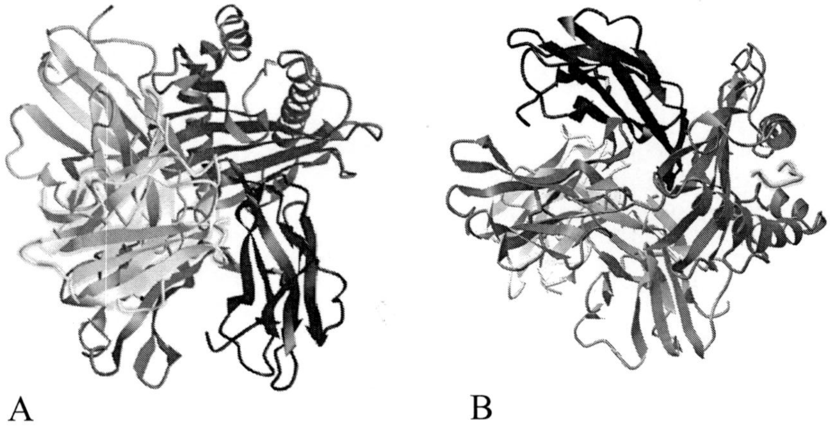

FIGURE 7.7 (A) The three dimensional protein structure of Class I MHC molecule. (b) The antigen binding site of Class I MHC molecule with the antigen is highlighted in yellow.

7.7.2 Class II MHC Molecule

The Class II MHC molecules are responsible for immune response and are found only on APCs. They are heterodimers and consist of two polypeptides known as α and β chains which are glycosylated. The two domains have about 230 and 240 amino acids and a molecular weight of 32–34 and 29–32 kDa respectively. The structure of both α and β chains resembles that of the α chain of the Class I MHC (Figure 7.4). Both chains similarly have

1. An extra-cellular hydrophilic region starts at the N-terminal and has 178 in the α chain and 192 amino acids in the β chain. This region in each chain has two extracellular domains known as alpha-I, alpha-II in the α chain and beta-I and beta-II in the β chain. Alpha-I has 84 and alpha-2 has 94 amino acid residues whereas beta-1 has 91 and beta-2 has 100 amino acid residues.
2. An immunoglobulin-like region which resembles the C- region of an immunoglobulin
3. A transmembrane region is hydrophobic and has 25 amino acids.
4. An intracellular region at the C-terminal which is hydrophilic is of variable length and anchors the polypeptide chain along with the transmembrane region.

The upper region of both α_1 and β_1 chains are helical whereas the lower part is made up of 8 beta-pleated sheets each supporting α_1 and β_1 chains respectively Figure 7.8. The helical portion is connected to the β-sheets by thin β-peptide strands. There is great variability in the amino acid sequence of both chains. This polymorphism is present in the region of the alpha-I and beta-I domains of both chains and is known as the hypervariable region. This is what forms the antigen binding site or the peptide binding cleft. This cleft is formed by, both the α_1 and β_1 chains, which lie very close to each other but are not bound together. The walls of the cleft are made by the α-helices and the bottom platform is made by the β-pleats of the two polypeptide chains. As these regions are not bound together the cleft is not closed like in Class I MHC but is open and can accommodate larger peptides.

Function: The Class II MHC molecule is tissue-restricted and is present only on the APCs such as macrophages, dendritic cells, activated T-cells, sperm cells, keratinocytes, epithelial cells, Langerhans cells and endothelial cells of blood vessels as well as myeloid stem cells. Exogenous antigens processed by the endocytic pathway are bound in the peptide cleft and these can extend beyond the cleft as it is open on both sides.

FIGURE 7.8 The structure of Class II MHC molecule containing the antigenic peptide held in the groove containing eight antiparallel beta pleated sheets connected through thin β -peptide strands to the α_1 and β_1 helical chains. The interaction between the antigenic peptide (in blue) and MHC molecule involves hydrogen bonds and van der Waals forces.

The antigens could be bacterial, viral, fungal or even self-peptides which are presented to the CD4 T_H cells which in turn activate B-cells to produce antibodies. This presentation of the exogenous peptide by Class II MHC is crucial for the activation of T_H-cells and in turn activation of adaptive immunity. As the MHC II molecules bind very stably to the antigenic peptide, T_H cell recognition is ensured leading to an antigen-specific immune response.

7.7.3 Class III MHC Molecule

The Class III MHC molecules are diverse and are not membrane-bound and are secreted proteins. They also include the functionally similar molecules responsible for inflammatory response known as MHC Class IV. These are

a. 21-hydroxylase (21 OHA) which is the enzyme involved in the glucocorticoid biosynthesis
b. Complement components mainly Factor B, C4a, C4b and C2
c. Cytokines such as
   1. Tumor necrosis factor α-TNFα
   2. Lymphotoxin α
      1. Heat shock proteins -HSP70

Function: The Class III MHC molecules play a role in immune function, inflammatory response, and auto-immune diseases.

7.8 MHC AND ANTIGEN PRESENTATION

The main function of MHC molecules is to present the antigen peptides to either the CD4 or CD8 T cells. The antigenic peptides that are presented, are either endogenously or exogenously processed with Class I MHC presenting the antigens of endogenous origin and Class II MHC the ones of exogenous origin.

7.8.1 The Endogenous Pathway of Antigen Presentation

The class I MHC glycoprotein molecules bind with intracellular antigenic peptides. The endocellular proteins from foreign sources such as viruses or tumour cells are processed into smaller peptides and these proteins are tagged with ubiquitin. The degradation process is done by passing these ubiquitin-protein conjugates in the cytoplasm through a multifunctional protease complex known as the proteasome. Each proteasome is a cylindrical large molecule consisting of four rings of subunits and central channel of 1–5 nm in diameter. Degradation of the ubiquitin-tagged foreign protein takes place in the hollow of the proteasome. The resulting smaller peptides are translocated into the rough endoplasmic reticulum (RER) by a process that utilises ATP. The loading mechanism requires a peptide loading complex which is made up of many components such as the transporter associated with antigen processing (TAP1 and TAP2), tapaisin, the oxidoreductase enzyme ERp57 and chaperone protein calreticulin. All these together are known as the antigen processing machinery (APM).

The TAP1 and TAP2 are ATP binding cassette proteins and have an ATP binding domain that projects into the cytosol and domains which project into the lumen of the rough endoplasmic reticulum. The TAP proteins have a high affinity for peptides the size of 8–10 amino acids which interact with the alpha-3 domain of the MHC in the cisternae of the ER. The TAP proteins have greater affinity to peptides with hydrophobic or carboxy-terminal amino acids which are also the preferred anchor residue for Class I MHC molecules. The peptide binding cleft can hold up to 8–10 amino acid long peptides but some are longer and a bulge is formed to accommodate them. The assembly and folding of the polypeptides of Class I MHC are facilitated by chaperone protein molecules. The α and β chains of Class I MHC are synthesised on the RER and are then assembled into a stable molecule. The Class I MHC molecule can exit the RER only after the binding of the processed antigenic peptide to its cleft. The chaperone protein calnexin, found in the membrane of the RER, associates with the α chain of Class I MHC and stabilizes its folding. This association breaks when the β microglobulin chain binds to the α chain. The Class I MHC molecule is now associated with calreticulin, also a chaperone protein and tapasin. The tapasin helps in the association of the TAP transporter with Class I MHC molecule along with an additional chaperone protein ERp57 and the resulting complex facilitates the binding of the antigenic peptide to the cleft. After the peptide binds to the cleft, the Class I MHC stabilizes and dissociates from calreticulin, tapasin and ERp57. It is then moved out of the RER into the Golgi and then presented on the surface of the cell membrane.

The CD8 T cell receptor recognizes the peptide presented on the Class I MHC and subsequently the T cell induces the destruction of the infected cell. This is the mechanism that the immune system uses to fight off organisms which have infected the cells (Figure 7.9).

7.8.2 The Exogenous Pathway of Antigen Presentation

The Class II MHC presents antigenic peptides to the CD4 T cells, with these peptides being exogenously derived. APCs are phagocytic in nature such as macrophages or dendritic cells. These ingest foreign material like invading organisms and ingest them by engulfing through phagocytosis. The phagosomes fuse with the lysosomes and the foreign proteins are broken down into peptides in this acidic environment. The internalised antigen proteins go through three different increasingly acidic environment and finally reach the lysosomes which have acid-dependent

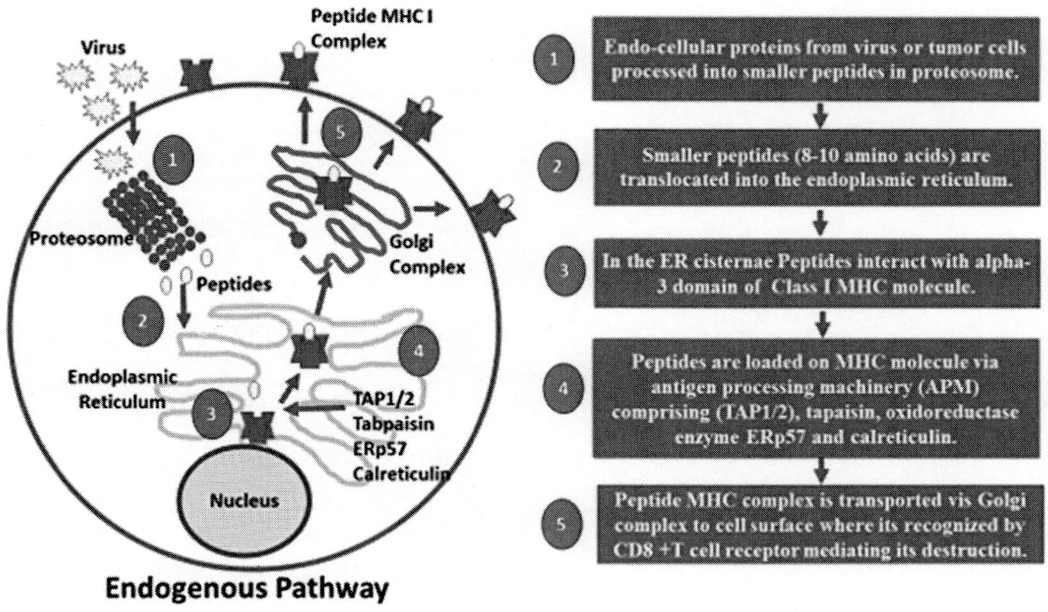

FIGURE 7.9 Endogenous pathway with the generation and processing of the antigenic peptide and its loading on MHC-I molecule.

hydrolases such as proteases and glycosides. The antigenic peptide is broken down into oligopeptides of 13–18 amino acid residues.

The Class II MHC molecules are transported into the ER after translation. The peptide binding cleft of the α and β chains is bound to a trimeric polypeptide known as the invariant chain and this complex is transferred to the endosomes via Golgi bodies. This invariant chain does not allow the binding of endogenously derived peptides to the peptide cleft but once this complex is inside the endosomes the invariant chain is digested by a protease in the acidic environment. Only a small peptide known as the Class II-associated invariant chain peptide (CLIP) is left which in turn blocks the binding of self-peptides to the cleft. Next, the CLIP is exchanged at peptide cleft with the exogenous antigenic proteins produced after phagocytosis in endosomes. The exchange of CLIP with the antigenic peptide is catalysed by HLA-DM molecule in the presence of HLA-DO. These loaded MHC molecules are then transported to the surface of the cells where they present the antigens to the T cell receptors (TCR) on the CD4 T_H cells (Figure 7.10).

7.9 HLA TYPING

Allografts or tissue transplant of the same species can either be accepted or rejected by the recipient. As each individual has a particular set of HLA molecules encoded by its specific set of genes, the immune system has learnt to recognize these as self. Cells which do not have these specific HLA molecules are identified as non-self and the immune system reacts to them as foreign. The HLA molecules act as antigens when tissue is transplanted into another individual and a transplant is successful if only there is compatibility between the donor and recipient tissue. Non-compatibility results in tissue rejection therefore, the identification of the HLA antigens of both donor and recipients is imperative and this matching or identification is known as HLA typing. The matching process tests and compares the HLA molecules of the donor and the recipient, identifying compatibility between the two, and increasing the chances of the success of the transplant. Genetic similarity helps the recipient's immune system tolerate transplanted tissue. The presence of incompatible

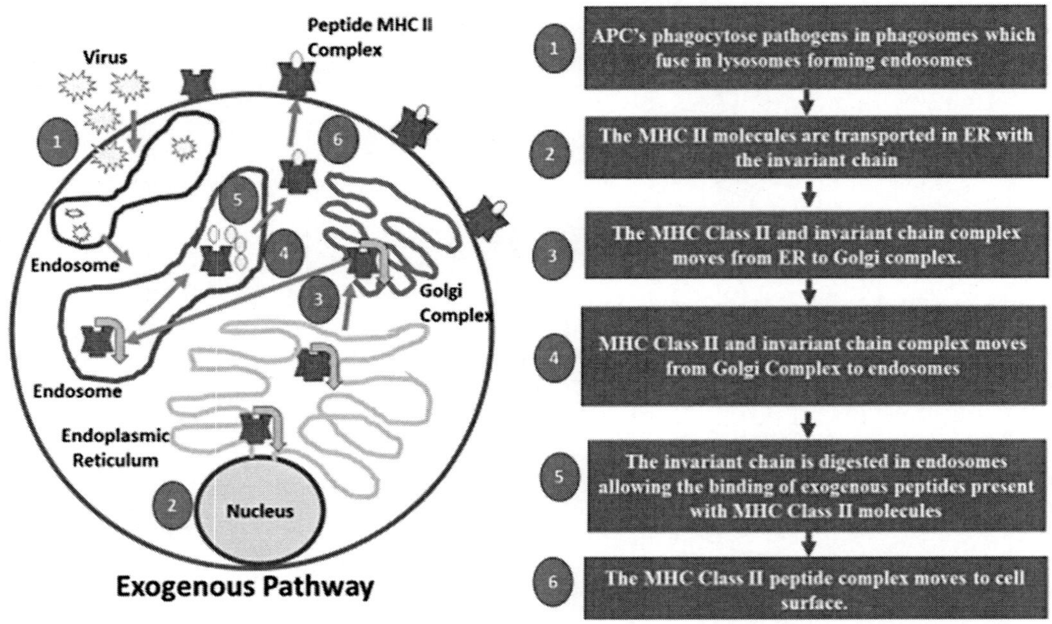

FIGURE 7.10 Exogenous pathway with the uptake of foreign protein and its processing into antigenic peptide followed by the loading of MHC-II molecule.

donor tissue leads to a cell-mediated immune response and tissue rejection can take place by either of two pathways, the direct or the indirect pathway.

The direct pathway is by which any nucleated cell or APC activates either the CD8 or the CD4 T cells respectively. The graft cell bears the MHC I molecules on its surface which are recognized as foreign by the recipient CD8 T_C cells and killed. Similarly, the donor APC of the graft tissue also carry the MHC II molecule which elicits a response in the CD4 T_H cells. These produce the cytokine IFN-γ which in turn activates macrophages to engulf and destroy the grafted cells.

In the indirect pathway, the donor MHC II molecules activate the recipient CD8 T_H cells which in turn activate the antibodies on the recipient B lymphocytes to recognize and bind to the antigens. The naïve B lymphocytes then will change into plasma cells which will generate a very large number of specific antibodies. These antibodies will bind to the MHC of the donor cells and signal the macrophages to destroy the antigen-bearing donor cells.

Both these pathways lead to tissue rejection and as MHC are very polymorphic there is a very big chance of the donor tissue to be recognized as foreign and consequently rejected. A very small pool of donors can have matching compatibility with the recipient apart from identical twins. Histocompatibility is the degree of similarity between the HLA donor genes and recipient genes. The chances of donor allograft being accepted increases manifold if HLA matching is done before the transplant is commenced.

A set of six markers have been identified out of which at least four markers have to match before a transplant can be planned. The best donors are siblings as they inherit one-half of the HLA haplotype from each parent and have one-fourth the probability of matching with each other. Donors can also be parents and children for achieving a better chance of survival of transplants.

HLA typing is done by the following methods:

7.9.1 Serological Method

Serological method is the conventional complement-dependent method and is done by the cytotoxicity assay. It is also known as the micro-cytotoxicity assay and was described for the first time by Gorer and Amos in mouse cells, with later improvements by Terazaki in humans. Unknown lymphocytes are exposed to antisera of known HLA haplotypes. Class I HLA antigens are tested using T lymphocytes and Class II HLA antigens by using B lymphocytes. These lymphocytes are incubated with anti-HLA antibodies and the complement in micro titre plates, with the incubation time being different for Class I and Class II HLA. If the antibodies bind to the HLA antigens the complement is activated, the lymphocytes are killed and can be visualized by using vital stains like Trypan Blue as their cell membranes are destroyed. On analyzing the wells, the HLA types can be determined. Antibodies of the recipient can also be screened by using solid phase ELISA and adsorbing HLA antigens on the walls and adding the recipient's serum. After washing, the walls are treated with anti-gamma globulin and a coloured substrate for detection. The appearance of colour indicates that the recipient has antibodies against the donor HLA leading to rejection.

7.9.2 Cellular Assays

The lymphocytes of two individuals either proliferate or become cytotoxic when cultured together according to the response of the HLA II (DR) antigens. The two types of cellular assays used are:

a. **Mixed leukocyte reaction (MLR):** It is carried out between the irradiated or mitomycin C treated lymphocytes of the recipient and CD4 cells of the donor. The treated recipient cells could be B cells, macrophages or dendritic cells and they behave as stimulator cells. The donor cells respond to the HLA II antigens present on these and can respond by synthesizing DNA and proliferating. The proliferative response is measured by the uptake of^3 H labelled thymidine. This uptake indicates the donor cells recognize the recipient as foreign and the proliferative response can be studied to assess histocompatibility.
b. **Cell-mediated lympholysis (CML):** It is seen when the donor responder cells give rise to cytotoxic cells when cultured together with the recipient stimulator cells. The HLA II of the stimulator cells stimulates the cytotoxic cells which can identify the HLA I antigens on the stimulator cells and then kill them. The CML is a variation of the MLR and its response also indicates the status of histocompatibility and the success of the transplant.

7.9.3 Flow Cytometry

Flow cytometry is a rapid method taking about less than an hour to indicate compatibility. Monoclonal antibodies against specific alleles are labelled with fluorescent probes and are added to freshly nucleated leukocytes that need to be tested. All the nucleated leukocytes which have antigens on their surfaces bind to the antibodies and can be detected as they emit fluorescence. This fluorescence can be measured and is used to assess histocompatibility.

7.9.4 Genetic or Molecular Methods

As the nature of the polymorphic alleles of the HLA locus became more apparent HLA typing is being done using molecular methods. DNA-based tests are more accurate and can pick up the presence of very low-level antigens as these tests analyze the genes and not the gene products increasing the success of transplantation. The serological methods were not always adequate, especially in cases where anti-sera were not available. The molecular methods are based on the sequence analysis protocols and are highly sensitive tests. Some of the methods employed are:

i. Sequence-specific primer amplification PCR (PCR-SSP)
ii. Sequence-specific oligonucleotide probe (PCR-SSO) and can be both dot blot or reverse dot blot method.
iii. Sequence-based typing methods (SBT)
iv. Restriction fragment length polymorphism (RFLP)
v. Microarrays
vi. Single stranded sequence conformation polymorphism (PCR-SSCP)

The molecular methods are rapidly replacing the serological methods due to the many advantages they present such as accuracy, time taken and the success of the transplant.

7.10 CLINICAL IMPORTANCE OF MHC

Apart from governing histocompatibility which becomes of utmost importance in the event of a transplant, the MHC molecules have other significant applications of clinical importance. Some of them are:

7.10.1 Susceptibility to Disease

Some specific HLA haplotypes are associated with certain diseases such as auto-immune diseases, immune deficiency diseases as well as susceptibility to infectious diseases, cancer, and neurological and metabolic diseases. This relationship between the HLA haplotypes and diseases can be explained by the similarities between antigens and the self-MHC molecules resulting in a weakened response by the Class II MHC. This mimicking of the self-peptides ensures that T cells are not activated. The immune response to different antigens also depends on the variability in the presentation of the antigens by the HLA II.

i. Celiac disease is associated with the presence of DR3 marker and patterns show intolerance to gluten. Intestinal damage takes place and leads to abdominal bloating, chronic diarrhea and abdominal pain. A gluten-free diet is the only relief from these symptoms.
ii. Ankylosing spondylitis occurs when the vertebral column shows inflammation and is a type of arthritis. The patient has the haplotype B27 and the disease peaks at around the age of 20 years of the individual.
iii. Reiter's disease is associated with the marker HLA B27 and is an auto-immune disorder affecting the genital mucosa. These patients show receptors for pathogens which cause diseases.
iv. Uveitis, psoriasis, Addison's disease, multiple sclerosis, insulin-dependent diabetes, Hashimoto's disease are some of the other auto-immune diseases caused by the presence of certain HLA haplotypes.

On the other hand, it is also known that certain HLA types provide protection from certain diseases such as the presence of HLA DR2 protects from insulin-dependent diabetes mellitus.

7.10.2 Immune Responsiveness

The Class II MHC plays a major role in immune responsiveness but different haplotypes show variability in this responsiveness. This variability directs the path of progress of infectious pathogens and two models propose to explain this variability.

i. Determinant-selection model proposes that it is the polymorphism which determines the specificity of the antigen determinants that will bind and be presented to the T cells. Any antigenic peptide which has no determinant which can bind to the MHC will not bring about an immune response. Thus, it is the peptide cleft region which allows an antigenic peptide of a matching determinant sequence to bind and then gets to trigger the T cells and as a result generate an immune response.

ii. "Hole in the T-cell" repertoire model states that functional T cells which can get triggered are absent and therefore the MHC may bind the antigen but no immune response takes place. The T cells are missing in the thymus due to clonal selection and processing. The MHC molecule may be present with the correct determinant which binds to the antigenic peptide but due to the lack of the T cell the immune response is not elicited.

Some more applications of the MHC are:

1. **Paternity testing:** Paternity or its absence can be established by matching two haplotypes. If haplotypes do not match then the two individuals do not share the same family lineage.
2. **Human evolutionary studies:** Human origins and migration patterns can be studied and relationships between human populations can be understood.

7.11 MINOR HISTOCOMPATIBILITY ANTIGENS

Minor histocompatibility antigens are proteins other than the MHC which can act as alloantigens and elicit a response from the immune system. These have implications in transplants especially of hematopoietic cell transplant (HCT) although the effects are more subdued than those seen by the MHC. These minor H antigens were first found in mice and later identified in humans and are derivatives of endogenous proteins which bind to the MHC. These are normal proteins which are polymorphic and these change in the amino acids leads to tissue rejection even if the MHC match between non-identical twin siblings.

7.11.1 Genetic Organization

The self-peptides of the minor H antigens can act as alloantigens because of the variability present in the genes. As many as 30 loci with more than 500 genes are located on the autosomes, the Y chromosome and the mitochondrial genome which encode for H antigens. The most studied antigens are HA-1, HA-2 which are found on the autosomes and HY-antigens which are found on the Y chromosomes

7.11.2 Graft Rejection: Role of H Antigen

The H antigens are found on the surface of cells including haemopoietic and epithelial cells where they can be recognized by the T cells of donor tissue, especially after HCT. Once these are recognized a graft versus host disease (GVHD) or graft versus leukaemia (GVL) develops. In the case of GVHD the transplant is not successful and tissue rejection takes place. The GVL is utilised to kill leukaemia cells and used as a therapy to prevent a relapse. The H antigens of the leukemia cells act as triggers for the donor T cells once the transplant is established and they are then subsequently not allowed to survive.

The HY antigens are instrumental in the rejection of tissue transplanted from males to females. The male Y chromosome has loci which codes for HY antigens and expressed on the cells of the male donor tissue. The female MHC recognize these male H antigens as foreign and an immune response ensues that targets the donor tissue.

7.11.3 Immunodominance

Certain H antigens are able to elicit a stronger response from the T cells as compared to other H antigens even when present on the same cell. This is known as immunodominance and can be

TABLE 7.1
Location of Minor Antigen Genes on Autosomes and Y Chromosome

Examples of mHA	Chromosome
HA-1	19
HA-2	6
HA-8	9
HB-1	5
HY-A1	Y
HY-A2	Y
HY-B7	Y
HY-B8	Y
HY-B60	Y
HY-DQ5	Y

Source: http://medweb4.unige.ch/immunologie.

due to the presence of a larger number of T cells responding to the dominant H antigen, or could be due to a more stable complex being formed between the MHC and the dominant peptide. The number of H antigens could be higher thus further stabilizing this complex of MHC-peptide and T cells (Table 7.1).

7.12 SUMMARY

- The mammals possess both the innate immunity and adaptive immunity. The adaptive immunity consists of humoral immunity which depends on antibody response and cell-mediated cytotoxicity which is produced by the T-cells and regulated by the major histocompatibility complex (MHC).
- The MHC is actually a tightly linked cluster of genes encoding for a set of glycoproteins that are expressed on cell surface and behave as antigens upon introduction into the body of another organism.
- The main function of MHC molecules are antigen presentation and lymphocyte recognition, therefore they help in distinguishing between self and non-self-molecules or cells.
- The MHC gene cluster belong to the immunoglobulin super gene family comprising of three classes called as MHC Class I, MHC Class II and MHC Class III molecules.
- In humans the MHC molecules are called as human leukocyte antigens- HLA Class I, Class II and Class III and the gene cluster is known as HLA complex.
- The MHC genes are responsible for rejection of transplanted tissue and were first recognized by Peter A. Gorer who characterized the MHC complex in mouse in 1936.
- The MHC gene cluster is known as the HLA system in humans and as H2 system in mice. This cluster of genes is found on the short-arm of human chromosome 6 and on chromosome 17 in mouse.
- Each class of MHC genes are polygenic and co-dominantly expressed on the surface of the cells which are tightly linked showing absence of recombination.
- Transcriptional activators and several cytokines at the transcriptional level regulate the differential gene expression of both MHC Class I and Class II molecules in different cell types and during different stages of cell differentiation.
- MHC Class I genes encode for glycoproteins found on the surface of all nucleated cells and which present antigens to Tc cells.

- The Class I MHC molecule comprises two dissimilar polypeptide chains. A glycosylated α-chain 350 amino acids long, having a molecular weight of 45 kDa, associates non-covalently with a small 12 kDa β2-micro-globulin molecule having 90 amino acids to form the functional molecule.
- MHC Class II genes encode for glycoproteins which are expressed on the surface of specialized APCs such as macrophages, dendritic cells, B cells. These APCs then process the antigens for presenting to TH cells.
- The Class II MHC molecule are dimers comprising of two glycosylated polypeptide chains known as the α and β chains having 230 and 240 amino acids and a molecular weight of 32–34 and 29–32 kDa respectively. Their structure resembles the structure of α chain of the Class I MHC.
- MHC Class III genes code for proteins that are structurally diverse, that are not membrane-bound and are secreted. Examples include proteins that are components of the complement system, soluble serum proteins and tumor necrosis factors (TNF).
- Allografts are either accepted or rejected by the recipient based on his/her inheritance of specific set of HLA molecules which the immune system recognize as self. Cells lacking these HLA molecules are recognized as non-self thereby eliciting the immune system against them.
- The Class I MHC molecules present antigens of endogenous origin derived from the processing of proteins from viruses or tumor cells through the endogenous pathway and present them to the T cell receptors of CD8 T cells.
- The Class II MHC molecules present antigens of exogenous origin processed through the exogenous pathway and present them to the T cell receptors of CD4 T cells.
- The HLA molecules act as antigens therefore, before transplantation of cells/tissue or organs HLA typing is done to identify the HLA antigens of both donor and recipients to prevent rejection of the transplant.
- HLA Typing is done using the serological method, the cellular assay methods, flow cytometry and the genetic or molecular methods.
- Research has shown the association of certain HLA haplotypes with some auto immune diseases such as Addison's disease, multiple sclerosis, insulin-dependent diabetes, Hashimoto's disease, and immune deficiency diseases. Also, certain HLA haplotypes makes an individual susceptible to infectious diseases, cancer, neurological and metabolic diseases.
- The presence of different haplotypes determines the variability in response of Class II MHC molecules. This has been explained through the two models namely the Determinant-selection model and the Hole in the T-cell repertoire model.
- Minor histocompatibility antigens are basically proteins but different from the MHC acting like allo-antigens and elicit a response from the immune system having implications in transplants.
- The H antigens are present on the surface of several cell types such as the haemopoietic and epithelial cells which upon recognition by the T cells of donor tissue can lead to graft versus host disease (GVHD) or graft versus leukaemia (GVL).
- Some H antigens elicit a stronger response from the T cells in comparison to other H antigens present on the same cell a phenomenon called as immunodominance. The reason can be greater number of T cells responding to the dominant H antigen, or stabler MHC and the dominant peptide complex formation.

QUESTIONS

1. Expand the following terms:
   a. MHC
   b. HLA
   c. PCR
   d. PCR-SSP

  e. PCR-SSO
  f. SBT
  g. RFLP
  h. PCR-SSCP
  i. APC
  j. TCR
  k. GVHD
  l. GVL
2. Mention the scientific contribution made by the following scientists:
  a. Peter A. Gorer
  b. George Snell
  c. Jean Daussct
  d. Peter C. Doherty
  e. Rolf M. Zinkernagel
3. Define the following terms:
  a. Allograft
  b. Transplantation
  c. Graft verses host disease
  d. Graf verses leukaemia
  e. Antigen
  f. Major histocompatibility antigens
  g. Immunodominance
4. Differentiate between the following terms:
  a. Class I MHC molecules and Class II MHC molecules
  b. Allograft and Autograft
  c. Host verses graft disease and Graft verses leukaemia
  d. Endogenous pathway and Exogenous pathway of antigen processing
5. Long answer questions:
  a. Describe the structure and function of Class I MHC molecules.
  b. Describe the structure and function of Class II MHC molecules.
  c. Describe HLA typing and the different methods by which it is carried out.
  d. Describe the endogenous pathway for antigen processing and presentation to Class I MHC molecules.
  e. Describe the exogenous pathway for antigen processing and presentation to Class II MHC molecules.
6. Write short notes on the following topics:
  a. Genetic polymorphism of MHC genes
  b. Clinical importance of MHC genes
  c. Methods of HLA typing
  d. Genetic organisation of Human Leukocyte Antigens (HLA) Complex
  e. Clinical importance of MHC molecules

BIBLIOGRAPHY

1. Thorsby E. A short history of HLA. *Tissue Antigens*. 2009. doi:10.1111/j.1399-0039.2009.01291.
2. Kindt, Thomas J., Osborne, Barbara A., Goldsby, Richard A. *Kuby Immunology*. 6th Edition. 2005. W. H. Freeman and Company, New York.
3. Taylor M, Hussain A, Urayama K, Chokkalingam A, Thompson P, Trachtenberg E, Buffler P. The human major histocompatibility complex and childhood leukemia: an etiological hypothesis based on molecular mimicry. *Blood Cells Mol Dis*. 2009;42:129–35.
4. https://www.immunopaedia.org.za/immunology/basics/4-mhc-antigen-presentation/

5. Hylenius S, Nybo Andersen AM, Melbye M, Hviid TV. Association between HLA-G genotype and risk of pre-eclampsia: a case-control study using family triads. *Mol Hum Reprod.* 2004;10(4):237–46.
6. https://www.slideshare.net/indiandentalacademy/topic-4-mhc
7. Kolb HJ. Graft-versus-leukemia effects of transplantation and donor lymphocytes. *Blood.* 2008;112(12):4371–83.
8. Ferrara JL, Levine JE, Reddy P, Holler E. Graft-versus-host disease. *The Lancet.* 2009;373(9674): 1550–61.
9. Wieczorek M, Abualrous ET, Sticht J, Álvaro-Benito M, Stolzenberg S, Noé F, Freund C. Major histocompatibility complex (MHC) Class I and MHC Class II proteins: conformational plasticity in antigen presentation. *Front Immunol.* 2017;8:292.
10. https://en.wikipedia.org/wiki/Major_histocompatibility_complex
11. Penn DJ, Ilmonen P. Major histocompatibility complex (MHC). In *Encyclopedia of Life Sciences.* 2005. doi: 10.1038/NPG.ELS.0003986.
12. Spierings E. Minor histocompatibility antigens: past, present, and future. *Tissue Antigens.* 2014;84(4):374–60.
13. Summers C, Sheth VS, Bleakley M. Minor histocompatibility antigen-specific T cells. *Front Pediatr.* 2020. doi: 10.3389/fped.2020.00284.
14. Thayer WP, Kraft JR, Tompkins SM, Moore IV JC, Jensen PE. Assessment of the role of determinant selection in genetic control of the immune response to insulin in H-2b mice. *J Immunol.* 1999;163(5):2549–2554.
15. Falkenburg JH, van de Corput L, Marijt EW, Willemze R. Minor histocompatibility antigens as target of graft-vs-leukemia reactions. *Curr Opin Hematol.* 2002;9:497–502.
16. Little AM, Marsh SG, Madrigal JA. Current methodologies of human leukocyte antigen typing utilized for bone marrow donor selection. *Curr Opin Hematol.* 1998;5:419–28.
17. Erlich HA, Opelz G, Hansen J. HLA DNA typing and transplantation. *Immunity.* 2001;14(4):347–56.
18. http://medweb4.unige.ch/immunologie/home/HSC/transplantation_barriers/structures_recognized/minor_histocompatibility_antigens/#
19. Gao GF, Tormo J, Gerth UC, Wyer JR, McMichael AJ, Stuart DI, Bell JI, Jones EY, Jakobsen BK. Crystal structure of the complex between human CD8alpha(alpha) and HLA-A2. *Nature.* 1997;387(6633):630–4. doi: 10.1038/42523.
20. Jiang W, Birtley JR, Hung SC, Wang W, Chiou SH, Macaubas C, Kornum B, Tian L, Huang H, Adler L, Weaver G, Lu L, Ilstad-Minnihan A, Somasundaram S, Ayyangar S, Davis MM, Stern LJ, Mellins ED. In vivo clonal expansion and phenotypes of hypocretin-specific CD4+ T cells in narcolepsy patients and controls. *Nat Commun.* 2019;10(1):5247. doi: 10.1038/s41467-019-13234-x.

8 Cytokines

Neha Niharika, Sunita K. Yadav, and Jyoti Singh
Daulat Ram College, University of Delhi

CONTENTS

8.1 INTRODUCTION

An effective immune response generation and coordination require a mechanism by which various types of immune cells can communicate with each other. This task is carried out by cytokines. The cytokines are a diverse family of peptides or glycoproteins (generally smaller than 30 kDa) that act as immunomodulation agents. Cytokines cannot cross the lipid bilayer membrane of cell to enter the cytoplasm, hence they act through cell surface receptors. The binding of the cytokine to the receptor triggers an intracellular signaling cascade that results in various gene expression in the target cell and the expression of genes lead to the biological effect of the cytokines. The effects of cytokines can be differentiation, proliferation or activation of other cells. Cytokines can influence innate as well as adaptive immune response and may stimulate the action in their own cell, in a nearby cell or in a distant cell depending on the requirement of the body at the time. Cytokines are classified based on their structure and function into different types. The important role of cytokines results in their therapeutic uses and recombinant cytokines are being made by using recombinant DNA technology for the benefit of the human race.

DOI: 10.1201/9781003286424-8

8.2 THEORY

Cytokines are a group of proteins (~5–20 kDa) that mediate an effective immune response by acting as a messenger between cells. Figure 8.1 shows that they are secreted by WBC and various other cells in the body like macrophages, B and T-lymphocytes, mast cells as well as endothelial cells, fibroblasts and various stromal cells. A given type of cytokine can be produced by more than one type of cell. The first cytokine called Interferon was discovered by Alick Issacs and Jean Lindemann in 1957.

Cytokines can be formed by different types of cells such as leucocytes that regulate immunity, inflammation and haematopoiesis. They can be classified based on their secretions into lymphokines, proinflammatory cytokines, anti-inflammatory cytokines, chemokines and_growth factors (see Table 8.1).

8.3 PROPERTIES OF CYTOKINE

Several properties have been attributed to cytokines like:

a. Pleiotropy
b. Redundancy
c. Synergy
d. Antagonism
e. Cascade Induction

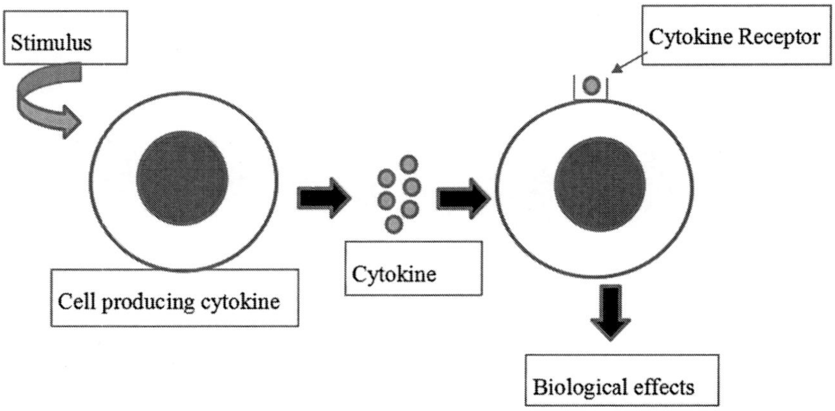

FIGURE 8.1 Overview of formation and function of cytokines.

TABLE 8.1
Different Cell Type and their Function

Cell Type	Function
1. Lymphokines	Regulate the immune response
2. Proinflammatory cytokines	Amplify and perpetuate the inflammatory process
3. Anti-inflammatory cytokines	Negatively modulate the inflammatory process
4. Chemokines	Chemotactic for inflammatory cells
5. Growth factors	Enhances cell survival and result in structural changes

a. **Pleiotropy:** A given cytokine performs a different effect on different target cells. eg: IL-4 (Interleukin-4) leads to activation, proliferation and differentiation of B-cell, proliferation of Thymocytes and Mast cells.
b. **Redundancy:** When two or more cytokines perform similar functions they are said to be redundant.eg: IL-2, IL-4 and IL-5 cause proliferation of B-cell.
c. **Synergism:** When the effect of two cytokines when combined shows greater activity than the effect of two cytokines. For example, IL-4 and IL-5 induce B-cell to class switch to IgE.
d. **Antagonism:** When the effect of one cytokine inhibits the activity of another cytokine. For example, IL-4 induces B-cell to class switch to IgE, whereas IFN-γ blocks class switch.
e. **Cascade induction:** When one cytokine induces the target cell to produce one or more cytokine. For example, the activated T_H cell produces IFN-γ which in turn induces macrophages to produce IL-12 which again activates T_H cell to produce a wide variety of cytokines like TNF, IL-2.
f. Initially, the lymphokine term was used for those cytokines that were secreted by lymphocytes, monokine for those which were secreted by macrophages and monocytes.

8.4 CYTOKINES AND HORMONES

8.4.1 Similarities

Cytokines have some properties similar to hormones and growth factors. Both cytokines and hormones are secreted factors. They affect cells at picomolar concentrations.

Differences between cytokines and hormones are shown in Table 8.2.

8.5 CLASSIFICATION OF CYTOKINES

In general, the main functionality of cytokines is to act as intercellular messenger molecules responsible for inducing particular biological activities post binding to a cytokine receptor on a cell. The two major producers of cytokines are T-helper cells and macrophages. The release of cytokines results in a vast array of actions ranging from evoking cytotoxic and humoral immune responses to hematopoiesis regulation and proliferation and differentiation of cells. Cytokines are antigen-nonspecific, which means that they are produced in response to antigen and bind to any cell non-specifically if the specific cytokine receptor is available for binding. The specificity of the immune system is achieved by the selection of the particular cytokine receptors that are expressed on the antigen-responsive cells. In addition, cytokines are produced by cells residing in the vicinity of the target cell after both cells interact with each other.

TABLE 8.2
Differences between Cytokines and Hormones

Cytokines	Hormones
They are produced by variety of cells	They are produced by specialized glands (endocrine glands)
Act over short distance in autocrine or paracrine fashion	Act in autocrine, paracrine and endocrine fashion.
Secretion is carefully regulated	Production is constitutive
Secretion is short lived (few hours to few days)	A hormone's half-life and duration of activity are limited and vary from hormone to hormone. For instance, the biological half-life of luteinizing hormone is_20 minutes, follicle-stimulating hormone (3–4 hours), and of human chorionic gonadotropin (24 hours).

8.5.1 Classification of Cytokines Based on Contribution in Innate Immunity

Cytokines can be classified based on their contribution in innate immunity or adaptive immunity:

1. **Innate immunity cytokines**
   These cytokines are secreted by cells that have a role in non-specific immunity and thereby act as the first line of defense against antigens. The various types of cytokines in this category are:
   - **Interleukin 1 (IL-1):** IL-1s are secreted by four different types of cells: Monocytes, macrophages, endothelial and epithelial. These molecules act on blood vessels and cause inflammation, bring fever by acting on hypothalamus cells and induce the generation of acute phase proteins by binding to receptors located on liver cells.
   - **Interleukin 6 (IL-6):** It is secreted by macrophages and endothelial cells and induces secretion of acute phase proteins by the liver (e.g. IL-1) and also induces T_H2 immune response (humoral) that produces and proliferates specific antibody secretion by B-cells.
   - **Interleukin 12 (IL-12):** Generally secreted by macrophages and dendritic cells and act on natural-killer cells. It also stimulates T_H1 (cytotoxic) type of adaptive immune response.
   - **Interferon α (IFN-α):** These are a group of molecules placed under IFN-α and secreted by macrophages. This class of molecules activates NK cells and increases the production of MHC class 1 expression. It also prevents virus proliferation in nucleated cells.
   - **Interferon β (IFN-β):** It is secreted by fibroblasts, and its functions are similar to those of IFN-α.
   - **Tumor necrosis factor-α (TNF-α):** Pro-inflammatory cytokines secreted by macrophages and having a role in systemic inflammation and the secretion of the liver's acute phase proteins. In addition, they are responsible for inducing cachexia body fat loss and necrosis (cell death) of cells and activating neutrophils. They also play an active role in combating viral proliferation.
2. **Adaptive immunity cytokines**
   - **Interleukin 2 (IL-2):** Induces proliferation of T cells and B cells and is secreted by T-Cells. Additionally, it promotes AICD. It also activates and proliferates NK cells.
   - **Interleukin 4 (IL-4):** It is produced by TH2 (humoral immunity cells) and mast cells. It helps in TH2 differentiation and immunoglobulin switching to IgE.
   - **Interleukin 5 (IL-5):** It is produced by TH2 cells and promotes proliferation and activation of eosinophils.
   - **Interleukin 25 (IL-25):** It helps in skewing the antibody response to TH2 (humoral) immunity by producing TH2-specific cytokines.
   - **Interferon γ (IFN-γ):** TH1 type of cells, CD8+ and NK cells secrete IFN-γ. It increases the expression of MHC-I and MHC-II by activating macrophages. Additionally, increased expression of MHC enhances both types of antigen presentations.
   - **Transforming growth factor β (TGF-β):** It is secreted by various types of cells, e.g., T cells and macrophages. It inhibits T-cell, B-cell and macrophage proliferation. Like IL-4, it promotes isotype switching to IgE.

8.5.2 Classification of Cytokines Based on Structure

Cytokines are also classified based on their structure. The structure of cytokines is evaluated by cloning and purifying them for X-ray crystallography and other techniques. Cytokines represent a diverse family of proteins or glycoproteins, usually less than 30 kDa. The structural cytokines characterized so far belong to five groups:

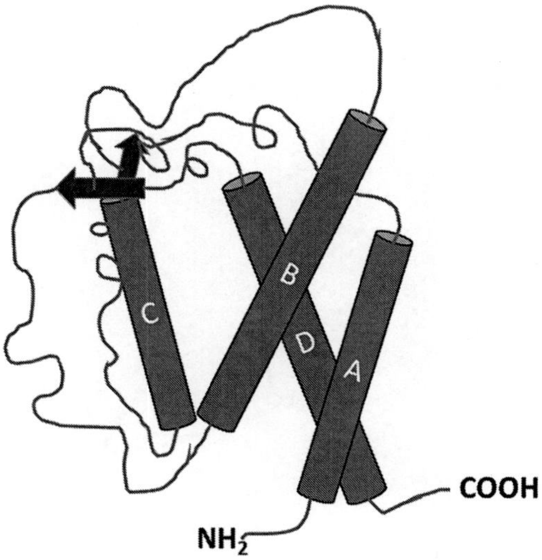

FIGURE 8.2 Structure of IL-4 cytokine of hematopoietin family.

1. **The hematopoietin family:** All members of this family possess a high degree of α-helical structure in spite of considerable differences in the amino acid sequences. β-sheets, if present, are very few. The cytokines such as IL6, IL11, IL12A, and IL23A are long-chain class I helical cytokines while other cytokines such as IL2, IL3, IL4, IL5, IL7, IL9, IL13, IL15, and IL21 are short-chain helical cytokines (Figure 8.2). The IL-2 and IL-4 share four parallel α-helical regions that are attached by loops.
2. **The interferon (IFN) family:** Interferon family consists of several interferon molecules which are slightly different from each other in structure and function. As shown in Figure 8.3 among various interferons, interferon-α has seven alpha helices and several loop regions with two disulfide bonds: one between the N-terminus and Helix E, and the other between Loop AB and Helix G. The helices A, C, and F run parallel to each other and anti-parallel to B, E, and G, however the latter run parallel to each other. Helix D runs at an angle of 45°–90° to them. Although interferon-β and interferon-α have 31% sequence homology (Figure 8.3), the former is a relatively simple biological response modifier, consisting of only five alpha helices, in addition to having multiple interconnecting loop regions. Helices A, B, and D run parallel to each other, and helices C and E run anti-parallel to them but parallel to one another.
3. **The chemokine family:** The chemokines are small size molecules between 8 and 10 kDa. They are 20%–50% identical to each other. An important characteristic of cytokines is that they have conserved amino acids that are required for forming their 3-D or tertiary structure (Figure 8.4 discusses the basic structure of chemokine), such as (in most cases) four cysteines which work together in sets to create a Greek key shape. Typical chemokine proteins are produced in the form of pro-peptides that begin with a signal peptide consisting

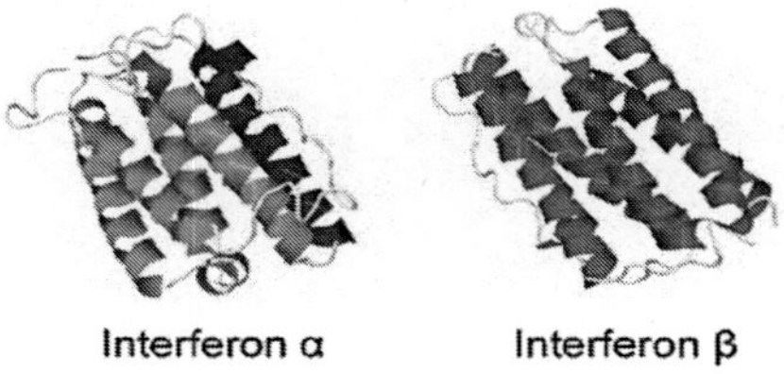

FIGURE 8.3 Structure of IFN-α and IFN-β.

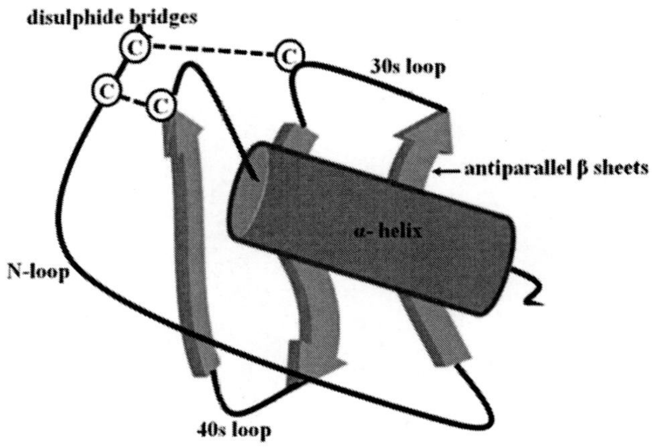

FIGURE 8.4 Structure of chemokine.

of ~20 amino acids that get cleaved from the active (mature) portion of the molecule during its secretion from the cell. In a chemokine, the first two cysteines are situated close to each other and near the N-terminal end of the mature protein, and the third cysteine resides in the center of the molecule and the fourth one close to the C-terminal end. First two cysteines are followed by a loop comprising approximately ten amino acids and is known as the N-loop. This is succeeded by a single-turn helix, termed as 3_{10}-helix, three β-strands and an α-helix at C-terminal. The turns connecting these helices and strands are named as 30, 40, and 50 seconds loops. The cysteine residues along the chemokine chain are numbered as they appear in the chain sequence and the third and fourth cysteines are located in the 30 and 50 seconds loops. The intramolecular disulfide bonds are present normally between the first and third, and the second and fourth cysteine residues.

4. **The tumor necrosis factor (TNF) family:** It's a 233-amino acid-long type II transmembrane protein arranged in stable homotrimers. The soluble 51 kDa trimeric TNF (Figure 8.5) dissociates at concentrations below the nano-molar range, thereby losing

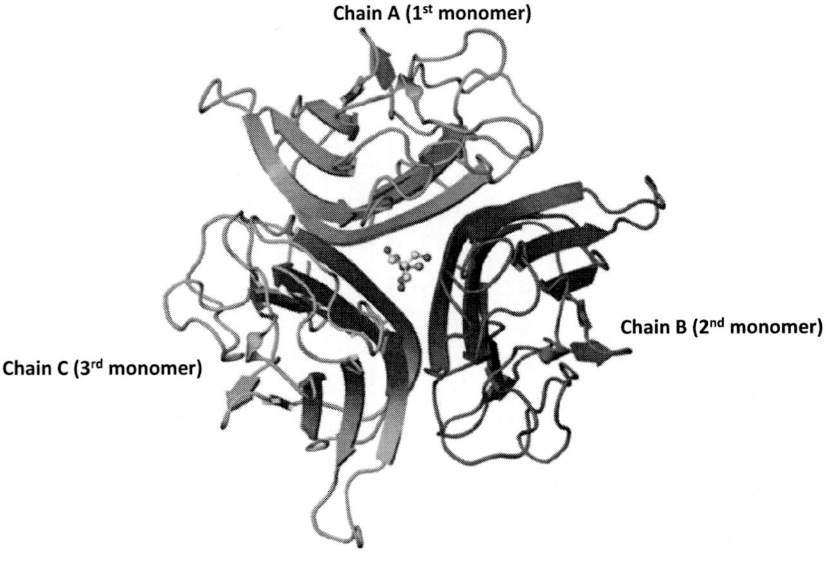

FIGURE 8.5 Ribbon diagram display of TNF-α homotrimer.

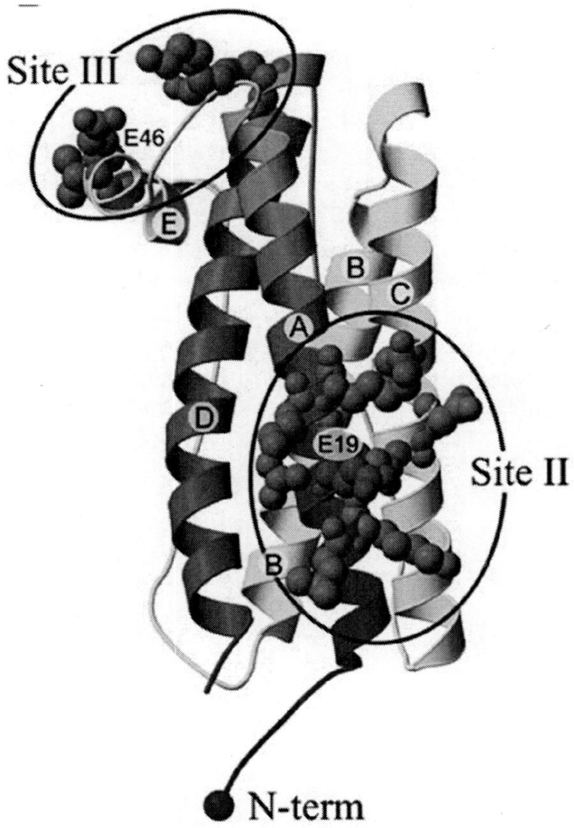

FIGURE 8.6 Ribbon diagram of G-CSF representing the main helices labeled A–D and additional E helix in the A–B loop. The residues involved in receptor binding (13–15) are shown in the space-filling representation, and the site II and III regions are indicated. The central site II residue, Glu 19 (in red), and the site III residue, Glu 46, are labeled. **Source: Layton et al. (2001).**

its bioactivity. The secreted form of human TNF-α takes a triangular pyramid shape and weighs about 17 kD. Both forms, the secreted and the membrane-bound are biologically active, although their specific functions are controversial. However, both forms have overlapping as well as distinct biological activities.

5. **Colony stimulating factor (CSF) family:** The structure is predominantly helical, with 104 of the 175 residues forming a four-alpha-helix bundle. The only other secondary structure is also helical. In the loop between the first two long helices, a 4-residue 3(10)-helix is immediately followed by a 6-residue alpha-helix. The small connection between the second and third bundle helices containing three residues creates nearly one turn of left-handed helix (Figure 8.6).

8.6 CYTOKINE RECEPTORS

Cytokines are released into the blood and transferred to the various parts of the body. The target organ that cytokines need to bind to in order to carry out the biological effect expresses cytokine receptors on their surface. The diverse function of cytokines on various cells is due to specific receptors that after binding to cytokines evoke signal transduction pathway or activate downstream proteins for the effects. Based on the structural diversity, cytokine receptors are divided into five families:

1. **Immunoglobulin superfamily receptors**
   This receptor family shares structural homology with immunoglobulins and cell adhesion molecules. Examples are IL-1, M-CSF, and C-Kit (Figure 8.7).
2. **Class I cytokine receptor family (or hematopoietin receptor family)**
   These cytokine receptors mainly function in immune and hematopoietic systems. They have conserved amino acid sequence motifs of tetra cysteine residues (CCCC) and tryptophan-serine-X-tryptophan-serine (WSXWS where X can be any non-conserved amino acid) located in the extracellular domain. An example is receptors for hematopoietins. They are connected to the Janus kinase (JAK) family (tyrosine kinase) of signaling pathway that phosphorylates and hence activates the downstream proteins to carry out the signal transduction (Figure 8.8).
3. **Class II cytokine receptor family (or interferon receptor family)**
   The members of this family (Figure 8.9) possess the conserved cysteine (CCCC) motifs, but WSXWS motif is absent unlike in class I cytokine receptors. Examples are receptors

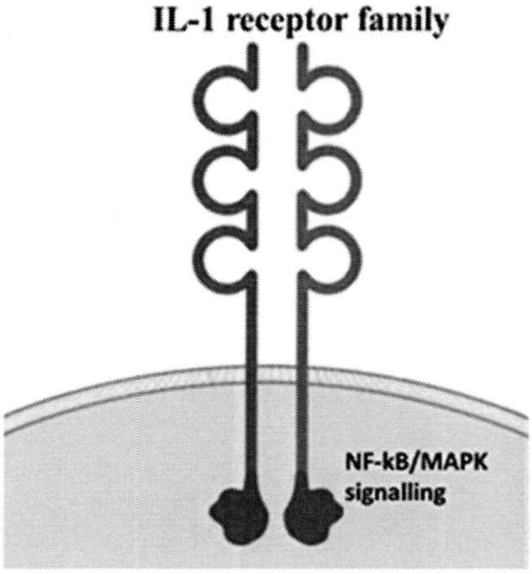

FIGURE 8.7 Structure of IL-1 recptor family.

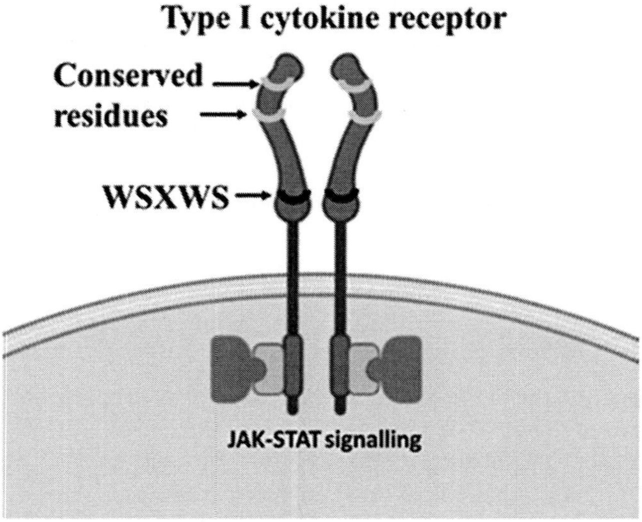

FIGURE 8.8 Structure of Type I cytokine receptor.

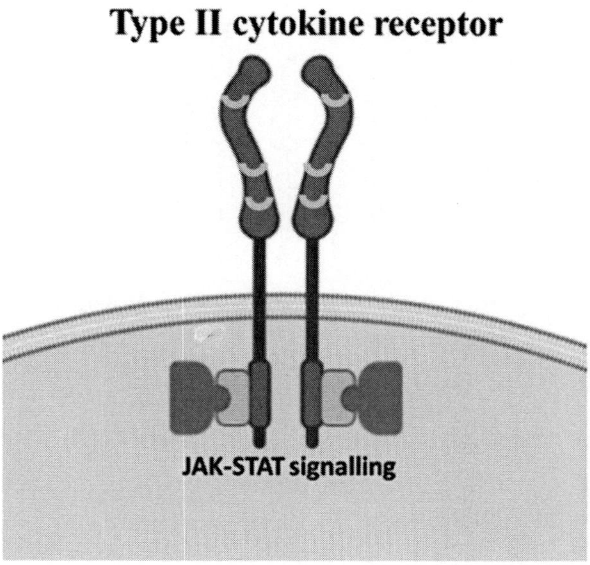

FIGURE 8.9 Structure of Type II cytokine receptor.

for IFNα, IFN-β, IFN-γ, and IL-10. These receptors also function in the same way as class I cytokine receptor involving the Janus kinase signal transduction pathway.

4. **TNF receptor family**

   These receptors have a central motif of CXXCXXC surrounded by a cysteine-rich domain (CRD) formed of three disulfide bonds generating an elongated molecule. Examples are TNF-α, TNF-β, CD40, NGF (nerve growth factor), and FAS (Figure 8.10). Adapter proteins, such as FADD and TRADD, help TNFR to associate with pro-caspases that can cleave other pro-caspases to activate them, and hence establish the caspase cascade resulting in irretrievable cell apoptosis.

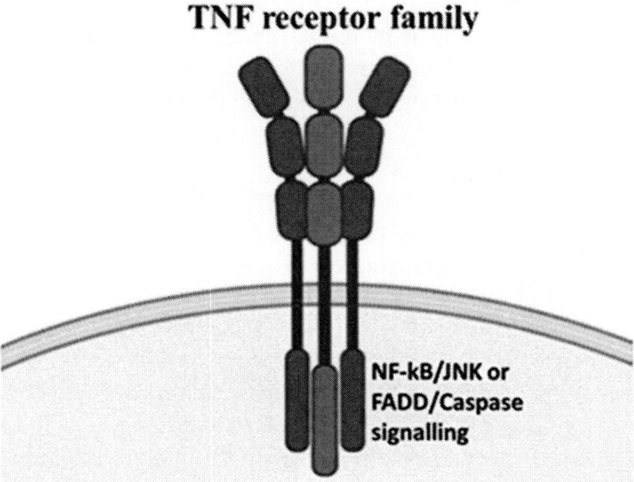

FIGURE 8.10 Structure of TNF receptor family.

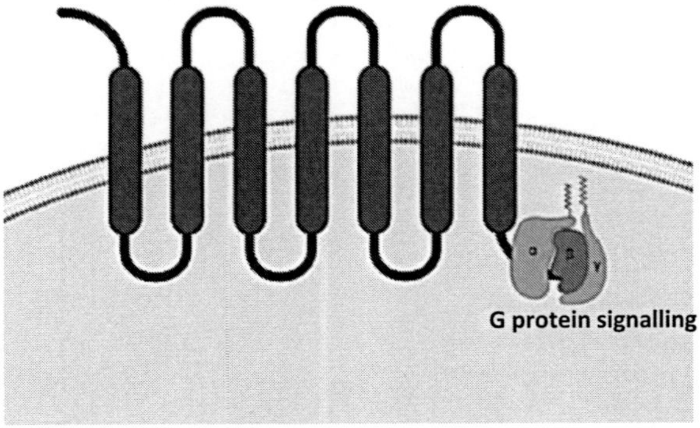

FIGURE 8.11 Transmembrane helical structure of chemokine receptor family.

5. **Chemokine receptor family**
   The members of this family have seven transmembrane helix structures and are coupled to G-protein for signal transduction. Examples include IL-8, RANTES, MIP-1, PF4, MCAF, and NAP-2 (Figure 8.11).

A new addition to the cytokine receptor family:

TGF-β receptors: These are transforming growth factor, beta receptor I and II (70/80kDa) that are membrane-bound. These are single-pass serine/threonine kinase receptors. Examples are TGF-β receptor 1 and TGF-β receptor 2.

8.7 CYTOKINE ANTAGONISTS AND AGONISTS

Cytokines carry out biological activities by binding to cytokine receptors. Several proteins or molecules have been identified that inhibit (called antagonists) or enhance (called agonists) the activities of cytokines by binding to the receptors or to the cytokines. Cytokine inhibitors and enhancers are present in the blood and extracellular fluid. During chronic inflammation, the body's defense mechanism detaches the binding motif of cytokine receptors by autolysis. These motifs are released in the body fluid in soluble form, allowing the traveling cytokines to bind to them, which results in the lowering of inflammation.

Viruses also adopt the property of anti-cytokines to restrict the generation of the immune response during viral entry and multiplication. A number of viruses have shown the characteristic of antagonists by producing molecules that mimic cytokines to bind the receptor but have no activity or produce cytokines binding protein, which renders the function of receptors inactive. An example of molecular mimicry is IL-10 homolog, a 17kDa protein produced by the Epstein-Barr virus. This protein shares 70% similarity with IL-10 of humans. Since it shares the structure with IL-10 of humans, it can easily bind to the IL-10 receptor and inhibit the signaling pathway that ultimately leads to the inhibition of T-cell proliferation and cytokine synthesis by Th1 cells. Another type of defense adopted by viruses is the production of inhibitors that bind to enzymes that convert cytokines into an active form. For example, the Cowpox virus produces an inhibitor that inactivates the IL-1 β converting enzyme, keeping IL-1 β from becoming functional. Additional mimicry molecules produced by viruses are receptor homologs that inhibit the interaction between cytokines and their receptors. An example of such mimicry is the soluble IFN-γ receptor encoded by the Myxoma virus. All the strategies adopted by viruses help them to evade the host immune response and aid

TABLE 8.3
Examples of Viral Protein Antagonists/Agonists to the Function of Cytokine

Viral Protein	Function of Cytokine
Vaccinia, cowpox virus	Soluble IL-1β receptor, inhibitor of IL-1β converting enzyme
Shope fibroma virus	Soluble TNF receptor
Poxviruses	Soluble TNF receptor and IFN-γ receptor
Fowlpox virus	IFN-γ receptor
Kaposi's sarcoma-associated Herpesvirus[a]	Cytokine receptor ORF74 (membrane protein that binds chemokines, constitutively activated, implicated in the development of Kaposi's sarcoma lesions
α-herpesviruses	Cytokine binding protein (neutralizes chemokines)

[a] An example of agonist.

in survival in the host environment. The other examples of viral protein antagonists/agonists to the function of cytokines are given in Table 8.3.

8.8 CROSS-REGULATION OF CYTOKINES

Cytokines are produced by various cells, including CD4, CD8, and macrophages, against the infection of antigens and micro-organisms. T-cells produce cytokines in response to the type of infection they encounter. T-cells can be distinguished based on the type of subsets they form and can be divided into the T_H1 and T_H2 subsets. The T_H1 subset is responsible for cell-mediated functions, such as the activation of cytotoxic T-cells, the promotion of phagocytosis and excessive inflammation. On the other hand, the T_H2 subset stimulates the humoral type of immune response that produces large amounts of IgM, IgE, and IgG isotypes, and also has a role in allergic reactions. The cytokines produced by these subsets have a cross-regulatory effect on the other subset. In other words, the cytokines produced by a particular subset inhibit the production of cytokines by another subset. In addition to this, the formed cytokines promote the growth of the subset that produces them. This effect is called cross-regulation. For example, the IL-12 cytokine is a critical cytokine for T_H1 subset differentiation. The IL-12 cytokine is down-regulated by the IL-4 and IL-10 (by acting on macrophages and dendritic cells) cytokines that are secreted by the T_H2 subset. Similarly, IFN-γ secreted by the T_H1 subset preferentially impedes the propagation of the T_H2 subset. B-cells are skewed to produce IgG2a in response to IFN-γ secreted by the T_H1 subset. IFN-γ also inhibits the production of IgG1 and IgE. The opposite occurs in the case of IL-4 produced by the T_H2 subset. IL-4 upholds the production of IgG1 and IgE; however, IgG2a production is suppressed. Hence, these immune responses are inversely proportional to each other, which means that an increase in antibody production decreases cell-mediated immunity propagation. Cytokines have an important role in the differentiation of T cell subsets. For example, IFN-γ favors the differentiation of T cells to the T_H1 subset by enhancing the production of molecules required for cell-mediated immunity. Likewise, IL-4 makes the T_H cells less responsive to cytokines that polarize them toward the T_H1 subset. As shown in Figure 8.12, the different sets of cytokines control the generation of the T_H1 and T_H2 types of immune response.

8.9 CYTOKINES IN HEALTH AND DISEASE (THERAPEUTIC CYTOKINES)

Cytokines are low molecular weight pleiotropic proteins. They can be formed by different types of cell types such as leucocytes which regulate immunity, inflammation and haematopoiesis. They can be classified based on their secretions into lymphokines, proinflammatory cytokines, anti-inflammatory cytokines, chemokines and growth factors.

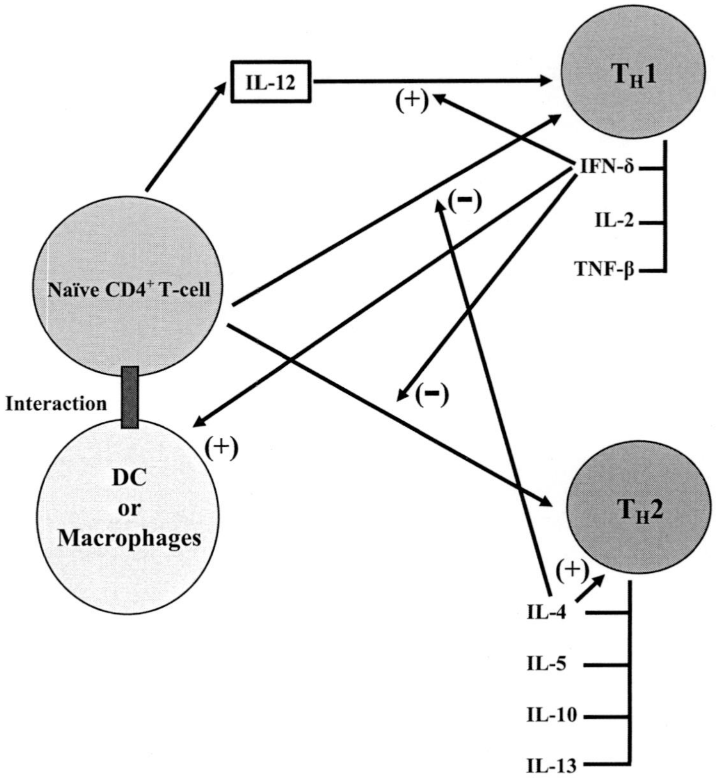

FIGURE 8.12 Cytokines regulate the development of T_H subsets.

To defeat an infection and other immune responses cytokines have a very crucial and important role. It also affects human health, diseases and many conditions like depression, stress, tumorigenesis, cancer, Alzheimer's disease, rheumatoid arthritis, reproduction, intestinal disease, pain, inflammation, respiratory disease, cytokines storms (a dangerous syndrome caused by excess secretion of cytokines) and viral pathogenesis. Cytokines produce a variety of effects on a number of cell types and help to maintain immune reaction and crucial role in altering and activating the innate response to an infection. Cytokines also have a prime role in macrophage activation and help the body in the initial process of fighting bacterial infection. Based on this cytokines can be used as biological markers (biomarkers) for disease and function as a therapeutic agent. Recombinant cytokines (cytokines have been developed by recombinant DNA technology) are being used in the medical industry as a drug.

Cytokine therapy is a new, advanced and innovative approach for the treatment of patients who are suffering from advanced malignancies and any manipulation of immune response. Diseases like cancer form a micro-environment (which includes cell stresses, injury caused by carcinogen, infection and inflammation) and different types of cytokines are produced in response to this. The aim of cytokines is to minimize cell damage, cellular stress, inflammation and promotes tissue repair. It is well known by clinical as well as behavioral observations that patients that were treated with cytokines probably show depression symptoms. 50% of the people who were treated with cytokine interferon-α show depression symptoms. Hypothalamo–pituitary–adrenocortical axis is activated by various cytokines in depressed patients. Certain inflammatory cytokines in dorsal root ganglion (DRG), in the spinal cord, injured nerve fibres and skin are not involved in pain but also in the persistence of pathologic pain behavior.

TABLE 8.4
Cytokines Function and Their Role in Health and Disease

Cytokines	Function
IFN type I and IL-7	Benificial cytokines: as they supress viral infection
IL-1B, IL-6 and TNF-α	Harmful cytoknes: as they can lead to cytokine strom
Cytokines	**Symptoms (in COVID Patients)**
IFN-γ	Headache, chill, fatigue and fever
IFN-γ, TNF-α	Flu like symptoms
IL-6	Blockage of small blood vessels
Cytokines	**Therapeutic Treatment**
IFN-α	Hepatitis B and C, lung cancer
IL-1	Sepsis, Rheumatoid arthritis, Alzheimer's disease, Cancer
IL-4, IL-5, IL-13	Asthma
IL-6	Rheumatic diseases, lung cancer
IL-10	Crohn's disease, Ulcerative colitis, Psoriasis, Rheumatoid arthritis, lung cnacer

Source: Modified from Gulati et al. (2016), Rabaan et al. (2021).

8.10 RECENT ADVANCES IN CYTOKINE

Cytokines are identified as assembly of integral membrane proteins. Some are permanently attached to membrane and may never be released from cell. Cytokines are biphasic in nature. Some molecules can be beneficial and at the same time harmful also, like interferon – γ causes several autoimmune diseases as well as it works in defense mechanism against several intracellular micro-organisms. Excessive release of inflammatory cytokines also called as cytokines strom (hypercytokinemia) can lead to many lung diseases like pneumonia, respiratory diseases and lung infections.

In bowel inflammatory diseases several cytokines and their pathways have been identified still anti-TNF- α is the basis of IBD treatment. The humanized IL-12/23 antibodies are an advance and promising therapy. Various cytokines are involved in the auto-immune disease rheumatoid arthritis (eg. cytokine is IL17) produced by the immune system. Cytokine therapy is a new research approach to treat patients suffering from cancer or other malignancies (Table 8.4).

8.11 SUMMARY

- Cytokines are small soluble proteins that are produced in response to an antigen. They are secreted in a controlled manner and secretion is short-lived.
- Cytokines act as chemical messengers for regulating the immune system including innate and adaptive immune response.
- Cytokines are pleiotropic (act on different types of cells), redundant (different cytokines can have the same function) and multifunctional (regulate different functions).
- Among various cytokines, tumor necrosis factor-alpha (TNF-α) and interleukin-1 (IL-1) are the chief cytokines that facilitate acute inflammation.
- Chemokines are a group of cytokines that directs the migration of leukocytes to the tissues at the site of inflammation.
- Interferons are involved in providing protection to cells (by activating them to produce defensive proteins) that are in vicinity to virus-infected cell. They are also involved in stimulating an inflammatory response.

- Cytokines perform diverse functions after binding to their specific receptors that are located on the cell surface. The cytokine-receptor binding leads to signal transduction pathway or activation of downstream proteins for effects.
- Since cytokines have a crucial role in defeating diseases, recombinant cytokines are being produced and are used in the pharmaceutical industry as a drug.
- Cytokine therapy is a new research approach to treat patients suffering from cancer or other malignancies.

QUESTIONS

Multiple Choice Questions

1. Which of the following is not a type of cytokine:
   a. Lymphokines
   b. Chemokines
   c. Tumour necrosis factor
   d. Hormones
2. Which of the following is incorrect about cytokines:
   a. Intercellular messengers
   b. Produced by T-helper cells
   c. Antigen-specific
   d. Hematopoiesis regulation
3. An example of chemokine receptor family is:
   a. TNF-α
   b. Rantes
   c. IL-1
   d. IFN-β
4. Cytokines that inhibit viruses from replication are:
   a. Chemokines
   b. Lymphokines
   c. Interferons
   d. Interleukins
5. Which of the following cytokines plays a role in terminating inflammatory responses:
   a. IL-2
   b. TGF-β
   c. IFN-α
   d. IL-4
6. Macrophage activation to kill pathogen is primarily mediated by
   a. IL-2
   b. IFN-γ
   c. GM-CSF
   d. IL-12
7. Which of the following type of cell signaling cytokines have been involved
   a. autocrine
   b. paracrine
   c. endocrine
   d. all of the above

Short Answer Questions

1. Define:
   A. Agonists
   B. Antagonists
   C. Cross-regulation
   D. Cytotoxic immunity
   E. Humoral immunity
   F. Cytokine therapy
   G. Pleiotropy
2. Expand following terms
   A. IFN
   B. TNF-α
   C. TGF-β
   D. CSF
3. Differentiate between the following
   A. Agonist and Antagonist
   B. Humoral immunity and Cytotoxic immunity
   C. T_H1 and T_H2 type immune response
   D. Interferons and Interleukins
   E. Pleiotropy and Redundancy
   F. Interferons and Interleukins

Long Answer Questions

1. Explain the properties of cytokines.
2. Explain different types of cytokines and their function.
3. Explain the role of cytokines in health and diseases and what are the recent advances in it?
4. What is cytokine receptor? Describe different types of cytokine receptors with diagram.
5. Explain cross-regulation among cytokines.

True/False

a. Each cytokine has a matching receptor inside the cell.
b. Cytokines have an important role in immune responses and in eliminating an infection.
c. Some cytokine molecules can be beneficial but at the same time can be harmful like IFN-γ.

BIBLIOGRAPHY

Brocker C., Thompson D., Matsumoto A., Nebert D.W. and Vasiliou V. 2010. Evolutionary divergence and functions of the human interleukin (IL) gene family. *Hum. Genomics* 5(1): 30–55.

Commins S.P., Borish L. and Steinke J.W. 2010. Immunologic messenger molecules: Cytokines, interferons, and chemokines. *J. Allergy Clin. Immunol.* 125: S53–S72.

Deverman B.E. and Patterson P.H. 2009. Cytokines and CNS Development. *Neuron* 64 (1): 61–78.

Diagrams drawn from biorender.com

Eck M.J. and Sprang S.R. 1989. The structure of tumor necrosis factor-α at 2.6Å resolution. *J. Biol. Chem.* 264: 17595–17605.

Fernandez E.J. and Lolis E. 2002. Structure, function, and inhibition of chemokines. *Annu. Rev. Pharmacol. Toxicol.* 42: 469–499.

Gulati K., Guhathakurta S., Joshi J., Rai N. and Ray A. 2016. Cytokines and their role in health and disease: A brief overview. *MOJ Immunol* 4(2): 1–9.

Hill C.P., Osslund T.D., and Eisenberg D. 1993. The structure of granulocyte-colony-stimulating factor and its relationship to other growth factors. *Proc. Natl. Acad. Sci. USA* 90: 5167–5171.

http://www.uniprot.org/uniprot/P00784

Javaid A., Jantan I. and Bukhari S.N.A. 2016. Rheumatoid arthritis: Recent advances on its etiology, role of cytokines and pharmacotherapy. *Biomed. Pharmacothe.* 92: 615–633.

Kindt T., Goldsby R., Osborne B., Kuby J. and Kuby J. 2007. *Kuby Immunology*. New York: W.H. Freeman.

Layton J.E., Hall N.E., Connell F., Venhorst J. and Treutlein H.R. 2001. Identification of ligand-binding site III on the immunoglobulin-like domain of the granulocyte colony-stimulating factor receptor. *J Biol Chem.* 276(39): 36779–36787.

Peter J.D., Seamus J.M., Dennis R.B. and Ivan M.R. 2011. *Roitt's Essential Immunology*. Hoboken, NJ: Wiley-Blackwell.

Rabaan A.A., Al-Ahmed S.H., Muhammad J., Khan A., Sule A.A., Tirupathi R., Mutair, A.A., Alhumaid S., Al-Omari A., Dhawan M., et al. 2021. Role of inflammatory cytokines in COVID-19 patients: A review on molecular mechanisms, immune functions, immunopathology and immunomodulatory drugs to counter cytokine storm. *Vaccines* 9: 436.

Ramachandran S., Verma A.K., Dev K., Goyal Y., Bhatt D., Mohammed A. Alsahli M. A., Rahmani A. H., Almatroudi A., Almatroodi S.A., Alrumaihi F. and Khan N.A. 2021. Role of cytokines and chemokines in NSCLC immune navigation and proliferation. *Oxid. Med. Cell. Longev.* 2021: 5563746.

Voet D., Voet J.G. and Pratt C. 2008. *Fundamentals of Biochemistry*, 3rd Edition. Hoboken, NJ: John Wiley and Sons.

Zhang J.M. and An J. 2007. Cytokines, inflammation and pain. *Int. Anesthesiol. Clin.* 45(2): 27–37.

9 The Complement System

Gouri Nandi
University of Delhi

CONTENTS

9.1 HISTORY

In a classical experiment, Jules Bordet led to the discovery of the complement protein system and was awarded with Nobel Prize in Physiology or Medicine in 1919. He demonstrated lyses of bacteria *Vibrio cholerae in-vitro* by sheep antiserum against it. He also showed that the antiserum lost its lytic function after heating while the serum's lytic function could be restored by adding the fresh sheep serum without antibodies. Based on the experiment, Bordet concluded that the lytic function of sheep antiserum depends upon two factors: first, the heat-resistant antibodies against the bacteria

DOI: 10.1201/9781003286424-9

and second, a heat-labile factor. He further confirmed his conclusions by red blood cell lysis (hemolysis), coated with antibodies on the addition of serum. Buchner in 1889, named the labile factor as Alexin (meaning "to ward off").

Later Paul Ehlrich coined another new term "Complement" after carrying out similar experiments independently. The term Complement means "the function of the serum that completes the action of the antibody"

9.2 COMPONENTS AND FEATURES OF THE COMPLEMENT SYSTEM

The complement system is composed of a series of more than 30 proteins (both plasma and cell surface) and glycoproteins, constituting 5% of serum globulin fraction by weight.

The characteristic features of the complement system are as follows -

1. Different complement proteins are designated by 'C' followed by a numeral, for example C1–C9, denoting the sequence of their discovery and not the sequence of action. A few other complement proteins are designated by letters (e.g., Factor B) or by protein names (e.g. Mannose-binding lectin).
2. Majority of the proteins of the complement system circulating in the blood serum are synthesized in the liver and are inactive precursor forms or Zymogens. After the proteolytic cleavage, these zymogens are activated by the removal of the inhibitory fragments, exposing the active site of the proteins. Usually, the small fragment generated is denoted as 'a' and the larger one as 'b'. For example, the complement protein C3 is cleaved into smaller C3a and larger C3b fragment. But C2 protein is an exception; C2a is larger and C2b is the smaller fragment. The smaller cleaved fragment diffuses and the larger protein fragments interact to form macromolecular complexes that have different functions, at the site of zymogen activation. These macromolecular complexes can act as enzymes or have cell membrane binding properties, designated by a bar overlying the symbols of the complement protein complex (e.g., $\overline{C4b2a}$).
3. The proteins of the complement system in activated form act in a predefined enzyme catalyzed cascade reaction. During Inflammation, a chain is triggered in which an initial active complement protein is generated from its precursor zymogen which acts as an enzyme catalyst cleaving the second zymogen substrate to convert it into an active complement protein product. The second active protein, thus formed will in turn catalyze the third step of the complement protein activation forming the third active protein in the cascade and hence the cascade of activation will proceed further in the same manner.
4. Another important feature of complement activation is the amplification of the products generated during the cascade reaction at C3 and C5 convertase steps.

9.3 THE THREE COMPLEMENT ACTIVATION PATHWAYS

The primary aim of the activated complement proteins is to mark and clear the pathogens from the body. Thus, the cascade activation of the complement proteins can occur in three ways, depending on the presence of the pathogen itself or due to antigen-antibody complexes (Flowchart 9.1).

1. **Classical pathway:** This is an antibody-dependent pathway for activation of the complement cascade. It is triggered due to the presence of antigen-antibody complexes (immune complexes) in serum or due to antibody molecules on large antigenic surfaces such as bacterial cells. Classical pathway is thus considered as a component of the Adaptive Immune system as it requires antibody molecules for its activation.
2. **Alternative pathway:** This is an antibody-independent pathway of activation and thus forms a part of the Innate Immune system. Many types of bacteria and their cell wall

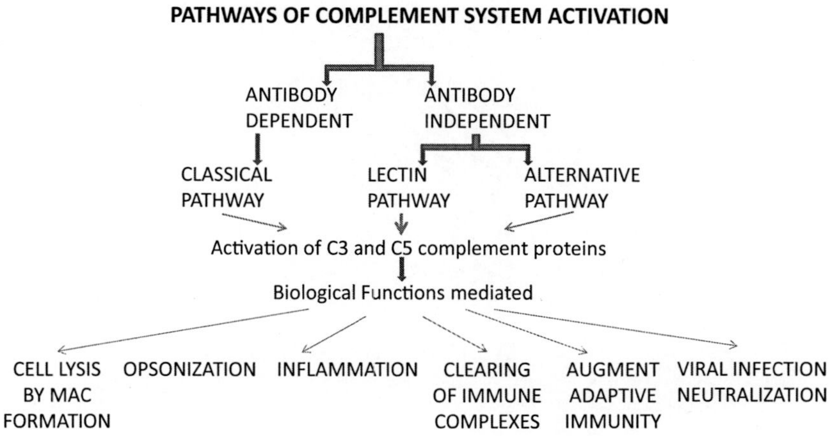

FLOWCHART 9.1 Overview of pathways of complement cascade activation and its biological functions.

components, yeast, virus-infected cells, parasites and several non-pathogenic substrates like heterologous erythrocytes, cobra venom, charged polymers etc. can activate this pathway.

3. **Lectin pathway:** This is also an antibody-independent pathway and thus can be called a component of the Innate Immune system, similar to the alternative pathway. It is the most recently discovered pathway and requires Mannose sugar present in glycoproteins or on the cell surfaces of microorganisms for its activation.

Some common characteristics of Complement System activation by the classical, alternative and lectin pathways are mentioned below.

9.4 ACTIVATION OF THE COMPLEMENT SYSTEM

9.4.1 Classical Pathway

9.4.1.1 C1 Macromolecule

The first protein to get activated in the classical pathway is C1. It is a large multimeric protein complex with a molecular weight of approximately 750 kDa. C1macromolecule is composed of one molecule of C1q and two molecules each of C1r and C1s subunits. All the subunits are stabilized by calcium ions in the complex.

The C1q subunit is composed of six collagen-like chains with globular heads (Figure 9.1). These are arranged in a radial manner like an inverted umbrella or a bunch of tulips. Each collagen-like chain of the hexamer consists of three polypeptides to form a triple helical structure with globular tips. The globular tips of the hexamer functions in binding to the CH_2 domain present on the F_c region of the antibody molecule. The C1q belongs to a category of carbohydrate-binding protein called Lectin. A subcategory of lectin with a characteristic collagen-like tail as observed in C1q are called collectin.

Each C1r and C1s monomer has a catalytic and an interaction domain. Both C1r and C1s are serine esterase enzymes, and the catalytic domains help in activating the C1 complex after it binds to antibody molecules. The C1s-C1r-C1r-C1s ($C1r_2$- $C1s_2$) tetramer is non-covalently associated with the C1q subunit with the help of their interaction domains.

9.4.1.2 Activation of Classical Complement Cascade System

The activation of the complement cascade system by the classical pathway is triggered due to the formation of soluble antigen-antibody complexes (immune complexes) or due to the binding of

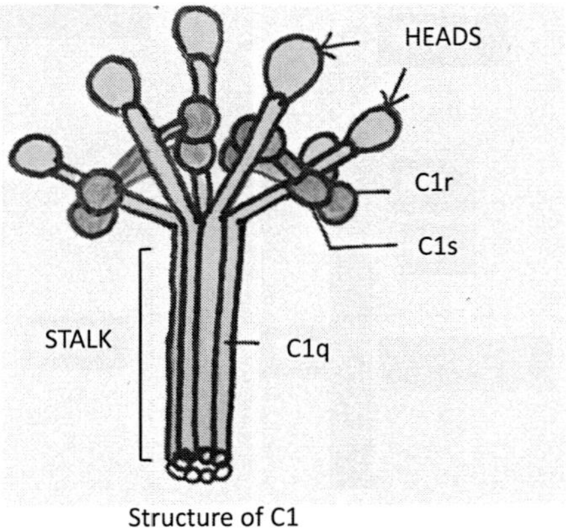

FIGURE 9.1 Structural details of C1 macromolecule.

antibody on a large antigen such as bacterial cell surface. The pentameric IgM and a few subclasses of IgG (human IgG1, IgG2 and IgG3) trigger the classical cascade pathway.

The binding of an antibody molecule to the antigen brings about conformational changes in the F_c region of the antibody exposing its C1 complement protein-binding sites. In the case of IgG bound to an antigen, a single C1-binding site is exposed, whereas an antigen-bound pentameric IgM has five such sites exposed. Specifically, the CH_2 domain located on the F_c portion of a bound antibody causes C1 complement protein activation. A single C1 macromolecule must bind to at least two of the F_c sites simultaneously with subunit C1q globular heads for a stable C1-antibody interaction and complement activation. This is the reason that a single molecule of secreted pentameric IgM, bound to a red blood cell and having five F_c regions exposed is enough for activating the complement system (also called Complement Fixation) and to cause lysis of the red blood cell. In contrast, several molecules of antigen-bound monomeric IgG must be aggregated, bringing multiple F_c regions close to C1 to bring about the complement activation (Figure 9.2).

The activation of the C1 molecule by binding of C1q globular heads to IgM or IgG causes a conformational change in the C1r subunit, converting it into an active serine protease enzyme $\overline{C1r}$, cleaving and activating the C1s subunit. The active serine protease $\overline{C1s}$ has two substrate proteins, C4 and C2 in the next step of the pathway.

The C4 molecule is the next protein to be activated in the cascade. A glycoprotein chemically, C4 consists of α, β and γ polypeptide chains. Activated $\overline{C1s}$ hydrolyzes a small fragment (C4a) from the amino terminus of α polypeptide of C4 thereby exposing the binding site on the larger protein fragment- $\overline{C4b}$. The $\overline{C4b}$ molecule covalently anchors to nearby cell surfaces or to bound antibody molecule in the vicinity of the C1 molecule. One $\overline{C1s}$ molecule can generate many $\overline{C4b}$ molecules. The smaller C4a diffuses away and functions as inflammation mediator called anaphylatoxin. The anaphylatoxin are small cationic peptides generated by the complement cascade, that are capable of binding to the receptors on the basophiles and the mast cells causing the release of mediators which cause inflammatory responses.

The third soluble serum component is the C2 complement protein composed of a single polypeptide. A C2 proprotein molecule binds to cell surface bound $\overline{C4b}$ molecules in presence of Mg^{2+} ions and is cleaved by nearby located $\overline{C1s}$ to produce a soluble C2b fragment of unknown importance which diffuses away. The larger $\overline{C2a}$ fragment remains physically associated to $\overline{C4b}$ on the bacterial cell surface. This results in formation of $\overline{C4b2a}$ molecular complex acting as classical pathway C3 convertase. The classical pathway C3 convertase binds and proteolytically cleave several C3 proenzyme molecules. The $\overline{C4b}$ component of the classical pathway C3 convertase, mediates binding and the $\overline{C2a}$ component catalyzes the proteolysis of C3 proenzymes.

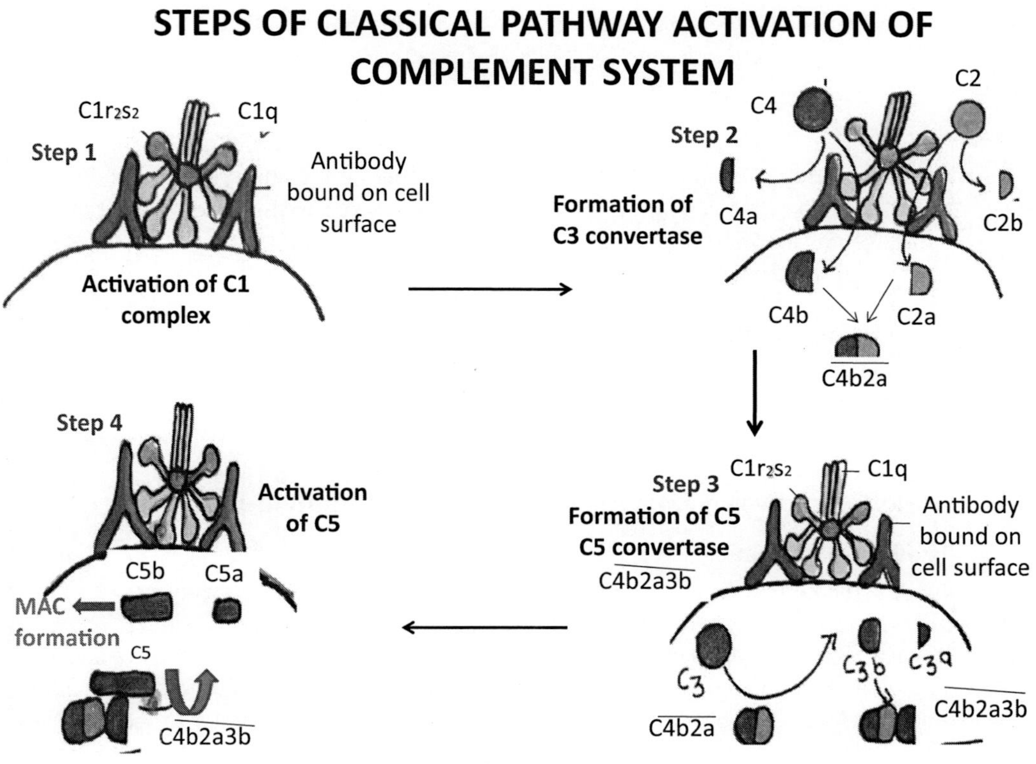

FIGURE 9.2 Steps of classical pathway activation of complement cascade system.

The C3 proenzyme molecule is sequentially the fourth soluble serum component participating in the classical pathway. The C3 proenzyme molecule has two, α and β polypeptide chains, A small fragment C3a, is hydrolyzed by classical pathway C3 convertase from the amino end of the α polypeptide chain generating a larger $\overline{C3b}$ fragment. The C3a molecule participates in mediating inflammation like C4a and also called anaphylatoxin. The $\overline{C3b}$ molecules contain unstable thio-ester bonds and can react with water yielding inactive $\overline{C3b}$ byproducts which cannot participate in the complement cascade. Only 10% of metastable $\overline{C3b}$ molecules generated by the classical pathway C3 convertase is able to bind to the $\overline{C4b2a}$ complex to form a trimolecular complex, $\overline{C4b2a3b}$, functioning as a classical pathway C5 convertase. To overcome the paucity of metastable $\overline{C3b}$ molecules, over 200 molecules of $\overline{C3b}$ are generated by a single molecule of classical pathway C3 convertase. This also causes a tremendous amplification of the complement cascade at this step. Another important function of activated $\overline{C3b}$ is $\overline{C3b}$ - mediated destruction of pathogens by phagocytes as $\overline{C3b}$ molecules act as opsonins (molecules which increase the rate of phagocytosis by coating the antigenic surfaces). Opsonization (the process of increase in phagocytosis rate, due to binding of small protein molecules to antigenic surfaces) by $\overline{C3b}$ occurs when it is deposited at a site away from bound C3 convertase of the classical pathway.

The next C5 proenzyme is the last molecule to be activated during the classical pathway. After this last enzymatic step, assembly of Membrane Attack Complex occurs. The serum protein C5 molecule has a disulfide-linked heterodimeric structure similar to C3 and C4 proteins but without a thioester bond internally. The $\overline{C3b}$ component of classical (or alternative) pathway C5 convertase, binds to C5 proprotein and alters its confirmation thus causing the $\overline{C4b2a}$ Component to degrade native C5 into smaller C5a and larger $\overline{C5b}$ fragments. The C5a fragment diffuses away and is called anaphylatoxin due to its function in inflammation, whereas the $\overline{C5b}$ fragment remains bound to the cell surface at a site near the complement reaction. The $C\overline{5b}$ molecule then attaches to the next complement protein in the cascade, the C6 molecule initiating assembly of the membrane attack complex.

Interesting facts: the complement function is dependent on the ability of $C\overline{4b}$ and $C\overline{3b}$ proteins to bind to cell surfaces. These molecules have structural homology that both have an internal thioester bond between a glutamine and a cysteine residue buried within the intact protein. After enzymatic activation and removal of C3a and C4a, the internal thioester bond is exposed, making it very unstable and susceptible to nucleophilic attack by hydroxy groups (-OH) and amine groups ($-NH_2$) present in membrane proteins and carbohydrates on cell surfaces. This allows the $C\overline{4b}$ and $C\overline{3b}$ to form a covalent bond, locking them onto the cellular antigens.

9.4.2 Antibody Independent Pathways of the Complement Activation

9.4.2.1 The Lectin Pathway

Lectin pathway is the most recently discovered complement system activation pathway.

The lectin proteins function to recognize and bind the specific carbohydrate residues. Those lectin proteins that activate the complement system bind specifically to mannose carbohydrate residues and thus, this pathway of complement activation is also called Mannose-Binding Lectin pathway (MBL Pathway) or MBLectin pathway. Unlike the classical pathway and similar to the alternative pathway for complement activation, the lectin pathway occurs independently of the antigen-antibody complex. But once activated, the mechanism of action of the lectin pathway is nearly similar to the classical pathway as compared to the alternative pathway.

Mannose-Binding lectin (MBL) is an acute phase protein. Most individuals have a low concentration of MBL in plasma and its synthesis increases during inflammatory responses. MBL, also known as defence collagen, has a collectin (C1q)-like structure with six heads.

The initiation of Lectin pathway occurs by binding of MBL to mannose residues in the glycoproteins or carbohydrate chains present on the surfaces of many pathogenic microorganisms. These microorganisms include bacteria such as *Salmonella*, *Neisseria and Listeria*; few fungi-*Cryptococcus neoformans* and *Candida albicans*; and some of the viruses like HIV-1. Human cell surfaces have sialic acid residues that masking the carbohydrate groups usually recognized by the MBL and hence no complement activation occurs during healthy conditions. After pathogen enters the body, the MBL binds to its target carbohydrate residue present on the pathogen, enabling MBL-associated serine proteases – MASP-1 and MASP-2, to bind to the pathogen-bound MBL. Both the MASP-1 and MASP-2 share structural and functional similarities with the C1r and C1s components of the C1 macromolecular complex of the classical pathway. The MASP-1 and MASP-2 proteases function to activate C4 and C2 component proteins by cleaving them and allowing the formation of $C\overline{4b2a}$ complex or classical pathway C3 convertase. This C3 convertase of the classical pathway gets deposited on the surface of the pathogens, further activating the downstream complement proteins in the cascade. Once the classical pathway C3 convertase is formed during the lectin cascade, the following reaction chain steps are exactly the same as the classical complement pathway.

Interesting facts: It has been suggested that the MBLectin pathway is an important innate defense mechanism for newborn children in the "gap period" of about 12 months, during which the protective cover of maternal antibodies is lost and the child's effective immune system develops.

9.4.2.2 The Alternative Pathway

The activation of the complement system by the alternative pathway was suggested by Pillemer and Ecker in 1941. This pathway is an alternative method of complement activation in the absence of antibodies and therefore considered a mechanism of innate immunity. The initiators of the alternative pathway include microbial pathogens such as bacterial polysaccharides, yeast, viruses and virus-infected host cells, tumor cells and parasites. Some nonpathogenic factors can also initiate this pathway such as heterologous erythrocytes, cobra venom, anionic polymers (dextran sulphate) and pure carbohydrates like agarose and inulin. Four blood serum proteins - C3, Factor B, Factor D and properdin are involved in complement activation by an alternative pathway. The proteins involved in this pathway are designated as Factors followed by a capital letter symbol. The $\overline{C5b}$ molecule is also generated during this pathway, initiating the assembly of the membrane attack complex, the same as described for classical and lectin cascade pathways (Figure 9.3).

A unique feature of the alternative pathway is that it is always in a state of low-level activation called "C3 tickover", caused by spontaneous hydrolysis of unstable internal thioester bond of C3 with fluid phase water. This generates another unstable intermediate called $\overline{C3b}$ (or C3[H_2O]) and C3a. The $\overline{C3b}$ thus generated binds to suitable targets such as bacterial cell walls, initiating the solid-phase of the alternative pathway. The bound $\overline{C3b}$ forms a complex with the next alternative pathway protein, Factor B. This allows proteolysis of bound factor B by the next protein component of the pathway, Factor D. The Factor D is a serine protease and circulates in the serum of healthy individuals at low concentrations, probably in its enzymatically active form. The proteolytic cleavage of bound Factor B by Factor D releases Ba (the small fragment) and $\overline{Bb}$, the larger fragment remains attached to $\overline{C3b}$ forming a very unstable complex $\overline{C3bBb}$.

The binding of the next alternative pathway component, Properdin or Factor P, stabilizes the $\overline{C3bBb}$ complex by preventing its spontaneous decay. The resulting stable $\overline{C3bBb}$ complex functions

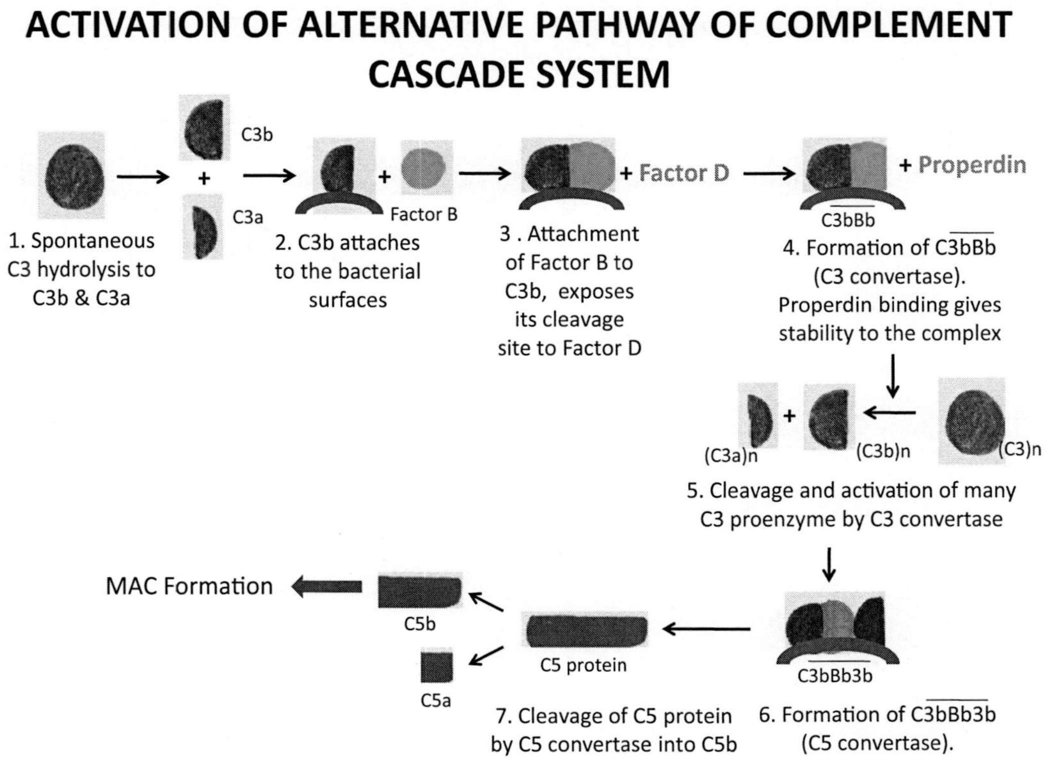

FIGURE 9.3 Steps involved in activation of alternative pathway of complement cascade system.

as the alternative pathway C3 convertase, with $\overline{Bb}$ fragment having serine protease function, capable of cleaving and activating numerous other C3 molecules to active $\overline{C3b}$. The newly formed $\overline{C3b}$. molecules get deposited on the microbial surfaces, recruiting factor B, factor D and properdin forming many new alternative pathway C3 convertase complexes. The cascade thus continues leading to positive feedback amplification at this step, generating several million $\overline{C3b}$ molecules within a few minutes of initiation of the alternative pathway.

Few $\overline{C3b}$ molecules combine with C3 convertase itself forming a new complex $\overline{C3bBb3b}$. This new complex is an alternative pathway C5 convertase, and proteolytically cleaves C5 to C5a and $\overline{C5b}$ fragments. The $\overline{C5b}$ molecule attaches to the microbial surfaces, initiating the assembly of membrane attack complex. At the assembly step, both the classical and the alternative pathway converge and the same terminal events occur, leading to the formation of MAC.

Interesting Facts – The $\overline{C3b}$ generated by the classical pathway, can also initiate the alternative pathway. The alternative pathway C3 convertase formed, functions by a similar mechanism of amplification as observed during the complement activation by the classical pathway.

9.4.3 Assembly of the Membrane Attack Complex

The last few steps of the complement cascade, either by Classical, Alternative or Lectin pathway involves complement proteins namely C5b, C6, C7, C8, and C9 interacting in a sequential manner to form the macromolecular cytocidal structure known as Membrane Attack Complex (MAC). This complex functions by forming large channels or pores in the target cell's plasma membrane, destroying the membrane integrity and killing the pathogen cells (Figure 9.4).

The initial step in MAC assembly by activated complement pathway is catalyzed by active C5 convertase. The C5 convertase proteolytically cleaves C5 zymogen to smaller C5a and larger $\overline{C5b}$ functional fragments. The C5 zymogen is a disulfide-linked heterodimer serum protein, consisting of α and β chains, homologous to C3 and C4 proteins, but is without the internal thioester bond.

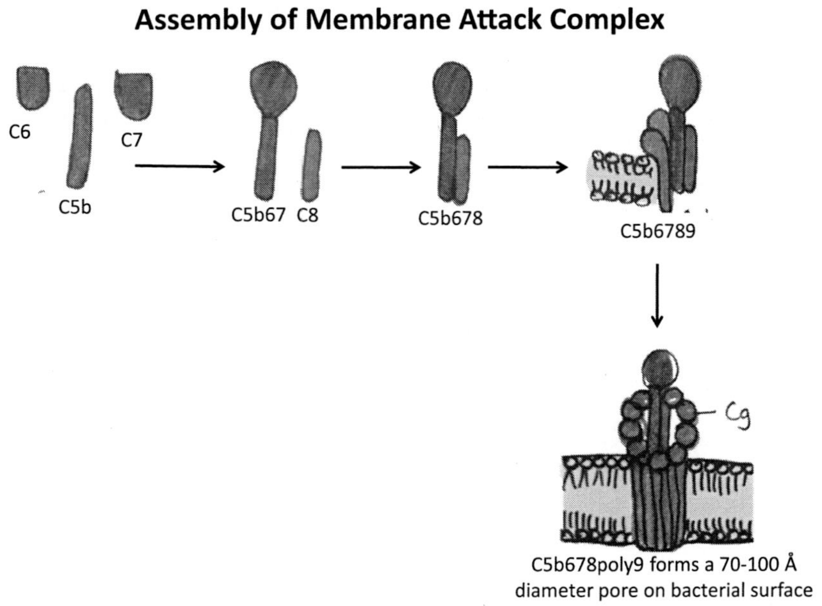

FIGURE 9.4 Assembly of Membrane Attack Complex (MAC) on bacterial cell surface.

The C5 zymogen binds to the nonenzymatic $\overline{C3b}$ component of the classical or the alternative pathway C5 convertase. It results in the cleaving of α chain amino terminus of C5 generating a small fragment C5a. The small active C5a fragment diffuses away from the site of reaction and participates in inflammation and is called as anaphylatoxin. The larger $\overline{C5b}$ fragment binds to the cell surfaces or immune complexes at the reaction site. The $\overline{C5b}$ component is very unstable and is quickly inactivated unless it is stabilized by complexing with the next protein in the cascade reaction, C6. The complement reaction occurs on hydrophilic cell surfaces or in the fluid phase immune complexes, up to this point in the cascade. As the next protein C7 binds to the C5b6 complex, the latter undergoes a structural transition exposing hydrophobic regions which are then inserted into the hydrophobic cell membranes. However if this reaction of C7 to C5b6 complex occurs on immune complexes or other noncellular surfaces, then the exposed hydrophobic sites of C5b67 cannot bind and get released. The released C5b67 having exposed hydrophobic binding sites can be inserted into the plasma membranes of the nearby nonpathogenic healthy self-cells. This could lead to killing of the host cells and the process is called innocent-bystander lysis, causing tissue damage. The regulatory proteins usually inhibit this effect but in the case of a few autoimmune diseases extensive damage to tissues occurs.

The binding of the next protein C8, of the cascade reaction to the membrane-bound C5b67 complex, causes a conformational change in the C8 so that now it becomes hydrophobic and also gets inserted into the cell membrane along with the bound complex. The C8 molecule is a trimer consisting of α, β and γ chains. The β chain of the C8 molecule binds to the C5b67 complex and the γ chain gets inserted into the cell membrane. The C5b678 membrane complex forms a small 10Å diameter pore which has a limited ability to lyse some types of cells. The formation of potent bactericidal and microbicidal MAC is accomplished by complexing and polymerization of C9 protein, the last complement cascade component. The C9 is a perforin-like monomeric serum protein that polymerizes and undergoes a structural transition to become inserted into the membrane at the site of bound C5b678 complex. As many as 10 to 17, C9 molecules become associated with the C5b678 complex, forming "poly C9" to complete MAC assembly. This increases the functional pore diameter to 70–100 Å. The small molecules and ions now freely diffuse through the central tubular channel of the MAC, and bacterial or host cell cannot regulate their osmotic stability. Thus, the MAC killing is caused by loss of the electrolytes and the influx of water into the cell.

9.5 FUNCTIONS OF THE COMPLEMENT SYSTEM

The complement cascade system serves to amplify the immune response forming an effective defense mechanism against the microbial infection. The main functions of complement activation are – Cell lysis, Opsonization, inflammation, immune complex clearance and augmenting the adaptive immune response.

9.5.1 Cell Lysis

The cell lysis function of the complement is brought about by formation of MAC. The MAC complex is capable of lysing the Gram-negative bacteria, viruses, parasites, erythrocytes and nucleated cells.

The complement cascade is very potent in killing Gram-negative bacteria, but a few of them have evolved complement evading mechanisms. For example, the complement-resistant strains of *Escherichia coli* and *Salmonella* possess long chain polysaccharide in their lipopolysaccharide (LPS) component of the cell wall. It has been suggested that these long chain polysaccharides interfere with the MAC insertion into the bacterial plasma membrane, conferring resistance to it. Some resistant strains of the *Neisseria gonorrheae* possess outer membrane proteins to prevent insertion of the MAC into bacterial cell membranes.

Generally, Gram-positive bacteria are resistant to MAC killing due to the thick peptidoglycan layer of their cell walls. The peptidoglycan layer inhibits the MAC insertion into bacterial plasma membranes.

Although the cell-mediated immune system has a major function in clearing the viral infections but anti-viral antibodies and complement system also have a vital role in host defenses against viruses. Most enveloped viruses are killed by complement-mediated lysis of the MAC. These include Retroviruses, Herpesviruses, Orthomyxoviruses (causing measles and mumps) and Paramyxoviruses (such as Influenza)

The assembly of single membrane attack complex on enucleated cell (RBC) membrane is enough to lyse it, but nucleated cell lysis requires formation of multiple MAC on its membrane. This could be very harmful for the host cells as the assembly of MAC can cause "Innocent bystander host cell damage". However, many host cells can endocytose the assembled MAC to prevent tissue damage. Unfortunately, the cancer cells also employ the same mechanism as the healthy cells, to evade the complement-mediated MAC lysis which otherwise would have been a potent weapon against the cancerous cells.

9.5.2 Opsonization

The complement activation generates $C\overline{3b}$, $C\overline{4b}$ and $iC\overline{3b}$ molecules (called Opsonins) which coat the surfaces of the antigens and immune complexes. The coated antigens and immune complexes bind to complement receptors present on the host cell surfaces including the phagocytic cells, increasing the phagocytosis (called Opsonization) and resulting in the killing of the microbial antigens or uptake of the immune complexes within the phagocytic cells.

The $C\overline{3b}$ fragment produced by classical, lectin and alternative pathways, is a major opsonin of the complement system. The amplification during C3 activation leads to the formation of several million $C\overline{3b}$ molecules, capable of coating the surfaces of microorganisms or immune complexes. In addition, $C\overline{4b}$ and i $C\overline{3b}$ (cleaved product of $C\overline{3b}$) also have opsonizing activity. Phagocytic cells like neutrophils and macrophages and few other cell types express the complement receptors like CR1, CR-3, and CR-4 capable of binding to $C\overline{3b}$, $C\overline{4b}$ and i $C\overline{3b}$ complement components. When C $\overline{3b}$ coated antigen binds to CR1 receptor on the phagocytic cell, the phagocytosis is enhanced. This in turn leads to augmented antigen processing pathway and specific antibody synthesis to clear the infection. In fact, $C\overline{3b}$ and $iC\overline{3b}$ dependent phagocytosis is a major defense mechanism against systemic bacterial and fungal infections.

9.5.3 Inflammation

Anaphylatoxins are small cationic peptides generated by the complement cascade, that are capable of binding to the receptors on the basophils and the mast cells causing the release of the histamine and many other pharmacologically active mediators that cause inflammatory responses. The C3a and C5a peptides are well known anaphylatoxins and called so because these fragments mediate inflammatory reactions characteristic of anaphylaxis. The anaphylatoxins C3a (molecular mass 9 Kilo Daltons) and C5a (molecular mass 11 Kilo Daltons) are generated by activation of complement cascade pathways. The C5a is 20-fold more potent anaphylatoxin as compared to C3a. Both anaphylatoxins, generated as a result of complement activation, induce contraction of smooth muscles and increase the permeability of blood vessels. This results in fluid influx carrying antibodies and the phagocytes to the site of antigen entry and infection.

Both C3a and C5a induce monocytes and neutrophils present in the bloodstream to adhere and extravagate through the vascular endothelial cells lining and migrate towards the complement activation site in the tissues.

The anaphylatoxins exert biological effects by binding on various cell types with the help of specific anaphylatoxin receptors. C5a receptor is characterized more thoroughly and its expression is reported on many cell types like neutrophils, basophils, eosinophils, monocytes, mast cells, macrophages, epithelial cells and astrocytes. Distinct C3a receptor are present on blood leukocytes and mast cells and are suggested to be G-protein linked.

The functions of anaphylatoxins are regulated by a plasma protease, carboxypeptidase-N, which functions to cleave arginine residues present on carboxyl-terminal of C3a and C5a. The carboxypeptidase-N thus generates slightly truncated forms of anaphylatoxins known as des-Arg forms. The C5a des-Arg form has only 10% of its chemotactic activity while C3a des-Arg is completely inactive.

9.5.4 Clearing of Immune Complexes from Circulation

An individual mounting vigorous humoral immune response against a circulating antigen will have a large amount of immune complexes in circulation. These immune complexes can get deposited on the blood vessel walls and can activate the complement system, leading to local inflammation damaging the surrounding tissues. The F_c regions of the antigen-bound immunoglobulins can interact and cause aggregation of immune complexes forming large lattices. The binding of complement components sterically blocks the F_c-F_c interactions of the bound antibody molecules, thus preventing lattice formation and dissociating the already formed lattices.

The second key function of the complement cascade system is rapid clearing of immune complexes from blood circulation. This is done with the help of complement receptor 1 (CR1) present on erythrocyte surfaces. The formation of immune complex activates the complement cascade, generating the $\overline{C3b}$ molecule which forms covalent bond with the bound antibody. The CR1 molecules binds to these $\overline{C3b}$ bound immune complexes with high affinity on erythrocyte surfaces. *In-vivo* studies suggest that phagocytic cells of liver and spleen remove and phagocytose the immune complexes from the erythrocyte as the latter passes through the sinusoids of these organs. This prevents the deposition of immune complexes on other tissues and the resulting tissue damage.

In the case of autoimmune disorder, Systemic Lupus Erythematosus (SLE) immune complexes are formed in large amounts in the body. The affected individuals suffer from pathologies caused by complement-mediated tissue damage and due to hypersensitivity. Although the major tissue damage is caused by the complement cascade system, it is a paradoxical finding that the predisposition of people to SLE is due to deficiencies of C1, C2, C4 and CR1 complement proteins and 90% of individuals lacking C4 complement protein suffer with SLE. This indicates that the deficiencies of complement proteins interfere with the immune complex clearance, reflecting the importance of complement in this process.

9.5.5 Augmenting Adaptive Humoral Responses

The complement protein system regulates humoral immunity in large part by CR2 complement receptor forming a co-receptor on B cell surfaces during antigenic activation. The CR2 also functions to retain antigens on follicular dendritic cells crucial for maintaining germinal centers. The display of antigens on the follicular dendritic cells is very important for the process of clonal selection and affinity maturation of selected B cells. The activated B cells secrete specific antibodies that bind and inactivate the target antigens.

9.5.6 Viral Infection Neutralization

The complement system can cause viral neutralization by a number of mechanisms. The $\overline{C3b}$ component, even at low concentrations facilitates antibody molecules, to form large viral aggregates reducing the load of pathogenic viruses.

The binding antibody and/or the complement protein on viral surfaces create thick protein coat. This coating blocks the attachment of viruses to susceptible cells. The antibody and complement deposits also cause binding of viruses to cells displaying Fc receptors or CR1 (Complement receptor 1). In the case of phagocytic cells, this can lead to phagocytosis and intracellular destruction of the viruses.

In addition, the complement system and MAC can lyse most enveloped viruses, helping in viral neutralization.

9.6 COMPLEMENT SYSTEM REGULATION

The cascade of complement proteins has evolved to become very rapid, self-amplified, with potent damaging effects necessary to combat the infection. Due to these properties, it is also indispensable that the complement system should be tightly regulated *in-vivo*. The general mechanisms of regulation include

1. Circulation of complement protein as zymogens which are inactive enzyme forms. These zymogens are activated only after binding to pathogenic surfaces.
2. The complement cascade after activation generates many labile components which undergo spontaneous inactivation within a short span of time or they diffuse away from the target site.
3. A few complement components can become activated spontaneously in the plasma and these could sometime bind to the host cell surfaces also. So, a series of complement regulating proteins are present that control the complement cascade at different steps. The regulatory proteins include soluble plasma protein regulators and the integral membrane protein regulators, to watch over both the fluid-phase and localized cell membrane events of complement activation.
4. The major regulatory mechanism control C3 convertase activity which is central to all the three pathways of complement activation. Two basic mechanisms are present for regulating the C3 convertase.
   A. First is the decay accerelation process, in which the dissociation of the convertase complex is enhanced.
   B. The second is cofactor-dependent proteolysis of $\overline{C3b}$ and $\overline{C4b}$ by Factor 1.
   Some regulatory proteins have both the decay accerelation and cofactor –dependent proteolysis properties, thus inhibiting both classical and alternative pathway.
5. Many discrete regulatory proteins control Cl activation of classical pathway, C3 convertase amplification in alternative pathway and MAC formation (Table 9.1).

9.6.1 Regulation of C1 Component

The activation of Cl is regulated by a heavily glycosylated plasma protein called Cl inhibitor (ClInh), which is a serine protease inhibitor that acts in two ways. The molar concentration of ClInh is much more than Cl, so majority of the Cl protein is non-covalently linked to ClInh in the blood. This inhibits the spontaneous activation of Cl. The bound ClInh is released when Cl is activated by immune complexes. ClInh can also bind to Clr_2s_2 subunits, resulting in its dissociation from Clq and thus inhibiting activation of downstream cascade proteins, the C4 and the C2.

9.6.2 Regulation of C3 Convertases

The C3 convertase activity and resulting generation of $\overline{C3b}$ in complement cascade pathway is monitored by a family of regulatory proteins. These regulatory proteins have characteristic repeated motifs spanning about 60 amino acid residues known as Short Consensus Repeats (SCRs). The genes encoding the C3 convertase regulatory proteins in humans are located at chromosome 1, in a single cluster, known as the Regulators of Complement Activation gene cluster abbreviated as RCA. The C3 convertase regulation can occur at different levels during the cascade as mentioned below.

TABLE 9.1
Regulation of the Complement Cascade System

Complement Protein	Regulator Protein	Mode of Action	Pathway Regulated
C1 component	C1 inhibitor (C1 Inh)	Binds to C1r2s2	Classical activation
C3 convertase	C4b-binding protein Complement Receptor Type 1 (CR1) Membrane Cofactor Protein (MCP)	Binds to C4b, preventing its complex formation with C2a	Classical activation
	Factor I	Cleaves C4b	Classical & Alternative
	Factor H	Binds to C3b	Alternative pathway
	Decay Accelerating Factor (DAF) or CD55	Decay of C3 convertase	Classical & Alternative
MAC	S protein	Binds to C5b67	MAC Insertion in membranes
	Homologous restriction factor (HRF)/ Membrane inhibitor of reactive lysis (MRL)/ Protectin	Inhibits poly-C binding to C5b678	MAC formation prevented

9.6.2.1 Inhibition of C3 Convertase Formation

The classical and lectin cascade pathways are regulated by three different RCA encoded proteins, all of which block C3 convertase assembly by inactivating $\overline{C4b}$, $\overline{C2a}$ or $\overline{C3b}$ subunits. Of these three regulatory proteins, one is soluble C4b-binding protein abbreviated as C4bBP and the rest are membrane-bound proteins namely- the Complement Receptor Type 1 (CR1) and the Membrane Cofactor Protein (MCP). The C4bBP, CR1 and MCP bind to $\overline{C4b}$ and prevent its complex formation with $\overline{C2a}$, thus no C3 convertase is formed. Once C4bBP, CR1 or MCP bind to $\overline{C4b}$; a fourth inhibitory protein Factor I degrades active $\overline{C4b}$ into C4d (complex bound form) and a soluble C4c form, thereby inactivating $\overline{C4b}$ completely.

A similar regulatory mechanism operates during an alternative pathway to inhibit the assembly of $\overline{C3bBb}$ to form C3 convertase. In addition to previously mentioned CR1, MCP, another regulatory protein known as Factor H complexes to $\overline{C3b}$ preventing its binding to factor B. In the next step, Factor I cleave $\overline{C3b}$ bound to CR1, MCP and factor H. This generates bound iC3b and soluble C3f fragments. Factor I further cleave iC3b to release C3c and C3dg which keeps bound to the membrane.

9.6.2.2 Inhibition after C3 Convertase Formation

The RCA proteins prevent C3 convertase activity by dissociating the assembled C3 convertase. These regulatory proteins include previously mentioned C4bBP, CR1 and Factor H. Another regulatory protein is Decay Accelerating Factor abbreviated as DAF or CD55. DAF is a glycoprotein in nature, covalently anchored to a peripheral membrane protein and can dissociate C3 convertase. All these regulatory proteins accelerate decay by setting free the enzymatic subunit ($\overline{C2a}$ or$\overline{Bb}$) of the cell membrane-bound C3 convertase. Following this, Factor I, cleaves the membrane-bound component rendering C3 convertase irreversibly inactive.

9.6.3 Regulation of Membrane Attack Complex Formation

Few other regulatory proteins prevent the entry of membrane attack complex (MAC) into plasma membranes of the healthy cells of host. The chance release of the C5b67 complex poses a threat to healthy host cells due to innocent bystander lysis. To prevent this, a regulatory serum protein called S protein binds to the C5b67 complex, and prevents its insertion in the membrane of the host cells.

Another intrinsic membrane protein known as Homologous Restriction Factor abbreviated as HRF or Membrane Inhibitor of Reactive Lysis (MRL) or CD59 or Protectin, inhibits complexing of poly - C9 subunits to C5b678 macromolecule thereby preventing the formation of fully functional membrane attack complex. Thus, CD59 protects host cells from nonspecific complement-mediated lysis. The HRF is called so because it inhibits complement components destroying the cells of the same species.

9.7 DEFICIENCIES OF COMPLEMENT SYSTEM

A. Silverstein in 1960 reported the first case of complement deficiency in adult human, following which many more incidences of such type have been described. The deficiencies of components of classical, alternative and terminal pathways or in soluble or membrane-bound regulatory proteins are being reported.

9.7.1 Genetic Complement Deficiencies

These involve mutations in gene coding for complement components and lead to congenital defects in functioning of the complement system.

1. Genetic deficiency of classical components including C1q, C1r, C1s, C2 and C4 proteins prevent classical cascade. These deficiencies cause the disease Systemic Lupus Erythematosus (SLE), highlighting the significant role of complement components in solubilization and clearing of the immune complexes from the body. These complement protein deficiencies are inherited as autosomal recessive disorders.
2. The deficiency of C3 complement zymogen and its regulatory proteins- factor H and Factor I leads to recurrent pyogenic bacterial infection which could prove fatal. The patients can additionally suffer from glomerulonephritis due to the accumulation of antigen-antibody complexes in the kidney. These deficiencies are autosomal recessive.
3. The deficiencies of the MAC components C5, C6, C7 and C8 prevent the formation of membrane attack complex. The individuals having these deficiencies suffer from recurrent *Neisseria* infections including *N. meningitides* and *N. gonorhoeae,* suggesting important role of complement –mediated bacteriolysis against these microbes. Individuals deficient of C9 are usually asymptomatic. Deficiencies of the proteins C5, C6, C7, C8 and C9 are inherited autosomally in recessive condition.
4. The deficiency of properdin and factor D inhibits activation of alternative pathway. This results in recurrent *Neisseria* infections. The deficiency of properdin is X-linked.
5. Some complementary deficiency diseases are caused due to defective regulatory proteins. The loss of the C1Inh protein causes Hereditary Angioneurotic Edema (HANE), an autosomal dominant disease. This disease manifests in form of extensive swelling of laryngeal, facial and intestinal mucosa. The patients suffer from abdominal pain and trachea choking which could be fatal.
6. Deficiency of regulatory protein CD59 or DAF reportedly cause Paraoxysmal Nocturnal Haemoglobulinuria (PNH). This deficiency leads to red blood cell lysis by complement leading to haemolytic anemia, pancytopenia, and thrombosis.
7. The complement receptors- CR3 and CR4 deficiencies results due to gene mutation and cause Leukocyte Adhesion Deficiency (LAD) disorder. LAD causes frequent pyogenic infections, due to inadequate adherence of phagocytes at the site of infection and lower phagocytosis of bacteria.

9.8 SUMMARY

- Jules Bordet led to the discovery of the complement protein system. The Complement cascade system consists of serum proteins that complete the action of the antibodies.
- The activated complement proteins mark and clear the pathogens from the body. These serum proteins are designated by 'C' followed by a numeral and are in the inactive zymogens form, activated after the proteolytic cleavage. Usually, the small fragment generated is denoted as 'a' and the larger one as 'b'.
- The cascade activation of the complement proteins can occur in three ways classical, alternative and lectin pathways.
- The classical pathway is activated by the presence of antigen-antibody complex, and includes activation of C1 macromolecule, C4, C2, C3 and C5 proteins in cascade manner.
- The alternative and mannose lectin pathways are antibody-independent pathways for complement cascade activation. The alternative pathway is activated by spontaneous hydrolysis of C3 proenzyme, then stabilized by binding to microbial surfaces. The alternative pathway involves C3, Factor B, Factor D and properdin.
- The initiation of the Lectin pathway occurs by binding of Mannose Binding Lectin (MBL) to mannose residues on the surfaces of many pathogenic microorganisms. This causes MBL associated serine proteases – MASP-1 and MASP-2, to bind to the pathogen bound MBL, activating C4, C2, C3 and C5 component proteins.
- The last few steps of complement cascade activation, either by Classical, Alternative or Lectin pathway involves C5b, C6, C7, C8 and C9 interacting in a sequential manner to form the macromolecular cytocidal Membrane Attack Complex (MAC).
- The main functions of complement activation are – cell lysis, opsonization, inflammation, immune complex clearance and augmenting the adaptive immune responses.
- The complement system is tightly regulated *in-vivo* to prevent nonspecific damage to host cells. These proteins circulate as zymogens and generate many labile components after activation. Also a series of complement regulating proteins exits which controls the complement cascade at different steps.
- Major regulatory mechanisms control C3 convertase activity which is central to all three pathways of complement activation.
- Many congenital defects in functioning of the complement system occur due to gene mutations and resulting complement protein deficiency. This includes Systemic Lupus Erythematosus (SLE), Hereditary Angioneurotic Edema (HANE), Paraoxysmal Nocturnal Hemoglobulinuria (PNH), Leukocyte Adhesion Deficiency (LAD) and repeated microbial infections.

QUESTIONS

1. Fill in the blanks:
    1. The term "Complement" was given by _________
    2. Majority of circulating complement proteins are in form of _________
    3. After proteolytic cleavage, C2a formed is _________fragment as compared to C2b.
    4. ________ and __________ complement activation pathways are components of innate immunity.
    5. ________ subunit C1 protein interacts with Fc region of antibody for its activation.
    6. The classical pathway C5 convertase and alternative pathway C5 convertase are represented by _________ and _______ respectively.
    7. The activation of Lectin pathway is similar to ________________ and its mechanism of action is very similar to ______________

8. The _________ and ______ peptides are anaphylatoxins produced during complement cascade activation.
9. _____________ bacteria are usually resistant to lysis by MAC due to their thick peptidoglycan layer.
10. The immune complexes are cleared out of the blood with the help of __________ present on the erythrocyte surface which binds to $\overline{C3b}$ associated with these complexes.

2. Select the correct answer of the given options:
   1. The following pathways of complement activation is considered as component of adaptive immunity
      a. Classical pathway
      b. Alternative pathway
      c. Lectin pathway
      d. None of the above
   2. The C1 protein is a large multimeric protein consisting of –
      a. $C1q\text{-}C1r_2\text{-}C1s_2$
      b. $C1q_2\text{-}C1r_2\text{-}C1s$
      c. $C1q_2\text{-}C1r\text{-}C1s_2$
      d. C1q-C1r-C1s
   3. The factor required to stabilize the C3 convertase formed during the alternative cascade pathway is –
      a. Factor B
      b. Factor D
      c. Properdin
      d. Lectin
   4. The deficiencies of complement proteins and their regulatory proteins causes the following diseases
      a. Hereditary angioneurotic edema (HANE)
      b. Paraoxysmal nocturnal haemoglobulinuria (PNH)
      c. Leukocyte adhesion deficiency (LAD)
      d. All of the above
   5. The C3 convertase formation and activity is tightly regulated by
      a. CR1
      b. Membrane Cofactor Protein
      c. Decay Acceleration Factor
      d. All of the above

3. Define the following terms:
   a. Complement fixation
   b. Anaphylatoxins
   c. Membrane Attack complex
   d. Opsonins
   e. Mannose binding lectin

4. Answer these questions briefly:
   i. Why a single pentameric IgM molecule is enough for classical cascade pathway activation?
   ii. Explain the activation of Lectin pathway.
   iii. What are the functions of Membrane Attack Complex?
   iv. Write a short note of systemic lupus erythromatosus caused due to complement protein deficiency.
   v. Explain the regulation of C1 complement protein.

5. Explain the following
   i. Explain the terminal step of complement cascade activation to form membrane attack complex.
   ii. Explain the different properties of complement proteins.
   iii. Briefly write the function of complement cascade activation.
   iv. Explain how C3 convertase regulation occurs in human body?
   v. Explain how antibody dependent activation of complement occurs?

BIBLIOGRAPHY

Books

1. Kindt, T. J., Goldsby, R.A., Osborne, B. A. and Kuby, J. 2006. *Immunology*, VI Edition. New York: W.H. Freeman.
2. David, M., Jonathan, B., David, R. B. and Ivan, R. 2006. *Immunology*, VII Edition. Philadelphia, PA: Mosby, Elsevier Publication.
3. Abbas, A. K. and Lechtman, A. H. 2003. *Cellular and Molecular Immunology*. V Edition. Philadelphia, PA: W.B. Saunders.
4. Owen, J. and Punt, J. 2013. *Kuby Immunology*. VII Edition. Basingstoke: W.B. Saunders Company.
5. Khan, F. H. 2009. *The Elements of Immunology*. Delhi: Pearson Education.

Journals

1. Walport, M. J. 2001. "Complement." *The New England Journal of Medicine* 344, no. 14 (April 5, 2001): 1058–66.
2. Walport, M. J. 2001. "Complement." *The New England Journal of Medicine* 344, no. 15 (April 12, 2001): 1140–44.

10 Tolerance and Autoimmunity

Charu Tripathi
CMP College, University of Allahabad

CONTENTS

10.1 INTRODUCTION

Although the function of the immune system is to target foreign organisms and altered self-cells, however, in some instances, it starts attacking healthy self-cells. Paul Ehrlich referred to this dysfunction as "horror autotoxicus". The pertinent question that arises here – one which bothered the minds of researchers for long – is what causes the immune system to turn onto itself? What components of the immune system differentiate the host's self cells from foreign components and in what circumstances can they go awry? The very components of the immune system that are responsible for recognizing foreign components i.e., antibodies and T-cell receptors (which confer antigen specificity to T cells) also hold the potential for recognizing and being activated in response to self-cells. Such antibodies are called auto-antibodies and such T cells are called self-reactive T cells respectively. There are mechanisms in place which ensure continuous removal of self-reactive T cells and B cells. These mechanisms are referred to as tolerance. In a healthy person, during T and B cell maturation in the thymus and bone marrow respectively, lymphocytes carrying self-reactive antibodies or receptors are eliminated through a process termed as Central Tolerance. Those lymphocytes having a low threshold for host cells are destroyed by the central tolerance mechanism. Self-reactive lymphocytes escaping central tolerance are rendered anergic or destroyed in the peripheral lymph organs by the process known as Peripheral Tolerance. However, it has been found that even healthy adults have circulating self-reactive components. Another safeguard to prevent

DOI: 10.1201/9781003286424-10

the activity of self-reactive lymphocytes is that activated lymphocytes have a small lifespan. After activation, they are destroyed soon by the process of apoptosis. Therefore, it had been understood that the lack of regulation of these normally circulating cells leads them to attack self-components. The failure of mechanisms, to check the proliferation and activity of self-reactive immune components leads to an immune attack on the host itself, this is what is known as autoimmunity. Examples include multiple sclerosis, Hashimoto's thyroiditis, tissue injury due to self-reactive antibodies, systemic lupus erythematosus (SLE), autoimmune hemolytic anemia and Rheumatoid arthritis, to name a few.

There are some factors which influence whether an antigen will be an immunogen or not. An antigen which fails to evoke an immune response due to the fact that the host is tolerant to its presence is known as a tolerogen. High dose, persistence of antigen in the host for a long time and absence of adjuvants, for example, may make a potent immunogen, a tolerogen. The route of entry also determines the immunogenicity of an antigen. An antigen administered orally is less likely to elicit an immune response, however, if the same antigen is delivered intravenously or subcutaneously, it becomes potent enough to evoke an immune response. In addition, the immune response is also dependent on the way an antigen is presented to lymphocytes, as in the case of T cells. An antigen presented to the T cell without proper co-stimulatory signals will fail to initiate an immune response causing a state of unresponsiveness or anergy. Tolerance mechanisms in healthy adults, act by causing cell death or anergy and also by means of a subpopulation of T cells known as T_{reg} (T regulatory) cells.

10.2 MECHANISMS OF TOLERANCE

10.2.1 Central Tolerance

Although a comprehensive understanding of the precise mechanism of action of central tolerance is still lacking, it is known that two processes are essentially responsible for maintaining tolerance, namely, clonal deletion or apoptosis and receptor editing (Figure 10.1).

EXPERIMENTAL FOCUS

C. Goodnow and colleagues first demonstrated experimentally in 1991 that even though self-reactive B cells are formed, however, they never complete maturation and are removed by apoptosis, thus do not appear in circulation. They crossed transgenic mice producing antibodies against hen egg-white lysozyme (HEL) to transgenic HEL mice expressing HEL protein. The F1 mice so produced were screened and found to lack B cells producing anti-HEL antibodies. This happened because the F1 mice produced HEL as well as anti-HEL antibodies, so the antibody-producing cells were deleted before they could enter circulation. Therefore, a central system exists that deletes self-reactive antibody-producing cells.

It is well understood that antibodies and T cell receptors (TCRs) are generated by random gene rearrangements in which the gene for the V region randomly associates with the D or J region (VDJ rearrangements). During the course of such random rearrangements, receptors and immunoglobulins are formed that have an affinity for self-components. The persistence of such cells will lead to auto-immune diseases; therefore, these cells are destroyed by a process known as negative selection or clonal deletion. The sites for negative selection are bone marrow and thymus and it involves clonal deletion of self-reactive clones by apoptosis. David Nemazee and colleagues have also provided experimental proof that apart from clonal deletion, receptor editing also takes place to maintain central tolerance. The specificity of receptors is due to the Variable (V) region, so in the

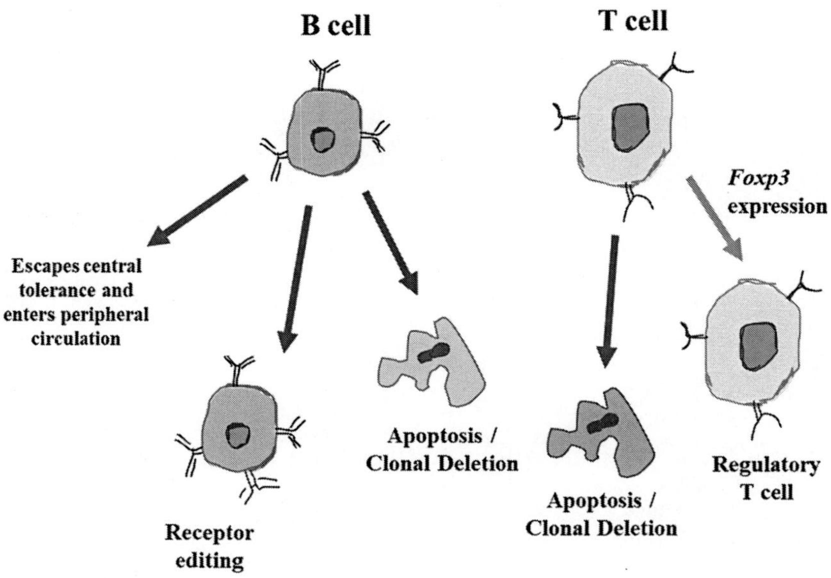

FIGURE 10.1 Schematic overview of central tolerance. Negative selection of B cells that harbor a high affinity to self-antigens takes place by apoptosis and receptor editing. In this process, some cells may be able to escape central tolerance and enter circulation. The negative selection of T cells involves the apoptosis of cells having receptors that exhibit a high affinity for host cells. Cells having intermediate affinity for self-antigens upregulate the expression of *Foxp3* gene and become T_{reg} cells.

process of receptor editing, the V region (specifically the V_L) of self-reactive antibodies is replaced. The same processes also occur in the case of T cells in order to remove self-reactive clones.

10.2.2 Peripheral Tolerance

Some lymphocytes may escape the central tolerance in bone marrow and thymus as it is not necessary that developing lymphocytes will be exposed to each antigen at these locations. Additionally, there is an affinity threshold, which renders the self cells and receptors having low affinity unrecognizable by central tolerance. Therefore, several self-reactive components can be found circulating in the periphery, where they are taken care of by the peripheral tolerance (Figure 10.2). To demonstrate this, an experiment was performed in which transgenic HEL expressing mice and anti-HEL Ig producing mice were used. It was found that when HEL was expressed on the membrane, anti-HEL antibodies were not detected in the periphery. However, when HEL was expressed as a secreted, soluble protein, anti-HEL B cells were detected in the periphery; however, they were non-responsive or anergic. Experiments were performed to investigate the reason for B cell anergy in peripheral tolerance. For this, anti-HEL Ig-producing B cells were introduced into HEL mice. As HEL mice have HEL as their self-antigens, their HEL-reactive T cells have already been clonally deleted by their central tolerance. Now, when anti-HEL antibody-producing B cells are introduced into their peripheral system, they enter a state of unresponsiveness or anergy due to the lack of activation by T cells.

Peripheral T cell tolerance relies on the stringent requirement of precise co-stimulatory signals for the activation of T cells. TCRs not only need antigens to be presented in conjugation with self-MHC cells but also CD28-B7 co-stimulation is required. The co-stimulatory signal is delivered by the binding of the CD28 receptor (on T cells) to B7 present on antigen-presenting cells (APCs). The absence of this necessary signal leads T cells to become unresponsive.

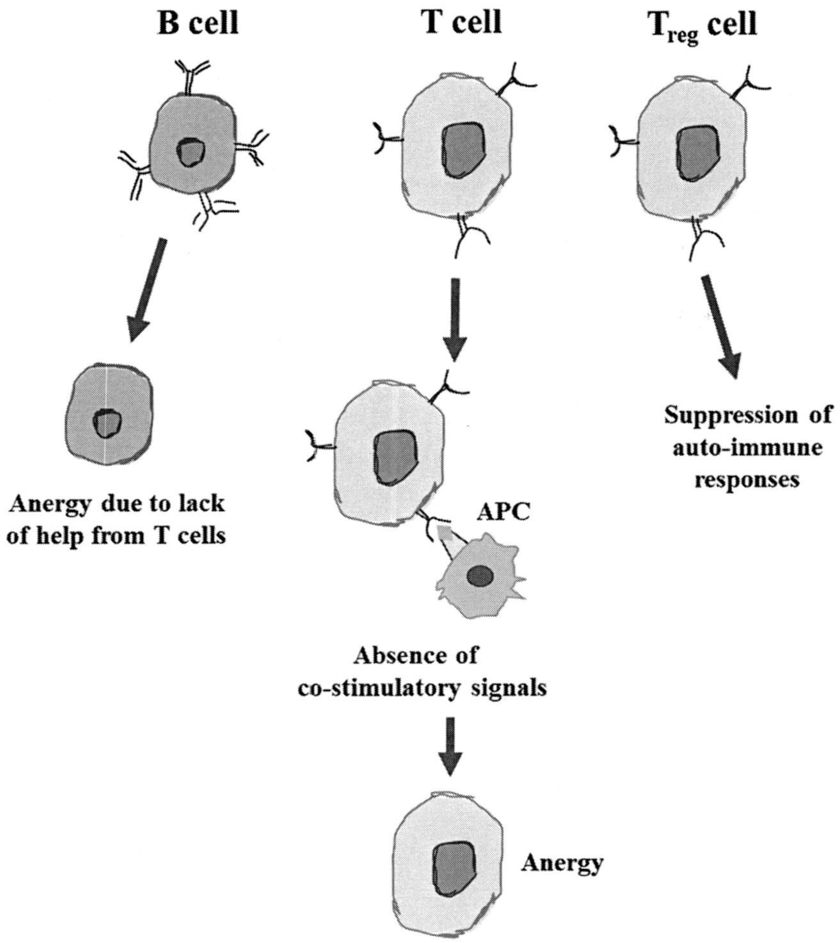

FIGURE 10.2 Schematic representation of peripheral tolerance which is responsible for preventing autoimmune responses occurring from self-reactive cells present in peripheral lymph nodes. B cells having self-reactive antibodies do not get activated in the absence of T cells and enter a state of anergy. T cells having receptors responsive to self-cells encounter self-antigens presented by an antigen-presenting cell on an MHC molecule, however, in the absence of co-stimulatory signals, they do not get activated.

INTERESTING TO KNOW

CTLA-4

CTLA-4 or CD152 is an inhibitory receptor on T cells. Similar to CD28, CTLA-4 binds to B7 on the APCs. However, CTLA-4-B7 interaction, instead of delivering an activation signal, delivers an inhibitory signal. It is important to note here that CTLA-4 expression on T cells only gets upregulated after T cell activation. The CTLA-4 mediated inhibitory signal prevents widespread cell destruction by inhibiting T cells and thus is a component of peripheral tolerance as well. CTLA-4 mediated inhibition is so vital to normal functioning, that mice carrying a genetic deletion of the CTLA-4 gene soon develop autoimmune diseases. Altered gene function of this gene has also been associated with a number of autoimmune diseases such as Hashimoto's thyroiditis, Systemic Lupus Erythematosus, Graves' disease, Celiac disease, Type I diabetes, and many others.

10.2.3 Role of T_{REG} Cells in Tolerance

T_{reg} (T regulatory) cells are cells that downregulate autoimmune reactions in secondary lymphoid organs. These are actually $CD4^+$ T cells that are distinguished by the presence of the IL-2R α chain (CD25) as one of the surface molecules. These cells are differentiated from other T-lymphocyte populations by their intermediate affinity to self-antigens during their maturation in the thymus. During T cell development in the thymus, those which have high affinity to self-antigens are deleted by clonal deletion/apoptosis, ones having low affinity are released and those having intermediate affinity mature to become T_{reg} cells. Once the precursor T_{reg} cells encounter a self-antigen to which they have an intermediate affinity, a transcription factor known as Foxp3 is upregulated, leading to T_{reg} cell development. T_{reg} cells have been known to be responsible for the suppression of autoimmune responses. Although gaps exist in the current understanding of the mechanism of T_{reg} cell-mediated immune suppression, it is known that they suppress immune responses by means of cytokines IL-10 and TGFβ and by inhibiting the proliferation of self-reactive T-cells.

INTERESTING TO KNOW

ANTIGEN SEQUESTRATION

Antigen sequestration can be considered one of the mechanisms of tolerance. If a self-antigen is sequestered and never encounters the immune system in normal conditions, it is protected by default from the immune system. However, if by accident, these antigens come in contact with immune system cells, they will be recognized as foreign and destroyed. Blood-brain barrier and Blood-testes barrier function in antigen sequestration. The brain and the testes are separated from the blood by means of a barrier through which only small molecules such as glucose, oxygen and amino acids can diffuse. One of the principal reasons for the presence of these barriers is to protect these sensitive organs from pathogens, which cannot cross this barrier. However, if this barrier is compromised by trauma, drugs, inflammation or any other agent, an autoimmune reaction can ensue and lead to the destruction of cells present in the CNS and testes.

10.3 AUTOIMMUNE DISEASES

The failure of tolerance mechanisms to protect the host from its own immune system results in an attack by the immune system on the host itself. This leads to tissue and organ damage, which can be fatal in severe cases. A number of autoimmune diseases are known. They are broadly categorized as organ specific or systemic depending on whether a single organ is affected or multiple organs respectively.

10.3.1 Organ Specific Autoimmune Diseases

In these diseases, antigens of only a single organ are targeted by the cell-mediated or humoral immune system. This affects the normal functioning of that organ and causes physical damage to the organ. Ultimately, the damaged tissues are replaced by connective tissue (fibrosis). In other cases, antibodies may cause inhibition or overstimulation of a particular organ function, causing functional impairment or over functioning of the organ. Some organ specific autoimmune diseases are described below:

10.3.1.1 Autoimmune Anemias

Autoimmune anemias are of three types, namely, pernicious anemia, autoimmune hemolytic anemia and drug-induced hemolytic anemia. The deficiency of vitamin B12 causes Pernicious anemia.

Vitamin B12 is required for the process of hematopoiesis. It is absorbed by a protein called intrinsic factor present on gastric parietal cells. In pernicious anemia, auto-antibodies to the intrinsic factor bind to the protein and block the absorption of vitamin B12 in the intestine, leading to anemia. The condition can be alleviated by the administration of vitamin B12 injections.

In patients suffering from autoimmune hemolytic anemia, an antigen present on self-RBCs is targeted by auto-antibodies. Antibodies agglutinate RBCs, after which complement-mediated lysis of RBCs or antibody-mediated phagocytosis occurs, leading to severe anemia. Drug-induced anemia is caused by certain drugs such as Penicillin, anti-hypertensive drug Methyldopa, Cephalosporins and NSAIDs (Nonsteroidal anti-inflammatory drugs) which cause the formation of auto-antibodies against RBCs, leading to their destruction.

INTERESTING TO KNOW

COOMBS TEST

This is a detection test for determining the occurrence of autoimmune hemolytic anemia. The test detects the presence of auto-antibodies to the RBC antigens, present in the blood. For performing this test, blood is added to anti-human immunoglobulin antiserum which contains antibodies against the antibodies which bind to RBC antigens. Agglutination of cells gives confirmation of the presence of auto-antibodies in the blood.

10.3.1.2 Goodpasture's Syndrome

In this syndrome, auto-antibodies are formed against basement membrane antigens. Kidney glomeruli and lung alveoli are the sites whose basement membranes are targeted, causing damage and deterioration over the long term. Subsequently, there is further deterioration due to complement activation and inflammatory responses. Over time, this causes damage to kidney functions and pulmonary hemorrhage. Death may result within months after the onset of this condition.

10.3.1.3 Hashimoto's Thyroiditis

The most affected group with Hashimoto's thyroiditis is middle-aged women. The disease occurs when auto-antibodies and sensitized T_H1 lymphocytes are directed against thyroid gland proteins. If self-reactive antibodies are generated against thyroid proteins thyroglobulin and thyroid peroxidase, then hypothyroidism may result, as these two proteins are responsible for iodine uptake in the thyroid gland. In this condition, the thyroid gland is infiltrated by lymphocytes, plasma cells and macrophages. A delayed-type hypersensitivity (DTH) reaction follows, causing the destruction of cells in germinal centers and lymphocytic follicles. Due to the destruction of cells, hypothyroidism results, leading to the enlargement of the gland (goiter).

10.3.1.4 Insulin Dependent Diabetes Mellitus (IDDM)

IDDM is an autoimmune disease targeting the β cells of the Islets of Langerhans located in the pancreas. As the β cells are the producers of insulin, the destruction of these cells leads to an increase in blood glucose levels. Sensitized CTLs (cytotoxic T lymphocytes) attack the β cells, leading to infiltration by activated macrophages and release of cytokines IFN-γ, TNF-α and IL-1. This stage of IDDM is known as insulitis. Insulitis is characterized by the lysis of β cells due to lytic enzymes secreted by macrophages. IDDM may also be caused by auto-antibodies alone or in conjugation with activated CTLs. In the presence of cytokines and auto-antibodies, a cell-mediated DTH, ADCC (antibody-dependent cell-mediated cytotoxicity) or antibody-mediated complement lysis may follow, causing more damage to the β cells.

10.3.1.5 Graves' Disease

The cause of Graves' disease is the formation of auto-antibodies known as long-acting thyroid-stimulating (LATS) antibodies which bind to TSH (thyroid stimulating hormone) receptors on thyroid cells. TSH is produced by the pituitary and binds to TSH receptors present on thyroid cells, stimulating them to produce adenylate cyclase which signals the secretion of the hormones thyroxine and triiodothyronine. When auto-antibodies bind to TSH receptors, there is a continued and unregulated stimulation of thyroid cells leading to excess secretion of thyroid hormones, a condition known as hyperthyroidism.

10.3.1.6 Myasthenia Gravis

Myasthenia gravis results due to the formation and binding of antagonistic antibodies that bind to acetylcholine receptors which are present on the motor end plate of muscles (Figure 10.3). These antibodies, when bound to acetylcholine receptors, block the binding of legitimate ligand acetylcholine and thus inhibit the stimulation of skeletal muscles when needed. Additionally, complement-mediated lysis ensues following the binding of auto-antibodies, leading to the deterioration of the skeletal muscle membrane and receptors, causing further damage. In the early stages of myasthenia gravis, patients have drooping eyelids and are unable to contract the corners of their mouths, giving a snarling appearance. With treatment, the condition can be managed, but without treatment, the disease progresses in such a way that even routine body movements become difficult.

10.3.2 Systemic Autoimmune Diseases

In systemic autoimmune diseases, autoimmunity targets not only a single antigen or organ, but multiple antigens present throughout many organs. Systemic autoimmune diseases are caused by a

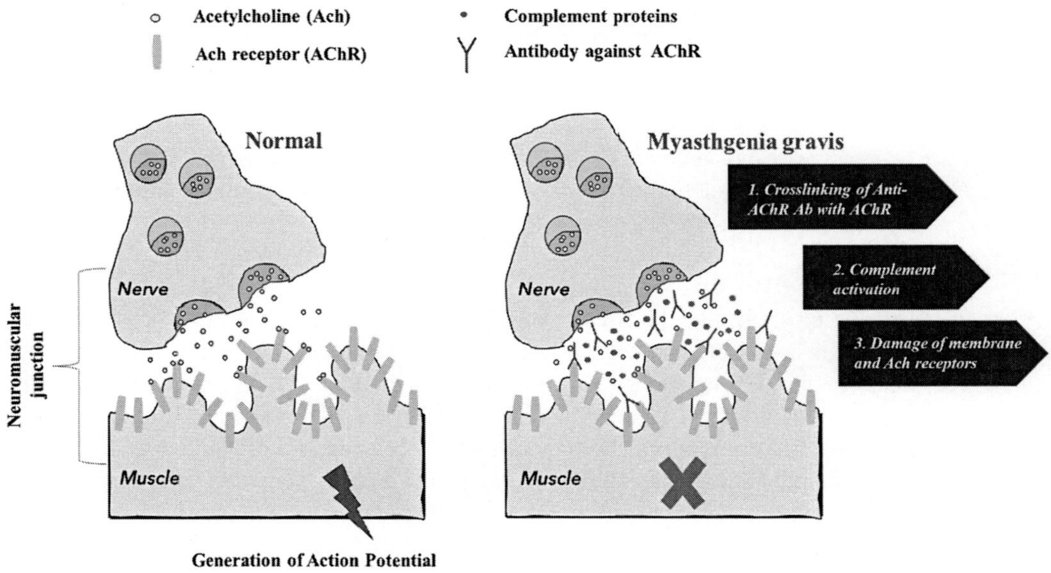

FIGURE 10.3 Schematic diagram showing the development of myasthenia gravis at a neuro-muscular junction. The action potential is generated by the binding of Acetylcholine released by the nerve to Acetylcholine receptors on the membrane of skeletal muscles (left panel). The auto-immune condition leading to myasthenia gravis is due to the generation of antibodies against Acetylcholine receptors which bind to and block the receptors, thereby hindering the generation of action potential (right panel). Henceforth, the complement system also gets activated, damaging the receptors and membrane.

general malfunctioning of the immune system. There is widespread tissue damage throughout the body due to direct damage by auto-antibodies and by cell-mediated immune responses.

10.3.2.1 Systemic Lupus Erythematosus (SLE)

SLE is most commonly observed in women between the ages of 20–40 years. It is more prevalent in women (ten times more than in men), especially of African American and Hispanic origin as compared to Caucasians, although the reason for this is not very clear. People with SLE experience weakness, fever, arthritis, pleuritis, kidney dysfunction and skin rashes. A number of tissue antigens, including red blood cells, platelets, DNA, histones, clotting factors, leucocytes and many others are targeted by auto-antibodies leading to different kinds of pathologies. Auto-antibodies bind to RBCs and platelets, thus forming immune complexes activating the complement system. Complement-mediated lysis then leads to hemolytic anemia due to the destruction of RBCs and thrombocytopenia due to the destruction of platelets. The walls of small blood vessels are lined with antigen-antibody complexes formed by binding to nuclear components of endothelial cells. Following this, a type III hypersensitivity reaction ensues, activating the complement system. Blood vessels gradually become damaged by inflammation of the vascular system, causing vasculitis. The glomerulus of kidneys is also affected causing glomerulonephritis. Increased complement activation leads to a three to four times increase in serum levels of complement split products C3a and C5a. C5a induces the expression of a receptor on neutrophils known as type 3 complement receptor CR3. This receptor causes the aggregation of neutrophils and their attachment to vascular endothelium causing a decrease in the number of neutrophils in circulation (neutropenia) and blockade of blood vessels (vasculitis). Over the long term, it leads to tissue damage. SLE can be diagnosed by detecting anti-nuclear antibodies targeting DNA, RNA, histones, nucleolar RNA and nuclear proteins. Immunofluorescent staining can be used for detection.

10.3.2.2 Multiple Sclerosis

Multiple sclerosis (MS) is one of the leading causes of neurological disorders in Western countries. It is usually diagnosed at the age of 20–40 years with mild symptoms such as numbness in limbs. The symptoms progress with the progression of the disease and may cause severe paralysis and blindness. In multiple sclerosis, activated T cells destroy the myelin sheath of nerves. As the myelin sheath of nerve fibers is responsible for its insulation and proper conduction of nerve impulses, therefore the insulation is lost leading to neurologic disorders. The cerebrospinal fluid of patients with MS has auto-reactive T cells which even enter the brain and cause lesions.

10.3.2.3 Rheumatoid Arthritis

Rheumatoid arthritis most commonly affects women aged 40–60 years causing chronic inflammation of joints. In this condition, auto-antibodies are generated against the Fc region of IgG antibodies. These auto-antibodies, referred to as Rheumatoid factors are IgM immunoglobulins. As IgMs are pentameric in nature, IgM-IgG complexes are bulky and get deposited in joints leading to complement activation, followed by type III hypersensitivity reaction, all of which lead to chronic inflammation of the joints. Sometimes, not only the joints but also the hematological, cardiovascular and respiratory systems may also get affected.

10.4 SUMMARY

- There are immune system failures which result in the immune system not being able to distinguish self from non-self, mounting an attack on its own tissues and organs, resulting in autoimmunity.
- Mechanisms that exist to prevent self-reactive lymphocytes from destroying self-cells are known as tolerance. Self-reactive lymphocytes are those T cells that have self-reactive

TCRs and those B cells that produce antibodies against self-cells. These are generated in primary lymphoid organs where the maturation of T and B cells take place, i.e., thymus and bone marrow respectively. Here, any such self-reactive cells having more than a low threshold affinity against self-antigens are destroyed by apoptosis. This is known as negative selection or clonal deletion and comprises Central Tolerance.

- The self-reactive lymphocytes that are not eliminated by central tolerance, are dealt with the peripheral tolerance in the secondary lymphoid organs. The lack of T cell-help to activate causes self-reactive B cells to become anergic in the absence of the former. Self-reactive T cells, having a stringent requirement of co-stimulatory signals, do not get activated in their absence, which checks T cell activation in the peripheral circulation.
- Increasing scientific studies have also pointed towards the crucial role of T_{reg} cells in suppressing auto-immune responses. These $CD4^+$ T cells, with CD25 as their distinguishing surface marker, have an intermediate affinity for self-cells.
- Antigen sequestration, wherein some antigens are never exposed to immune cells in primary lymphoid organs is another strategy for tolerance.
- Autoimmune diseases occur when tolerance mechanisms fail and auto-antibodies or self-reactive T cells are generated attacking the host's own cells and tissues causing and accelerating the damage.
- Organ-specific autoimmune disorders result when self-reactive cells targeting a single self-antigen are generated such that damage is restricted to a single organ or tissue. In some cases, auto-antibodies act as agonists or antagonists, binding to certain receptors and stimulating or blocking them respectively. Examples of organ-specific autoimmune diseases are Hashimoto's thyroiditis, autoimmune anemias (pernicious anemia, autoimmune hemolytic anemia and drug-induced hemolytic anemia), Goodpasture's syndrome, insulin-dependent diabetes mellitus, Graves' disease and Myasthenia gravis.
- Systemic autoimmune diseases are those in which self-reactive T cells and auto-antibodies targeting various organs and tissues are generated, such that they cause damage to multiple organs. Examples of systemic autoimmune diseases include systemic lupus erythematosus, multiple sclerosis and rheumatoid arthritis.

QUESTIONS

1. What is the significance of tolerance?
2. Define: central tolerance, peripheral tolerance, autoimmunity.
3. Describe the mechanisms by which central tolerance functions.
4. What is the role of peripheral tolerance and how is it achieved?
5. What are T_{reg} cells? What is their significance in tolerance?
6. What is blood-brain barrier? What will happen if it is compromised?
7. Differentiate between organ specific and systemic autoimmune diseases.
8. With the help of the following statements, identify the autoimmune disorders:
   i. Auto-antibodies bind to intrinsic factor, interrupting the absorption of vitamin B12.
   ii. Auto-antibodies are formed against an antigen present on self-RBCs.
   iii. Auto-antibodies are produced that bind to TSH receptors on thyroid cells.
   iv. Auto-antibodies and sensitized T_H1 lymphocytes are directed against thyroid gland proteins.
   v. An autoimmune disease in which activated CTLs and/or auto-antibodies targeting the β cells of pancreatic Islets of Langerhans are formed.
   vi. Activated T cells destroy the myelin sheath of nerves.
   vii. Auto IgM antibodies are generated against the Fc region of IgG antibodies, forming IgM-IgG complexes.

BIBLIOGRAPHY

1. Kindt, T.J., Goldsby, R.A. and B.A. Osborne. 2006. *Kuby Immunology.* Sixth edition. W.H. Freeman and Company, New York.
2. Delves, P.J., Martin, S.J., Burton, D.R. and I.M. Roitt. 2011. *Roitt's Essential Immunology.* Twelfth edition. Wiley-Blackwell, USA.
3. Janeway, C.A. Jr., Travers, P. and M. Walport. 2001. *Immunology: The Immune System in Health and Disease.* Fifth edition. Garland Science, New York.

11 Hypersensitivity

Vatsala Tripathi and Ritu Rai
University of Delhi

CONTENTS

11.1 INTRODUCTION

The immune system present in our body ensures protection from various infections. There are times when the same immune system shows exaggerated responses which prove to be detrimental to the body. These abnormal reactions may be damaging or cause various degrees of discomfort, or at times are fatal. When the normal immune system exhibits these kinds of unwanted reactions they are referred to as Hypersensitivity Reactions.

DOI: 10.1201/9781003286424-11

Table 11.1
Different Types of Hypersensitivity Reactions

Type	Immune Reactant	Method of Detection
I	IgE	Provocation test, Skin test, IgE RAST
II	IgM or IgG, Complement, MAC	IgG serum test
III	IgG, IgM, Complement, Neutrophils	IgG serum test
IV	T-cells	Skin test, MELISA®

Allergies and several autoimmune disorders are covered under the term hypersensitive reactions. The distinction between these two types is that the immune reaction in response to antigens or allergens (exogenous) are referred to as allergies and those that occur in response to autoantigens (endogenous) are referred to as autoimmune diseases. These reactions develop when the individuals have been at least once exposed to the antigen earlier. Subsequent encounter with the same antigen leads to a symptomatic reaction in the sensitized individuals.

P. G. H. Gell and R. R. A Coombs in 1963 recommended that these reactions are classified into four groups (Table 11.1) which is well established. Based on the types of effector molecules generated, drug allergies are distinguished into four pathophysiological types. Each kind involves well-defined mechanisms, cells and mediator molecules. Hypersensitive reactions (Types I II and III) occur by a humoral branch which is mediated by antibody or antigen-antibody complexes and immune complex reaction whereas type IV hypersensitive reactions occur by cell-mediated branch (Figure 11.1).

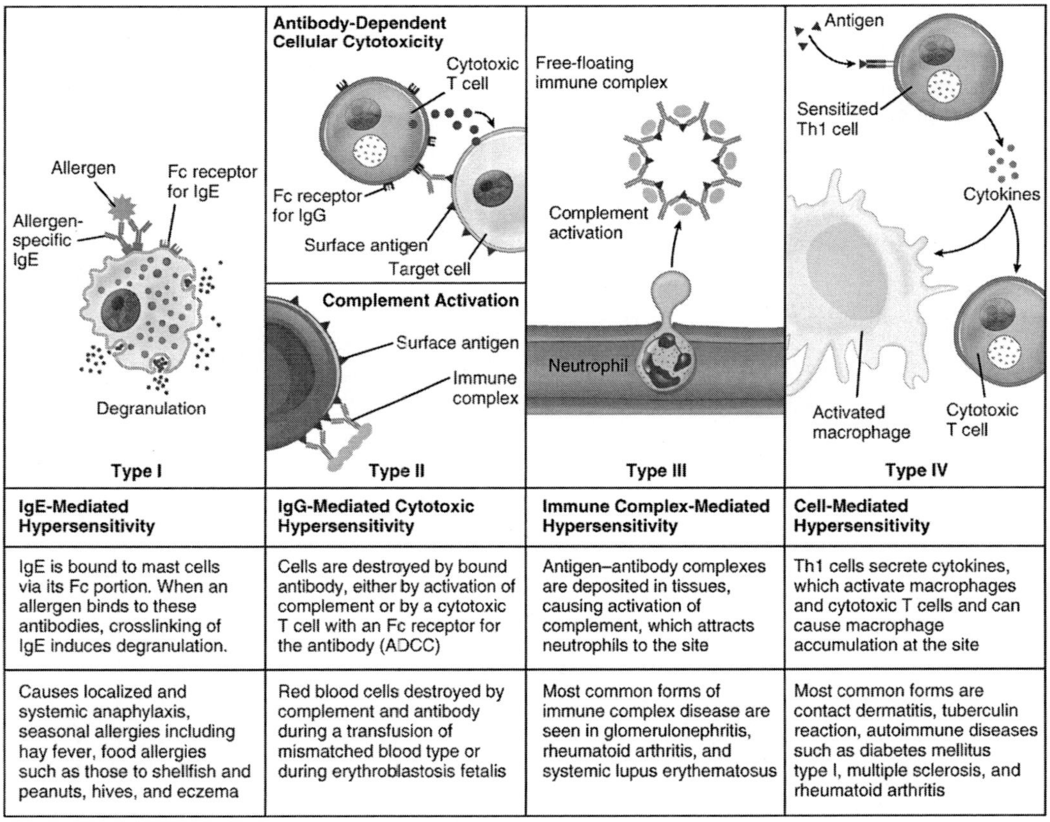

FIGURE 11.1 Different types of hypersensitivity reactions. (Reproduced from "File:2228 Immune Hypersensitivity new.jpg" by OpenStax College is licensed under CC BY 3.0.)

11.2 TYPE I HYPERSENSITIVITY

Type I hypersensitivity is additionally referred to at times as immediate hypersensitivity. This is an allergic reaction which occurs instantly on exposure to an antigen to which the individual is sensitized earlier. The time of reaction may vary from minutes to hours. Antigens eliciting these allergic reactions are called allergens e.g. dust, pollen, drugs, certain foods, proteins (vaccines), insect products (bee venom), microbes etc. Individuals may get exposed to these allergens by direct contact or by ingestion, inhalation and injection. As soon as the individual encounters an allergen for the first time, the body elicits an IgE antibody response. This sensitizes an individual to a Type I hypersensitivity reaction as soon as it gets a future exposure to the same allergen. Type I hypersensitivity reactions are quite common.

This type of hypersensitive reaction is elicited when the individual is exposed to an allergen. B cells get activated and synthesize IgE-secreting plasma cells instead of IgG, IgA or IgM. IgE has the ability to attach to F_c receptors with a great affinity. These receptors are located on the surface of tissue mast cells and blood basophils. The IgE-coated mast cells and basophils are referred to as sensitized. Crosslinking of IgE on sensitized mast cells and basophils occur on subsequent exposure to identical allergen promoting their degranulation. This leads to the freeing of pharmacologically active mediators which leads to vasodilation, increased vascular permeability and smooth-muscle contraction. The magnitude of the mediator released determines whether the effect would be systemic or localized. The development of wheal and flare reaction is the most classic manifestation of immediate hypersensitivity (Figure 11.2).

11.2.1 Components Involved in Type 1 Reaction

11.2.1.1 Allergen

The serum levels of IgE usually increase in humans as a defense against parasitic infections and decrease as they are removed from the body. However, few individuals show a hereditary predisposition to the development of immediate hypersensitivity reactions. These individuals have increased serum levels of IgE even on exposure to non-parasitic antigens (common environmental antigens). These individuals are referred as atopic individuals and have inappropriate IgE production which is due to genetic predisposition. These persons are at greater risk to allergic disorders e.g., allergic rhinitis (hay fever), atopic dermatitis (eczema) and asthma. The response seen in the affected individuals by these allergens is dependent not only on the genetic constitution but also on dosage and the route of sensitization.

11.2.1.2 IgE Antibody

IgE antibody is short-lived in serum. The half-life is somewhere between 2 and 3 days. However, receptor-bound IgE is stable for weeks eventually leading to significant clinical implications. The IgE bind with great affinity to the mast cells and basophils.

11.2.1.3 Mast Cells and Basophils

Mast cells and basophils have pharmacologically active mediators stored within them which are set free in type I hypersensitivity reaction. Basophils release histamine and mast cells release large diversity of cytokines (GM-CSF, TGF-β, TNFα, IL-1, IL-5, IL-6) that elicit diverse immunological responses.

11.2.1.4 IgE-Binding Fc Receptor

Depending on the capability to tether to receptor that is specific for the Fc region of the € heavy chain the activity of IgE is defined. There are two types of these receptors, FcεRI and FcεRII of which FcεRI binds with much greater affinity to IgE as compared to FcεRII.

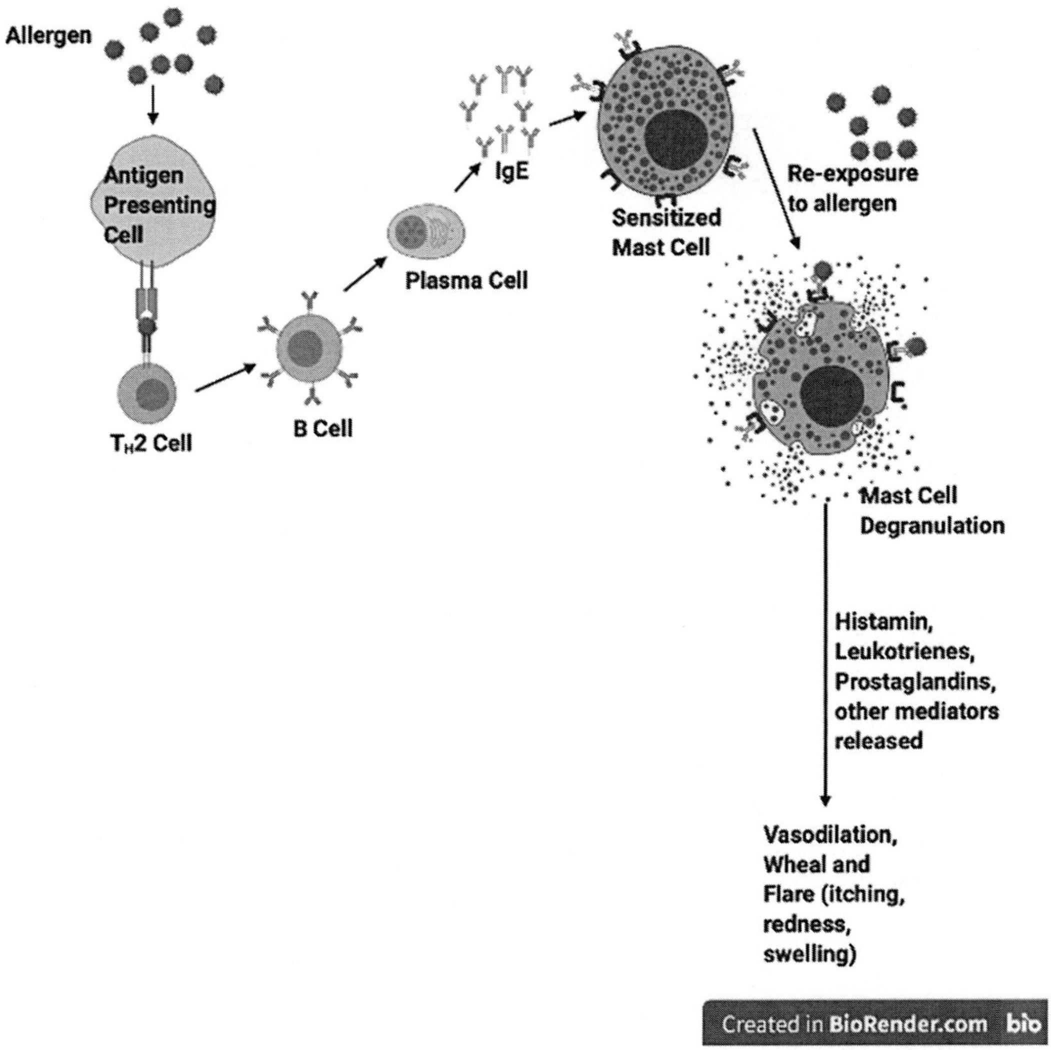

FIGURE 11.2 Diagrammatic illustration of Type I Hypersensitivity. Whenever an individual is first exposed to an allergen, the Antigen Presenting Cells processes and presents it to T_H cells. T_H cells produce interleukins which activate the B cells. Activated B cells produce IgE, which is the main effector molecule in Type I Hypersensitivity. IgE bind to mast cells with their F_c region, sensitizing the mast cells. On re-encounter with similar allergen the IgE molecules crosslink leading to mast cell activation and degranulation. The chemicals released during degranulation are responsible for the clinical manifestations seen during type I hypersensitivity. (Adapted from *Elsevier's Integrated Review Immunology and Microbiology*, 2nd Ed., 2012.)

11.2.1.5 IgE Cross-Linkage Initiates Degranulation

When two or more FcεRI molecules crosslink which occurs on binding of allergen to IgE-receptor complex it facilitates degranulation of mast cells and leads to:

- Activation of protein tyrosine kinase leading to tyrosine phosphorylation
- This induces the production of a number of secondary messengers.
- Membrane phospholipids phosphatidylethanolamine is methylated to phosphatidylcholine which increases the fluidity of the membrane.

- This favours the opening of Ca^{2+} channels. Due to the freeing of Ca^{2+} from the intracellular storage site (endoplasmic reticulum) and uptake of extracellular Ca^{2+}, the total concentration of calcium ions increases which is very important for the degranulation of mast cells.
- The enzyme phospholipase A_2 gets activated because of an increase in Ca^{2+} ions which leads to the genesis of arachidonic acid which is transformed to inflammatory mediators such as prostaglandins and leukotrienes.
- Microtubules assembly and microfilament contraction are also assisted because of the rising level of calcium ions. This is essential for the movement of granules to plasma membrane.
- Along with phospholipid methylation and an increase of calcium ions, a short-term rise of cAMP is also observed around 1 minute after the FcεRI crosslink.
- This alters the permeability of the granules to water and calcium ions.
- The granules therefore increase in size and this enables their fusion with the plasma membrane and allows the mediators to leave. For degranulation to proceed the decrease in cAMP is necessary which is mediated by kinases.

11.2.2 Mediators of Type I Reactions

Type I hypersensitivity shows different expressions across different species or different tissues due to variations in the release of mediators. Various known mediators are histamine, heparin, leukotrienes, prostaglandins, platelet-activating factor, cytokines such as IL-5, IL-8, TNF, tryptase, and kininogenase. These mediators are classified into two types: primary and secondary. These mediators affect local tissues and effector cells such as monocytes, T lymphocytes and platelets.

The pharmacological effects seen due to the release of these mediators are vasodilatation, capillary permeability, bronchoconstriction, platelet aggregation or may act as a chemo-attractant.

11.2.2.1 Primary Mediators

Primary mediators are deposited in granules and are generated prior to degranulation e.g. histamine, serotonin, proteases (tryptase, chymase), eosinophil chemotactic factor, neutrophil chemotactic factor and heparin. Histamine, a crucial contributor in allergic diseases is majorly stored in mast cells and basophils. Within minutes of mast cell activation, the biological effects of the release of histamines can be observed. H_1, H_2, H_3 and H_4 are four different kinds of receptors for histamine that have been recognized. When mast cells release histamine, it binds to the target cells that have specific receptors that produce clinical allergic symptoms. Effects such as smooth muscle contraction, increased vascular permeability and mucus secretion are usually seen when histamine binds to the H_1 receptor whereas stimulation of exocrine glands, increased vaso-permeability and dilation are observed when it binds to the H_2 receptor. Vasodilation and decreased peripheral vascular resistance are seen on stimulation of both receptors. Binding to H_2 receptors leads to feedback inhibition on the release of histamine as it suppresses the degranulation of mast cells and basophils.

11.2.2.2 Secondary Mediators

Platelet-activating factors, leukotrienes, prostaglandins, bradykinin and varied cytokines, are secondary mediators that are produced and liberated by activated mast cells and basophils. The effects caused are bronchoconstriction, increased vascular permeability and increased mucus productions, which are more sustained in comparison to histamine. Secondary mediators are synthesized only after mast cells degranulation and phospholipids in the plasma membrane are degraded by enzymes such as phospholipases.

The complexity of type I reaction is enhanced by cytokines. Human mast cells secrete IL-1, IL-3, IL-4, IL-5, IL-6, TNF and colony-stimulating factors (CSF). These cytokines have the ability to alter the microenvironment. The increase in IgE production is attributed to IL4. Inflammatory cells such as neutrophils and eosinophils are engaged and activated by the cytokine IL-5. Mast cells secrete large concentrations of TNF which activate endothelial expression of adhesion molecules and may give shock in systemic anaphylaxis.

11.2.3 Clinical Manifestations of Type I Reactions

Systemic or localized are two types of clinical manifestations seen in type I reactions. Localized reactions are put up in one area whereas a systemic reaction progresses from one organ to other organ systems in an individual. They show diverse effects ranging from being fatal, such as systemic anaphylaxis and asthma to merely annoying such as eczema. Inflammation is a common response seen in individuals due to contact with toxins, allergies or infections.

11.2.3.1 Systemic Hypersensitive Reactions

Systemic anaphylaxis occurs when mast cells and basophils release mediators that elicit a systemic response. These allergic events are many times life-threatening. The chief targets are usually the cardiovascular, integumentary, respiratory and gastrointestinal as here the number of mast cells is profuse. During anaphylactic shock, many systems are affected together. For instance, the integumentary system experiences discomfort in the form of inflammation itching, redness etc. During anaphylaxis, the respiratory system may show shortness of breath and/ or the circulatory system may show a drop in blood pressure. These reactions are seen in individuals susceptible to antigens such as bee venom, ant stings, etc.; drugs such as penicillin, insulin and antitoxins; seafood and nuts. The key to effective treatment is to identify the systemic reaction quickly and seek medical attention instantly. The treatment of such systemic reaction is based on whether the reaction is allergic, toxic or septic. All systemic reactions are not lethal, but care needs to be taken as the condition may turn critical if the infection affects different organs simultaneously.

The most effective drug for the treatment of systemic reaction is epinephrine as it stalls the effects of primary and secondary mediators by increasing the concentration of cAMP in mast cell. This prevents degranulation and vascular collapse which is a characteristic feature of an anaphylactic reaction by improving cardiac output.

11.2.3.2 Localized Hypersensitive Reactions

11.2.3.2.1 Localized Anaphylaxis (Atopy)

An immediate (Type I) hypersensitivity is restricted to only some tissue/organ resulting in manifestations like reddening of the skin, watery eyes, hives, etc. and they may also be hereditary. Some of the examples are mentioned below:

Allergic rhinitis or hay fever is an allergic reaction to airborne allergen such as pollen, dust, mould, etc. Binding of allergen to sensitized mast cells prompts the liberation of pharmacologically active mediators from them which leads to localized vasodilation and increased capillary permeability. Common symptoms include runny nose, congestion along with sneezing and coughing.

Asthma triggers vary between individuals. Asthmatic attack may be triggered in some individuals on exposure to specific allergens like pollens, dust, fumes, viral antigens etc. while in some it can occur due to strenuous exercise or inhaling cold dry air. Degranulation of mast cells occurs in asthma just as in hay fever which results in the release of mediators and cytokines. Bronchoconstriction occurs in asthma and so people experience difficulty in breathing. The treatment is done with bronchodilators which relax bronchial muscles, expectorants that help in the discharge of mucous which is accumulated in alveoli and anti-inflammatory steroids to provide relief.

Food allergies: Localized anaphylaxis reaction is observed in individuals who are allergic to specific food. These allergies are mediated by IgE and therefore fall under type I hypersensitivity reactions. Symptoms may range from mild (vomiting) to severe.

Atopic dermatitis is also referred to as allergic eczema. It is associated frequently with family history. It is an inflammatory disease of the skin and common symptoms are itching and reddening of the skin. It is quite common in young children and affected individuals show elevated levels of IgE in serum.

11.2.3.2.2 Late-Phase Reactions Induce Localized Inflammatory Reactions

The late phase reaction begins after the type I hypersensitive reaction starts receding. This is also referred to as the "second" round of inflammatory reaction. It starts from 4 to 6 hours after the initial type I reaction and continues for 1–2 days. The mediators released are cytokines and platelet-activating factors. The reaction is characterized by the infiltration of neutrophils, eosinophils, macrophages, lymphocytes and basophils.

11.2.3.3 Detection Methods

Type 1 hypersensitive reaction can be identified by skin test or by measuring serum IgE levels. The skin tests can be performed by pricking or scratching the skin with the help of a needle that is exposed to a little quantity of antigen. A patch test can also be done by applying a patch exposed to the known antigen on the surface of the skin. A positive skin test usually occurs within 30 minutes in the form of a wheal (swelling due to fluid accumulation) and flare (reddening of the skin) at the site of injection. This is more widely used as it is cost-effective, non-invasive and also permits to screen various allergens in a single test. However, the limitation is that sometimes it may sensitize the individual to a new allergen which may be detrimental.

Another method to detect Type I hypersensitivity reaction is via radioallergosorbent tests (RASTs) or radioimmunosorbent test (RIST). In this test, the concentration of IgE antibody which reacts with allergen is quantified. An increased level of IgE to a specific antigen shows that the person is allergic to that particular antigen. Leukocyte histamine release assays and surface markers for basophil activation are various other tests used to detect a Type I hypersensitivity reactions.

11.2.3.3.1 Prausnitz–Küstner Test (P-K TEST)

Carl Prausnitz and Heinz Küstner in 1921 did an experiment that generated an immediate wheal and flare reaction. Küstner after eating fish developed symptoms of clinical manifestations. Prausnitz injected Küstner's serum in his arm intracutaneously and later injection of the same fish extract evoked an immediate wheal and flare response. This test was used by physicians to identify if a person is allergic to a particular antigen but now it is no longer in practice as it may lead to blood-borne diseases such as variant Creutzfeldt–Jakob disease, AIDS, etc.

11.2.3.4 Therapeutic Measures for Type I Hypersensitivities

- Stay away from known allergens.
- The intensity of type I reactions can be reduced or eliminated completely by allergy shots (hyposensitization). In this immunotherapy technique, the allergens are injected subcutaneously at certain durations which vary from weeks to months.
- Drug therapy e.g., antihistamines, cromolyn sodium, theophylline, epinephrine, cortisone.

11.3 TYPE II HYPERSENSITIVITY

Type II hypersensitivity reaction is additionally referred to as cytotoxic reaction. The cell destruction is mediated by antibody (IgG or IgM). Destruction of cells can occur by complement-mediated cell lysis, antibody-dependent cell-mediated cytotoxicity (ADCC) or by opsonisation. These reactions usually take between 2 and 24 hours to develop e.g. Blood transfusion, erythroblastosis fetalis, drug-induced hemolytic anemia.

11.3.1 ABO Blood Transfusion Reaction

A, B, AB and O are the four major blood types. The presence or absence of the antigen A and B determines the blood group of an individual. There are some minor antigens also such as Rh, Kell, Duffy, etc. These antigens play an important role in blood transfusion reactions. When a person receives blood from another individual having a different blood type then it will show an antibody response as it considers the transfused blood as foreign due to difference in the allelic form of a blood-group antigen. Isohemagglutinin are antibodies to ABO antigen and they belong to the IgM class but on the contrary antibodies to the minor antigen are of the IgG class. In order to prevent a transfusion reaction, there must be compatibility between the blood of donor and the recipient. For example, individual whose blood type is B can only receive blood from a donor having either blood group B or O and not from individuals whose blood type is either A or AB. On transfusion mismatch, incompatible red blood cells will be coated by the isohemagglutinins and cause severe reactions. Lysis of RBC occurs intravascular and free hemoglobin can be detected in urine (hemoglobinuria). Hemoglobin may be transformed into bilirubin but it is extremely toxic to tissues. Clinical manifestations can be immediate (IgM class) or delayed (IgG) which occur between 2 and 6 days. Upon mismatch of minor antigen such as Rh a delayed reaction is observed. The transfused RBCs undergo lysis by the process called opsonisation (IgG antibody) and no free hemoglobin is detected in urine. In case a mismatch is discovered the transfusion should be instantly terminated and diuretic should be prescribed to increase the flow of urine to prevent disastrous outcomes.

11.3.2 Hemolytic Disease of New Born (Erythroblastosis Fetalis)

Erythroblastosis fetalis is a condition that arises due to blood group incompatibility between fetus and the mother. This is a severe autoimmune condition that involves an inherited protein called Rh factor. This is present on the surface of erythrocytes. The presence or absence of this protein determines if the individual is Rh positive or Rh negative respectively. Complications are bound to occur in a pregnancy in which the Rh factor of fetus and mother is not identical. Placenta, the feto-maternal organ prevents mixing of fetal and maternal blood with each other. Therefore, if Rh^+ woman has Rh^- fetus or vice versa during first pregnancy the complications do not arise it does not encounter enough fetal red blood cells to activate antibodies as it takes time. But during parturition, a copious amount of blood from the umbilical cord enters into mother's body. The body of the mother considers this blood as a foreign substance because of the difference in the Rh factor and forms antibodies in the mother. These antibodies cause problems in subsequent pregnancies with Rh^+ fetus as the memory cells get activated during second pregnancy and form IgG anti-Rh antibodies, which may cross the placenta and destroy the blood cells of fetus (Figure 11.3). If preventive measures are not taken, complications such as mild to severe anemia are seen to develop in the baby. It also may have fatal consequences. Treatment involves administration of RhoGAM antibodies to prevent Rh immunization to the mother within 24–48 hours after the first childbirth.

11.3.3 Drug-Induced Hemolytic Anemia

Drug-induced hemolytic anemia is a disorder in which the defense system detects its own red blood cells as foreign and leads to its destruction by a process hemolysis. Some antibiotics (e.g., penicillin, cephalosporin and streptomycin) can get non-specifically adsorbed to proteins on erythrocyte membrane, leading to the formation of a complex closely resembling a hapten-carrier complex. In a few cases, these drug-protein complexes influence the synthesis of antibodies, which then attach to the drug adsorbed on the membrane of erythrocytes. This initiates complement-mediated lysis of RBC leading to anemia. Hemolytic anemia disappears when the drug is withdrawn. In certain individuals, this disorder is also due to a lack of enzyme glucose-6-phosphate dehydrogenase.

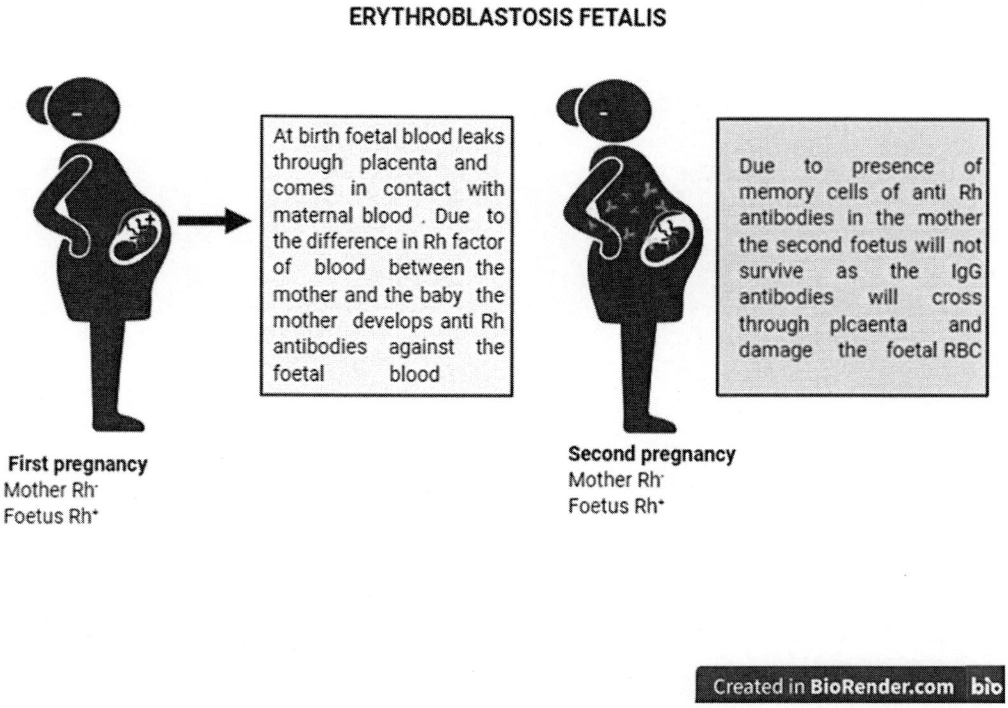

FIGURE 11.3 Diagmrammatic representation showing the development of erythroblastosis fetalis.

11.4 IMMUNE COMPLEX–MEDIATED (TYPE III) HYPERSENSITIVITY

Immune complexes are formed when the antibody reacts with antigen and are continuously eliminated from blood circulation by phagocytes, macrophages or by the complement system. Protection to host cells from complement attack is imparted with the help of proteins that block the genesis of membrane attack complexes and quicken the inactivation of complement components. However, if the host cells do not generate these inactivator proteins at an optimum rate then the complement system gets activated via a classic pathway due to the generation of immune complexes. Formation and aggregation of a large amount of immune complexes result in tissue damage along with inflammation leading to Type III hypersensitivity reactions. These reactions are more commonly found in areas such as blood vessel endothelium, glomerular basement membrane of kidney, synovial membrane of joints and choroid plexus of the brain. The Type III hypersensitivity reactions are moderated by mainly IgG and IgM antibodies. The reactions may develop within a few hours to several days of antigen exposure. The immensity of the reaction depends on the amount formation of immune complexes and the nature of its distribution. The reaction results in edema, pain and inflammation.

The activation of various components of complement systems (like C3a, C4a and C5a which are lymphotoxins) by the immune complex leads to the development of Type III hypersensitivity reaction. These are responsible for localized mast cell degranulation which releases histamine. The presence of histamine induces vascular hyper-permeability and aids in the accumulation of immune complexes in the blood vessel wall.

C5a, C3a, and C5b67 act as chemotactic factors that cause the aggregation of neutrophils at the place of immune complex deposition. The neutrophils then try to phagocytose the immune complex by the release of lytic enzymes but are unsuccessful in their attempt as the immune complexes are deposited on the basement membrane. These enzymes cause maximum tissue damage in the areas surrounding the immune complex deposits (Figure 11.4). The complement proteins too aid

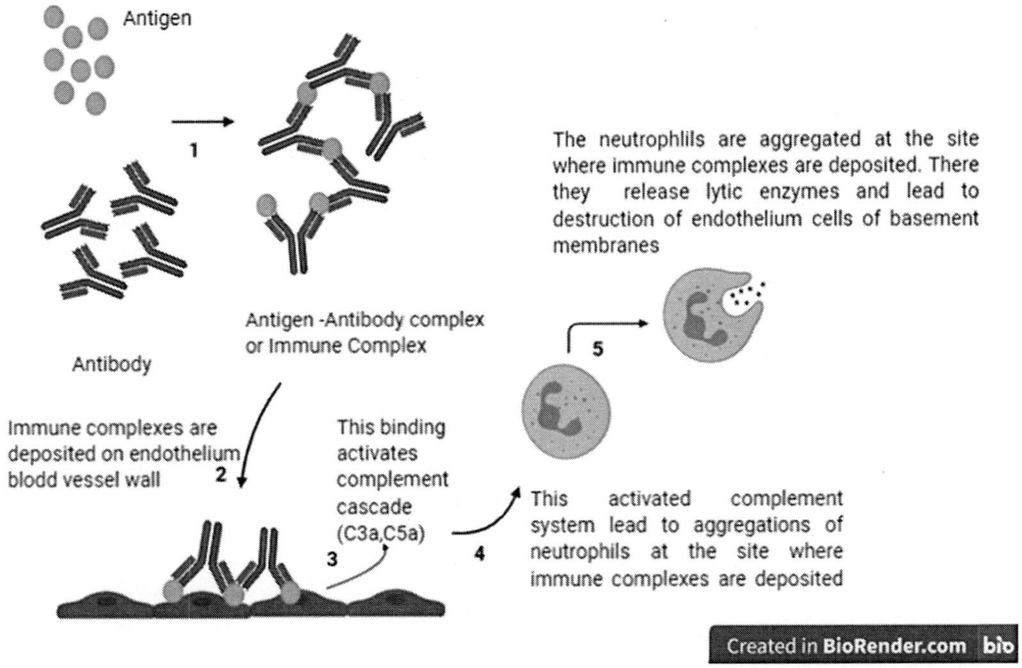

FIGURE 11.4 Diagrammatic representation of Type III Hypersensitivity. (Created by authors using Biorender.com from the YouTube Channel PhysioPathoPharmaco (2018).)

in the destruction of tissue. The C3b binds to the immune complex with the help of a C3b-specific receptors and thereby functions as an opsonin. Neutrophils attach themselves to these C3b-coated immune complexes.

11.4.1 Localized Type III Hypersensitivity Reaction

- The most common example of localized Type III hypersensitivity reaction is Acute Arthus reaction.
- In acute Arthus reaction, when an antigen is injected subcutaneously it binds within four to eight hours to the antibody to generate localized immune complexes.
- These immune complexes precipitate and bind to the complement system and anaphylatoxin is generated causing mast cell degranulation.
- As the reaction progresses, there is the damage of localized tissue. Edema and reddening of skin are observed at the point from where the antigen enters due to vascular damage.
- The intensity of reaction ranges from slight inflammation and redness to tissue necrosis (Figure 11.5).

11.4.2 Generalized Type III Hypersensitive Reaction

Circulating immune complexes are formed as a result of the interaction of antigens, which enter bloodstream, with the antibody. When the amount of antigen is large vis a vis to antibody, the immune complexes produced are minute and dissolvable. These immune complexes do not get phagocytosed by the phagocytic cells and result in Type III hypersensitivity reaction. The common example of generalized Type III hypersensitivity reaction is serum sickness. It develops 8 to 12 days after exposure to a foreign protein such as anti-diphtheria serum. The receiver forms antibodies

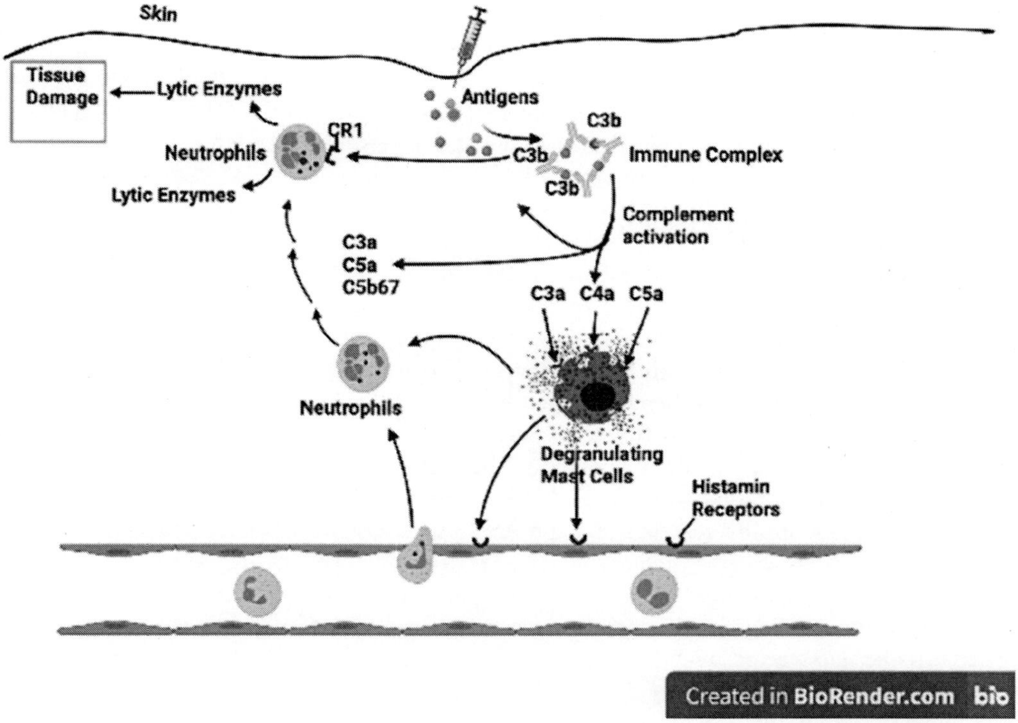

FIGURE 11.5 Diagrammatic representation of a localized Arthus reaction. (Adapted from *Kuby Immunology*, 7th Ed.)

specific to the foreign serum resulting in the formation of immune complexes. The common symptoms manifested include fever, rashes, redness, inflammation, weakness, lymphadenopathy, arthritis and sometimes also glomerulonephritis. The degree of manifestation is not only dependent on the amount of immune complexes generated but also on the site where they get accumulated. The accumulation of complexes happens at site of blood filtration. The symptoms may persist from two days to two weeks.

The common examples of Generalized Type III hypersensitivity reactions are:

- **Autoimmune diseases:** Systemic Lupus Erythematosus (SLE), Rheumatoid Arthritis, Goodpasture Syndrome (GPS)
- **Drug-induced serum sickness:** Allergies to penicillin and sulphonamides
- **Inhalation of antigens into lungs:** Farmer's Lung, Pigeon Fancier's Lung, and Humidifier fever

11.5 TYPE IV OR DELAYED-TYPE HYPERSENSITIVITY (DTH)

Type IV hypersensitivity reaction takes longer than 12 hours to develop. Usually, the maximum reaction time is 2–3 days. Type IV reactions are carried out by direct action of T cells and not by antibodies. Inflammatory reactions are either exogenous or due to autoantigens. T cells and antigen-presenting cells (APC) such as macrophages and dendritic cells are involved in hypersensitivity reactions that arise due to exposure to exogenous antigens. They produce cytokines which instigate a local inflammatory response.

Delayed-type hypersensitivity is further classified into three sub-types based on kind of skin reaction produced when the antigen is encountered. The reaction may be local (thickening of the skin) or systemic (T cell division and cytokine synthesis). Contact dermatitis, tuberculin and granulomatous are the three subtypes of Type IV hypersensitivity.

Contact sensitivity (48–72 hours): In contact sensitivity/dermatitis, haptens (exogenous agents) invade the skin and interact with tissue protein resulting in inflammatory reactions. Antigen-presenting cells (Langerhans cells and dendritic cells) are associated in recognition of hapten-tissue protein complexes and its presentation to T cells. The sensitized T cells (mainly CD4 and CD8 cells) secrete cytokines and other chemicals which recruit other immune cells at the site. Other cells like keratinocytes (those that exhibit MHC class II molecules and ICAM-1) secrete cytokines (IL-1, IL-6, IL-8) resulting in skin inflammation. Various chemical agents and plant products that induce contact sensitivity include nickle, formaldehyde, turpentine, poison ivy, poison oak, etc.

Tuberculin (48–72 hours): Tuberculin-type hypersensitivity is seen upon intradermal injection with tuberculin (derived from *Tuberculosis bacillus*). It is seen either in tuberculosis patients or in individuals vaccinated against tuberculosis. It produces local reactions including skin induration, swelling and redness within 48–72 hours. Sensitized individuals may exhibit similar reaction on exposure to antigens from different pathogens (like *Leishmania tropica* and *Mycobacterium leprae*).

Granulomatous (21–28 days): Granulomatous-type hypersensitivity results from a variety of antigens, including beryllium, talc, silica etc. It results because the macrophages which engulf the antigen are unable to digest them or present them. The persistence of these intracellular substances within macrophages, which the cell is neither able to process nor damage, leads to the accumulation of more macrophages resulting in granuloma formation. Sarcoidosis disease, where the antigen is unknown, is an example of granulomatous hypersensitivity.

11.5.1 Phases of the DTH Response

DTH response can be categorized into two phases: sensitization (primary contact) and effector phase (secondary contact). During the initial sensitization phase (1–2 weeks), i.e., as soon as there is a primary contact with an antigen, the $CD4^+$ (T helper cells) are activated essentially. However, in some cases, it has been observed that the DTH response may be also induced by $CD8^+$ cells. When the same antigen is encountered in future (secondary contact), the effector phase of the DTH response is induced. The response of effector phase is the maximum observed after 48–72 hours. During the effector phase, numerous kinds of cytokines that engage and activate macrophages, non-specific inflammatory cells are secreted by T helper cells. Response is delayed because aggregation and activation of macrophages by cytokines needs time. The principal effector cells of the DTH response are macrophages. Blood monocytes attach to vascular endothelial cells and move into the neighbouring tissues due to the presence of cytokines. In this process, the monocytes differentiate into activated macrophages. Due to the presence of different cytotoxic mediator molecules, the activated macrophages exhibit phagocytosis and an enhanced ability to kill microorganisms. In addition, activated macrophages also function as antigen-presenting cells as they show high levels of class II MHC molecules, TNF receptors and other cell-adhesion molecules (Figure 11.6).

11.5.2 Mechanism of Type IV Hypersensitivity

The Type IV hypersensitivity or DTH response plays a crucial role in host defence against parasites and bacteria that reside within cells where the circulating antibodies cannot interact with them. The infected area displays enhanced phagocytic activity of the macrophages and increased presence of lytic enzymes. This not only leads to the pathogen elimination but also nonspecific destruction of the cell. Usually, little damage to the tissues is incurred when the pathogen is destroyed. However, occasionally the antigen does not get cleared easily and in such circumstances, a prolonged DTH

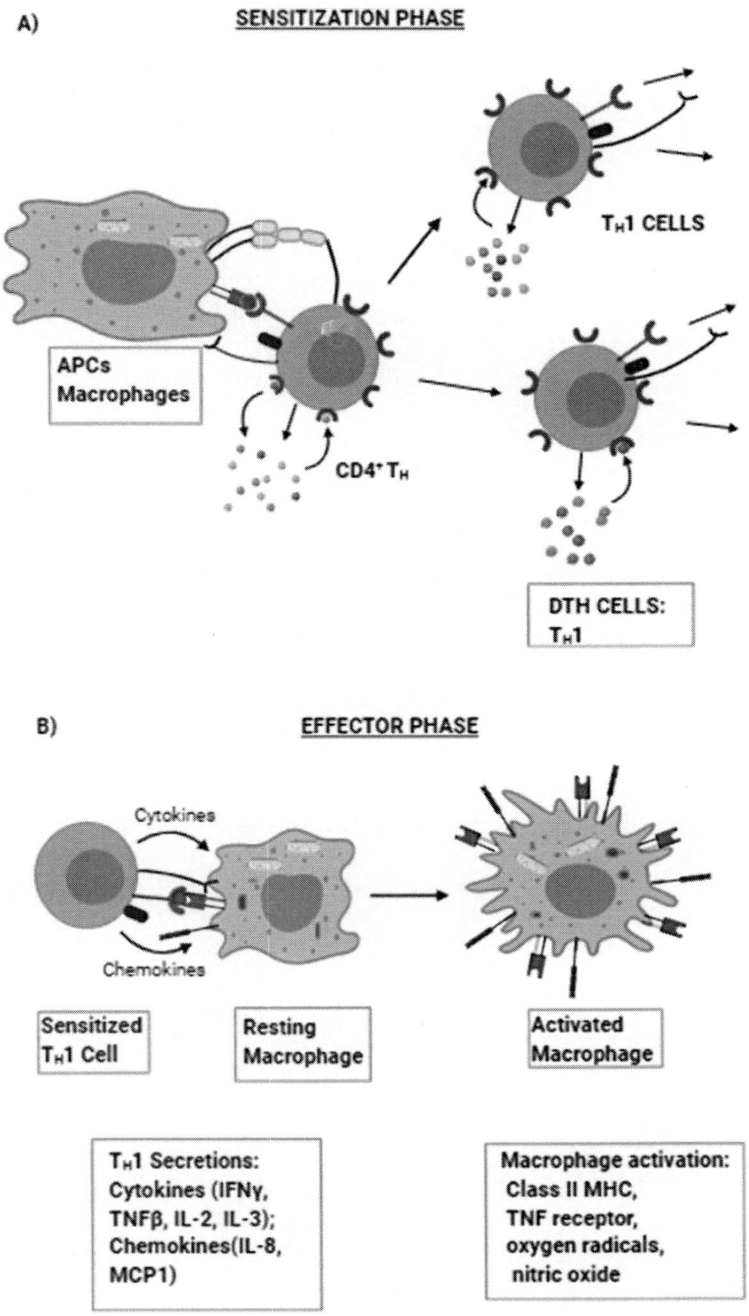

FIGURE 11.6 Diagrammatic representation of Type IV Hypersensitivity (a) Sensitization Phase and (b) Effector Phase. (Adapted from *Kuby Immunology*, 7th Ed.)

response is observed. This sustained DTH response is damaging for the host and inflammatory response observed emerges as a visible granulomatous reaction.

A granuloma forms when macrophages are activated continuously and cohere with one another. Under these conditions, they assume an epithelioid shape and often fuse to form multinucleated giant cells. These giant cells form palpable nodules by displacing normal tissue cells. They may also damage the neighbouring tissues by the release of an increased quantity of lytic enzymes. This also leads to substantial tissue necrosis and damage to blood vessels. DTH response is also seen when *Mycobacterium tuberculosis* infects the lung. Immune response is initiated on its re-exposure by activation of macrophages that contain it within a granuloma-type lesion called a tubercle. These activated macrophages release lytic enzymes that start damaging the lung tissues that the immune response hoped to preserve.

11.5.3 Examples of Type IV Hypersensitivity

Type IV hypersensitivity reaction can occur in many parts of the body. Various diseases associated include Atopic dermatitis, Tuberculosis, Hypersensitivity pneumonitis, Type I diabetes mellitus, Leprosy, Granulomatous inflammations observed in Sarcoidosis, Crohn disease, Hashimoto's thyroiditis, Allograft rejection, Graft versus Host Disease (GvHD), Fungal diseases (like Candidiasis, Dermatomycosis, Histoplasmosis etc), Parasitic diseases (like leishmaniasis, schistosomiasis).

11.5.4 The Tuberculin Reaction/Mantoux Reaction

Koch in the year 1890 developed tuberculin skin test (TST). The presently used intradermal technique was described by a French physician, Charles Mantoux, in 1912. TST is the most standard technique for detecting tuberculosis (TB). In this technique purified protein derivative (PPD) tuberculin, derived from *M. tuberculosis* is injected into the skin of the forearm and monitored. An individual with latent TB infection will develop a small hard lump in 48–72 hours at the place of injection (Figure 11.7). Redness and swelling appear at the injection site and then increase through the next day, and disappear with time. The tissue samples from the site of the positive reaction on examination show infiltration by lymphocytes and monocytes. Multinucleate large cells are formed due to the fusion of macrophages in severe cases. Granuloma is a result of the aggregation of activated macrophages. Sensitized T cells are recruited to the skin site where they release chemicals like lymphokines. These lymphokines induce various reactions like local vasodilatation, edema, fibrin deposition and recruitment of other inflammatory cells to the area. A similar kind of sensitive

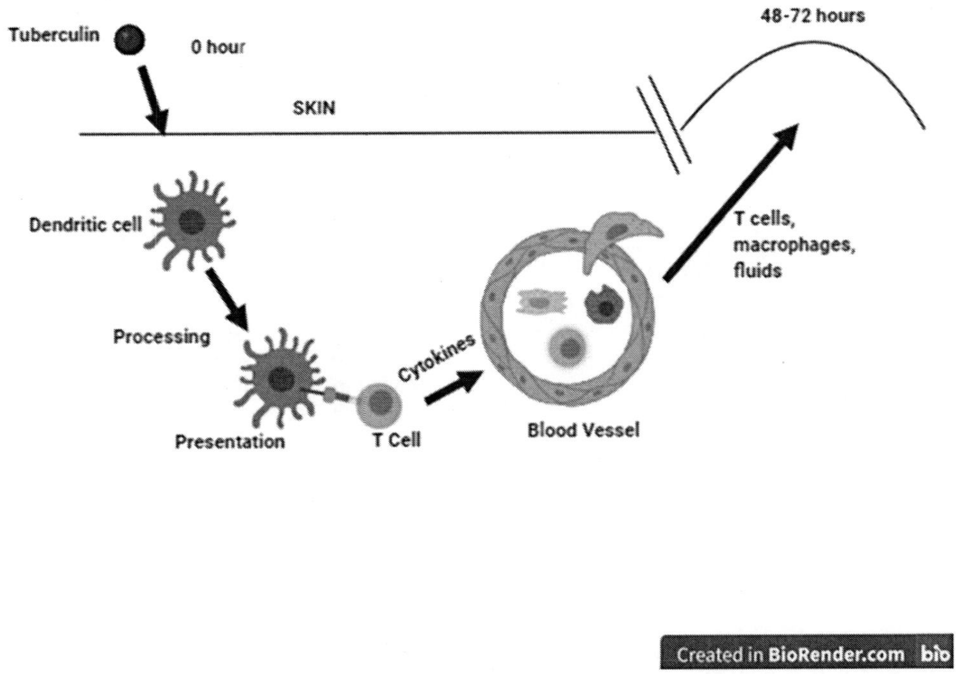

FIGURE 11.7 Diagrammatic representation of tuberculin reaction. (Adapted from *BIOS Notes Immunology*, 3rd Ed., 2011.)

test can also be used to check immunity for other diseases such as leprosy, leishmaniasis, coccidiosis, etc. by injecting individuals with appropriate antigen.

11.6 SUMMARY

- Hypersensitivity is defined as the exaggerated immune response to an antigen that rather than providing protection leads to damage. The severity of the reaction ranges from mild to severe.
- There are four different types of hypersensitive reactions, which can be distinguished from each other based on their mechanisms or time of onset. These reactions only occur in pre-sensitized individual meaning the individual has been exposed to the antigen earlier which led to the development of antibodies against these antigens.
- Type I hypersensitive reaction is IgE mediated. On subsequent exposure to the allergen, IgE binds to its receptors located on basophils and mast cells leading to its crosslinking. This crosslinking promotes degranulation of mast cells and basophils by triggering a signalling cascade. The pharmacological active mediators (e.g. histamine) thus releases lead to vasodilation, smooth muscle contraction and lead to the development of rashes, edema, etc. Type I reactions are immediate in nature, i.e., it develops within minutes of exposure to antigen and may be mild (allergy to food, pollen etc) or life-threatening (anaphylactic shock).
- Type II hypersensitive (antibody-dependent cell-mediated cytotoxicity) reactions are mediated by IgG and IgM. These antibodies bind to the particular cells in sensitized individual. This binding leads to lysis of cells by activation of complement system or phagocytosis. These reactions are observed in autoimmune disorders and also in Hemolytic Disesase of Newborn (HDN).
- Type III hypersensitive reactions are also carried out by IgG and IgM antibodies. The antibody binds with free floating antigen. This leads to the formation of insoluble immune complexes which are deposited on the wall of blood vessels. It activates the complement proteins causing damage to the area where these immune complexes are deposited e.g. serum sickness and arthritis.
- Type IV hypersensitive reaction is mediated by T-cells. It is also a type of delayed hypersensitive reaction. The reaction is seen after 48–72 hours. Pre-sensitized T cells when re-exposed to the antigen the T_H cells release inflammatory cytokines and T-killer cells release cytotoxic reactions e.g., Contact dermatitis.

QUESTIONS

1. Define
   a. Hypersensitivity
   b. Isohemagglutinin
   c. Histamine
   d. Arthus Reaction
   e. Mantoux Reaction
2. Expand the following
   i. RAST ii. DTH iii. ADCC iv. PK Test
3. Write short notes on
   i. Erythroblastosis fetalis
   ii. Allergens
   iii. Cell-mediated Hypersensitivity
   iv. Anaphylactic Reaction
   v. Primary and secondary mediators

4. Give a detailed account of mechanism of Type 1 Hypersensitivity
5. Explain the mechanism and variants of cell-mediated hypersensitivity.

BIBLIOGRAPHY

Actor, J. K. Adaptive immune response and hypersensitivity. In *Elsevier's Integrated Review Immunology and Microbiology*, 2nd Ed., 2012. Philadelphia, PA: Elsevier Saunders, pp. 53–59.

Dean, L. *Blood Groups and Red Cell Antigens*, 2005. Bethesda, MD: National Center for Biotechnology Information (US).

Karki, G. Type II hypersensitivity reaction: Mechanism and examples. 2018. https://www.onlinebiologynotes.com/type-ii-hypersensitivity-reaction-mechanism-and-examples/.

Lovenberg, T. W., Roland, B. L., Wilson, S. J., Jiang, X., Pyati, J., Huvar, A., Jackson, M. R., & Erlander, M. G. Cloning and functional expression of the human histamine H3 receptor. *Mol Pharmacol.*, 1999; 55, 1101–1107.

Lydyard, P., Whelan, A., & Fanger, M. *BIOS Instant Notes in Immunology*, 3rd Ed., 2011. London: Taylor & Francis; Amsterdam, The Netherlands: Elsevier B.V.

Marc, D. & Olson, K. *Hypersensitivity Reactions and Methods of Detection*, 2009. Osceola, USA: NeuroScience, Inc., pp. 1–4.

Oda, T, Morikawa, N, Saito, Y, Masuho, Y, & Matsumoto, S. Molecular cloning and characterization of a novel type of histamine receptor preferentially expressed in leukocytes. *J Biol Chem.*, Rockville, Maryland, 2000; 275, 36781–36786.

Owen, J. A., Punt, J., & Stranford, S. A. *Kuby Immunology*, 7th Ed., 2013. New York: W.H. Freeman and Company.

PhysioPathoPharmaco (10 April, 2018). Type III Hypersensitivity, PhysioPathoPharmaco Youtube Channel, retrieved on 23 august 2021 from https://www.youtube.com/watch?v=-_yUsZO6Vio

Primeau, M. N. & Adkinson, N. F., Jr. Recent advances in the diagnosis of drug allergy. *Curr. Opin. Allergy Clin. Immunol.*, 2001; 1, 337–341.

Roitt, I. *Essential Immunology*, 8th Ed., Oxford, UK: Blackwell Science.

Schwartz, L. B. *Systemic Anaphylaxis, Food Allergy, and Insect Sting Allergy Goldman's Cecil Medicine*, 24th Ed., 2012. Philadelphia, United States: Saunders, Vol. 2, pp. 1633–1638.

Smith, T. F. Allergy testing in clinical practice. *Ann. Allergy*, 1992; 68, 293–301.

White, M. V. The role of histamine in allergic diseases. *J. Allergy Clin. Immunol.*, 1990; 86, 599–605.

Williams, P. B., Dolen, W. K., Koepke, J. W., & Selner, J. C. Comparison of skin testing and three in vitro assays for specific IgE in the clinical evaluation of immediate hypersensitivity. *Ann. Allergy*, 1992; 68, 35–45.

12 Transplantation Immunology

Priya Singh and Nidhi Garg
University of Delhi

CONTENTS

DOI: 10.1201/9781003286424-12

12.1 INTRODUCTION

Transplantation is a medical procedure involving the removal of either cells, or tissues, or organs, from an individual and its placement on another site within him/her or to another individual. The individual donating the cells, tissues or organ/s is called the donor and the one receiving them is known as recipient. When a part of the body gets irreversibly damaged owing to either disease or injury, transplantation with cells, tissues or organs that are in good physical shape is a way of restoring normal functioning. Organ transplantation being expensive and complicated remains the only choice when no other treatment works. Thus, making it a life-saving procedure which actually saves thousands of lives every year. According to the Organ Procurement and Transplantation Network (OPTN), about 750,000 transplantation surgeries have been performed in the United States since 1988. The biggest hurdle in performing successful transplant is the immune system, due to its recognition of cells/tissues or organs as non-self or foreign, thereby eliciting an immune response against them. Strategies have been devised to reduce the rejection of transplants by identifying the components of the immune system responsible for rejection and developing suitable immunosuppressive drugs.

The first step involves carrying out the matching process between the organ donor and the recipient. The matching process comprises the following

1. Matching the blood group of donor and recipient.
2. Tissue typing.
3. Testing the reaction of the recipient's blood serum towards donor cells.

Following transplantation, the recipient is given immunosuppressive drugs to suppress the action of the immune system toward foreign cells/organs. The use of immunosuppressive drugs has several disadvantages in being non-definite in action and the recipients become highly prone to diseases along with several side effects. The field of transplantation is fast evolving through intensive research in discovering the pathways and components of the immune system that will facilitate improved donor and recipient matching and prevention of rejection of transplant.

12.2 HISTORY OF TRANSPLANTATION

If we believe that the oldest known transplantation existed in the pre-historic era as many human mythological scriptures state, one of them is well known as the Lord Ganesha, with an elephant head transplanted onto the human torso with multiple forelimbs. Such legends lack the evidences of the medical procedures carried out then until recent history covering the last 70 years. Records prior to these states unsuccessful attempts of transplantation or it seems more like miracles. The idea of replacing dead/diseased body organs is doing rounds in human history for more than a millennium. The experiments for organ transplantation (mostly xenografts) began in the early years of the twentieth century, most of them did not see the light of the day due to immediate failure and most importantly lack of evidences to prove the findings as many immunology facts were unfounded then. In the year 1903, Paul Ehrlich and Danish biologist Carl Jensen studied tumor transplantation in mice, but they failed to establish the kind of immune response that overwhelmed recipients leading to graft rejection. Then in 1912, Georg Schöne, studying skin grafts at Ehrlich's laboratory, established that a second graft (homograft) followed by rejection of the first graft was surely rejected and that too in less time as compared to the first graft (years later this phenomenon was established as second set graft rejection). At Rockefeller Institute, scientists worked to establish some additional knowledge on transplant experiments, most notable was the central role of lymphocytes deduced by James B. Murphy's work in 1914. He furthered his findings by eliminating lymphocytes by irradiation, splenectomy, or benzol (first chemical immunosuppressant) buying some extra time for graft survival. These findings were well noted, published in renowned scientific journals, but not regarded as breakthrough achievements in transplantation immunology as none resulted in a successful transplant.

The pioneering work of Alexis Carrel leading to vascular suturing and its role in organ transplants won him the 1912 Nobel Prize. Although this feat was achieved by him under the guidance and help of Mathieu Jaboulay and Charles Guthrie respectively as he trained with the former and collaborated with the later. Before Carrel several others, Emerich Ullmann (in 1902), Jaboulay (in 1906), Ernst Unger (in 1909) tried and succeeded to a certain extent in xenograft renal transplants using dog-to-goat, pig, goat-to-human, monkey-to-human respectively. But these transplants were regarded unsuccessful as the recipients survived only for a few days after receiving the transplants. In the 1930s, Leo Loeb successfully determined that genetic distantness between recipient and donor resulted in quick and strong graft rejections. He also found the involvement of lymphocytes in the graft rejections, but his work was strongly ridiculed by his colleagues and fellow immunologists as the humoral basis of graft rejection was a more sought-after answer to transplantation immunology then. From 1933 to 1949, Yu Yu Voronoy carried out five human-to-human kidney transplants that failed eventually probably due to a major blood group mismatch and the findings were lost in history. Sometime during World War II, Peter Brian Medawar, while working in the burn unit at Glasgow's Royal Hospital confirmed homografts rejection providing insights into the acquired immunological tolerance. Then he worked on inducing immunological tolerance by neonatal injection of allogeneic cells in mice, leading to acceptance of allogeneic grafts in the 1950s and went on to win the 1960's Nobel Prize for his contributions. With his findings began the modern era of actual transplantation immunology summarized in Figure 12.1.

The work in transplantation furthered after considering brain dead people as potential organ donors with their heart's functioning which was regarded unacceptable prior to the 1960s and successful preservation procedures for tissue/organs for transplants. Other recent developments include the use of synthetic organs for transplant into humans for which the successful procedure was carried out in the year 2011 for the first time by transplanting artificial windpipe into the host after coating it with the host's stem cells in Sweden. More recently, we are heading on the road to 3-D bioprinting of human organs with the first successful transplant of the 3-D printed human bladder that was developed from the host's bladder tissue. Techniques that have been developed lately aim to ensure near zero probability of graft rejection.

12.3 TYPES OF TRANSPLANTATION

The following are the different types of transplantation (12.2).

12.3.1 Autograft

It is defined as the removal of either cells or tissues or organs and its transplantation at another site or place in the same organisms. An example of this is the skin graft or rhinoplasty.

12.3.2 Isograft

It is a type of allograft, which involves transplanting cells, tissues or organ/s between donor and recipient both of which happen to be genetically identical, due to which there is no rejection of transplant by the immune system.

12.3.3 Allograft

It is the transplantation of either cells, or tissues, or even organs from a genetically non-identical individual to another of the same species. The transplanted cells, tissues, or organs are called as an allograft. Due to the absence of genetic identity of transplanted cells and the host cells the immune system is activated, causing rejection, the risk for which is known through Panel reactive antibody level.

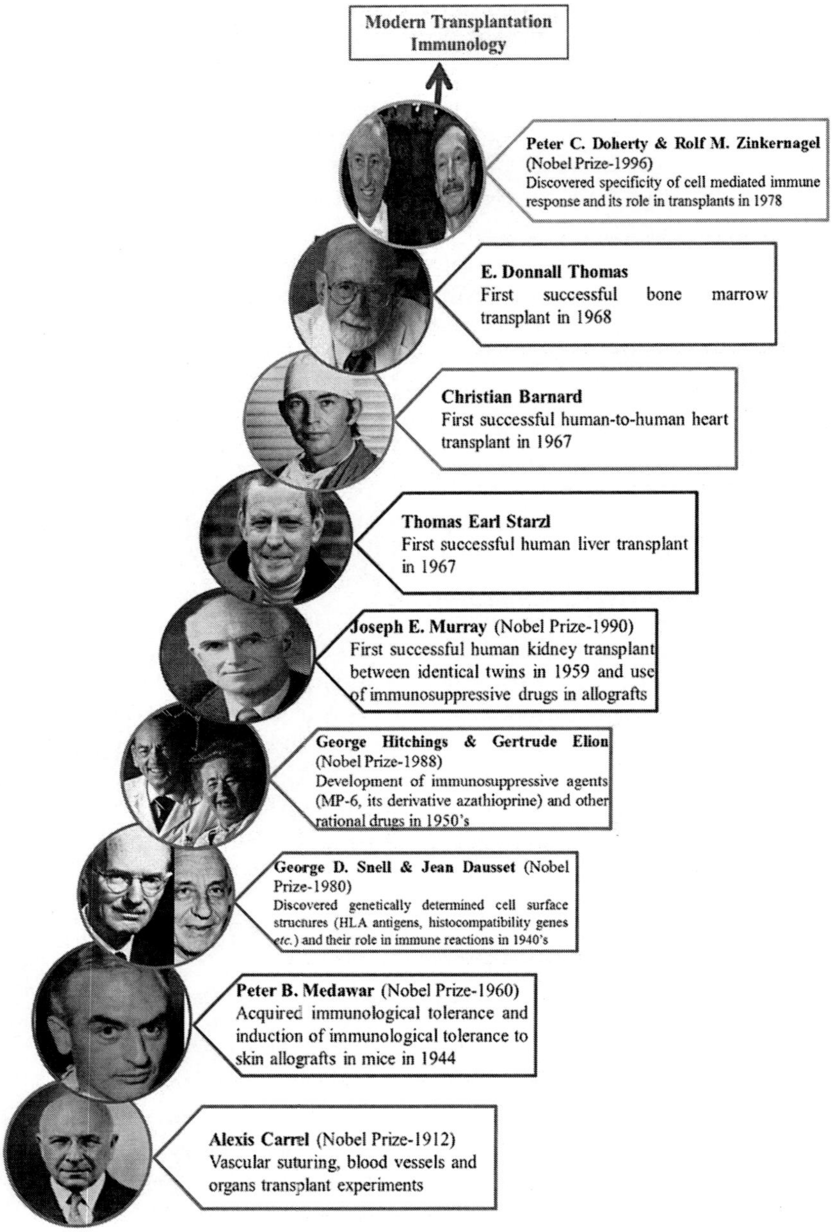

FIGURE 12.1 History of Transplantation: List of scientists, researchers and/or doctors who are regarded for their contributions in the development of immunological knowledge base of transplantation experiments and establishment of tissue/organ transplant as feasible procedures to cure some severe disease states in humans. This also enlists their major contributions in the development of modern transplantation immunology and the year they went on to win the Nobel Prize in Physiology or Medicine for their pioneering work. **Source: Authors.**

12.3.4 XENOGRAFT

It is a type of transplantation called xenotransplantation where either an organ or tissue is transplanted to an individual belonging to another species. An example of this could be the transplantation of a Pig's valve in the human heart for the replacement of a defective valve (Figure 12.2).

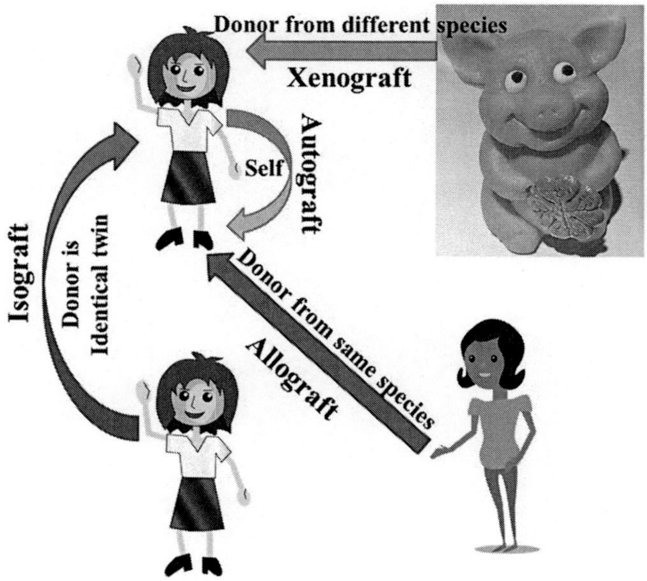

FIGURE 12.2 Types of Transplantation: Autograft is when the donor and recipient is the same individual; Isograft is done between the identical twins; Allograft is from a donor who is not identical twin, but is some relative or other individual from the same species and Xenografts are accepted from the tissue/organ donors of the different species *e.g.* pig. **Source: Authors.**

12.3.5 ABO Incompatible Transplantation

ABO denotes the blood grouping in humans, whose determination of donor and recipient is done to reduce transplant rejection. But an exception is seen in young kids with underdeveloped immune systems which offer the advantage of transplantation without the need of matching the blood groups between the donor and transplant which is known as ABO incompatible transplantation.

12.3.6 Stem Cell Transplantation

Stem cells are endowed with the ability to differentiate into different cell types. The hematopoietic stem cells are present in bone marrow in infants and adults and in the blood of the umbilical cord and placenta. These cells are capable of giving rise to all the different types of blood cells. Hence, they are of immense value through transplantation from a healthy donor to recipient for treating leukemia and other diseases of blood due to faulty bone marrow.

12.4 CELLS, TISSUES AND ORGANS FOR TRANSPLANTATION

May both refer to organs and tissues that are relatively often or routinely transplanted (here "main organs and tissues"), as well as relatively seldom transplanted organs and tissues and the ones on the experimental stage (Figure 12.3).

12.5 TYPES OF DONORS

12.5.1 Living Donors

In living donors, the donor remains alive and donates a renewable tissue, cell, or fluid (e.g., blood, skin), or donates an organ or part of an organ in which the remaining organ can regenerate or take on the workload of the rest of the organ (primarily single kidney donation, partial donation of liver,

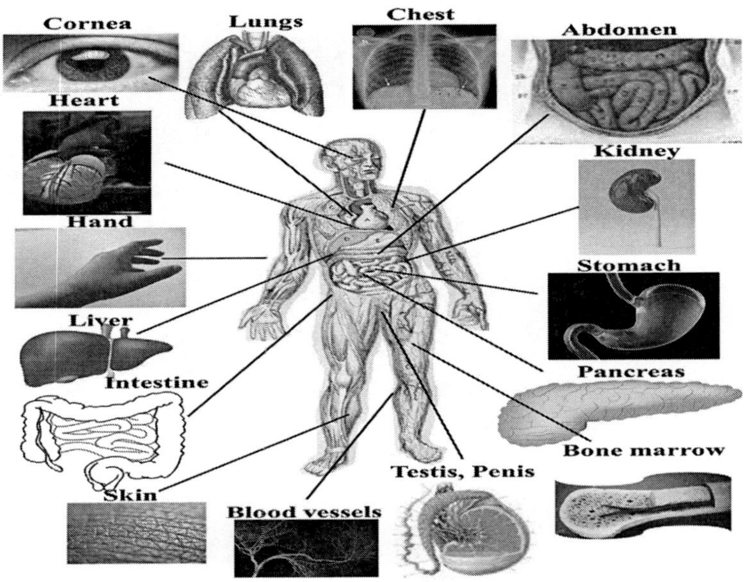

FIGURE 12.3 A pictorial representation of the tissues and organs used for transplant in humans. Chest, Heart (deceased-donor only), Lung (deceased-donor and living-related lung transplantation), Abdomen, Kidney (deceased-donor and living-donor), Liver (deceased-donor, which enables donation of a whole liver; and living-donor, where donation from only one person cannot provide a whole liver, if an entire liver is needed), Pancreas (deceased-donor only; a very severe type of diabetes ensues if a live person's entire pancreas is removed), Intestine (deceased-donor and living-donor; normally refers to the small intestine), Stomach (deceased-donor only), Testis (deceased-donor and living-donor), Penis (deceased-donor only), Hand (deceased-donor only), Cornea (deceased-donor only), Skin, including face replant (autograft) and face transplant (extremely rare), Islets of Langerhans (pancreas islet cells) (deceased-donor and living-donor), Bone marrow/Adult stem cell (living-donor and autograft), Blood Vessels (autograft and deceased-donor). Others not represented in the figure include Bone (deceased-donor and living-donor), Heart-valve (deceased-donor, living-donor and xenograft [porcine/bovine]), Tissues, cells and fluids, Blood transfusion/Blood Parts Transfusion (living-donor and autograft). **Source: Authors.**

lung lobe, small bowel). Regenerative medicine may one day allow for laboratory-grown organs, using person's own cells via stem cells, or healthy cells extracted from failing organs.

12.5.2 Deceased Donors

Deceased donors (formerly cadaveric) are people who have been declared brain-dead and whose organs are kept viable by ventilators or other mechanical mechanisms until they can be excised for transplantation. Apart from brain-stem dead donors, who have formed the majority of deceased donors for the last 20 years, there is increasing use of donation-after-circulatory-death-donors (formerly non-heart-beating donors) to increase the potential pool of donors as demand for transplants continues to grow. Prior to the recognition of brain death in the 1980s, all deceased organ donors had died of circulatory death. These organs have inferior outcomes in comparison to organs from a brain-dead donor.

12.6 POSSIBLE REASONS FOR ORGAN DONATIONS

12.6.1 Living Relative Donors

Living-related donors donate to family members or friends with whom they have an emotional investment. The risk of surgery is offset by the psychological benefit of not losing someone related to them, or not seeing them suffer the ill effects of waiting on a list.

12.6.2 Paired Exchange

A "paired-exchange" is a technique of matching willing living donors to compatible recipients using serotyping. For example, a spouse may be willing to donate a kidney to their partner, but cannot since there is not a biological match. The willing spouse's kidney is donated to a matching recipient who also has an incompatible but willing spouse. The second donor must match the first recipient to complete the pair exchange. Typically, the surgeries are scheduled simultaneously in case one of the donors decides to back out and the couples are kept anonymous from each other until after the transplant. The first pair exchange transplant in the U.S. was in 2001 at Johns Hopkins Hospital.

12.6.3 Good Samaritan

Good Samaritan or "altruistic" donation is giving a donation to someone that has no prior affiliation with the donor. The idea of altruistic donation is to give with no interest in personal gain, it is out of pure selflessness. On the other hand, the current allocation system doesn't assess a donor's motive, so altruistic donation isn't a requirement. Some people choose to do this out of a personal need to donate. Some donate to the next person on the list; others use some method of choosing a recipient based on criteria important to them. Web sites are being developed that facilitate such donation. It has been featured in recent television journalism that over half of the members of the Jesus Christians, an Australian religious group, have donated kidneys in such a manner.

12.6.4 Financial Compensation

Now monetary compensation for organ donors is being legalized in Australia, and strictly only in the case of kidney transplant in the case of Singapore (minimal reimbursement is offered in the case of other forms of organ harvesting by Singapore). Kidney disease organizations in both countries have expressed their support. In compensated donation, donors get money or other compensation in exchange for their organs. This practice is common in some parts of the world, whether legal or not, and is one of the many factors driving medical tourism. In the illegal black market, the donors may not get sufficient after-operation care, the price of a kidney may be above $160,000, middlemen take most of the money, and the operation is more dangerous to both the donor and receiver, and the receiver often gets hepatitis or HIV. In the legal markets of Iran, the price of a kidney is $2,000–$4,000.

In the United States, The National Organ Transplant Act of 1984 made organ sales illegal. In the United Kingdom, the Human Organ Transplants Act 1989 first made organ sales illegal and has been superseded by the Human Tissue Act 2004 (Figure 12.4).

12.6.5 Forced Donation

Forced labor has always stayed there in a human civilization based on the power one group had over the other. None of the continents is spared from the tradition of slavery in recent past. Such roots have given way to modern day exploitation of the deprived and needy human beings by those in power (monetary/discriminatory/traditional/arms *etc.*). The history of vaccine development is full of inhuman exploitation of mentally weak, financially feeble, uninformed human adults and majorly children. Such is the case of forced organ donation, there are reports surfacing time and again about the human organ trafficking. There are organized crimes against the underprivileged/prisoners for forced organ harvest to supply organs to the rich who can afford. It is estimated to be a much bigger nexus of forced organ harvest than it appears in China, the Indian subcontinent, African countries and poor communities all over the world. The agents working at ground level either deceive the poor or forcefully abduct them for harvesting organs. There have been reports of uninformed organ harvesting from patients coming for minor health issues to doctors, such patients are operated on without properly informing them or their family. Kidney rackets have been busted every

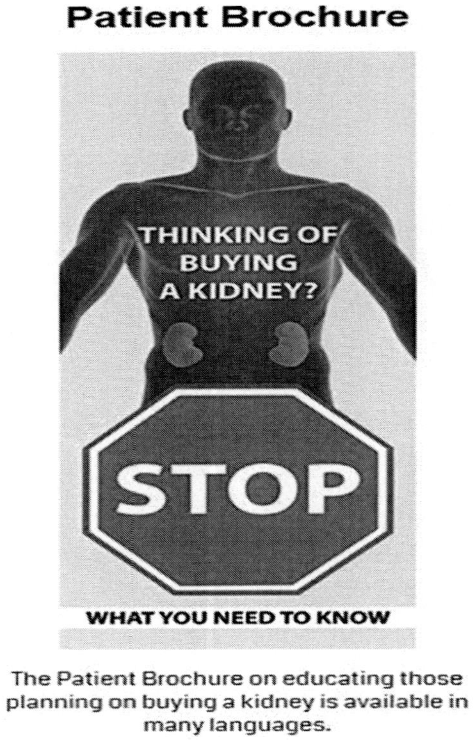

FIGURE 12.4 The information brochure available for the primary reference to the patients and/or their acquaintances prior to look for buying organs for transplantation bypassing the legal procedures to make an informed choice. (http://www.declarationofistanbul.org/.)

now and then in India and elsewhere, even after the sharp eye of the ethics committee and medical vigilance. The shortage of available tissues and organs for transplantation, even after the expansion of deceased organ donation is increasing, day by day with the improvement of medical technology related to rejection of transplantation. This has prompted countries to frame guidelines to check the commercial trafficking of organs, especially from unrelated donors. WHO guiding principles on Human cell, tissue and organ transplantation endorses The Declaration of Istanbul on Organ Trafficking and Transplant Tourism formulated by the guidance of the Transplantation Society and the International Society of Nephrology in May 2008 and urges professionals worldwide to follow the norms laid in the declaration that is constantly updated.

12.7 IMMUNE RESPONSE TO TRANSPLANTATION

The degree and type of immune response a transplant elicits depends on the similarity or differences in the histocompatibility of recipient and host. Whether a graft gets rejected or accepted also depends on the organ, for example, eye and brain are immunologically spared to a certain extent and grafts therein are readily accepted while in the case of highly vascularized organs grafts get rejected more easily after vascularization establish, resulting in cell-mediated immune response (Figure 12.5).

Autografts and isografts are usually accepted with no or minimal immune response, owing to the genetic similarity between donor and recipient (Figure 12.5a). But allografts face a spectrum of immune response due to their genetic dissimilarity to the host and for obvious reasons they are recognized as foreign by the host immune system subsequently leading to graft rejection. Xenografts exhibiting great genetic dissimilarity receive a massive immune response and rapid rejection.

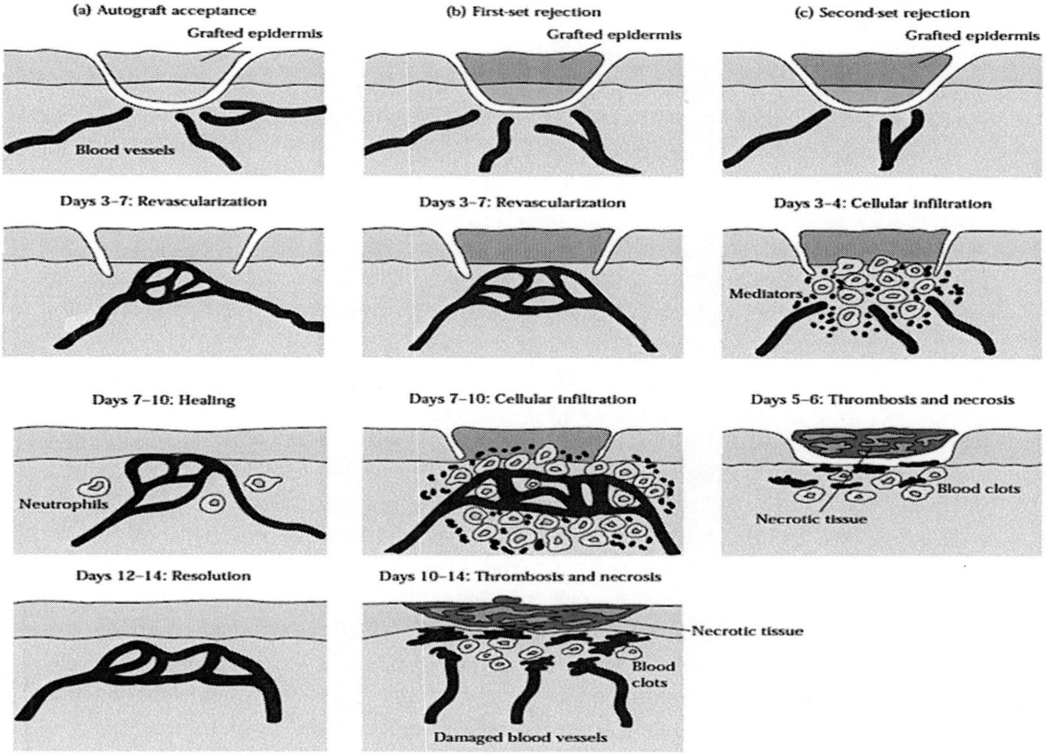

FIGURE 12.5 Schematic representation of the process of acceptance and rejection of graft. (a) Autograft is accepted within 12–14 days; (b) First set rejection of allograft start at 7–10 days after transplant, and full rejection occur around 10–14 days after grafting; (c) Second set rejection of graft begins on third or fourth day and complete rejection is done by fifth or sixth day. The immune cells that infiltrate at the site of allograft include phagocytes, lymphocytes, macrophages and other inflammatory cells. (Kindt, T.J., Goldsby, R.A. and Osborne, B.A. (2007). *Kuby Immunology*. WH Freeman; New York, 6th Edition.)

Allograft rejection, as stated varies depending on the tissue transplanted. Skin grafts are readily rejected as compared to kidney or heart. But there is always specificity and memory of immune system involved in precise graft rejection response. As indicated in Figure 12.5, skin graft first-set rejection (primary graft rejection) begins around the seventh day after revascularization of transplant between 3 and 7 days and subsequent infiltration of various immune cells of the host. Vascularization reduces after the seventh day and tissue necrosis begin around the tenth day with complete rejection by 12–14 days.

But in the case of second-set rejection (when tissue from the same donor is transplanted to the same host again) immune response is very quick leading to complete rejection within 5–6 days. This rapid action is attributed to the immunological memory. The specificity of second-set rejection may be tested by grafting from another donor at the same time in the same host. This will be rejected as first-set rejection only, while the previous graft will be rejected rapidly.

T-cells play an important role in the rejection of allografts, and this ability of T-cells is transferrable. This is detailed in the next paragraph of this chapter.

12.8 MECHANISMS OF GRAFT REJECTION

Both humoral (cell-mediated) and cellular (cell-mediated) immune reactions play a part in graft rejections. T-cells play a major role in graft rejection along with other immune cells. Graft rejection is carried out in two stages, sensitization and effector stage.

Sensitization stage: Immune reactions begin with sensitization of the CD8 and CD4 T-cells on recognizing non-self with the help of receptors. It involves two signals, interaction of TCR with MHC presenting the antigen and costimulatory interaction of receptors and ligands on T-cells and APCs. Costimulatory interactions include CD28 on the T-cells and its B-7 ligands on the APC surface, CD40 and its ligand CD40L (CD154), whereas the cytotoxic T lymphocyte-associated antigen-4 (CTLA-4) binding to B-7 ligands provides an inhibitory signal.

Through the central and peripheral tolerance mechanisms leading either to clonal deletion or anergy, self-antigens are spared from action by T-cells. Two basic ways of presenting an antigen from graft tissue (allo-antigen) exist *viz.* direct and indirect pathway.

Direct pathway: Here the host T-cells recognize peptides presented by allo-MHC molecules on the surface of the donor cells. In other words, host T-cells perceive allo-MHC molecule presenting allo-peptide (allo-antigen) as being equivalent to self-MHC presenting foreign peptide. Thus, it readily identifies donor tissue as non-self. Direct pathway is a dominant pathway in the early immune response in the allograft. The allo-MHC molecules are present on the interstitial dendritic cells or other APCs in donor tissue that has a good capacity to stimulate host T-cells. Also, the number of T-cells proliferating clonally in response to allo-MHC + allo-antigen complex is much higher as compared to antigen presented by self-MHC. Thus, this pathway is more prevalent in acute allograft rejection.

Indirect pathway: In this, the host T-cells recognize donor cell peptides (allo-antigen) presented by the self-MHC molecules. Along with self-MHC + allopeptide-primed T-cells, accelerated alloantibodies production play part in the acute vascular type rejection. Chronic allograft vasculopathy is also mediated by T-cells primed by an indirect pathway.

Effector stage: Effector mechanisms are controlled both by the allo-antigen and other independent factors. It starts with a non-immunologic injury response resulting in inflammatory reactions that subsequently result in more increased antigen presentation by MHC to T-cells, increased T-cell proliferation, overexpression of chemokines and cytokines. MHCs with bound allo-peptide are shed in response leading to increased proliferation of antibody-producing B-cells, that are also stimulated by macrophage-mediated delayed-type hypersensitivity (DTH) response initiated by the activated CD4 T-cells.

T-cells and the related cytokines such as IL-2 and IFN-γ are up-regulated in transplantation early on. Subsequently, β-chemokines like IP-10, MCP-1 and RANTES (Regulated upon Activation, Normal T-cell Expressed and Secreted) are over-expressed along with IL-6, TNF-α, inducible nitric oxide synthase (iNOS) and growth factors leading to intense macrophage infiltration at the transplant site and causing intimal thickening, smooth muscle proliferation, sclerosis, interstitial fibrosis *etc.*

T-cell-derived cytokines activate endothelial cells and macrophages express adhesion molecules, class II MHC and co-stimulatory molecules. These present antigens excessively, leading to the recruitment of more T-cells increasing rejection immune response. CD8 T-cells result in cell-mediated cytotoxicity reactions either by killing target cells or inducing apoptosis. Apoptosis of target cells for transplant is either induced by the granules secreted by activated CTLs that contain perforin and granzyme or, alternatively, through a Fas-dependent pathway leading to cytolysis.

Additionally, natural killer (NK) cells are active immune cells leading to acute and chronic rejection of solid graft tissue. NK cells do not require MHC for sensitization towards non-self ('missing self' hypothesis) rather the recognition is meditated through NK inhibitory receptors receiving signals from MHC-I antigens on cell surfaces. NK cells carry out their function by stimulatory signals including, cytokine release, increased cytotoxicity through granzyme and perforin, TNF-related apoptosis-inducing ligand (TRAIL) and Fas ligand (Fas L). They help CD8 T-cells in their graft rejection response.

The role of innate immune responses involving TLR agonists *etc.* in transplantation is being studied and knowledge of its role will help in the optimization of adaptive immune response for more successful transplantations in future.

Types of graft rejection in the clinical setup are as follows:

Hyperacute rejection: Graft rejection is very quick, occurring within minutes to hours of transplant and the grafted tissue never gets vascularized. Hyperacute rejection is mediated by the humoral immune system with the help of pre-existing antibodies specific for the graft antigens. The Ag-Ab complexes in large number activate the complement system that results in excessive infiltration of neutrophils and other immune cells in the graft tissue. This inflammatory reaction that ensues post infiltration results in obstruction of blood vessels by way to blood clot formation within capillaries and prevents any chances of vascularization of graft. The reason for the presence of pre-existing antibodies can be repeated blood transfusions, pregnancies and/or previous graft in host, as these expose the person to allogenic MHC antigens.

Acute rejection: In acute rejection, graft tissue can be rejected anytime between the tenth day and six months of transplantation surgery. It is usually cell-mediated allograft rejection that occurs after the massive infiltration of macrophages, lymphocytes and other immune cells at the site of the transplant. As it leads to tissue destruction carried out by T-cells. In some cases of acute allograft rejection, the humoral system plays an important role as allograft dysfunction and destruction are mediated by antibody and complement system. Acute rejection of graft tissue makes use of effector mechanisms stated previously.

Chronic rejection: Chronic allograft rejection occurs months or years after the transplant after the subsiding of acute rejection reactions. Here also mechanisms include both humoral and cell-mediated immune responses from the host. In the first few years after transplant, the use of immunosuppressive drugs and tissue typing, PRA for finding the best suitable donor contributes to increasing graft acceptance and chances of host survival. But in the long-term chronic rejection can occur as little progress has been done in this regard as immunosuppressive drugs fail to prevent chronic rejection. In some cases, another transplant is warranted timely to save the patient. Certain factors listed below increase the chances of chronic rejection:

- Previous case of acute rejection
- Improper immunosuppression
- Injury to transplanted organ leading to reperfusion
- Delay in graft function initially
- Long cold ischemia time
- Post-transplant infection (*e.g.* CMV)
- Donor-related factors (old age, high blood pressure *etc.*)
- Host-related factors (diabetes, high blood pressure, hyperlipidemia)

12.9 TRANSPLANTATION ANTIGENS

The grafted tissue is rejected by the immune response to the cell surface molecules capable of inducing antigenic stimulus. The number of cell surface molecules has been described as the transplantation antigens, including major histocompatibility antigens, minor histocompatibility antigens, ABO antigens on RBCs and monocytes and endothelial cell surface antigens.

12.9.1 Major Histocompatibility Complex (MHC)

MHCs were discovered back in 1967 by Gorer & Snell, and since then they have been shown to play an important part in the overall immune response. These cell surface antigens (HLA-human leucocyte antigens) are coded by a closely linked cassette of genes termed as MHC. These genes are present on chromosome 6 in humans. Details of MHC function and their role in immune reactions is detailed in the chapter concerning major histocompatibility complex and it can be referred. It is important to pre-match the MHC of donor and recipient during transplantation, as MHC compatible organs are readily accepted and organ transplant between MHC mismatched individuals is rejected which can be fatal as well. MHCs hone in the cytotoxic activity of the T-cells by primary signalling through the presentation of graft antigens as foreign to host lymphocytes or vice versa. This is well established now that T-cells play a key role in the allograft rejection and T_C- or T_H-cells act only after they are sensitized by primary signals of exogenous or endogenous antigen presence on MHC I or II on nucleated cells or the APCs respectively. It has been found in mice experiments that both CD4+ and CD8+ T-cells participated in graft rejection. Dendritic cell signaling (presentation of exogenous Ag on MHC I to CD8 T_C-cells) play a significant role in raising tolerance or rejection, as their inhibition was found to aid in graft acceptance.

12.9.2 Minor Histocompatibility Antigens

Minor histocompatibility antigens are very much different from the MHCs and they are recognized as peptides bound to self MHC. Minor incompatibilities result in T_C- cell activation, meaning that they are presented on class I MHC that is they are a type of endogenous antigens derived from intra-cellular peptides. These minor antigens can be encoded by either of the sex chromosome, autosome or mitochondrial genes. It is notable that these minor antigens-T_C interactions require the recipient to have the restricting MHC class I allele if a given antigen has to act as a transplantation antigen. This means that recipient T_C-cells can bind the minor donor peptide, and the donor cells have a shared MHC antigen on their surface that presents the same peptide. Although the targets for $CD8^+$ T_C-cells are class I MHCs that present peptide, it is clear that a $CD4^+$-mediated T_H-cells response is needed during immunization. Also, minor antigens do not provoke antibodies.

Based on the observations based on bone marrow graft-versus-host reactions and organ graft rejections in cases of genetically matched HLA antigens, it is established that minor histocompatibility antigens can be immunogenic in humans. A few cases of donor-directed, MHC class I-restricted, T_C-cell responses have been confirmed.

12.9.3 ABO Blood Group Antigens

The blood transfusion reactions are known since blood transfusions are being carried out and ABO blood group antigens on RBCs surface were initially held responsible for those reactions completely. A- and B-antigens have different types of glycosylations while O-Ag lacks it completely and these antigens are well identified by hemagglutinins (natural Ab) that agglutinate RBCs upon non-identity. In addition to RBCs, ABO antigens are expressed on the surface of other cells as well, *e.g.* on endothelial cells, thus they play important role in transplantation. Their mismatch results in hyperacute rejection of allografts through the action of preformed antibodies on cells expressing either of the mismatched ABO antigens. Graft donation follows the same pattern of donor-recipient as is done in blood transfusions pertaining ABO-blood groups like graft from the donor of O-blood group can be transplanted in either A, B or O, AB can donate to only AB and so on. Although ABO mismatching is of less importance as compared to MHC compatibility for transplant survival but ABO incompatibility can lead to hyperacute rejection of vascularized transplants, such as kidney and heart. Rhesus (Rh) factor and other blood cell antigens are not present on endothelial cells and are not considered during transplants due to this. Recent advancements in desensitization techniques and immunosuppressive agents have made it possible to consider graft donation between HLA-incompatible (even ABO-incompatible) individuals.

12.9.4 Monocytes and Endothelial Cell Antigens

In a few cases, the transplant fails even after ABO- and HLA-compatibility and these rejections are attributed to some still unknown non-ABO-antigens. They are presumed to be expressed on the surface of endothelial cells and/or monocytes. Researchers are still in the phase of finding out such antigens and their attributes. So, such antigens are poorly understood for now. Also, there are no methods to do their typing before transplantation as there is no Ab known for these, and we lack proper reagents/procedures as well.

12.10 GRAFT-VERSUS-HOST DISEASE (GVHD) AND HOST-VERSUS-GRAFT DISEASE (HVGD)

In a graft, rejection can be both ways depending on the tissue/organ used for transplantation. Graft-versus-host disease (GvHD) is a serious complication of allogeneic organ transplantation and major cause of morbidity and mortality post-transplant as the donor lymphocytes show alloimmune response against the normal host cells. There are medicines and treatments to manage GvHD but in some people, it shows up post-transplant. Effects of GvHD may vary from mild to severe and many patients show one or the other symptom of GvHD after receiving transplants. GvHD occurs in the case of bone marrow, stem cells and sometimes solid organ transplants. It is believed that having some GvHD is not bad in certain cases rather it indicates that the new immune cells are doing their job right *e.g.* in the case of bone marrow transplant for curing blood cancer, GvHD indicates transplant cells will take care of cancer cells if any. GvHD may affect different areas of the body, but it commonly affects skin, liver and gut (including stomach and bowel). Thus, based on the symptoms and their relative time of occurrence GvHD is majorly of two types:

Acute (aGvHD): It occurs in the early weeks or months after the transplant. It is late acute GvHD if it develops after 3–4 months of transplant. Symptoms include skin redness or rashes in case of skin aGvHD; yellow discolouration of eyes/skin with abnormal blood test results indicates liver aGvHD; and diarrhea, nausea, vomiting, abdominal cramps signal GI tract/gut aGvHD.

Chronic (cGvHD): It usually develops after a year or so of transplant surgery. Symptoms include skin rashes, discolouration, tightening and thickening in cGvHD of skin; abdominal swelling, yellow disclouration of skin/eyes with abnormal blood tests in cGvHD of liver; dryness in eyes, vision change in cGvHD of eye; dryness, increased sensitization, and white patches with pain in mouth indicate cGvHD of the mouth; shortness of breath and changes in chest x-rays indicate pulmonary cGvHD of lungs; decreased joints activity in cGvHD of connective tissues; fatigue, muscle weakness with pain in neuromuscular cGvHD; vaginal dryness and pain in cGvHD of vagina/vulva; difficulty swallowing, pain and weight loss in gut cGvHD.

The overlap syndrome may show features of both acute and chronic GvHD.

Host-versus-graft disease (HvGD) occurs when the host immune cells target the graft cells after recognizing them as non-self and mount an immune response. This can also be of two types, acute HvGD and chronic cHvGD. The symptoms of aHvGD include hepatic infarcts, gastrointestinal bleedings, anemia, severe thrombocytopenia and leucocytosis. Symptoms of cHvGD include thrombocytopenia, autoimmune antithymocyte antibodies (ATA) *etc.* In a nutshell transplantation immunology is all about the level to which a host can tolerate a graft from the donor and how clinicians can reduce the risk of graft rejection.

12.11 DONOR-RECIPIENT MATCHING

12.11.1 ABO Blood Group Compatibility

ABO blood group testing is the first step in determining the compatibility of host and potential donor. The common A, B, AB and O blood antibodies that react with other blood groups are matched. The Rh factor (positive or negative) does not matter. The donor-recipient compatibility is as listed in Table 12.1 below:

TABLE 12.1
The Table Enlists The Donor-Recipient Compatibility

Blood Group Compatibility Table

	Donor			
Recipient	**A**	**B**	**AB**	**O**
A		×	×	
B	×		×	
AB				
O	×	×	×	

Source: Authors.

Even if the blood group matching is not there, the transplant can still be done by treatments to lower the host blood antibodies. Another way is to look for another donor via the paired exchange, where somebody else's intended donor also does not match with them, but the other recipient whose donor can donate to that somebody else. The transplant coordinator at the transplant centres can be reached out for such help. If the donor's blood matches with the host/recipient, then the donor takes the next blood test, i.e., tissue typing.

12.11.2 Tissue-Typing

As the name suggests in this test, tissue (HLA) type of the potential donor and recipient are matched to see how well they go with each other so that no or minimum immune reaction is there after transplant. Everybody has different tissue types except in the case of identical twins. Genes for HLA are received as haplotype from each parent, so have about 50% match between child and parents and is about 25% similar in the case of the siblings. But in the case of siblings the match can dwindle between 0% and 100% match due to the polymorphic nature of MHC genes and crossing over (rare, once in 200 meiotic cycles). It is less likely that an unrelated donor shows a match. Transplant immunologists have established that the better the HLA match, the more successful the transplant over a longer period of time with eliciting immune response of any kind leading to its failure. Gross HLA mismatch leads to sudden transplant rejection by the host immune system. The best way is to match 12 out of 12 antigens of the recipient with that of the donor (showing a zero mismatch). In case a patient has a very common HLA type, it is possible for all 12 markers to match, even with an unrelated donor.

The next blood test is done to measure the antibodies to HLA in recipient's blood and it is done for the patient only repeatedly (based on transplant policy). The presence of HLA Ab is deemed harmful for transplant and their quantities may vary over time, so need to be measured multiple times and immediately before surgery and if required after surgery as well. Person with HLA Ab in their blood are HLA 'sensitized' and require a donor with HLA type that do not react with HLA Ab of the recipient. Notably, HLA Ab levels show changes following blood transfusion, severe infections, miscarriage and some small surgery (*e.g.* dental) and need to be taken care of before transplant surgery.

12.11.3 Cross-Matching

Serum cross-match test is done between donor cells and recipient serum multiple times and right before transplant surgery. It is done by mixing the serum of the recipient with cells from the donor and observed under the fluorescent microscope to detect the presence of any antibodies binding with donor cells. If such antibodies are found high in number, it is a positive cross-match. It indicates that the transplant will not be successful due to destroying of donor cells and rejected consequently.

12.11.4 Panel Reactive Antibody (PRA) Test

It is another blood test done prior to transplantation. PRA percentage is determined by testing the serum of a prospective host against a panel of cells from prospective donors with different HLA proteins. PRA is the amount of anti-HLA antibody present in the blood of prospective recipient's serum and it is expressed as a percentage. Such antibodies can result from the exposure of the recipient to different foreign HLA proteins because of prior transplant, blood transfusion and/or pregnancy. If one tests positive for HLA Ab, they are 'sensitized; and his/her PRA will be greater than 0. The PRA percentage can vary from 0 to 100. It can also vary from time to time. If one has a very high number of HLA Ab in their serum with a high PRA percentage, it is difficult to find a suitable donor for such people. In these, plasmaphereses can be carried out to reduce Ab in the blood of the recipient.

12.11.5 Serology Screening

A proper pre-transplant serological screening for many viral infections, including cytomegalovirus (CMV), human immunodeficiency virus (HIV), hepatitis B virus (HBV), hepatitis C virus (HCV), Epstein-Barr Virus (EBV), and others is important. In addition, serological testing for detection of prior infection or immunity (with IgG) for *varicella*, measles, mumps, and rubella is suggested prior to transplantation surgery. These test results and vaccination records help in the decision to update the vaccination status of recipients. The recipients would be more susceptible to common infections like tuberculosis (TB) due to the weakened immune system. So, it is important to chart the history of TB risk factors. The epidemiological survey will search the risk of TB by tracing close contact with the active TB patient, birth, residence or work in areas highly under TB burden *etc.* Transplant recipients are supposed to be tested for latent syphilis and TB along with other probable bacterial infections.

12.12 IMMUNOSUPPRESSIVE DRUGS

In order to increase the chances of graft acceptance, immunosuppressive drugs are used. These drugs are pharmaceutical agents that suppress the immune response to foreign tissue or organ and reduce the chances of its rejection by the host. These agents are used for the treatment of autoimmune diseases and other immune mediated-diseases along with transplantation. Roy Calne, working under the guidance of Peter B. Medawar in the 1060s, was the pioneer to use pharmaceutical drugs for suppressing the immune system of the host to aid graft acceptance. He worked with whole body irradiation in his earlier experiments and later used the immunosuppressive agents 6-MP and its derivative azathioprine (AZA) developed by Gertrude Elion and George Hitchings. Then in the 1070s, cancer chemotherapy agent cyclophosphamide was adapted for immune suppression, followed by the use of anti-lymphocyte serum. The advancing knowledge of the immune system components (Ab, mAbs, cytokine, chemokine B and T cell, *etc.*) and their functions furthered the use of targeted immunosuppressive agents. Immunosuppressive effects of cyclosporin A (derived from extracts of *Tolypocladium inflatum* fungus) and sirolimus (rapamycin) were worked upon and applied in transplantation procedures successfully. Research for finding suitable agents is ongoing and see new developments every now and then. In clinical transplantation blocking of the co-stimulatory signals has shown to develop anergy. Such inhibitors include anti-CD40L, anti-CD3 mAb, CTLA-4 Ig and anti-CD25 mAB that block co-stimulatory signals, while cyclosporine A (FK506) and JAK3 inhibitors restrict signalling pathways, thus blocking immune cells interactions and drugs like rapamycin, azathioprine, cyclophosphamide, mycophenolate mofetil and methotrexate interfere with cell cycle at some point keeping immune cells proliferation in check. More details are summarized in the following text.

There are some immunosuppressive drugs that target antigen APC and T-cell interaction and early stages of T-cell activation by APC in particular. Such as the agents abatacept and belatacept CTLA4 Ig that blocks B7 interaction with CTLA4 (cytotoxic T lymphocyte-associated protein 4), ASKP1240 α-CD40 mAb (anti-CD40 mAb) that block CD40 and CD40 ligand interaction, efalizumab α-LFA-1 (lymphocyte function-associated antigen-1) mAb block ICAM-LFA interaction, cyclosporine tacrolimus (CNI- calcineurin inhibitor) block signal transduction through calcineurin, α-CD3 mAb block CD-3 signal, Alemtuzumab block CD52 signal. In summary, these immunosuppressive agents block signals 1 and 2 hindering the T-cell activation through APC interaction.

Additionally, there are immunosuppressive drugs that target T-cell signal 3 (differentiation and proliferation) and few selected cytokines. Such as basiliximab or α-IL-2R block IL-2 and IL-2 receptor signalling, TNFR Ig and α-TNF mAb block TNF-α and TNF receptor signalling, tofacitinib block the JAK-STAT pathway, corticosteroids block the NF-κB signal, azathioprine and mycophenolate arrest T-cells in S-phase restricting their proliferation, sirolimus and everolimus blocks mTOR signalling and thus check G1-phase to S-phase transition. In summary, these agents inhibit later stages of T-cell differentiation and proliferation. Additionally, α-IL-17 mAb, secukinumab block IL-17 and α-IL-6 mAb, tocilizumab block IL-6 function.

Lastly are the immunosuppressive drugs that target B- and T- cell interaction, plasma cell function, and complement-system function. α-CD40 mAb (ASKP 1240) restrict CD40 and CD40 ligand interaction crucial for B- and T-cell interaction, Alemtuzumab block CD52, α-CD22 mAb (epratuzumab) block CD22, α-CD20 mAb (rituximab, ocrelizumab, ofatumumab, veltuzumab) block CD20 function, α-BAFF mAb (belimumab) inhibit BAFF from binding to its receptors on B-cell, TACI (transmembrane activator and CAML interactor) Ig (anacicept) block APRIL (A proliferation-inducing ligand) ligand from binding to its receptor and also inhibit BAFF (B cell–activating factor) interaction with BAFF-R on B-cell, Bortezumib acts on proteasome function thus checking antibody production by plasma cell, α-C5 mAb (eculizumab) block C5 convertase action on compliment protein, C5 eventually blocking MAC formation (C5b-C9).

There are different immunosuppressive agents developed and in the process of being developed. They have some specific target in the immune response cascade that involves cytokines, chemokines, B-and T- cell, plasma cells, complement system proteins and other immune system cells and signalling/effector molecules.

12.13 SUMMARY

- Tissue and organ transplants are life-saving for millions of people worldwide helping them to lead a healthy and normal life.
- But a miss-matched graft can have a serious manifestation in their life and many a times it can be fatal.
- Graft rejection is an immunologic process that uses the properties like specificity, memory and self-nonself recognition of the immune system.
- Graft rejection can be quick and rapid (hyperacute rejection), within 10–200 days (acute graft rejection) or delayed beyond a year or so (chronic graft rejection).
- Rejection can involve humoral as well as cell-mediated immune reactions against the graft tissue.
- Other than rejection there can be graft vs host disease (GvHD) of acute or chronic types with varying symptoms, most common in cases of a bone marrow transplant.
- To prevent and minimize the chances of immune response to graft tissue, various tests such as ABO blood group matching, tissue typing, cross-matching PRA and serological screening are done to find the best suitable organ donor.

- Other than these immunosuppressive drugs are also used in clinical situations to suppress the host immune system to reduce the chances of graft rejection.

QUESTIONS

1. Define the following terms:
   a. Transplantation
   b. Graft
   c. Donors
   d. Autograft
   e. Allograft
   f. Xenograft
   g. Isograft
   h. Cross-matching
   i. PRA test
2. Differentiate between the following terms:
   a. Autograft and allograft
   b. Donor and recipient
   c. Sensitization and effector stage of graft rejection
   d. Acute and hyperacute graft rejection
   e. Chronic and acute GvHD
   f. First set and second set graft rejection
   g. Living Donor and Deceased Donor
   h. GvHD and HvGD
3. Write the major contribution of the following scientists in transplantation immunology:
   a. Joseph E. Murray
   b. Zinkernagel and Doherty
   c. George Hitchings and Gertrude Elion
   d. Snell and Dausset
   e. Donnall Thomas
   f. Thomas E. Starzl
   g. Alexis Carrel
   h. Peter Medawar
4. Expand the following:
   a. CTLA-4
   b. PRA
   c. GvHD
   d. HvGD
   e. RANTES
   f. mTOR
   g. HLA
   h. EBV
   i. CMV
   j. HBV
   k. HAV
   l. BAFF
   m. APRIL
   n. OPTN
5. Why is organ donation important? Describe the reasons for the same.
6. Describe the Immune Response to Transplantation.
7. What are the different mechanisms involved in Graft Rejection?

8. What are the different methods used for Donor- Recipient Matching?
9. Mention the different types of cells, tissues and organs that can be transplanted.
10. Write short notes on the following:
    a. Host verses Graft Disease
    b. Immuno-suppressive drugs given to patients undergoing transplantation
    c. ABO- Incompatible Transplantation
    d. Stem cell Transplantation

BIBLIOGRAPHY

Abbas, A.K., Lichtman, A.H. and Pillai, S. (1994). *Cellular and Molecular Immunology*. Elsevier Health Sciences, W.B. Saunders, Philadelphia, United States..

Barker, C.F. and Markmann, J.F. (2013). Historical overview of transplantation. *Cold Spring Harbor Perspectives in Medicine*, *3*(4), p. a014977.

Chinen, J. and Buckley, R.H. (2010). Transplantation immunology: solid organ and bone marrow. *Journal of Allergy and Clinical Immunology*, 125(2), pp. S324–S335.

Fishman, J.A. (2009). Introduction: infection in solid organ transplant recipients. *American Journal of Transplantation*, *9*(s4), pp. S3–S6.

Kindt, T.J., Goldsby, R.A. and Osborne, B.A. (2007). *Kuby Immunology*. WH Freeman; New York.

Wiseman, A.C. (2016). Immunosuppressive medications. *Clinical Journal of the American Society of Nephrology*, *11*(2), pp. 332–343.

https://www.ncbi.nlm.nih.gov/pmc/articles/PMC2857984/pdf/nihms179519.pdf

https://www.immunology.org/sites/default/files/Transplant%20Immunology.pdf

https://www.immunopaedia.org.za/immunology/advanced/1-transplantation/?print=pdf

https://www.immunopaedia.org.za/immunology/advanced/1-transplantation/

https://en.wikipedia.org/wiki/Transplantable_organs_and_tissues

https://en.wikipedia.org/wiki/Organ_transplantation

http://firm.uni-frankfurt.de/education-transplantation

https://www.immunopaedia.org.za/immunology/advanced/1-transplantation/

https://www.immunology.org/sites/default/files/Transplant%20Immunology.pdf https://www.who.int/transplantation/Guiding_PrinciplesTransplantation_WHA63.22en.pdf?ua=1

http://www.declarationofistanbul.org/

13 Vaccines

Nidhi Garg
University of Delhi

CONTENTS

DOI: 10.1201/9781003286424-13

13.1 INTRODUCTION

Vaccination is defined as a process of injecting an antigen into the body of an individual, so his/her immune system can be stimulated leading to the development of adaptive immunity towards that antigen. The injected antigen may be a complete pathogen like a bacterium or virus or it could be a part of the pathogen and is termed as vaccine. Vaccines have been extremely effective in preventing and even eradicating infectious diseases. Vaccination of a large number of people in a population results in a phenomenon called as herd immunity. It has been estimated, that on an average 2.5 million child deaths are prevented globally, through vaccination every year. Additionally, the death of 2 million children each year can be prevented by immunization. Immunization is defined as the process which protects humans and animals from a particular disease through a heightened immune response. Immunization is not as same as vaccination. Vaccination is a part of immunization that enables the body to identify a specific antigen which is known as active immunization. Passive immunization can be conferred through the administration of external antibodies, which for the short term fortify body's response to a particular antigen without inducing immunological memory of the antigen. The worldwide immunization programmes had a remarkable effect on global health management through a drastic reduction in both mortality and morbidity of millions of infants and children due to infectious diseases. Large-scale vaccination has led to the universal elimination of smallpox and partial eradication of polio, measles, and tetanus.

If one asks what is the importance of vaccination or immunization programmes, it can be simply answered by assessing the effect of vaccination on the health of public health globally, which is incomparable to anything else, not even antibiotics. The vaccines happen to be extremely important as they:

a. **Promotes health:** Vaccines aid in human development by helping people stay healthy.
b. **Extensive reach to people:** Vaccines protect not only individuals or particular communities but entire populations. An example of this is smallpox, which in 1980 was declared eradicated by the WHO.
c. **Quick impact:** There is an instant effect of maximum vaccines on people of different populations. An example of this is the worldwide reduction in deaths by 78% due to measles from 2000 to 2008, due to vaccination.
d. **Lifesaving and cost-effective:** A group of economists have placed extended immunization coverage for children in the fourth position in a list of 30 cost-effective methods of advancing global welfare.

13.2 HISTORICAL BACKGROUND

Even though the importance of vaccines has been highlighted in the last century, its story is rather old due to the long history of infectious diseases worldwide and the use of cowpox pustules for giving immunity against it. The story of vaccines doesn't start with the development of first vaccine by Edward Jenner, rather it dates back to 10th century when the Chinese physicians practiced inoculation of cowpox pustules. There are well documented examples of practice of vaccination in India and China in the 17th century. Vaccination was performed even in Africa and Turkey. Later the practice of vaccination reached Europe and America.

Nevertheless, the British physician, Edward Jenner is credited for pioneering the development of the concept of first vaccine in 1796. He inoculated live and virulent cowpox virus from the cowpox pustules in humans to show that exposure to this virus will induce cross-protective immunity against smallpox. In fact, the term "vaccine" has been given by Edward Jenner, and is derived from the Latin word "vacca", meaning cow. The act of immunization started by Edward Jenner became popular and turned out to be a widespread practice. Even though this method protected the individuals from smallpox, there was still risk of death or epidemic outbreak.

As time went by, Jenner's original procedure went through both medical and technological changes. There was universal case detection as well as vaccination programmes against smallpox which finally, led the World Health Assembly in 1979, to officially declare smallpox as eradicated. This has been till date the hallmark of human success over one of the deadliest diseases. After Jenner, it's Louis Pasteur whose work led to the development of the next vaccine against rabies in 1885. For more than a century people followed Pasteur's principle, which involved isolation, purification and injection of the pathogen for inducing protective immunity.

By 1900, the researchers had developed five vaccines, two were against viruses, smallpox and rabies and three were against bacteria causing typhoid, cholera, and plague. The other diseases that became targets of vaccine development were diphtheria, tetanus, anthrax and tuberculosis to name a few, they were developed during the 1930's. The time from 1950 to 1960's saw tremendous research and development of vaccines. Researchers developed suitable methods for growing viruses in laboratory, leading to several discoveries and innovations, including the development of vaccine targeting polio. This was followed by creation of vaccines targeting diseases that commonly affect children like mumps, measles, and rubella. This greatly helped in reducing infant mortality rate.

Thus, the 20th century witnessed the development of vaccines targeting the common fatal infections such as pertussis, diphtheria, tetanus, etc. along with certain communicable diseases. With the availability of vaccines, the developed countries started recommending routine vaccination for children. Once the World War II was over there were widespread and systematic childhood vaccination programs which led to great improvement in public health. With the successful eradication of smallpox, the WHO in 1974, started the "Expanded Programme on Immunization (EPI)". The principal objectives of EPI were

a. Ensuring vaccination of each child from birth till 1 year against six diseases *viz.* tuberculosis, polio, diphtheria, pertussis, tetanus and measles.
b. To vaccinate pregnant women with tetanus toxoid to protect them and the new-borns from tetanus.

By 1990, vaccination could protect more than 80% children worldwide from the six main EPI diseases. In addition to this, new vaccines were added to the EPI programmes in several nations. The Global Alliance for Vaccines and Immunization (GAVI) was formed in 1999, for two reasons:

a. For spreading the range of the EPI throughout the world.
b. For helping the underprivileged nations to add novel and lesser used life-saving vaccines in their immunization schedule.

In spite of the efforts of WHO, more than 24 million infants are unable to receive all the EPI vaccines in the first year of their life. But nevertheless, it may be said that the EPI has been successful due to universal decrease in the cases of measles and poliomyelitis. Together measles and poliomyelitis are among the list of several diseases that have been targeted for elimination by WHO through vaccines. The 20th century also saw the development of innovative techniques such as the use of recombinant DNA technology and new delivery techniques driving vaccine research ahead. Currently, there are more than 20 diseases which can be prevented through vaccines. Some of the recent WHO recommended vaccines used globally are the hepatitis B, rotavirus, *Haemophilus influenzae* type b (Hib) and pneumococcal vaccines. Vaccines against yellow fever is recommended in only those countries which show a high incidence of this disease. Over the years, the disease targets have expanded due to viral and bacterial outbreaks in different parts of the world such as the Zika virus, Ebola virus and the Nipah virus outbreak. Vaccine research is also focusing on non-infectious conditions like drug addiction and allergic reactions. As of now, the WHO guidelines mention routine vaccination against 12 different diseases (Table 13.1). In addition to these, more vaccines are required for people that are at a great risk or in areas with different needs.

Table 13.1
The Main Vaccine-Preventable Diseases Targeted by the EPI and the Associated Vaccines

Pathogen	Disease	Vaccine
Tubercle bacillus	Tuberculosis	Bacillus Calmette-Guérin (BCG) vaccine
Polio virus	Polio	Oral polio vaccine (OPV) vaccine Inactivated polio vaccine (IPV) vaccine
Corynebacterium diphtheria	**Diphtheria:** It is a vaccine preventable disease caused by *Corynebacterium diphtheria.* The throat of the infected individuals shows presence of a false membrane.	Diphtheria toxoid vaccine: It consists of diphtheria toxoid, providing immunity against diphtheria which is available as combination vaccine with tetanus toxoid.
Clostridium tetani	**Tetanus:** It's caused by the toxins produced by *Clostridium tetani* which is characterized by aching muscular contractions.	Tetanus toxoid (TT) vaccine: The tetanus toxoid provides immunity against tetanus. The tetanus toxoid is also given twice to pregnant women that protects both the mother and new-born infants against tetanus.
Bordetella pertussis	**Pertussis:** A communicable disease whose causative agent is the bacterium *Bordetella pertussis.* The disease is characterized by vehement and irregular coughing.	The pertussis vaccine is present in two forms, i.e., whole-cell pertussis (wP) vaccine and acellular (cell-free) pertussis (aP) vaccine. This vaccine is usually administered as combination vaccine with Diphtheria and Tetanus toxoid to infants and young children.
Measles virus	It is a highly infectious disease characterized by fever. Outbreak of red circular spots on skin proving to be fatal to young and debilitated people.	Measles vaccine
Hepatitis B virus	This virus causes liver infection. Its transmission is by contact with infected blood or other body fluids from an infected individual to a healthy individual.	Hepatitis B vaccine (HepB): it is a subunit protein-based recombinant vaccine.
Rotavirus	It refers to a group of viruses leading to gastroenteritis or diarrhoea commonly affecting children.	Rotavirus vaccine
Haemophilus influenzae type B (Hib)	*Haemophilus influenza*	Hib conjugate vaccine: It is a subunit polysaccaride-conjugate vaccine providing immunity against Hib disease.
Streptococcus Pneumoniae (Pneumococcal infection)	*S. pneumoniae* is responsible for causing pneumonia	Pneumococcal vaccines
Yellow fever virus	A viral disease transmitted through mosquitoes and common in the tropics. Common symptoms are high fever, jaundice, and gastrointestinal bleeding.	Yellow fever vaccine
Human papilloma virus (HPV)	Cervical cancer develops in the cervix due to infection with Human papillomavirus infection (HPV). It should be noted that not all HPV infections result in cervical cancer.	HPV vaccine

Source: https://vaccine-safety-training.org/vaccine-preventable-diseases.html.

13.3 ROUTES OF VACCINE ADMINISTRATION

Vaccines targeting different diseases are given through different routes or ways such that they enter the body and then reach the site of action through the body's transport mechanisms to show their desired effects. Thus, the success of immunization is dependent on the correct route of administration. Table 13.2 mentioned below describes the different routes of vaccine administration.

13.4 COMPONENTS OF A VACCINE

Each vaccine is made up of several components such

a. Antigens
b. Stabilizers
c. Adjuvants
d. Antibiotics
e. Preservatives
f. Left over by-products accumulated during the process of vaccine production.

The knowledge of precise composition of each vaccine is essential, for one is able to investigate adverse events following immunization (AEFIs). Following vaccination, a person might develop allergy or certain complications, thus the knowledge of components of vaccines will help to determine which component is the causative agent. This will help in choosing alternative products to prevent further complications.

13.4.1 Antigens

Antigen is a molecule that elicits an immune response in the body of an individual because it is considered as foreign. The antigen is a component of the pathogen.

13.4.2 Stabilizers

They are chemical substance which help maintain the efficacy of vaccines during their storage. The stability of vaccines is extremely important under conditions of unreliable cold chain. Instability

Table 13.2
Different Routes of Vaccine Administration

Routes of Vaccine Administration	
Intramuscular (IM) injection	The vaccine is injected directly into the muscle mass. Usually, those vaccines that comprise of adjuvants are injected intramuscularly for reducing adverse local effects. Example: Tetanus toxoid
Subcutaneous (SC) injection	The vaccine is injected into the subcutaneous layer of the skin. Example: Influenza and MMR
Intradermal (ID) injection	The vaccine is injected in the dermis of the skin. Example: BCG vaccine is injected intradermally for reducing the danger of neurovascular injury. It's difficult to inject the BCG vaccine into the new born baby's arms due to their small size.
Oral administration	It's the easiest and painless way of vaccinating children. Example: Polio vaccine
Intranasal spray application	It's another easy and painless way of administering the vaccine to the patient via the nasal mucosa. Example: Influenza

Source: https://vaccine-safety-training.org/route-of-administration.html.

can result in loss of antigenicity and diminished infectivity of Live Attenuated Vaccines. Changes in temperature and pH jeopardize the stability of vaccines. Instability of bacterial vaccines results from hydrolysis and accumulation of protein and carbohydrate moieties. Commonly used stabilizing agents are Magnesium Chloride for OPV, Magnesium Sulphate for measles, lactose-sorbitol and sorbitol-gelatine.

13.4.3 Adjuvants

Adjuvants comprise a heterogeneous group of chemical substances that enhance the immunogenicity of antigens in the vaccines thereby enhancing their efficacy. Adjuvants are commonly added to inactivated (killed) vaccines prepared by traditional and biosynthetic methods, recombinant DNA technology and other modern technology, the reason being that the synthetic or subunit vaccines have poor immune response. Currently, hundreds of diverse kinds of adjuvants are used during vaccine preparation or else are under consideration.

13.4.4 Antibiotic

It is a chemical substance produced by organisms which either kill or inhibit bacterial growth. In the manufacture of certain vaccines, minor amounts of antibiotics are used to avoid bacterial contamination of viruses in tissue culture. Generally, trace amounts of antibiotics are present in vaccines. An example of this is the MMR vaccine and IPV, both of which have less than 25 µg of neomycin per gram/dose. People allergic to neomycin are kept under observation after vaccination so they can be immediately treated for any type of allergic reaction.

13.4.5 Preservatives

Preservatives are chemical substances such as thiomersal, formaldehyde, or phenol derivatives that are added to multi-dose vaccines for preventing bacterial and fungal growth.

Did You Know?

Preservatives used in Vaccines

A huge variety of preservatives are in use in multi-dose vaccines to prevent microbial growth. The most commonly used preservatives, thiomersal and formaldehyde are mentioned below:

A) Thiomersal

- It is one of the oldest preservatives used since the 1930s and contains ethyl mercury.
- The biggest advantage of using thiomersal is the absence of detrimental effects for doses added to vaccines, with the exception of minor reactions such as redness and swelling at the site of injection.
- In several nations, it is added to both multi-dose vials and single-dose vials to decrease the costs of storage and administration.
- The use of Thiomersal, owing to the presence of ethyl mercury has been a subject of controversy. Thiomersal has been under continuous scrutiny by the Global Advisory Committee on Vaccine Safety. As of now, there exists no known case of toxicity of Thiomersal in infants due to the presence of trace amounts of it in vaccines on their neurological development.

B) Formaldehyde

- It is used for inactivating viruses such as the polio virus in IPV and for detoxifying bacterial toxins, like diphtheria and tetanus toxins for the production of toxoid vaccines.
- During the course of vaccine production, there is a purification step for the removal of even trace amounts of formaldehyde.
- The quantity of formaldehyde present in vaccines is several hundred times lower than the prescribed levels that can cause harm to neither human adults nor infants. For example, the DTP-HepB+Hib is a "5-in-1" vaccine that consists of less than 0.02% formaldehyde per dose (<200 ppm).

Source: http://courses.washington.edu/medch401/pdf_text/RT_vaccines_2.pdf.

13.5 ACTIVE AND PASSIVE IMMUNITY

Immunity is defined as "the state of protection against infectious disease conferred either through an immune response generated by immunization or previous infection or by other non-immunological factors." The immune system consists of both innate and adaptive immunity (Figure 13.1).

13.5.1 Innate Immunity

It is also referred to as native immunity. It is dependent upon the genetic constitution of the organism, independent of external stimulation or a prior infection. For example, the innate immune system in mammals consists of primitive bone marrow cells capable of recognising foreign substances and reacting. Innate immunity is of the following two types:

a. **Non-specific innate immunity:** It is responsible for the resistance that the body shows to any type of infection.
b. **Specific innate immunity:** It helps the body show resistance towards a particular type of pathogen. Therefore, certain races, individuals or plant varieties are not affected by specific infectious diseases.

13.5.2 Adaptive Immunity

It is of two kinds subject to the mode of attainment of immunity.

a. **Passive immunity:** It is attained by transferring the antibodies or activated T-cells to the individual. This type of immunity is for a short duration of time, say a few months.
b. **Active immunity:** This type of immunity develops by exposure of individual to an antigen such that an immune response is evoked thus, the immunity attained lasts for the lifetime.

Both passive and active immunity is attained either naturally or artificially:

- **Natural immunity:** It is acquired by an organism by chance contact with a pathogen.
- **Artificial immunity:** It is developed by deliberately exposing an organism to a pathogen as in the case of vaccination.

Passive immunity: It is the transmission of readymade antibodies derived from an immune person to another person lacking immunity. Passive immunity gives instantaneous

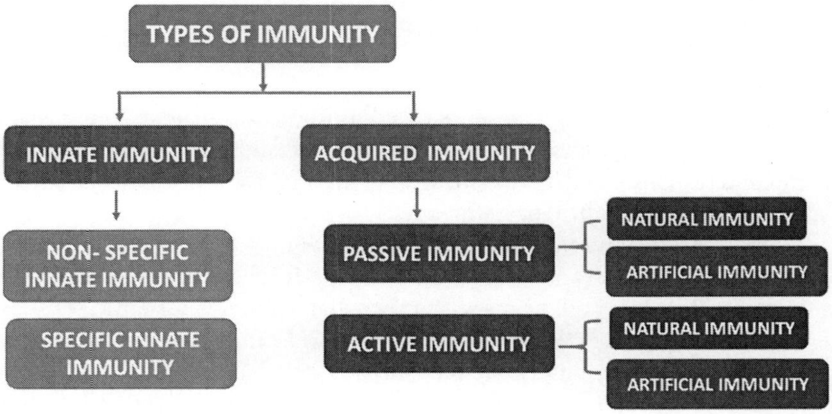

FIGURE 13.1 The flowchart depicts the different types of immunity. (Copyright Author.)

protection, although there is no generation of immunological memory, thus the individual can get infected again by the same pathogen. Despite the disadvantages, passive immunization is beneficial under the following circumstances:

a. In case of a great possibility of infection.
b. For treatment of several acute infectious disease and food poisoning.
c. When the body does not have sufficient time for developing the immune response.
d. For reducing the symptoms of current or immunosuppressive diseases.

Natural passive immunity is acquired as maternal antibodies pass through placenta, during pregnancy, from mother to foetus. During the third month of gestation, maternal antibodies (MatAb) enter the foetus by binding to the FcRn receptor present on placental cells. Only the IgG antibody is capable of passing through the placenta. Breastfeeding the baby is another way of providing natural passive immunity to the newborn. The breast milk contains the IgA antibodies which enter the gut of the newborn, which protects the baby from bacterial infections, till the baby is able to produce his/her own antibodies.

Artificial passive immunity can be obtained by injecting an individual with antibodies of either human or horse towards a particular pathogen or toxin. Artificially acquired passive immunity is administered in many forms as mentioned below:

a. In the form of human or animal blood plasma.
b. In the form of pooled human immunoglobulin for intravenous (IVIG) or intramuscular (IG) injection.
c. By means of monoclonal antibodies (MAb).

The side effects of artificial passive immunization involve the danger of hypersensitive reactions, and serum illness, particularly from the use of gamma globulin derived from non-human species. In spite of the side effects, the artificial method of obtaining passive immunity is an effective means for the treatment of communicable illness, and in the past, it served as the sole means of treatment of some infections until the advent of antibiotics.

Artificial passive immunity can also be achieved through the transmission of activated T-cells (cell-mediated immunity), from the donor immune individual to the recipient. Cell-mediated immunity is seldom used in humans as there is difficulty in finding histo-compatible donors, otherwise there will be a problem of graft versus host disease. In spite of these problems, cell-mediated immunity is used for curing some kinds of cancer and immunodeficiency. The bone marrow transplant differs from cell-mediated immunity because undifferentiated hematopoietic stem cells are transferred whereas, the cell-mediated immunity consists of differentiated cells.

13.5.2.1 Active Immunity

The pathogenic invasion of the body causes activation of the B cells and T cells leading to the elimination of pathogen along with the generation of both memory B cells and T cells, which is the primary immune response. During the lifespan of an organism, the memory cells remember every pathogen encountered, and upon reinfection, by the same pathogen a strong secondary response is evoked. Thus, the presence of an immunological memory causes a fast rise in the titre of antibody and effector T cells following reinfection late in life, which may appear either mild or even imperceptible. In 1921, both primary and secondary immune response was described by English immunologist Alexander Glenny for the first time. The immunity described above is both adaptive and active since the immune system of the body is organizing itself for further pathogenic invasions. Active immunity consists of B cells and T cells and takes help from the innate immune system.

If an individual comes in contact with a live pathogen, the immune system is activated which later leads to the development of immunological memory, this is how active immunity develops naturally. This is called as "natural" since the organisms are not deliberately introduced to it. Both immunodeficiency and immunosuppression readily affect the development of active immunity. Vaccination is the method of artificially inducing active immunity.

13.6 TYPES OF VACCINES

There exist diverse kinds of vaccines, which are classified based on the antigen present in the preparation. The preparation of the vaccine greatly determines the way they are used, stored, and administered. The following are the goals of vaccine manufacturers:

- The vaccine was developed to be effective enough in either preventing or counteracting the infectious disease.
- The efficacy of vaccine is such that long-term immunity is achieved with minimal doses.
- To develop vaccine containing maximal antigens conferring total protection against infectious disease.
- To develop vaccines with minimal or no side effects.
- To develop vaccines formulations conferring prolonged stability to vaccines extremes of storage conditions.
- Mass availability of vaccines at affordable rates through large-scale production.

In the past, vaccines developed contained either live attenuated pathogens or whole inactivated organisms or inactivated bacterial toxins. They induced effective immune response leading to the development of long-term protective immunity. One disadvantage of using live systems is the adverse effects which are sometimes mild but can range from severe to fatal. Safety is the main apprehension limiting the applicability of traditional approaches during the development of new vaccines. There exists the possibility of disease condition or virulence of pathogenic strains in traditional vaccines in immune-compromised individuals. Keeping in mind these concerns, the new parenteral vaccines will in most likelihood won't be live attenuated vaccines. Newer approaches for vaccine development thus arose, leading to much safer technology of subunit vaccine. Further developments occurred in the field of vaccinology with an important role played by recombinant DNA technology, solving the problem of antigen manufacturing. Conjugate vaccines, subunit vaccines, and the non-replicating recombinant virus-like particles (VLPs) have tremendously contributed to vaccine development and success.

Vaccines are categorized based on the number of antigenic strains in the formulation (Figure 13.2).

13.6.1 Mono- and Polyvalent Vaccines

A monovalent vaccine refers to a formulation containing only one pathogenic strain of an antigen, for example the Measles vaccine. A polyvalent vaccine consists of two or more strains/serotypes of the same antigen for example the Oral Polio Vaccine.

13.6.2 Combination Vaccines

Combination vaccines containing several antigens are pooled in one formulation that either targets different diseases or provides protection against pathogens that have multiple strains but cause the same disease. For example, the DPT vaccine protects against three diseases namely diphtheria, pertussis and tetanus. The advantage of combination vaccines is the absence of multiple boosters, which decrease the fear and pain of vaccination in children. Combination vaccines go through rigorous testing so that the national regulatory authorities approve them ensuring that they are safe, of high quality, and effective.

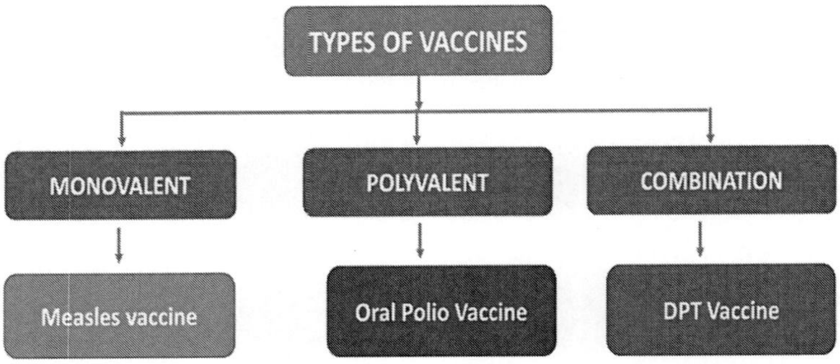

FIGURE 13.2 Classification of vaccines based on the number of antigenic strains present in them. (Copyright Author.)

The following are some of the advantages that combination vaccines confer:

a. There is a reduction in the cost of stocking separate vaccines for different diseases.
b. It is cost-effective for patients, as it reduces the number of visits to the healthcare providers and multiple injections.
c. Reduction in fear and pain in kids due to administration of separate vaccines.
d. It has simplified vaccine administration by allowing the addition of new vaccines to the immunization programmes.
e. There is an improvement in timelines of vaccination, the reason being the apprehension or objection of certain parents and doctors for multiple injectable vaccines in one visit. This is because of fear of needles and pain, in the child's mind along with concerns about safety.

Extensive testing of the combination vaccines is extremely important before commercial approval. For example, the adjuvants in the formulation might lead to the drastic reduction in the activity of one antigen while the reactivity of the other antigen is disproportionately increased. Cross-reactivity might occur between vaccine components like buffers, stabilizers and preservatives. While developing combination vaccines the makers are required to evaluate the following criteria:

a. Effectiveness of every antigenic constituent.
b. Effectiveness of vaccine components in combination to induce immunity.
c. The possibility of reversion to virulent forms.
d. Cross-reactivity among vaccine components.

Figure 13.3 depicts globally recommended vaccines and vaccines that are still in various stages of development for the treatment of infectious diseases.

13.6.3 Live Attenuated Vaccines (LAV)

The live attenuated vaccines (LAV) came into existence in the 1950s. The LAVs are prepared from laboratory weakened pathogenic viral or bacterial strains. Upon injection of LAV, the weak pathogens can replicate in the body of the vaccinated child, but since they are weak, it would lead to either very mild or no disease at all. LAVs are capable of arousing an immune response equivalent to an infection caused by a wild-type pathogen. The logic behind the use of LAVs is their ability to mimic the natural infection, which leads to long-term protection, usually achieved with a single inoculation. The live pathogens offer continuous antigenic stimulation that provides the immune system sufficient time to generate memory cells. The LAVs provide the advantage of generating a robust

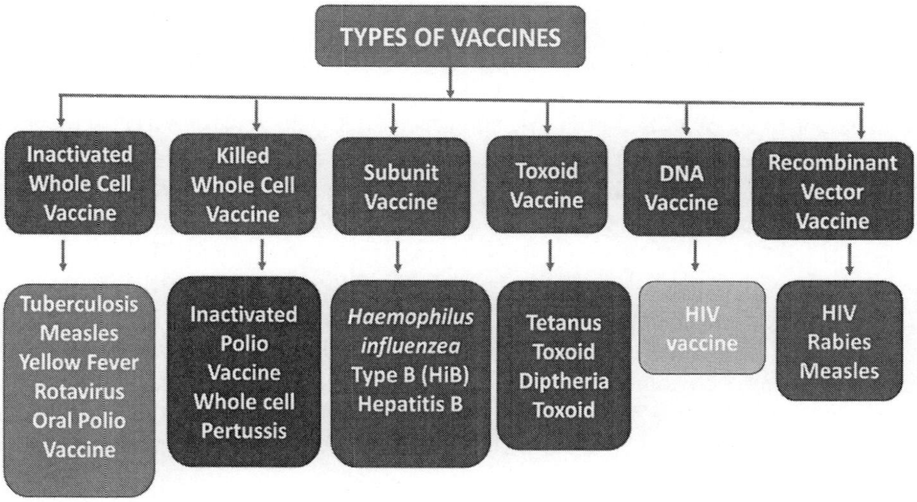

FIGURE 13.3 A flowchart depicting the diverse kinds of vaccines presently used and the ones that are being developed to treat diseases such as cancer and AIDS. (Copyright Author.)

cellular and humoral response. Examples of LAVs include the BCG vaccine against Tuberculosis and vaccines against viruses that cause polio, measles, yellow fever and dysentery such as the rotavirus.

13.6.3.1 Methods of Attenuation

Attenuation of pathogenic strains is attained through different methods, some of which are mentioned below:

a. The methodology adopted by Edward Jenner involved the use of the cowpox virus that was pathogenic for cow but wasn't pathogenic to humans. Therefore, he isolated pus from cows suffering with cowpox, which led to the development of the smallpox vaccine.
b. Virulent stains that have undergone natural attenuation are used for vaccine development. For example, OPV contains type 2 poliovirus.
c. Attenuation can be achieved by subjecting pathogenic viruses to harsh conditions such as the cold adaption of influenza virus. Another notable example is of Bacillus Calmette Guérin (BCG) vaccine developed against Tuberculosis (TB). Presently, the BCG vaccine in use consists of strains that have been derived from the original *M. bovis* strain which was attenuated by Calmette and Guérin passing it through many cycles. Several new BCG strains showing both phenotypic and genotypic differences have been generated by subjecting them to different laboratory conditions. The strain 1331 was produced using this procedure by the Danish Serum Institute. The transmission of TB occurs from adults suffering from lung TB thus, BCG vaccination of children is unable to make a huge impact on the global TB epidemic. BCG vaccine is ineffective against two billion people already suffering from TB. Thus, there is an urgent need to develop new vaccines targeting TB.
d. The measles, mumps, and rubella vaccine (MMR) contain attenuated virus strains. The Schwarz strain targets measles, the RIT 4385 strain is against mumps and the Wistar RA 27/3 strain is against rubella. All three strains were obtained by separately growing them in tissue culture. The viruses conferring immunity against mumps and measles are grown in chick embryo tissue cultures while the virus targeting rubella is cultured on MRC5 human diploid cells.

13.6.3.2 Mechanism of Action

To elucidate how immunity has been achieved, the example of measles is chosen. During vaccination the individual is injected deep subcutaneously from there the virions enter the body and then into different types of cells through receptor-mediated endocytosis. Within the cytoplasm, viral proteins are processed via the endogenous pathway giving rise to peptides e laden on MHC type I molecules followed by the display of the complex on cell surface. Circulating cytotoxic T cells (Tc) displaying suitable TCRs which identify the complex thereby releasing the cytokines which induce the infected cell to apoptosis. It seems that some cytotoxic T cells develop into memory cells. The virus vaccine is phagocytosed by immature dendritic cells which initiates a process similar to the above-mentioned process which generates plasma cells, IgG antibodies and memory B cells. If someone has been vaccinated for measles and then inhales wild measles virus, it activates both the protective mechanisms so the virus undergoing multiplication in cells at the site of infection is killed by the Tc. In case the virus eludes this and enters the blood stream the IgG antibody present in the blood plasma will neutralize the virus thereby, preventing the occurrence of measles.

13.6.3.3 Safety and Stability of LAV

Due to the presence of live pathogens in LAVs, there exists some amount of unpredictability, giving rise to both safety and stability concerns. There exists the rare possibility of attenuated organisms reverting into their pathogenic form thereby causing disease in vaccinated individuals or in people who come in contact with them. The following are two examples of very rare vaccine-related complications:

- Vaccine-associated paralytic poliomyelitis (VAPP)
- Disease-causing vaccine-derived poliovirus (VDPV) associated with the oral polio vaccine (OPV).

Normally, the immune systems are capable of eliminating the attenuated pathogens but people who are immune-compromised, like individuals infected with HIV, are unable to effectively eliminate the attenuated antigens. For example, if an individual is suffering from TB for a long period of time, then BCG vaccination might lead to local lymphadenitis or to disseminated infection. If the tissue culture cells are contaminated and the attenuated virus is grown on them, then there exists the possibility of contamination by other viruses like the retroviruses with the measles vaccine. Thus, to avoid complications, LAVs aren't given to pregnant women, although the actual potential for foetal damage is yet to be completely assessed. Another disadvantage of LAVs are the errors that can occur during immunization:

- Certain LAVs are manufactured in powdered or lyophilized form. Thus, before vaccination, they have to be reconstituted using a specified diluent. At this step manual error like the use of a wrong diluent can happen.
- Some LAVs have a low-temperature requirement to maintain their activity and if the temperature is not properly maintained then it could lead to failure of the immunization programme.

13.6.4 Inactivated Whole-Cell Vaccines

Inactivated whole cell vaccines are prepared from pathogens such as viruses or bacteria, which are killed by physical or chemical methods. It offers the advantage of safety as killed pathogens won't cause disease, the reason being that killed or inactivated organisms, cannot revert to virulence. The disadvantage of inactivated vaccines is that they won't continuously induce an immune response and the response might be short-lived. This is because dead organisms cannot replicate leading to speedy clearance from the body thereby, decreasing the efficiency of inactivated vaccines in comparison to living attenuated vaccines. Many doses of inactivated whole-cell vaccines are required to induce an adequate immune response. In comparison to LAVs, the inactivated whole-cell vaccines are considered more stable.

In general, the inactivated vaccines lead to either a highly complex or grander inflammatory immune response when compared to the newer subunit vaccines because many pathogenic components are preserved. In spite of the disadvantages, inactivated vaccines have been used world over. Hepatitis A vaccine is an example of inactivated whole-cell vaccine containing strain RG-SB of hepatitis A virus which is inactivated through treatment with formalin. The inactivated vaccine is adsorbed on an adjuvant system which is the virosome formulation.

13.6.4.1 Mechanism of Action

After injection, the complete pathogen is phagocytosed by the immature dendritic cells. The pathogen is digested inside the phagolysosome giving rise to several diverse antigenic fragments. These fragments are displayed on the surface of the cell as discrete MHC II: antigenic fragment complexes. In the lymph node, several T_H2 cells, each having a specific TCR for each antigenic fragment, are activated by the activated mature dendritic cell. The B cells possess specific BCRs for different antigenic fragments, which bind to the antigens that are draining along the lymphatic channels. Thus, each of the antigens is internalized followed by presentation as MHC II: antigenic fragment, and then recognized by suitable T_H2 cells. The T_H2 cells release different types of interleukins such as IL2, IL4, IL5 and IL6 which bring about B-cell activation, differentiation and proliferation. This is followed by isotype switching from IgM to IgG with subsequent formation of memory B cells. This complete process takes about 10–14 days. But in a vaccinated individual, contact with the pathogen will elicit a secondary response by activation of the memory B cells, leading to elevated levels of the different IgG molecules in just 24–48 hours.

13.6.5 Subunit Vaccines

The subunit vaccine consists of a portion of the pathogen rather than the complete pathogen as in the case of attenuated or inactivated whole cell vaccine. Subunit vaccines contain inactivated antigenic parts of the pathogen which are necessary for provoking a suitable immune response. Subunit vaccines contain either one or many recombinant peptides/proteins or polysaccharides which are structural components of the pathogen against which the vaccine is to be developed. The subunit vaccines are comparatively more advantageous than traditional vaccines in terms of safety and manufacturing cost, as they contain well-known components of high purity. These vaccines are safer because of the absence of live organism; therefore, there is the absence of replication and also materials which can invoke unwanted responses of the immune system are absent. This accuracy in vaccine development is a costly affair because the antigenic properties of different possible subunits of a pathogen are scrutinized at length to know the specific combinations that can elicit not only effective but also correct immune response. At times, although a response is elicited, there is no surety that immunological memory will be generated in the right way.

The subunit vaccines targeting virus are split vaccines because the structure of viruses is disrupted, giving rise to a mixture of different viral constituents. On the other hand, subunit vaccines might contain a single or many viral or bacterial peptides or proteins. Sometimes the antigens are sufficiently immunogenic, for example, the influenza vaccine consists of two surface antigens namely hemagglutinin (HA) and neuraminidase (NA). To develop the seasonal flu vaccine, the above-mentioned two proteins were isolated from three different virus strains. These two proteins were combined in a trivalent vaccine, which may not necessarily contain an adjuvant. Another example is the hepatitis B vaccine, containing the surface antigen, HBsAg, which is adequately immunogenic. The first genetically engineered vaccine contained the recombinant HBsAg which was commercially manufactured and used globally.

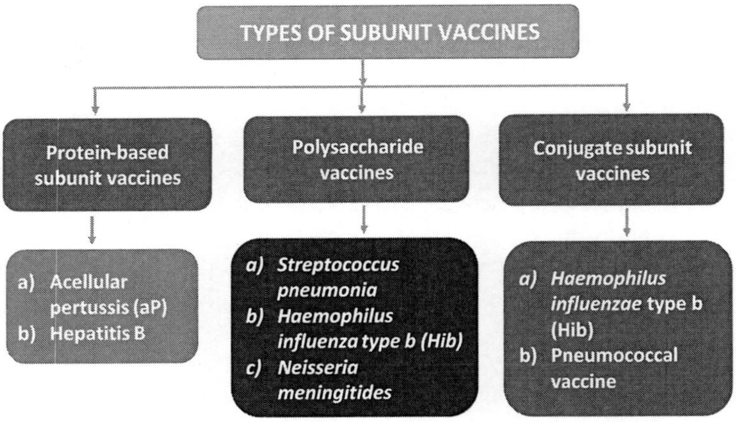

FIGURE 13.4 Different types of subunit vaccines along with their target diseases. (Copyright Author.)

Many a times, the very purified subunit antigens are inadequately immunogenic as they lack the intrinsic pathogenic features, thus, co-administration of adjuvants is necessary. The presence of adjuvants helps in the induction of an effective immune response, along with its modulation. Addition of adjuvants can lead to a reduction in the number of doses of an injection or a single dose might be sufficient. Subunit vaccines are categorized as (Figure 13.4):

a. Protein-based subunit vaccines
b. Polysaccharide vaccines
c. Conjugate subunit vaccines

13.6.5.1 Protein-Based Subunit Vaccines

These vaccines do not contain viral particles rather they contain a highly specific and purified protein of the pathogenic organism. The disadvantage of these vaccines is that in case the isolated proteins get denatured; they might bind to other antibodies rather than binding to the original antibody to which the protein of the pathogen binds. Examples of protein-based subunit vaccines are mentioned below:

a. **Acellular pertussis (aP) vaccine:** It consists of inactivated pertussis protein which is actually a toxin and might have just one or many other bacterial components. The detoxification of pertussis toxin is done by treatment through chemical means or by genetic engineering.
b. **Hepatitis B vaccines:** It consists of the surface antigen produced by the hepatitis B virus (HBsAg), which is a protein. Formerly, the vaccine was prepared using the proteins obtained from the plasma of infected people by thorough purification. The current technology of vaccine production involves the use of recombinant DNA technique which produces the viral antigen HBsAg thus, eliminating the use of human plasma. This is a better and safe method of vaccine development because now there is no risk of probable contamination from human plasma.

13.6.5.2 Polysaccharide-Based Subunit Vaccines

Certain bacteria upon infection in humans are able to escape an immune response because of the presence of a polysaccharide capsule that protects the bacteria, particularly in new-born and young children. The polysaccharide vaccines target the components of the capsule to generate an effective immune response. The polysaccharide molecules are small in size, and usually not very immunogenic; therefore, they are

1. Ineffective in newborns and young children that are 18–24 months old.
2. Capable of inducing merely short-term immunity because there is an extremely slow increase in the titre of antibody levels due to slow immune response and no immune memory.

Mentioned below are some of the polysaccharide vaccines developed against diseases produced by

- *Streptococcus pneumoniae*- Pneumococcal disease.
- *Neisseria meningitides*- Meningococcal disease caused by *Neisseria meningitides* groups A, C, W135 and Y
- *Haemophilus influenza* type b (Hib)

13.6.5.3 Conjugate Subunit Vaccines

These vaccines are capable of generating an immune response against certain molecules present in the capsule of the pathogen. In contrast to the polysaccharide vaccines, these vaccines contain polysaccharides attached to a carrier protein capable of eliciting a lasting immune response. Different types of protein carriers are used for conjugating to the polysaccharide, which also includes diphtheria and tetanus toxoid. Conjugate subunit vaccines are effective in preventing common bacterial infections against which the polysaccharide vaccines prove to be useless, as in the case of infants, or elicit just short-term protection. The development of conjugate subunit vaccines has ushered immunization programmes into a new era. Effective vaccines are developed that can target diseases caused by meningococcus, pneumococcus and *Haemophilus influenzae* type b (Hib), all of which contain a protective capsule. The conjugate vaccines currently recommended by WHO, to children as part of a normal immunization programme, are the *Haemophilus influenzae* type b (Hib) and pneumococcal vaccine. Additionally, in Africa, the meningococcal A vaccine was included in immunization programme due to the prevalence of the disease.

13.6.5.3.1 Mechanism of Action

The adaptive immune response towards the subunit vaccine will be different based on the antigen present in the vaccine i.e., whether it has protein or polysaccharide or both as in the case of conjugate vaccine. The protein-based subunit vaccines such as the hepatitis B and influenza vaccine generate a T cell-dependent response, whereas the polysaccharide-based subunit vaccine gives rise to a T cell-independent response.

Pneumovax is a subunit vaccine that generates T cell-independent immune response. It is given to workers suffering from chronic respiratory, heart, renal and liver disease, immunosuppression etc. Pneumovax contains capsular polysaccharides from 23 different pneumococcal serotypes as the antigen. The vaccine is injected deep either into the subcutaneous tissue or intramuscularly. At the site of injection, the immature dendritic cells and macrophages phagocytose a few polysaccharide molecules. They then later move to local lymph nodes and encounter naive T_H2 cells. But the TCR identifies only the protein molecules, which are also presented by mature dendritic cells and displayed on the surface with the help of Class-MHC II molecules, still the T_H2 cells are not activated. At the same time, non-phagocytosed polysaccharide molecules through the lymph vessels reach the draining lymph nodes encountering the B cells, possessing specific BCRs. The antigen comprises tandem repeats of high molecular weight capsular polysaccharide therefore, it binds with great affinity to several specific receptors on the surface of the B cell. This multivalent binding activates the B cell, eliminating the requirement for the T_H2 cell and generating IgM. Since T_H2 cells are not involved, therefore very less isotype switching is there thus, minor quantities of IgG and a small number of memory B cells are generated. In the case of a vaccinated person, the entry of *Streptococcus pneumoniae* through the mucosa will activate specific IgM antibody leading to complement-mediated lysis. An important property of IgM is its ability to activate the complement system effectively but is unable to behave as neutralizing or opsonizing antibody.

By combining a protein molecule with the T-independent vaccines through conjugation they are transformed into effective T-dependent vaccines. The immature dendritic cells phagocytose the conjugated protein and polysaccharide molecules after which they are displayed on the cell surface as MHC II:protein and MHC II: polysaccharide complexes. These mature and activated dendritic cells migrate to the draining lymph node which is a T-cell-rich zone leading to the activation of T_H2 cells which are highly specific for the carrier protein. The antigen in the vaccine concurrently passes through the draining lymph channels into the lymph nodes which is a B-cell-rich zone, resulting in binding of polysaccharide:protein conjugate to polysaccharide specific BCR on the B cell surface. This is followed by internalization of the polysaccharide:protein complex and phagocytosis, leading to the generation of protein that will bind to MHC class II molecules to be displayed as complex on the cell surface. There exists a link amongst the activated T_H2 cells, having great specificity towards the carrier protein and the activated B cell. T_H2 cells play a crucial role as they co-stimulate and cause the release of cytokines which leads to the generation of first IgM antibodies and then IgG antibodies and lastly the memory cells.

13.6.6 Toxoid Vaccines

The development of toxoid vaccines was based on the toxins produced by some bacteria such as tetanus and diphtheria. Once the bacteria invade the body it releases the toxin that enters the bloodstream and is principally accountable for the disease symptoms. The toxin is composed of proteins which are made harmless, after that it is used in the form of antigen for vaccine development to provoke an immune response. The inactivation of toxins is done by chemical methods such as the use of formaldehyde thereby, converting it to nontoxic toxoids which are safe for vaccine development. It is the similarity of the toxoid to the toxin that leads to the generation of an immune response similar to that elicited by the use of natural toxins. Thus, the anti-toxoid antibodies are produced thereby neutralizing the toxins. Adjuvant such as aluminium or calcium salts are used for increasing the immune response of the toxoid by adsorbing the toxoid onto the adjuvant. Examples of toxoid vaccines include vaccines for diphtheria, tetanus, and pertussis.

Clostridium tetani produces a toxin called tetanospasmin which binds to specific receptors located in the membrane of pre-synaptic motor nerve cells. The toxin is internalized after which it migrates to the central nervous system, thereby blocking glycine metabolism necessary for the usual functioning of gamma amino butyric acid (GABA) neurons. The function of GABA neurons is to inhibit the functions of motor neurons. Thus, if the GABA neurons become non-functional it leads to excessive activity of motor neurons which innervates the muscles producing continuous contractions causing muscular spasms. This is a distinctive feature of tetanus. The development of tetanus toxoid vaccine involves culturing the *Clostridium tetani* in a semi-synthetic medium. Following the growth of bacteria and successive lysis, the toxin is released in the supernatant. Treatment of the toxoid with formaldehyde converts the toxin to a toxoid through the alteration of certain amino acids and minor change in conformation. Using the technique of ultrafiltration, the unwanted proteins are separated from the toxoid in the pellet to get purified toxoid for vaccine preparation.

13.6.6.1 Mechanism of Action

After injecting the tetanus vaccine deeply in subcutaneous/intramuscular (sc/im) tissue, the toxoid molecules are internalized by the immature dendritic cells. In the cytoplasm, the toxoid molecules undergo processing by the endosomal pathway after which they bind to MHC type II molecules forming the MHC II:toxoid complex which will migrate to the surface of the cell. The dendritic cells undergo maturation and move along lymph channels to the draining lymph node come across naive T_H2 cells, each possessing a unique T-cell receptor (TCR). Identification of MHC II:toxoid and subsequent binding of it to particular T_H2 receptor lead to activation of naive T cell, initiating their multiplication.

At the same time, some toxoid molecules simply pass along lymph channels to the draining lymph nodes coming in contact with B cells, each possessing a unique B-cell receptor (BCR). The B cell binds through their specific BCR to tetanus toxoid leading to the internalization of toxoid, where it is processed via the endosomal pathway, followed by cell surface presentation as MHC II:toxoid complex. Since both the processes are happening in the same part of the lymph node, the B cells displaying MHC II:toxoid complex encounter specific activated T_H2 cells. This process is called as linked recognition, which causes T_H2 cells to activate B cells thereby converting them to plasma cells that in the beginning produce IgM, followed by isotype switch to IgG. Additionally, a small number of B cells differentiate into memory cells. This is the adaptive immune response towards protein-based antigen such as the tetanus toxoid. These antigens are called as T cell-dependent vaccines because of the significant role played by T-helper cells in eliciting an immune response. The rationale behind the development of tetanus vaccine is the generation of antibodies targeting the toxoid having heightened capability for binding to toxin in comparison to toxin receptor binding sites on nerve cells. Thus, in a vaccinated individual accidental exposure to *C. tetani*, will lead to the generation of a giant toxin:antibody complex which cannot bind to the receptor; therefore, the toxin will be neutralized thereby, preventing tetanus.

13.6.6.2 Advantages

Toxoid vaccines are considered safe since they lack both live and killed pathogens. Therefore, there is no possibility of pathogen multiplication in the vaccinated individual and thus cannot spread to the unimmunized people. Only the products of pathogens such as their toxins are used for vaccine preparation in the inactivated form thus probability of disease and reversion to virulence is negligible. The vaccine formulations are stable as they are less vulnerable to variations in light, temperature and humidity.

13.6.6.3 Disadvantages

The first disadvantage is the need for an adjuvant and administration of many doses of vaccines. This leads to an increase in the cost of vaccination. Secondly, at the site of injection, local reactions are commonly observed. The reason is the use of adjuvant or a type III Arthus reaction which usually begins as redness and induration at the site of injection many hours following vaccination and subsides within 48–72 hours. The reaction occurs due to the presence of the abundant amount of antibody forming a complex with toxoid molecules thus, activating the complement system and causing an acute local inflammatory reaction.

13.6.7 DNA Vaccines

The principle of development of DNA vaccine is their ability to induce immunity by transforming the host cells with plasmid DNA containing the gene encoding the target antigen (Figure 13.5). This strategy is different in comparison to injecting the desired antigen as peptide or protein. After DNA vaccination, the host cells that gets transformed produce the desired protein antigen from the DNA present in the plasmid. The presence of a new protein in the body elicits an immune response against it. DNA vaccine is one of the new generation vaccines offering several advantages over the traditional attenuated or inactivated vaccines.

13.6.7.1 Advantages

Some of the Advantages of DNA Vaccines are:

a. Their manufacturing costs are relatively low.
b. They can elicit both humoral and cellular immune responses.

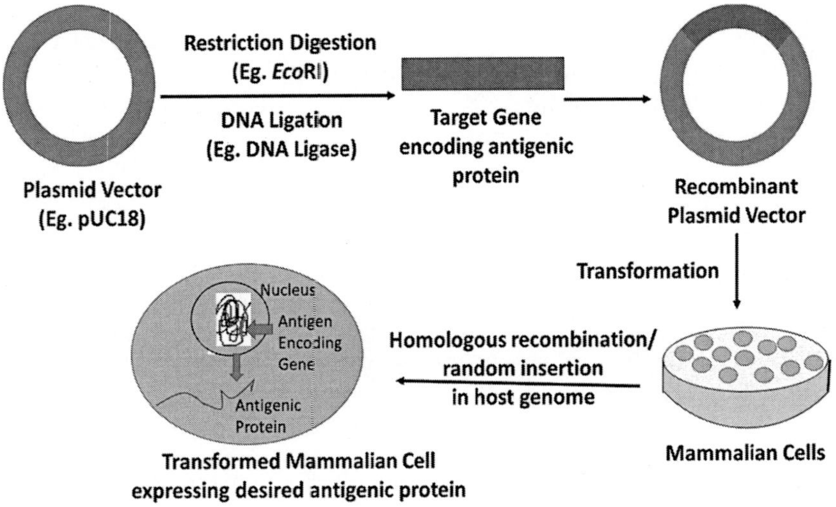

FIGURE 13.5 Development of DNA vaccine. (Copyright Author.)

c. The pDNA is quite stable at room temperature, therefore the requirement for cold storage of DNA vaccine is reduced. This is an extremely important factor for the success of immunization programs in developing nations.

The DNA vaccines are also called as naked DNA vaccines when DNA is directly injected into the body. The administration of DNA vaccines is done either through needle and syringe or through a needle-less device using high-pressure gas for shooting microscopic DNA-coated gold particles straight into the cells. Occasionally, certain molecules are mixed with DNA helping its uptake in cells. As of now, DNA vaccines are not approved for human use. Currently, the clinical trials are being done for developing cure for different types of cancers and HIV/AIDS. Presently, researchers are testing the efficacy of naked DNA vaccines on humans against influenza and herpes viruses.

13.6.8 Dendritic Cell-Based Vaccines

The function of dendritic cells (DCs) involves acquiring, processing and presenting antigens to the T-cells. Not only this, the DCs generate stimulatory signals and release cytokines which lead to the production of T-cells followed by their differentiation into effector cells. The vaccination strategy involves loading the antigens on the *in vitro*-generated DCs and their administration into the patient's body. This leads to the generation of T-cell-mediated immune response, particularly in the case of cancer because the function of DCs *in vivo* is quite disrupted by the factors that are released by the tumour cells. The preclinical studies have time and again revealed that DC-based vaccines are capable of delaying or even preventing the progress of tumour formation. But the disappointment came with human clinical trials giving just minimal help to patients. Therefore, there exists a possibility for an increase in the stimulatory capability of the injected dendritic cells.

13.6.9 Recombinant Vector Vaccines

These vaccines fall under the category of a new generation of vaccines that are under development. They are quite similar to DNA vaccines, but differ from them because they require either an attenuated virus/bacterium for introducing microbial DNA into the cells of the body. The term "vector" denotes the use of virus or bacterium as the carrier of DNA. Under natural conditions, the virus attaches to the cell surface and inserts its genetic material into it. The researchers utilize this natural process for vaccine development. Certain portions of the genomes of some harmless or attenuated viruses are deleted to generate space for inserting microbial genes encoding the desired antigens in them. This recombinant carrier virus is used for transferring microbial DNA to cells. Since these vaccines are capable of mimicking natural infection, they are able to stimulate the immune system. In case the attenuated bacteria are used as vectors the microbial genes are inserted in their genome thus, the bacteria now display the foreign antigens on their cell surface. Thus, a harmless bacterium is converted by genetic engineering into a harmful microbe capable of provoking an immune response. Scientists are trying to develop both bacterial and viral-based recombinant vector vaccines for curing HIV, rabies, and measles.

13.7 LIST OF CURRENTLY AVAILABLE VACCINES AND PIPELINE VACCINES

Table 13.3 enlists vaccines under two headings: available and pipeline vaccines. "Available vaccines" refers to a list of diseases or pathogens against which vaccines are available in the market. If one clicks on the link http://www.who.int/immunization/diseases/en/ then, the list of internationally available vaccines is given. Additional information regarding each vaccine is given as a hyperlink regarding the WHO policy recommendations, alongside other key resources. The "Pipeline vaccines" refers to a list of diseases or pathogens for which vaccines are currently under development. The WHO's PDVAC, which stands for Product Development for Vaccines Advisory Committee is responsible for developing Pipeline vaccines. If one clicks on the link http://www.who.int/immunization/diseases/en/ then, for every disease or pathogen, a hyperlink is given which provides information in brief about the current status of research and development of vaccine for it.

13.8 DIFFICULTIES IN VACCINE DEVELOPMENT

Although, the development of vaccines is a huge medical success story, but there lie several challenges before the present vaccination programmes. There are almost three million deaths worldwide every year due to lack of vaccines against vaccine-preventable diseases, and infectious diseases. The past decade has witnessed a slowdown in universal vaccination coverage in comparison to the progress seen in the 1970s and 1980s. Owing to inefficient vaccine coverage there is a re-emergence of many childhood diseases, like measles, pertussis and polio in some industrialized and developing nations. The probable reasons are

1. Perception of public towards vaccination- people might refuse vaccination for avoiding any type of side effects
2. Political reasons.

The lack of effective vaccines leads to death from several infectious diseases, because traditional methods of vaccine development prove to be unproductive owing to factors such as

1. Antigenic drift
2. Diseases that are difficult to treat such as tuberculosis, human deficiency virus-acquired immune deficiency syndrome (HIV-AIDS), malaria etc.

Antigenic shift refers to changes in surface antigens arising from mutations in the genome of the pathogen. For example, the influenza virus shows annual changes in surface antigens. The problem of antigenic variability is solved by changing the composition of the vaccine every year. Currently, antigenic shifts possess the biggest challenge for vaccine development. Vaccine development in the past was successful because the pathogens they targeted showed little or no antigenic drift. Diphtheria and tetanus-causing pathogens are the best examples of micro-organisms whose toxins show no antigenic drift. On the other hand, the human immunodeficiency virus mutates faster giving

Table 13.3
The Table Below Enlists the Disease or Pathogens Against Which Vaccines are Currently Available to Public for Immunization and Vaccines that are Under Development as per WHO Guidelines

Available Vaccines	Pipeline Vaccines
Cholera	*Campylobacter jejuni*
Dengue	Chagas Disease
Diphtheria	Chikungunya
Hepatitis A	Dengue
Hepatitis B	Enterotoxigenic *Escherichia coli*
Hepatitis E	Enterovirus 71 (EV71)
Haemophilus influenzae type b (Hib)	Group B Streptococcus (GBS)
Human papillomavirus (HPV)	Herpes Simplex Virus
Influenza	HIV-1
Japanese encephalitis	Human Hookworm Disease
Malaria	Leishmaniasis Disease
Measles	Malaria
Meningococcal meningitis	Nipah Virus
Mumps	Nontyphoidal Salmonella Disease
Pertussis	Norovirus
Pneumococcal disease	Paratyphoid fever
Poliomyelitis	Respiratory Syncytial Virus (RSV)
Rabies	Schistosomiasis Disease
Rotavirus	Shigella
Rubella	*Staphylococcus aureus*
Tetanus	*Streptococcus pneumoniae*
Tick-borne encephalitis	*Streptococcus pyrogenes*
Tuberculosis	Tuberculosis
Typhoid	Universal Influenza Vaccine
Varicella	

Source: http://www.who.int/immunization/diseases/en/s.

rise to newer antigens, thus conventional techniques of vaccine development are futile. Traditional vaccinology has also been successful in the case of vaccines targeting pathogens where protection is through the production of antibodies. Vaccine development again becomes difficult in the case of pathogens against which antibodies are unable to give adequate protection such as *Mycobacterium tuberculosis* the causative agent of tuberculosis which in 2012, infected 8.6 million people of which about 1.3 million people died. Thus, it's extremely important to develop new prophylactic and therapeutic vaccines targeting diseases such as cancer, malaria, HIV-AIDS and tuberculosis which are responsible for maximum death globally. Changes in conventional vaccines can hugely affect vulnerable population groups like the elderly, because they are immunologically hypo-responsive.

13.9 KEY CONCEPTS

- The immune system has two components namely innate and adaptive immunity.
- Innate or native immunity is an outcome of the genetic constitution of an individual.
- The vaccine comprises an antigen in either an alive or dead state or its components which provoke an immune response causing the development of immunity against a particular pathogen.

- Immunity develops in two ways- either by natural contact with a pathogen or by vaccination.
- Based on the mode of attainment adaptive immunity is of two types: passive and active each of which can be achieved either naturally or artificially.
- When a vaccine containing preformed antibodies is administered to an individual passive immunity develops which is useful in fighting infection but does not generate immunological memory.
- Passive immunity is short-lived and attained by transferring the antibodies or activated T-cells to the individual.
- Active immunization occurs either by contacting the infection directly or through vaccine. It is time taking process but generates immunological memory.
- Active immunity lasts for a lifetime as it develops through the exposure of individual to the antigen initiating an immune response.
- There are several different routes of administration of vaccine such as intradermal, intramuscular, intravenous, oral and intranasal spray to facilitate its efficient entry and transport to the site of action in the body.
- Every vaccine has a defined composition whose knowledge is necessary to enable the investigation of adverse effects after immunization.
- Most vaccines are either live attenuated (bacterial or viral) or killed/inactivated (bacterial or viral), toxoid or subunit vaccines. The reason behind the diversity of vaccines are the differences in the chemical and genetic composition of different pathogens hence, their modus operandi is different.
- Development of vaccines efficient in combating infectious disease is faced with challenges such as antigenic drift, lack of sufficient knowledge, high costs of development, personal and political reasons and difficulties in development of vaccines against certain pathogens.

13.10 SUMMARY

- Vaccines have been extremely effective in preventing and even eradicating infectious diseases, such as smallpox.
- The vaccine consists of an antigen which can be either alive or dead or its components are capable of eliciting an effective immune response. This leads to the development of immunity which is a state of resistance towards the particular pathogen in the infected person if contracted again.
- Immunity can be developed both by natural exposure to a pathogen or through vaccination.
- The terms immunization and vaccination are often used interchangeably, although the two are different. Immunization is the process of development of the state of immunity which can occur naturally or through vaccination.
- Vaccination involves the administration of vaccine which might contain preformed antibodies or the antigen or its components. The former is known as passive immunization which is fast and gives the body an instant ability to combat the infection but unfortunately there is no immunological memory, while the latter, is known as active immunization which takes time to activate the body's immune system and generates immunological memory in most cases. Thus, each type has its own advantages and disadvantages.
- The vaccines promote health by having a quick impact and are life saving along with being cost-effective due to which they have mass reach. All these factors make vaccines extremely important.
- The first vaccine was developed by Edward Jenner the British physician, in 1796 against smallpox and also coined the term vaccine. This gave impetus to the development of different types of vaccines with a significant contribution from Louis Pasture.
- The immune system comprises both innate and adaptive immunity.
- Innate immunity is also referred to as native immunity. It is dependent upon the genetic constitution of the organism independent of external stimulation or a prior infection.

- Adaptive immunity is of two types passive and active depending upon how the immunity was acquired. Passive immunity is attained by transferring the antibodies or activated T-cells to the individual. This type of immunity is for a short duration of time say a few months. Active immunity develops by exposure of the individual to the antigen such that an immune response is evoked thus, the immunity attained lasts for a lifetime.
- Both passive and active immunity can be attained either naturally or artificially.
- The route of administration of the vaccine can be intradermal, intramuscular, intravenous, oral and intranasal spray. The main reason for the difference in the route of administration is that it can enter the body effectively and then reach the site of action through the body's transport mechanisms to show their desired effect.
- Each vaccine is made up of several components such as antigens, stabilizers, adjuvants, antibiotics, preservatives and left-over by-products accumulated during the process of vaccine production. It's important to know the precise composition of each vaccine so that one is able to investigate adverse events after immunization.
- As of now, most of the vaccines administered are either live attenuated (bacterial or viral) or killed/inactivated (bacterial or viral), toxoid or subunit vaccines. There exist so many different types of vaccines because of the differences in the chemical and genetic composition of different pathogens that leads to differences in their modus operandi.
- Vaccines are either monovalent or polyvalent. A monovalent vaccine refers to a formulation containing only one pathogenic strain of an antigen, for example, the Measles vaccine. A polyvalent vaccine consists of two or more strains/serotypes of the same antigen for example the Oral Polio Vaccine.
- Combination vaccines contain several antigens that are pooled in one formulation that either target different diseases or provide protection against pathogens that have multiple strains but cause the same disease. For example, the DPT vaccine protects against three diseases namely diphtheria, pertussis and tetanus.
- The live attenuated vaccines (LAV) are prepared from laboratory weakened pathogenic viral or bacterial strains. Upon injection, the weak pathogens replicate in the body, but since its weak, it would lead to either very mild or no disease at all but, arouse an immune response equivalent to an infection caused by a wild-type pathogen. There is long-term protection, usually achieved with a single inoculation. The LAVs provides the advantage of generating a robust cellular and an antibody response. Examples of LAVs include vaccines against bacteria such as the BCG vaccine against Tuberculosis and viruses that cause polio, measles, Yellow fever and dysentery such as the Rotavirus.
- Inactivated whole cell vaccines comprise of pathogens such as viruses or bacteria, which are killed by physical or chemical means. It is safe as killed pathogens cannot be reverting to virulence. The disadvantage is the short-lived immune response. This decreases the efficiency of inactivated vaccines in comparison to live attenuated vaccines. Example includes the Hepatitis A vaccine.
- The subunit vaccine consists of a portion of the pathogen rather than a complete pathogen. Subunit vaccines contain either one or many recombinant peptides/proteins or polysaccharides having antigenic properties and are the structural components of the pathogen against which the vaccine is to be developed. The subunit vaccines prove to be more advantageous than traditional vaccines in terms of safety and manufacturing cost, as it contains well known and components of high purity.
- The Protein-based subunit vaccines contain a highly specific and purified protein of the pathogenic organism. This protein binds to the specific antibody and elicits the immune response. The main disadvantage of these vaccines is the denaturation of isolated proteins which renders them ineffective. Examples include acellular pertussis vaccine and Hepatitis B vaccine.

- The polysaccharide vaccines target the components of the capsule to generate an effective immune response against certain pathogenic bacteria, that evade immune response because of the presence of a polysaccharide capsule that protects them, particularly in new-borns and young children.
- The conjugate subunit vaccines comprise of polysaccharide bound to a carrier protein capable of eliciting long-term immune response. Different types of protein carriers are used which also includes the diphtheria and tetanus toxoid. The conjugate vaccines currently recommended by WHO are the *Haemophilus influenzae* type b (Hib) and pneumococcal vaccine.
- These vaccines are capable of generating an immune response against certain molecules present in the capsule of the pathogen.
- The toxoid is an inactivated toxin. Certain bacteria such as tetanus and diphtheria invade the body of the host and release toxins which are responsible for the disease symptoms. These toxins are proteins that are made harmless, for vaccine development either by heat or chemical treatment. They are safe and capable of eliciting immune response.
- The DNA vaccines are the new generation vaccines currently under development, which can induce immunity by transforming the host cells with plasmid DNA containing the gene encoding the target antigen.
- Dendritic Cell-Based Vaccines use dendritic cells (DCs) which acquire, process and then present antigens to the T-cells along with the generation of stimulatory signals and release of cytokines. The vaccine consists of antigens loaded on the *in vitro*-generated DCs that are injected into the patient's body, which generates T-cell-mediated immune response, particularly in cancer cells.
- Recombinant Vector vaccines belong to the new generation of vaccines that are still being developed. Although similar to DNA vaccines, they differ from them owing to the use of an attenuated virus or bacterium as vector for introducing microbial DNA into the cells of the body.
- There exist several difficulties in the development of effective vaccines against infectious diseases such as antigenic drift, high costs of development, both personal and political reasons and difficulties in the development of vaccines against certain pathogens.

QUESTIONS

1. Expand the following terms
   a. LAV
   b. WHO
   c. HIV-AIDS
   d. NIP
   e. PDVAC
2. Differentiate between the following terms:
   a. Live attenuated vaccines and Inactivated vaccines
   b. Subunit vaccines and Toxoid vaccines
3. Long answer questions:
   a. Why are immunization programmes important?
   b. Describe the different types of vaccines.
   c. Mention the various components of the vaccines.
4. Write short notes on the following:
   a. Challenges in vaccine development
   b. Different routes of vaccine administration
   c. Antigenic shift
   d. Toxoid vaccines

BIBLIOGRAPHY

1. https://vaccine-safety-training.org/vaccine-preventable-diseases.html
2. https://vaccine-safety-training.org/route-of-administration.html
3. http://courses.washington.edu/medch401/pdf_text/RT_vaccines_2.pdf
4. http://www.who.int/immunization/diseases/en/
5. https://www.niaid.nih.gov/research/vaccine-types
6. http://www.who.int/vaccine_safety/initiative/tech_support/Part-2.pdf
7. https://en.wikipedia.org/wiki/Vaccination
8. https://en.wikipedia.org/wiki/Immunity_(medical)
9. Baxter, D. (2007). Active and passive immunity, vaccine types, excipients and licensing. *Occupational Medicine*, Vol. 57 (8), pp. 552–556. https://doi.org/10.1093/occmed/kqm110.
10. Kindt, T. J., Goldsby, R.A., Osborne, B. A. and Kuby, J (2006). *Immunology*, VI Edition. W.H. Freeman and Company, New York, United States.

14 Generation of Receptor Diversity

Garima Nagar, Pooja Mittal, and Archana Singh
University of Delhi

CONTENTS

14.1 INTRODUCTION

In all jawed vertebrates, an acquired immune response is generated whenever the body encounters any foreign antigen. It has remarkable properties like high specificity towards the antigen and long-lasting immunological memory. The host immune system generates a large receptor diversity to encounter frequently evolving pathogens. Identification of the antigen and binding with its antigenic determinant is one of the crucial events in the defense mechanism that is carried out by the B cell. A host can produce a wide array of antibodies against a single antigen molecule, provided it has multiple recognition sites. This diversity in antibody production is required to identify various surface molecules of antigens, which also possess a frequent mutational tendency. T cells and B cells express a distinctive and unique receptor. T cells recognize specific small peptide antigens bound to the major histocompatibility complex classes I and II on the membrane of antigen-presenting cells. At the same time, the highly specific paratope of antibody molecule on the surface of B cells identifies the complementary epitopes on the surface of invading antigen (Cooper and Alder, 2006; Ahmed and Rouse, 2006; Laver et al., 1990). In 1965, J. Claude Bennett and William Dryer proposed that the antigen receptor's diversity results from DNA recombination. They suggested that two separate genes encode both the light and heavy chain of Immunoglobulin, in which one encodes the constant region and the other encodes the variable region. This theory contradicted the one coding gene, one polypeptide hypothesis, and due to the unavailability of modern techniques, they were unable to prove it (Dreyer and Bennett, 1965). Later in 1976, Susumu Tonegawa and Hozumi reported the role of somatic recombination in the rearrangement of the immunoglobulin genes. They discovered the genetics behind the generation of antibody diversity. They experimentally

DOI: 10.1201/9781003286424-14

proved that the V and C regions are encoded by three separate genes that undergo rearrangement at maturation and B cell differentiation. They confirmed that antigen receptor loci contain well-organized coding elements – variable, diversity, joining, and constant in a distinct genomic region (Hozumi and Tonegawa, 1976). Susumu Tonegawa was awarded Noble Prize to discover the genetic principle for the generation of antibody diversity in 1987 (Tonegawa et al., 1987).

14.2 IMMUNOGLOBULIN MULTI-GENE ORGANIZATION

Immunoglobulin (Ig) is a hetero-dimeric glycoprotein exclusively expressed by B lymphocytes. Ig comprises two identical light chain isotypes (lambda or kappa) and two identical heavy chains (Townsend et al., 2016). Gene segments encoding heavy chain, kappa, and lambda light chain are organized at IgH, Igκ, and Igλ locus on the distinct chromosome as shown in Figure 14.1.

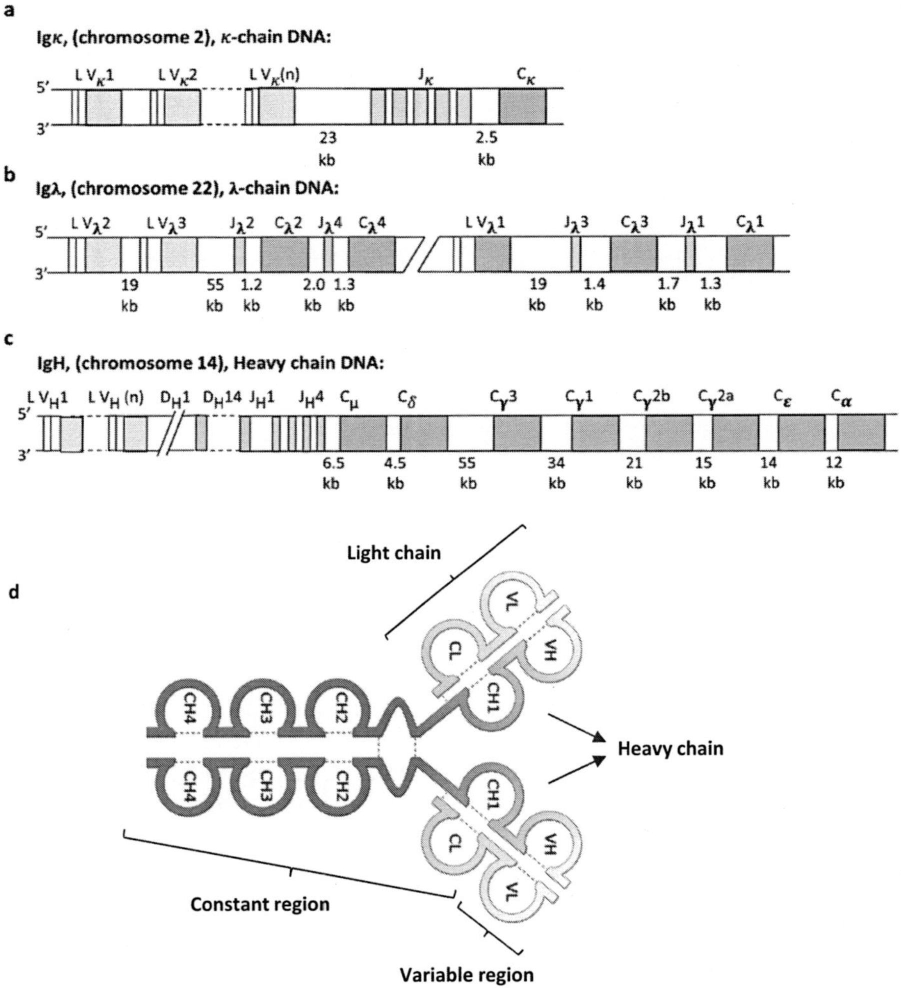

FIGURE 14.1 (a) Immunoglobulin kappa chain loci are formed by three families of germline DNA segments. In the kappa chain, DNA upstream of a small cluster of J_k segments, a large cluster of V_k segments is found. From the 3′ end at the downstream of the J_k segment, a C_k segment is present. (b) Immunoglobulin lambda chain loci are formed by three families of germline DNA segments. In lambda chain DNA downstream of the V_λ segment cluster, pair of J_λ and C_λ segment cluster are present in a sequential manner (c) Immunoglobulin heavy chain locus is formed by four families of germ-line DNA segments. At the downstream of the cluster of V_H segment, D_H segment cluster is presently followed by C_H segment cluster's first exon C_μ.

Variable (V) and Joining (J) gene segments encode the light chain variable region. In contrast, the heavy chain variable region is encoded by Variable (V), Joining (J), and Diversity (D) gene segments (Chi et al., 2020). Joining these multiple gene segments through somatic recombination results in the formation of the variable final exon (Hwang et al., 2015). In humans, the heavy chain locus (IGH) is a single gene locus at chromosome 14 q32.33 which consists of 44 V_H, 23D_H, and 6 J_H functional gene segments (Lange et al., 2014; Watson and Breden, 2012; Watson and Breden, 2012). The light chain has two separate gene loci, Kappa and Lamda. Immunoglobulin kappa locus (IGK) is present on chromosome 2 at 2p11.2, with 52 V_k and 5 J_k functional gene segments, whereas the Immunoglobulin lambda locus (IGL) is present on chromosome 22 at 22q11.2 with 30 V_λ and 7 J_λ functional gene segments (Townsend et al., 2016). Only one type of light chain gene loci encodes both the chains in a complete immunoglobulin (Nakano et al., 2011).

14.3 V REGION DNA REARRANGEMENT

In humans, the exon that encodes the V (variable) region is the outcome of the somatic recombination of multi-gene segments – D (diversity), J (joining), V (variable) in the variable region of the IgH (Immunoglobulin heavy chain) and IgL (Immunoglobulin light chain). In the V region of the IgH chain - D, V and J gene segments reorganize, whereas in the IgL chain –V and J gene segments reorganize (Bassing et al., 2002). This somatic recombination in the V region of immunoglobulin plays a crucial role in the adaptive immune response (Tonegawa, 1983; Flajnik and Du Pasquier, 2004). A recombination signal sequence (RSS) is a non-coding sequence that guides the recombination mechanism. Recombination activated gene (RAG1/2 Recombinase) creates double-stranded breaks that initiate the recombination. RSS flanks a coding sequence; it directs the recombination process and acts as a DNA cleavage site (Bassing, Alt, Hughes, Auteuil, et al., 2000; Gellert, 2002). Heptamer, spacer, and nanomer constitute an RSS (Swanson and Desiderio, 1998). A heptamer is a highly conserved sequence of seven base pairs (5′-CACAGTG-3′); it's an important element for cleavage recombination. A spacer is a non-conserved sequence of 12 or 23 bp. It is required for efficient recombination. A nonamer is a less conserved sequence of 9 base pairs (5′-ACAAAAACC-3′); it is an important element for initial recognition and RAG protein binding (Arnal et al., 2010). The VDJ recombination in the IgH chain follows the so-called "12/23 rule". According to the 12/23 rule, the Recombination signal sequence having a space length of 12–23 mer (amino acids) flanks the VDJ gene segments. These RSS guide the process of recombination. It states that if different spacer length is present between RSS, then only the efficient recombination of gene segments will occur (Eastman et al., 1996; Van Gent et al., 1996).

14.3.1 Mechanism of VDJ Recombination

DNA rearrangements assemble the V region exon, encoding the antigen-binding domains of T cell and B cell receptors. It contributes to the adaptive immune system's infinite antigen recognition property in jawed vertebrates (Schatz et al., 1992).

14.3.1.1 First Phase – Recognition of Specific DNA Sequence and Cleavage

Two exclusive genes are expressed during lymphoid development- RAG-1 (Recombination activating gene 1) and RAG-2 (Recombination activating gene 2). As shown in Figure 14.2a proteins encoded by these genes initiate the formation of double-strand break (DSBs), which is essential for VDJ recombination (Schatz et al., 1989; Oettinger et al., n.d.; Carmona and Schatz, 2017). The RAG complex cuts specifically between the variable, diversity, and joining coding segments and their flanking RSS (Bassing, Alt, Hughes, D'Auteuil, et al., 2000; Ru et al., 2015). High mobility group box proteins like HMGB1 or HMGB2 help in DNA binding and bending, facilitate RAG binding with RSS, and result in a Signal complex formation. The signal complex seizes another unbound RSS and forms a paired complex. Nicking is followed by a trans-esterification reaction in which

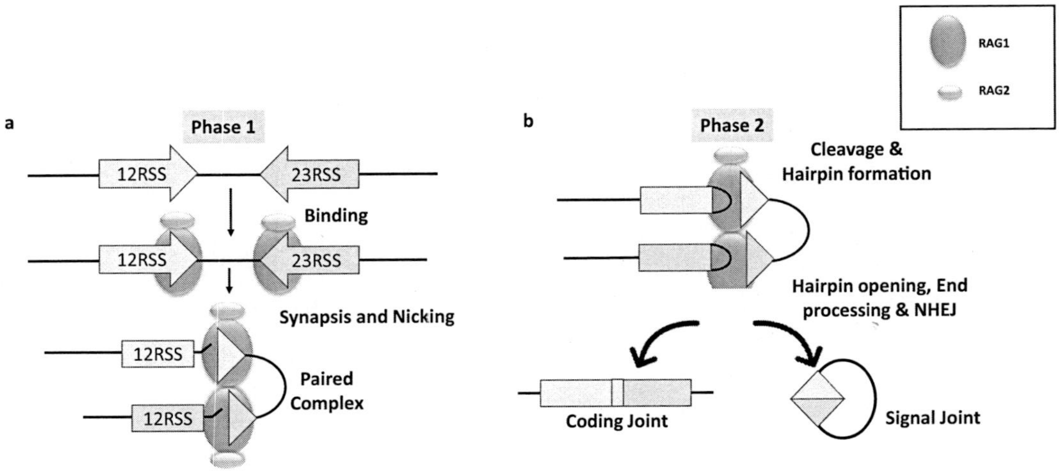

FIGURE 14.2 Mechanism of VDJ recombination (a) Phase 1–12 or 23 signal complex is formed by binding 12 or 23 RSS with RAG1 and RAG2 in the antigen receptor gene segments. Synapsis occurs, and a paired complex is formed by arresting another RSS. In this complex, DSBs are introduced between the RSS and the gene segments. (b) Phase 2 – NHEJ and RAG are involved in hairpin opening and coding-end processing. The addition and deletion of nucleotide occur to form a coding joint, and unprocessed RSS ends join, forming a signal joint.

the free 3′ hydroxyl group forms a cleaved RSS signal end and a coding hairpin end by attacking the opposite strand (Lieber, 2011; Zagelbaum et al., 2016; Schatz and Swanson, 2011; De Villartay, 2015; Jones and Gellert, 2002; McBlane et al., 1995; Hanisch and Baiasu, 2016).

14.3.1.2 Second Phase – Processing Followed by Joining of Ends

In the second phase of VDJ recombination, as shown in Figure 14.2b the coding-end opening occurs by recruiting NHEJ (Non-homologous end joining) DNA double-strand break repair factors (Lieber, 2011; Panchakshari et al., 2018; Lescale et al., 2016). In the repair mechanism of the coding end, nucleotide deletion and addition occur. Due to this, the indefinite joining of coding ends in the variable region attains a junctional diversity (Ma et al., 2002; Moshous et al., 2001; Rooney et al., 2002). After this, an exonuclease deletes some of the nucleotides on the coding end. P and N nucleotide addition occurs in the junction, which adds to the diversity in the variable region. P Nucleotide – Random cleavage of coding hairpin and addition of nucleotide with the help of template results in the formation of a palindromic sequence. N Nucleotide – To the coding end, when TdT adds a nucleotide without using a template, N nucleotides are formed (Repasky et al., 2004; Desiderio et al., 1984). (N represents that any of the four nucleotides could be added).

14.3.2 DNA Repair Machinery

NHEJ process is the dominating pathway of re-joining of double-stranded DNA break. NHEJ involves two pathways competing with each other, Canonical NHEJ and Alternative NHEJ (Chiruvella et al., 2013).

14.3.2.1 Canonical NHEJ (c-NHEJ) Pathway

A heterodimer Ku70/Ku80 formed by two proteins Ku70 and Ku80 is the first DSB (Double-strand breaks) binding component that prevents the digestion of DNA end. Ku: DNA complex modifies and prepares each end of DNA for rejoining. As shown in Figure 14.3a it also recruits DNA-PKcs (DNA-dependent protein kinase catalytic subunit) and Artemis (both 5′ and 3′ endonucleases) (Chang and

Lieber, 2016; Spagnolo et al., 2006; Maezawa et al., 2016; Chang et al., 2017). Auto-phosphorylation of DNA –PKcs triggered by DNA hair pinned DNA ends result in Artemis activation. This activation of Artemis turns on its nuclease activities, which confirms the compatibility of both ends. The hairpin of the coding end is opened by this Artemis (Rooney et al., 2005; Meek, 2020; Goodarzi et al., 2006; Pryor et al., 2015). The final ligation and filling steps of the coding and signal joints are carried out by the "End ligation complex." This complex is formed by binding the scaffold proteins like XRCC4 (X-ray repair cross-complementing protein 4) and XLF (XRCC4-like factor) with DNA ligase IV. After processing, Ku recruits Pol μ/λ, which fills the existing gaps. A newly discovered protein, PAXX, interacts with Ku and promotes the double-strand DNA break repair. PAXX also shows a structural resemblance to XRCC4 and XLF (Tadi et al., 2016; Ochi et al., 2015; Kumar et al., 2016; Pryor et al., 2015).

14.3.2.2 Alternative NHEJ (Alt-NHEJ) Pathway

Poly ADP Ribose Polymerase 1 (PARP 1) is an important enzyme involved in the single-stranded DNA break repair mechanism. As shown in Figure 14.3b for recruitment of the factors of DNA damage, PARP 1 acts as a flexible scaffold (Wei and Yu 2016). MRN complex (MRE 11-RAD50-NBS1) and CtIP (C terminal binding protein-interacting protein) form single-stranded DNA overhangs by cleaving 5′ terminated strand near the double-strand DNA break site (Truong et al. 2013). Resection reveals the terminated strands of ssDNA and break sites and exposes them to the microhomology internal, resulting in instant ssDNA annealing. Annealing creates a hindrance in

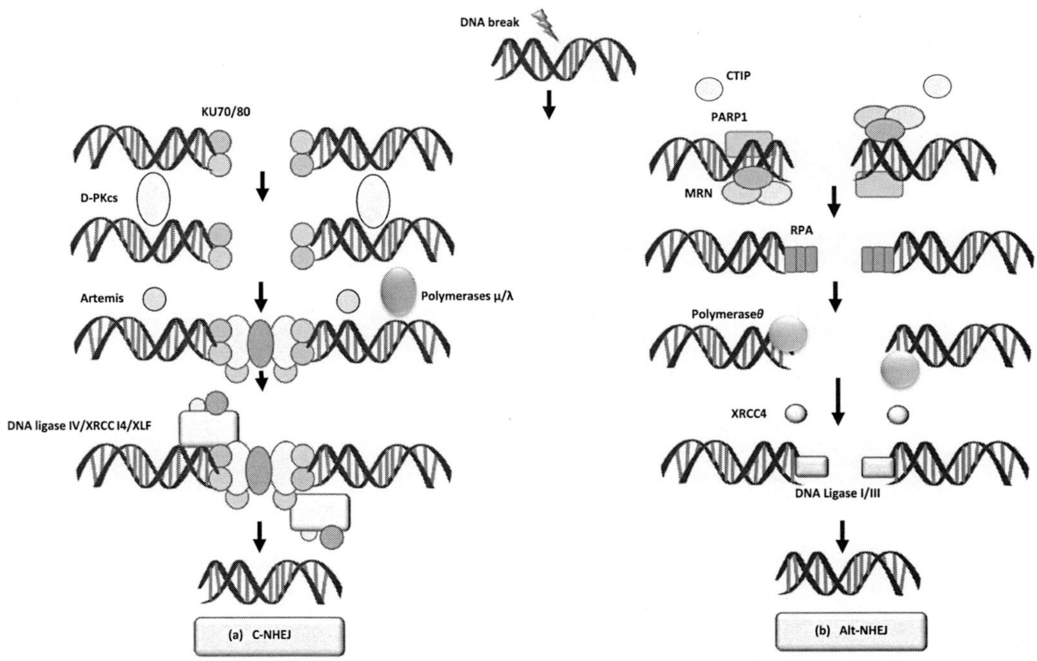

FIGURE 14.3 (a) In canonical nonhomologous end-joining: Ku 70 and Ku 80 bind with DNA DSB and recruit DNA PKcs along with a nuclear protein Artemis which has both exo- and endonuclease activity. The DSB gap filling and ligation are completed by DNA polymerase μ (Pol μ), DNA polymerase λ (Pol λ) and a complex, DNA ligase IV/XRCC4/XLF. (b) In alternative non-homologous end-joining: Poly {ADP-ribose} polymerase-1 binds with DNA DSB and recruits MRN complex and CtIP for creating single-stranded DNA overhangs. RPA prevents the instantaneous single-stranded DNA annealing, which is initiated by internal microhomology. Eventually, the DNA annealing and Alt- NHEJ process is triggered by RPA, which itself is dissociated by Pol helicase. Extension and rejoining of ends are done by Pol polymerase and DNA ligase (I or III) XRCC1 complex.

the NHEJ mechanism; therefore, Replication protein A (RPA) prevents the overhangs from annealing, which binds with ssDNA protrusions and takes off the secondary structure (Deng et al., 2014, 2015). ATP-dependent dissociation of RPA by Pol Θ helicase (annealing helicase) activates DNA annealing and alternative-NHEJ machinery. Further, Pol 1Θ polymerase extends the paired single-stranded DNA overhangs. A complex - DNA ligase I (lig I) or DNA ligase III (lig 3)/XRCC1 accomplishes the process of end re-joining (Deng et al., 2015; Doksani and de Lange, 2016; Sfeir and Symington, 2015).

14.4 DIVERSIFICATION OF ANTIBODY REPERTOIRE

Diverse antibody repertoires have significant importance in the adaptive immune response. After VDJ recombination, alterations like SHM (somatic hypermutation) and class switch recombination (CSR) in the subset of B cells result in the formation of antibodies of high affinity. An enormous diversity is formed within the antigen encountering hypervariable Complementarity determining regions (CDRs) (Briney and Crowe, 2013).

14.4.1 Somatic Hypermutation (SHM)

Though the V-D-J rearrangement results in a primary repertoire of antibodies with different specificities (~order of 10^5–10^6 in humans), it is followed by the diversification of this repertoire by the process known as somatic hypermutation (SHM). B cells within the germinal centers in secondary lymphoid organs undergo SHM following the antigen-binding; thus, it is antigen-dependent antibody affinity maturation. SHM involves a key enzyme termed Activation-induced deaminase (AID), which is also known to trigger CSR. In the process of SHM, as shown in Figure 14.4, single nucleotide substitutions take place in the V (variable) region with the frequentness of

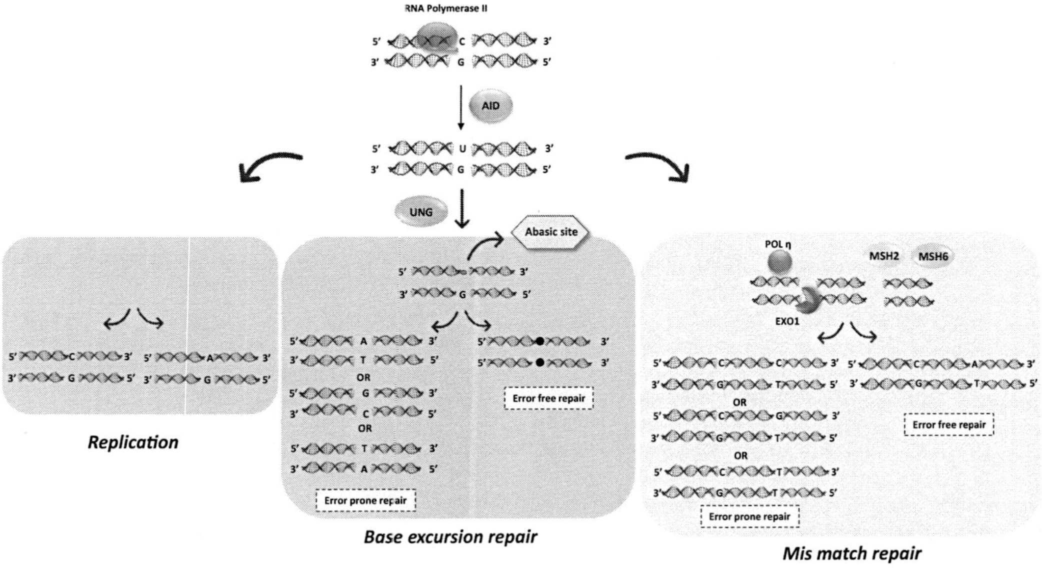

FIGURE 14.4 Schematic for Somatic Hypermutation Mechanism. AID-mediated deamination of cytosine in single-stranded DNA template results in uracil formation, which can be further processed by base excursion repair or mismatch repair pathway. Abasic site is formed by the UNG binding with U:G mismatch in BER. The cleavage of the abasic site and resynthesis of DNA is mediated by APE1 and Polβ. In contrast, MSh6, MSH2, DNA Polη, and EXO1 mediate the error-prone synthesis of nucleotide in MMR.

approximately 10^{-3} mutations per base pair per generation (cell division). Nucleotide base targeting is non-random with a C:G biased predominant phase of mutation. AID catalyzes the deamination of deoxycytidine into deoxyuridine, thus converting the C:G base pair into U:G mispair. Depending on the downstream processing, the U:G mismatch has different fates. Firstly, during replication, U:G mispair is replicated as U:A and U to T conversion takes place by the process of base excision repair (BER), thus leading to C:G–T:A transitions (Di Noia and Neuberger, 2007; Odegard and Schatz, 2006). Secondly, the uracil excision repair (UER) enzyme and uracil DNA glycosylase (UNG) may identify the U:G mismatch to create an abasic site by removal of uracil, which is then rectified by the trans-lesion synthesis mediated by Rev1 and Polη enzymes, leading the C:G into G:C/T:A transitions or A:T transversions (Methot and Di Noia, 2017). Lastly, mutations in the A:T base pairs near the U:G mismatch may occur by the MSH2/MSH6 patch repair complex, also known as the mismatch repair (MMR) machinery (Teng and Papavasiliou, 2007). Thus, targeted nucleotide modification and deamination of deoxycytidines at the Ig loci represents a key mechanism for the generation of antibody diversity.

14.4.2 Class Switch Recombination

Class switch recombination (CSR) mainly occurs before the beginning of somatic hypermutation (Roco et al., 2019; Flemming, 2019). CSR is an intra-chromosomal region-specific DNA deletion recombination of the Immunoglobulin heavy chain (IgH) locus (Figure 14.5) (Feng et al., 2020; Stavnezer et al., 2008). The constant region of heavy chain initially expressing δ and μ are replaced with α, ε, and ϒ regions (Stavnezer and Schrader, 2014). As a result, antibodies having the same variable region with higher antigen-binding specificity but different effector functions are expressed (Dunnick et al., 1993; Kracker and Durandy, 2011).

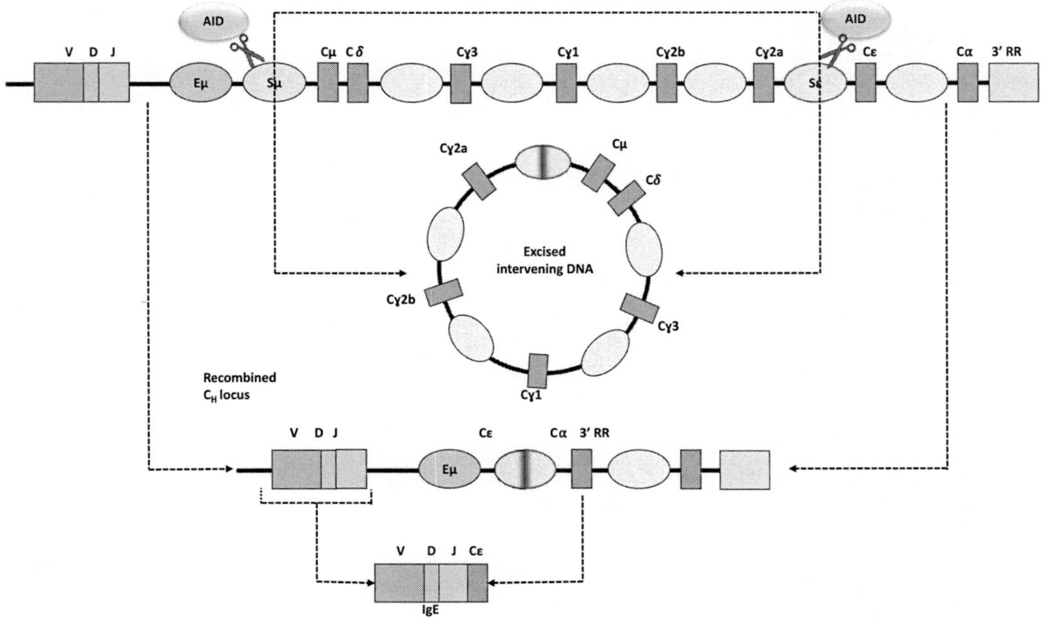

FIGURE 14.5 Schematic diagram for Class Switch Recombination. In the class switch recombination heavy chain, the constant region is interchanged by the cluster of constant region genes, which are located at its downstream. AID initiates a double-stranded DNA break at the S (switch region). Subsequently. DNA repair mechanism occurs, leading to the generation of a recombined constant heavy chain locus and an excised intervening DNA episomal circle.

Except Cδ, transcriptionally active G-rich repetitive S (switch) regions are present at the upstream of all C_H regions (Dunnick et al., 1993; Min et al., 2005). The switch regions have high-density RGYW motifs, which serve as the target sites for the AID binding (Bonaud et al., 2015). AID is an ssDNA-specific cytidine deaminase that deaminates the deoxycytidines of the switch region into deoxyuracils (Dickerson et al., 2003; Di Noia and Neuberger, 2002; Maul et al., 2011).

DNA uridine glycosylase removes dU bases due to which abasic residues are formed in the DNA backbone. A vital base excursion repair enzyme AP endonuclease, identifies and nick these abasic sites; it results in several single-stranded breaks in the acceptor and donor S site (Stavnezer et al., 2008). DNA mismatch repair enzymes like MSH6 and MSH2 convert DNA single-strand breaks into the double-strand break required for CSR by recruiting an exonuclease EXO1 (Stavnezer, 2011). In between the S region of acceptor and donor site, end-joining recombination occurs. Re-joining of double-strand break is done through alternative non-homologous end-joining or classical non-homologous end-joining pathway (Pulendran and Ahmed, 2006; Stavnezer et al., 2008).

IGHV MUTATIONAL STATUS

IgH gene mutational status is considered a powerful tool in the prognosis of Chronic lymphocytic leukemia (CLL). IGHV mutation status is determined by the proliferation rate of CLL cells that undergo somatic hypermutation. In slowly dividing cells, the IGHV mutation rate is low, whereas, in fast-dividing cells, the IGHV mutation rate is high. Therefore, the proliferation rate predicts the clinical outcome and overall survival chances with chemoimmunotherapy. IGHV mutational status suggests that CLL patient's survival rate is worse in unmutated IGHV patients than in mutated IGHV patients (Gladstone et al., 2012; Ghia et al., 2007).

14.5 SUMMARY

- To encounter a large variety of molecules, present on foreign antigens, the immune system requires diverse repertoires of B and T lymphocytes, which specifically bind to their corresponding antigens. This binding leads to proliferation and differentiation into the effector cells to eliminate the microbes (Cooper and Alder, 2006). Tonegawa discovered the genetic basis of Ig receptor diversity. He stated that the recombination of the multiple gene segments – V, D, and J in the heavy chain variable region and V and J in the light chain variable region result in the receptor diversity (Tonegawa, 1983). RSS, a non-coding sequence, follows the 12/23 rule and guides this VDJ recombination.
- Recombination occurs by forming a double-strand break followed by a repair mechanism (Bassing et al., 2000; Gellert, 2002). At the time of antigen selection after immunization, high-affinity antibodies are formed when receptors undergo somatic hypermutation (Meffre et al., 2001).
- Before the beginning of Somatic hypermutation, Class switch recombination (CSR) occurs and results in forming a different class of antibodies with higher affinity. Intrachromosomal deletion recombination occurs in CSR initiated by a potential and well-recognized RNA editing enzyme - Activation-induced cytidine deaminase (AID) (Stavnezer et al., 2008; Roco et al., 2019; Muramatsu et al., 2000).
- Regulatory error in the AID enzyme results in DNA damage in Ig genes, resulting in tumor formation. AID is a kind of mutant enzyme, which in Burkitt lymphoma, causes c-Myc chromosomal translocation. AID dysregulation causes tumorigenesis by inducing DNA damage in various Ig genes, B cell, and non-B cell genes.
- Association of B cell lymphoma is seen with the chromosomal translocation, which deregulates the oncogene's expression. In the germinal centers, B cells proliferate, and

the immunoglobulin variable region (IgV) genes get diversified by somatic hypermutation (SHM).

- Aberrant expression of AID in B cells is linked with the c-Myc, RhoH, Pim1and Pax 5 oncogene mutation, and diffuse large B cell lymphoma growth. The AID-induced mutation is a characteristic attribute of Burkitt lymphoma in which the c Myc and IgH loci in B cells undergo a reciprocal translocation (Gu et al., 2012).
- Understanding the complexities of antibody diversity and affinity maturation is necessary to progress in evolving vaccine development strategies.

QUESTIONS

1. Antigen receptor diversity is a result of……………………
2. Name the gene which initiates recombination by a double-stranded break.
3. DSB end processing, gap-filling, and ligation in the canonical nonhomologous end-joining (C-NHEJ) are carried out by – (a)…………………(b)……………(c)…………
4. Extension and rejoining of ends in alternative non-homologous end-joining (Alt- NHEJ) are done by – (a)………………… (b)………………
5. How do T cells and B cells recognize antigens?
6. Draw a schematic diagram of an immunoglobulin molecule.
7. Briefly explain the phases of VDJ recombination.
8. Write the steps involved in high-affinity antibody formation.
9. Explain the different fates of U:G mismatch repair in Somatic hypermutation (SHM).
10. What is the significance of Class switch recombination (CSR)?

BIBLIOGRAPHY

Ahmed, Rafi, and Barry T. Rouse. 2006. "Immunological Memory." *Immunological Reviews*. doi:10.1111/j.0105-2896.2006.00424.x

Arnal, Suzzette M., Abigail J. Holub, Sandra S. Salus, and David B. Roth. 2010. "Non-Consensus Heptamer Sequences Destabilize the RAG Post-Cleavage Complex, Making Ends Available to Alternative DNA Repair Pathways." *Nucleic Acids Research* 38 (9): 2944–54. doi:10.1093/nar/gkp1252

Bassing, Craig H., Frederick W. Alt, Maureen M. Hughes, Margaux D'Auteuil, Tara D. Wehrly, Barbara B. Woodman, Frank Gärtner, J. Michael White, Laurie Davidson, and Barry P. Sleckman. 2000. "Recombination Signal Sequences Restrict Chromosomal V(D)J Recombination beyond the 12/23 Rule." *Nature* 405 (6786): 583–86. doi:10.1038/35014635

Bassing, Craig H., Frederick W. Alt, Maureen M. Hughes, Margaux D. Auteuil, Tara D. Wehrly, Barbara B. Woodman, J. Michael White, Laurie Davidson, and Barry P. Sleckman. 2000. "Nature_VDJ.Pdf," no. X: 3–6.

Bassing, Craig H., Wojciech Swat, and Frederick W. Alt. 2002. "The Mechanism and Regulation of Chromosomal V(D)J Recombination." *Cell* 109 (2 SUPPL. 1): 45–55. doi:10.1016/S0092-8674(02)00675-X

Bonaud, Amelie, Fabien Lechouane, Sandrine Le Noir, Olivier Monestier, Michel Cogne, and Christophe Sirac. 2015. "Efficient AID Targeting of Switch Regions Is Not Sufficient for Optimal Class Switch Recombination." *Nature Communications* 6 (May). doi:10.1038/ncomms8613

Briney, Bryan S., and James E. Crowe. 2013. "Secondary Mechanisms of Diversification in the Human Antibody Repertoire." *Frontiers in Immunology* 4 (MAR): 1–7. doi:10.3389/fimmu.2013.00042

Carmona, Lina Marcela, and David G. Schatz. 2017. "New Insights into the Evolutionary Origins of the Recombination-Activating Gene Proteins and V(D)J Recombination." *FEBS Journal* 284 (11): 1590–1605. doi:10.1111/febs.13990

Chang, Howard H.Y., and Michael R. Lieber. 2016. "Structure-Specific Nuclease Activities of Artemis and the Artemis: DNA-PKcs Complex." *Nucleic Acids Research* 44 (11): 4991–97. doi:10.1093/nar/gkw456

Chang, Howard H.Y., Nicholas R. Pannunzio, Noritaka Adachi, and Michael R. Lieber. 2017. "Non-Homologous DNA End Joining and Alternative Pathways to Double-Strand Break Repair." *Nature Reviews Molecular Cell Biology* 18 (8): 495–506. doi:10.1038/nrm.2017.48

Chi, Xiying, Yue Li, and Xiaoyan Qiu. 2020. "V(D)J Recombination, Somatic Hypermutation and Class Switch Recombination of Immunoglobulins: Mechanism and Regulation." *Immunology* 160 (3): 233–47. doi:10.1111/imm.13176

Chiruvella, Kishore K., Zhuobin Liang, and Thomas E. Wilson. 2013. "Repair of Double-Strand Breaks by End Joining." *Cold Spring Harbor Perspectives in Biology* 5 (5): 1–22. doi:10.1101/cshperspect.a012757

Cooper, Max D., and Matthew N. Alder. 2006. "The Evolution of Adaptive Immune Systems." *Cell* 124 (4): 815–22. doi:10.1016/j.cell.2006.02.001

Deng, Sarah K., Huan Chen, and Lorraine S. Symington. 2015. "Replication Protein A Prevents Promiscuous Annealing between Short Sequence Homologies: Implications for Genome Integrity." *BioEssays* 37 (3): 305–13. doi:10.1002/bies.201400161

Deng, Sarah K., Bryan Gibb, Mariana Justino De Almeida, Eric C. Greene, and Lorraine S. Symington. 2014. "RPA Antagonizes Microhomology-Mediated Repair of DNA Double-Strand Breaks." *Nature Structural and Molecular Biology* 21 (4): 405–12. doi:10.1038/nsmb.2786

Desiderio, Stephen V., George D. Yancopoulos, Michael Paskind, Elise Thomas, Michael A. Boss, Nathaniel Landau, Frederick W. Alt, and David Baltimore. 1984. "Insertion of N Regions into Heavy-Chain Genes Is Correlated with Expression of Terminal Deoxytransferase in B Cells." *Nature* 311 (5988): 752–55. doi:10.1038/311752a0

De Villartay, Jean Pierre. 2015. "Congenital Defects in V(D)J Recombination." *British Medical Bulletin* 114 (1): 157–67. doi:10.1093/bmb/ldv020

Dickerson, Sarah K., Eleonora Market, Eva Besmer, and F. Nina Papavasiliou. 2003. "AID Mediates Hypermutation by Deaminating Single Stranded DNA." *Journal of Experimental Medicine* 197 (10): 1291–96. doi:10.1084/jem.20030481

Di Noia, Javier, and Michael S. Neuberger. 2002. "Altering the Pathway of Immunoglobulin Hypermutation by Inhibiting Uracil-DNA Glycosylase." *Nature* 419 (6902): 43–48. doi:10.1038/nature00981

Di Noia, Javier M., and Michael S. Neuberger. 2007. "Molecular Mechanisms of Antibody Somatic Hypermutation." *Annual Review of Biochemistry* 76: 1–22. doi:10.1146/annurev.biochem.76.061705.090740

Doksani, Ylli, and Titia de Lange. 2016. "Telomere-Internal Double-Strand Breaks Are Repaired by Homologous Recombination and PARP1/Lig3-Dependent End-Joining." *Cell Reports* 17 (6): 1646–56. doi:10.1016/j.celrep.2016.10.008

Dreyer, W. J., and J. C. Bennett. 1965. "The Molecular Basis of Antibody Formation: A Paradox." *Proceedings of the National Academy of Sciences of the United States of America* 54 (3): 864–69. doi:10.1073/pnas.54.3.864

Dunnick, W., G. Z. Hertz, L. Scappino, and C. Gritzmacher. 1993. "Corrigenda: Dna Sequences at Immunoglobulin Switch Region Recombination Sites." *Nucleic Acids Research* 21 (9): 2285. doi:10.1093/nar/21.9.2285

Eastman, Quinn M., Thomas M. J. Leu, and David G. Schatz. 1996. "Initiation of V(D)J Recombination in Vitro Obeying the 12/23 Rule." *Nature*. doi:10.1038/380085a0

Feng, Yuqing, Noé Seija, Javier M. Di Noia, and Alberto Martin. 2020. "AID in Antibody Diversification: There and Back Again." *Trends in Immunology* 41 (7): 586–600. doi:10.1016/j.it.2020.04.009

Flajnik, Martin F., and Louis Du Pasquier. 2004. "Evolution of Innate and Adaptive Immunity: Can We Draw a Line?" *Trends in Immunology* 25 (12): 640–44. doi:10.1016/j.it.2004.10.001

Flemming, Alexandra. 2019. "Class-Switch Recombination Revised." *Nature Reviews Immunology* 19 (10): 596–97. doi:10.1038/s41577-019-0214-8

Gellert, Martin. 2002. "V(D)J Recombination: RAG Proteins, Repair Factors, and Regulation." *Annual Review of Biochemistry* 71 (D): 101–32. doi:10.1146/annurev.biochem.71.090501.150203

Ghia, Paolo, K. Stamatopoulos, C. Belessi, C. Moreno, S. Stilgenbauer, F. Stevenson, F. Davi, and R. Rosenquist. 2007. "ERIC Recommendations on IGHV Gene Mutational Status Analysis in Chronic Lymphocytic Leukemia." *Leukemia* 21 (1): 1–3. doi:10.1038/sj.leu.2404457

Gladstone, Douglas E., Amanda Blackford, Eunpi Cho, Lode Swinnen, Yvette Kasamon, Christopher D. Gocke, Constance A. Griffin, Javier Bolaños-Meade, and Richard J. Jones. 2012. "The Importance of IGHV Mutational Status in Del(11q) and Del(17p) Chronic Lymphocytic Leukemia." *Clinical Lymphoma, Myeloma and Leukemia* 12 (2): 132–37. doi:10.1016/j.clml.2011.12.005

Goodarzi, Aaron A., Yaping Yu, Enriqueta Riballo, Pauline Douglas, Sarah A. Walker, Ruiqiong Ye, Christine Härer, et al. 2006. "DNA-PK Autophosphorylation Facilitates Artemis Endonuclease Activity." *EMBO Journal* 25 (16): 3880–89. doi:10.1038/sj.emboj.7601255

Gu, Xiwen, Velizar Shivarov, and Matthew P. Strout. 2012. "The Role of Activation-Induced Cytidine Deaminase in Lymphomagenesis." *Current Opinion in Hematology* 19 (4): 292–98. doi:10.1097/MOH.0b013e328353da3a

Hanisch, Christoph, and Sorin Baiasu. 2016. "Constitutivism and Kantian Constructivism in Ethical Theory: Editorial Introduction." *Philosophia* 44 (4): 1125–28. doi:10.1007/s11406-016-9743-6

Hozumi, N., and S. Tonegawa. 1976. "Evidence for Somatic Rearrangement of Immunoglobulin Genes Coding for Variable and Constant Regions." *Proceedings of the National Academy of Sciences of the United States of America* 73 (10): 3628–32. doi:10.1073/pnas.73.10.3628

Hwang, Joyce K., Frederick W. Alt, and Leng-Siew Yeap. 2015. "Related Mechanisms of Antibody Somatic Hypermutation and Class Switch Recombination." In *Mobile DNA III*, pp. 325–48. doi:10.1128/9781555819217.ch15

Jones, Jessica M., and Martin Gellert. 2002. "Ordered Assembly of the V(D)J Synaptic Complex Ensures Accurate Recombination." *EMBO Journal* 21 (15): 4162–71. doi:10.1093/emboj/cdf394

Kracker, Sven, and Anne Durandy. 2011. "Insights into the B Cell Specific Process of Immunoglobulin Class Switch Recombination." *Immunology Letters* 138 (2): 97–103. doi:10.1016/j.imlet.2011.02.004

Kumar, Vipul, Frederick W. Alt, and Richard L. Frock. 2016. "PAXX and XLF DNA Repair Factors Are Functionally Redundant in Joining DNA Breaks in a G1-Arrested Progenitor B-Cell Line." *Proceedings of the National Academy of Sciences of the United States of America* 113 (38): 10619–24. doi:10.1073/pnas.1611882113

Lange, Miles D., Lin Huang, Yangsheng Yu, Song Li, Hongyan Liao, Michael Zemlin, Kaihong Su, and Zhixin Zhang. 2014. "Accumulation of VH Replacement Products in IgH Genes Derived from Autoimmune Diseases and Anti-Viral Responses in Human." *Frontiers in Immunology* 5 (JUL). doi:10.3389/fimmu.2014.00345

Laver, W. Graeme, Gillian M. Air, Robert G. Webster, and Sandra J. Smith-Gill. 1990. "Epitopes on Protein Antigens: Misconceptions and Realities." *Cell* 61 (4): 553–56. doi:10.1016/0092-8674(90)90464-P

Lescale, Chloé, Vincent Abramowski, Marie Bedora-Faure, Valentine Murigneux, Gabriella Vera, David B. Roth, Patrick Revy, Jean Pierre De Villartay, and Ludovic Deriano. 2016. "RAG2 and XLF/Cernunnos Interplay Reveals a Novel Role for the RAG Complex in DNA Repair." *Nature Communications* 7. doi:10.1038/ncomms10529

Lieber, Michael R. 2011. "The Mechanism of DSB Repair by the NHEJ." *Annual Review of Biochemistry* 79 (3): 181–211. doi:10.1146/annurev.biochem.052308.093131

Ma, Yunmei, Ulrich Pannicke, Klaus Schwarz, and Michael R. Lieber. 2002. "Hairpin Opening and Overhang Processing by an Artemis/DNA-Dependent Protein Kinase Complex in Nonhomologous End Joining and V(D)J Recombination." *Cell* 108 (6): 781–94. doi:10.1016/S0092-8674(02)00671-2

Maezawa, So, Saori Nakano, Takaaki Kuniya, Osamu Koiwai, and Kotaro Koiwai. 2016. "Double-Strand Break Repair Based on Short-Homology Regions Is Suppressed under Terminal Deoxynucleotidyltransferase Expression, as Revealed by a Novel Vector System for Analysing DNA Repair by Nonhomologous End Joining." *FEBS Open Bio* 6 (1): 16–23. doi:10.1002/2211-5463.12001

Maul, Robert W., Huseyin Saribasak, Stella A. Martomo, Rhonda L. McClure, William Yang, Alexandra Vaisman, Hillary S. Gramlich, et al. 2011. "Uracil Residues Dependent on the Deaminase AID in Immunoglobulin Gene Variable and Switch Regions." *Nature Immunology* 12 (1): 70–76. doi:10.1038/ni.1970

McBlane, J. Fraser, Dik C. van Gent, Dale A. Ramsden, Charles Romeo, Christina A. Cuomo, Martin Gellert, and Marjorie A. Oettinger. 1995. "Cleavage at a V(D)J Recombination Signal Requires Only RAG1 and RAG2 Proteins and Occurs in Two Steps." *Cell* 83 (3): 387–95. doi:10.1016/0092-8674(95)90116-7

Meek, Katheryn. 2020. "Activation of DNA-PK by Hairpinned DNA Ends Reveals a Stepwise Mechanism of Kinase Activation." *Nucleic Acids Research* 48 (16): 9098–9108. doi:10.1093/nar/gkaa614

Meffre, E., N. Catalan, F. Seltz, A. Fischer, M. C. Nussenzweig, and A. Durandy. 2001. "Somatic Hypermutation Shapes the Antibody Repertoire of Memory B Cells in Humans." *Journal of Experimental Medicine* 194 (3): 375–78. doi:10.1084/jem.194.3.375

Methot, S. P., and J. M. Di Noia. 2017. "Molecular Mechanisms of Somatic Hypermutation and Class Switch Recombination." In *Advances in Immunology*. 1st ed. Vol. 133. Elsevier Inc. doi:10.1016/bs.ai.2016.11.002

Min, Irene M., Lisa R. Rothlein, Carol E. Schrader, Janet Stavnezer, and Erik Selsing. 2005. "Shifts in Targeting of Class Switch Recombination Sites in Mice That Lack μ Switch Region Tandem Repeats or Msh2." *Journal of Experimental Medicine* 201 (12): 1885–90. doi:10.1084/jem.20042491

Moshous, Despina, Isabelle Callebaut, Régina De Chasseval, Barbara Corneo, Marina Cavazzana-Calvo, Françoise Le Deist, Ilhan Tezcan, et al. 2001. "Artemis, a Novel DNA Double-Strand Break Repair/V(D) J Recombination Protein, Is Mutated in Human Severe Combined Immune Deficiency." *Cell* 105 (2): 177–86. doi:10.1016/S0092-8674(01)00309-9

Muramatsu, Masamichi, Kazuo Kinoshita, Sidonia Fagarasan, Shuichi Yamada, Yoichi Shinkai, and Tasuku Honjo. 2000. "Class Switch Recombination and Hypermutation Require Activation-Induced Cytidine Deaminase (AID), a Potential RNA Editing Enzyme." *Cell* 102 (5): 553–63. doi:10.1016/S0092-8674(00)00078-7

Nakano, Takanari, Masanori Matsui, Ikuo Inoue, Takuya Awata, Shigehiro Katayama, and Takayuki Murakoshi. 2011. "Free Immunoglobulin Light Chain: Its Biology and Implications in Diseases." *Clinica Chimica Acta* 412 (11–12): 843–49. doi:10.1016/j.cca.2011.03.007

Ochi, Takashi, Andrew N. Blackford, Julia Coates, Satpal Jhujh, Shahid Mehmood, Naoka Tamura, Jon Travers, et al. 2015. "PAXX, a Paralog of XRCC4 and XLF, Interacts with Ku to Promote DNA Double-Strand Break Repair." *Science* 347 (6218): 185–88. doi:10.1126/science.1261971

Odegard, Valerie H., and David G. Schatz. 2006. "Targeting of Somatic Hypermutation." *Nature Reviews Immunology* 6 (8): 573–83. doi:10.1038/nri1896

Oettinger, Marjorie A., David G. Schatz, Carolyn Gorka, and David Baltimore. n.d. "RAG-I1," no. D.

Panchakshari, Rohit A., Xuefei Zhang, Vipul Kumar, Zhou Du, Pei Chi Wei, Jennifer Kao, Junchao Dong, and Frederick W. Alt. 2018. "DNA Double-Strand Break Response Factors Influence End-Joining Features of IgH Class Switch and General Translocation Junctions." *Proceedings of the National Academy of Sciences of the United States of America* 115 (4): 762–67. doi:10.1073/pnas.1719988115

Pryor, John M., Crystal A. Waters, Ana Aza, Kenjiro Asagoshi, Christina Strom, Piotr A. Mieczkowski, Luis Blanco, and Dale A. Ramsden. 2015. "Essential Role for Polymerase Specialization in Cellular Nonhomologous End Joining." *Proceedings of the National Academy of Sciences of the United States of America* 112 (33): E4537–45. doi:10.1073/pnas.1505805112

Pulendran, Bali, and Rafi Ahmed. 2006. "Translating Innate Immunity into Immunological Memory: Implications for Vaccine Development." *Cell* 124 (4): 849–63. doi:10.1016/j.cell.2006.02.019

Repasky, Jamie A. E., Elizabeth Corbett, Cristian Boboila, and David G. Schatz. 2004. "Mutational Analysis of Terminal Deoxynucleotidyltransferase- Mediated N-Nucleotide Addition in V(D)J Recombination." *The Journal of Immunology* 172 (9): 5478–88. doi:10.4049/jimmunol.172.9.5478.

Roco, Jonathan A., Luka Mesin, Sebastian C. Binder, Christian Nefzger, Paula Gonzalez-Figueroa, Pablo F. Canete, Julia Ellyard, et al. 2019. "Class-Switch Recombination Occurs Infrequently in Germinal Centers." *Immunity* 51 (2): 337–350.e7. doi:10.1016/j.immuni.2019.07.001

Rooney, Sean, Frederick W. Alt, Jo Ann Sekiguchi, and John P. Manis. 2005. "Artemis-Independent Functions of DNA-Dependent Protein Kinase in Ig Heavy Chain Class Switch Recombination and Development." *Proceedings of the National Academy of Sciences of the United States of America* 102 (7): 2471–75. doi:10.1073/pnas.0409857102

Rooney, Sean, Jo Ann Sekiguchi, Chengming Zhu, Hwei Ling Cheng, John Manis, Scott Whitlow, Jeff DeVido, et al. 2002. "Leaky Scid Phenotype Associated with Defective V(D)J Coding End Processing in Artemis-Deficient Mice." *Molecular Cell* 10 (6): 1379–90. doi:10.1016/S1097-2765(02)00755-4

Ru, Heng, Melissa G. Chambers, Tian Min Fu, Alexander B. Tong, Maofu Liao, and Hao Wu. 2015. "Molecular Mechanism of V(D)J Recombination from Synaptic RAG1-RAG2 Complex Structures." *Cell* 163 (5): 1138–52. doi:10.1016/j.cell.2015.10.055

Schatz, David G., Marjorie A. Oettinger, and David Baltimore. 1989. "The V(D)J Recombination Activating Gene, RAG-1." *Cell* 59 (6): 1035–48. doi:10.1016/0092-8674(89)90760-5

Schatz, David G., Marjorie A. Oettinger, and Mark S. Schlissel. 1992. "V (D) J Recombination : Molecular Biology and Regulation" *Annual Review of Immunology* 10: 359–383.

Schatz, David G., and Patrick C. Swanson. 2011. "V(D)J Recombination: Mechanisms of Initiation." *Annual Review of Genetics* 45 (D): 167–202. doi:10.1146/annurev-genet-110410-132552

Sfeir, Agnel, and Lorraine S. Symington. 2015. "Microhomology-Mediated End Joining: A Back-up Survival Mechanism or Dedicated Pathway?" *Trends in Biochemical Sciences* 40 (11): 701–14. doi:10.1016/j.tibs.2015.08.006

Spagnolo, Laura, Angel Rivera-Calzada, Laurence H. Pearl, and Oscar Llorca. 2006. "Three-Dimensional Structure of the Human DNA-PKcs/Ku70/Ku80 Complex Assembled on DNA and Its Implications for DNA DSB Repair." *Molecular Cell* 22 (4): 511–19. doi:10.1016/j.molcel.2006.04.013

Stavnezer, Janet. 2011. "Complex Regulation and Function of Activation-Induced Cytidine Deaminase." *Trends in Immunology* 32 (5): 194–201. doi:10.1016/j.it.2011.03.003

Stavnezer, Janet, Jeroen E.J. Guikema, and Carol E. Schrader. 2008. "Mechanism and Regulation of Class Switch Recombination." *Annual Review of Immunology* 26: 261–92. doi:10.1146/annurev.immunol.26.021607.090248

Stavnezer, Janet, and Carol E. Schrader. 2014. "IgH Chain Class Switch Recombination: Mechanism and Regulation." *The Journal of Immunology* 193 (11): 5370–78. doi:10.4049/jimmunol.1401849

Swanson, Patrick C., and Stephen Desiderio. 1998. "V(D)J Recombination Signal Recognition: Distinct, Overlapping DNA- Protein Contacts in Complexes Containing RAG1 with and without RAG2." *Immunity* 9 (1): 115–25. doi:10.1016/S1074-7613(00)80593-2

Tadi, Satish K., Carine Tellier-Lebègue, Clément Nemoz, Pascal Drevet, Stéphane Audebert, Sunetra Roy, Katheryn Meek, Jean Baptiste Charbonnier, and Mauro Modesti. 2016. "PAXX Is an Accessory C-NHEJ Factor That Associates with Ku70 and Has Overlapping Functions with XLF." *Cell Reports* 17 (2): 541–55. doi:10.1016/j.celrep.2016.09.026

Teng, Grace, and F. Nina Papavasiliou. 2007. "Immunoglobulin Somatic Hypermutation." *Annual Review of Genetics* 41: 107–20. doi:10.1146/annurev.genet.41.110306.130340

Tonegawa, Susumu. 1983. "Somatic Generation of Antibody Diversity." *Nature* 302 (5909): 575–81. doi:10.1038/302575a0

Tonegawa, Susumu, Nobel Prize, and Thomas Waldmann. 1987. "Research Garners."

Townsend, Catherine L., Julie M.J. Laffy, Yu Chang Bryan Wu, Joselli Silva O'Hare, Victoria Martin, David Kipling, Franca Fraternali, and Deborah K. Dunn-Walters. 2016. "Significant Differences in Physicochemical Properties of Human Immunoglobulin Kappa and Lambda CDR3 Regions." *Frontiers in Immunology* 7 (SEP): 1–12. doi:10.3389/fimmu.2016.00388

Truong, Lan N., Yongjiang Li, Linda Z. Shi, Patty Yi Hwa Hwang, Jing He, Hailong Wang, Niema Razavian, Michael W. Berns, and Xiaohua Wu. 2013. "Microhomology-Mediated End Joining and Homologous Recombination Share the Initial End Resection Step to Repair DNA Double-Strand Breaks in Mammalian Cells." *Proceedings of the National Academy of Sciences of the United States of America* 110 (19): 7720–25. doi:10.1073/pnas.1213431110

Van Gent, Dik C., Dale A. Ramsden, and Martin Gellert. 1996. "The RAG1 and RAG2 Proteins Establish the 12/23 Rule in V(D)J Recombination." *Cell* 85 (1): 107–13. doi:10.1016/S0092-8674(00)81086-7

Watson, C. T., and F. Breden. 2012. "The Immunoglobulin Heavy Chain Locus: Genetic Variation, Missing Data, and Implications for Human Disease." *Genes and Immunity* 13 (5): 363–73. doi:10.1038/gene.2012.12

Wei, Huiting, and Xiaochun Yu. 2016. "Functions of PARylation in DNA Damage Repair Pathways." *Genomics, Proteomics and Bioinformatics* 14 (3): 131–39. doi:10.1016/j.gpb.2016.05.001

Zagelbaum, Jennifer, Noriko Shimazaki, Zitadel Anne Esguerra, Go Watanabe, Michael R. Lieber, and Eli Rothenberg. 2016. "Real-Time Analysis of RAG Complex Activity in V(D)J Recombination." *Proceedings of the National Academy of Sciences of the United States of America* 113 (42): 11853–58. doi:10.1073/pnas.1606721113

15 Biology of the B Lymphocyte

Renu Solanki and Shailly Anand
Deen Dayal Upadhyaya College, University of Delhi
Monisha Khanna
Acharya Narendra Dev College, University of Delhi

CONTENTS

15.1 INTRODUCTION

B cell or B lymphocyte is a lymphocyte sub-type that circulates in the blood and lymph and constitutes the important part of the adaptive humoral immune system. B cells are responsible for the production of antigen-specific antibodies against invasive pathogens.

Early B cell development and the commitment thereof to the B cell family occurs prenatally in the fetal liver, prior to its continuation in the bone marrow throughout life. B cell development can broadly be divided into three stages:

i. Generation of mature B cells
ii. Activation of mature B cells after their interaction with antigen
iii. Formation of plasma cells and memory B cells from activated B cells.

On the basis of their dependence on antigen, B cell development can also be classified into two phases namely:

i. **Antigen-independent phase:** This phase is accomplished in the bone marrow by hematopoiesis. This occurs without any interaction with an antigen and therefore comprises the antigen-independent phase of B cell development.
ii. **Antigen-dependent phase:** B cell which leaves bone marrow has single antigenic specificity. These cells have membrane-bound Immunoglobulin M and Immunoglobulin D (mIgM and mIgD) because of clonal selection. B cell which has not encountered an antigen is called a naïve cell. The interaction of B cell with antigens takes place in the secondary lymphoid

DOI: 10.1201/9781003286424-15

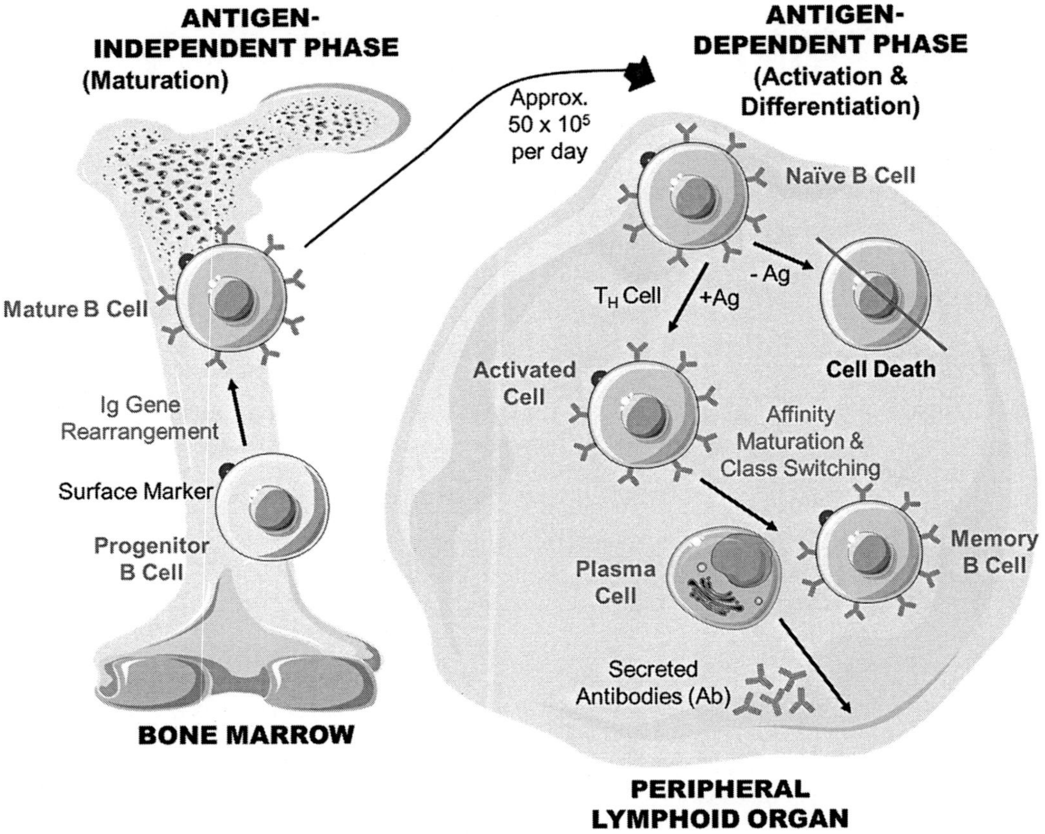

FIGURE 15.1 Overview of B cell development (Author.)

organs mainly the spleen and lymph nodes. After interaction with antigen, B cell undergoes antigen-dependent phase of B cell development. During this phase, affinity maturation and class switching from IgM to other classes of antibodies takes place (Figure 15.1).

15.2 ANTIGEN-INDEPENDENT PHASE OF B CELL DEVELOPMENT

15.2.1 B Cell Maturation

During embryonic development, B cell generation occurs in the yolk sac, then in the fetal liver and finally in fetal bone marrow but after birth bone marrow takes over the function. All the types of blood cells including B cells arise from a multipotent stem cell called hematopoietic stem cell (HSC). Early in hematopoiesis, HSC differentiates either into common lymphoid or myeloid progenitor cell. Progenitor cells subsequently lose the capacity for self-renewal and are committed to a particular cell lineage. B, T, and NK (natural killer) cells arise from common lymphoid progenitor cells. On the other hand, common myeloid progenitor stem cells generate progenitors of red blood cells, white blood cells including neutrophils, eosinophils, basophils, monocytes and also mast cells, dendritic cells, and platelets.

In the bone marrow, hematopoiesis takes place on a meshwork of stromal cells that include fat cells, endothelial cells, fibroblasts, and macrophages. Although stromal cells themselves are nonhematopoietic, they support the growth and differentiation of hematopoietic stem cells. Stromal cells provide a hematopoietic-inducing microenvironment (HIM) that comprises a cellular matrix and various growth factors that promote both the growth and differentiation of hematopoietic stem cells. The stromal cells play two important roles in the maturation of B cells. Firstly, they interact directly with progenitor B cells (pro-B) and precursor (pre-B cells) and secrete various cytokines, mainly IL-7 (Interleukin 7), that support the developmental process of B cells. Secondly, during the early stages of B cell development, direct contact between pro-B cells and stromal cells is necessary. This interaction is mediated by a number of adhesion molecules including VLA-4 (very late antigen 4) present on the pro-B cell and its ligand, VCAM-1 (Vascular cell adhesion molecule 1), on the stromal cell. After the initial association between VLA-4 and VCAM-1, an interaction between a receptor c-Kit (tyrosine kinase) on pro B cell and stem-cell factor (SCF) on the stromal cell occurs due to which activation of c-kit takes place. As a result, pro-B cells first divide and then differentiate into pre-B cell which after that express IL-7 receptor. Stromal cells secrete IL-7 which bind to its corresponding receptor and downregulates the expression of adhesion molecules on B cells. Consequently, proliferating cells lost their association with the stromal cells. There is no requirement of direct contact at this stage between pre B cells and stromal cells but IL-7 is required for growth and maturation (Figure 15.2).

15.2.1.1 Production of Immature B Cells

Maturation of B-cell depends on the rearrangement of the immuno-globulin DNA in the common lymphoid stem cells. Immunoglobulins are made up of two similar heavy and light chains, which are linked by disulphide bonds. During the development of B cell, the first Ig heavy chain gene rearranges from D_H to J_H which is followed by a V_H-to-D_HJ_H rearrangement resulting in the formation of a functional heavy chain protein (Ig) in large pre-B cells. If the heavy chain rearrangement is not successful, the V_H -D_H -J_H rearrangement proceeds on the other chromosome. The developing cell is classified as a pre-B cell only when the heavy chain rearrangement is accomplished. A productive light chain gene rearrangement is also required for the continued development of pre-B

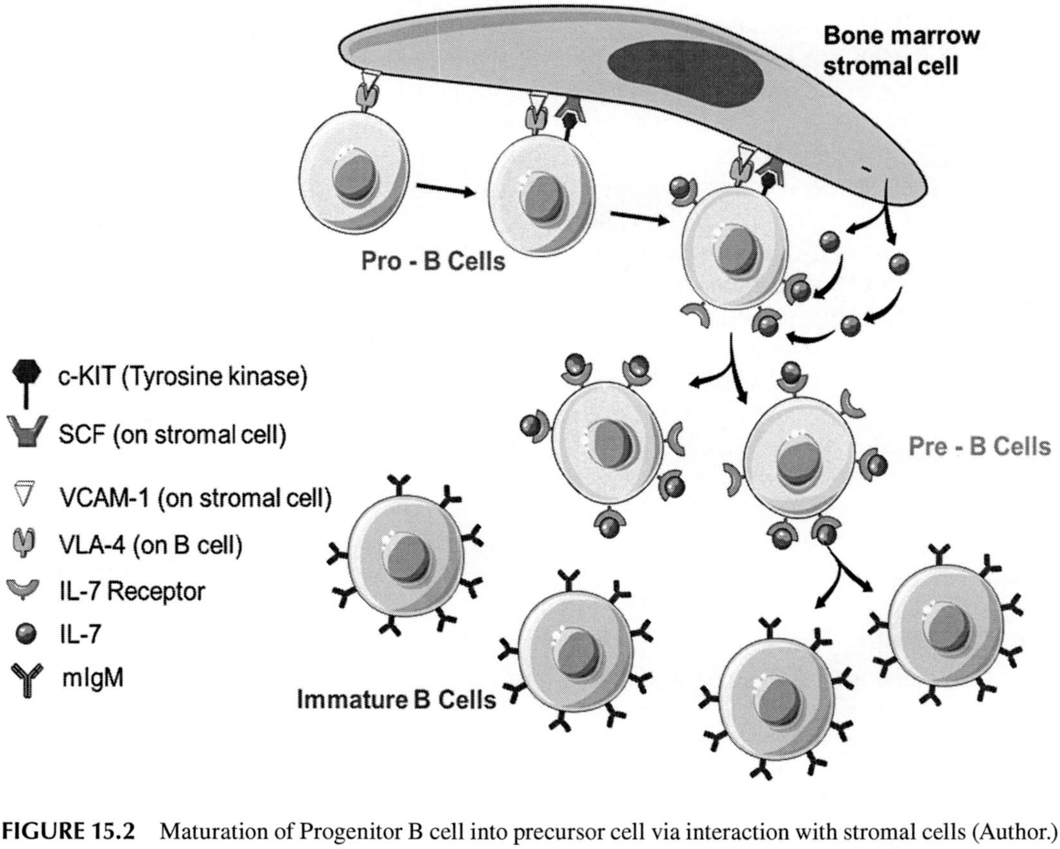

FIGURE 15.2 Maturation of Progenitor B cell into precursor cell via interaction with stromal cells (Author.)

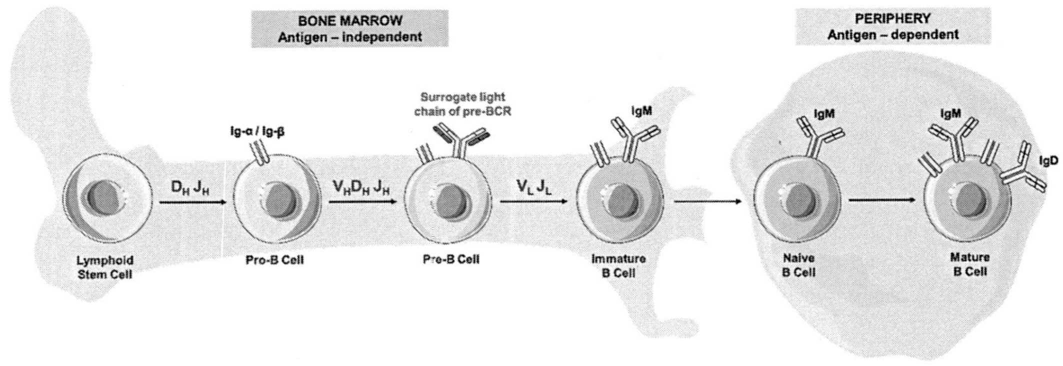

FIGURE 15.3 Sequence of events of B cell maturation in the bone marrow and secondary lymphoid organs (Author.)

cell into an immature B cell. Only a single light chain isotype is expressed on the B cell membrane because of allelic exclusion. Immature B cell commits to an antigenic specificity only after productive light chain gene rearrangement. The recombined heavy chain is then associated with the surrogate light chains and the Ig/dimer thereby forming the pre-B cell receptor (Pre-BCR). Pre-BCR is then expressed on the cell surface and in turn drives proliferation and differentiation into small pre-B cell stage intensively. Inactive small pre-B cells subsequently undergo V-J rearrangement of the Ig light chain. This helps in the production of a completely functional B cell Receptor which expresses on the surface of immature B cells as IgM (Figure 15.3).

Immature B cell commits to an antigenic specificity only after productive light chain gene rearrangement. Antigenic specificity is determined by heavy-chain VDJ sequence along with light-chain VJ sequence. mIgM (membrane IgM) is expressed on the cell surface of immature B cells. In pro and pre-B cell there is an expression of RAG1 and RAG2 (recombinase enzymes) which play an important role in the heavy and light chain gene arrangements. The enzyme terminal deoxyribonucleotidyl transferase (TdT) is found only in the pro-B cell stage not in pre-B cell stage because N-nucleotides are not usually found in the V_L-J_L coding joints.

During B cell development, the receptor of pre-B cell has a surrogate light chain comprising of a Vpre-B polypeptide (V-like structure) and a lambda 5 (C-like sequence) polypeptide which are associated noncovalently. In immature B cells, surrogate light chain is replaced by kappa or lambda light chain. Pre-BCR has the membrane-bound complex of heavy chain and surrogate light chain along with the Ig-α/Ig-β heterodimer. This is required for the maturation of pre-B cell. Four factors including E2A, EBF (early B-cell factor), BSAP (B-cell specific activator protein) and Sox-4 are particularly important for B-cell development. During the developmental progression of the pro-B cell to pre-B cell and then finally to the mature B cells. There is a difference in the profile of cell surface markers (Table 15.1).

15.2.1.1. Different Types of B Cells

There are two lineages of B cells i.e. B-1 and B-2 cells. Different types of antigens are encountered by these two types of cells. This is because of the specificity and requirement for help of T cells. B1 cells are predominantly present in the peritoneum and very less number in secondary lymphoid organs like lymph nodes and spleen. B-1 B cells in present during fetal life, marked by the display of CD45. B-1 and B-2 cells comparison is explained in Table 15.2.

15.2.1.2 Selection of Immature B Cells

During B cell development, auto-reactive B cells undergo apoptosis as target self-antigens. These are immature B cells which express autoantibodies against- self-antigens. This is also called as negative selection.

TABLE 15.1
Cell Surface Markers on the Pro B Cell, Pre B Cells, Immature B Cells and Mature B Cells

	Pro-B Cells	Pre-B Cells	Immature B Cells	Mature B Cells
CD 19	Yes	Yes	Yes	Yes
HSA(CD24)	Yes	Yes	Yes	Yes
CD45R	Yes	Yes	Yes	Yes
CD 43	Yes	No	No	No
Ig-alpha/Ig-beta	Yes	Yes	Yes	Yes
c kit	Yes	No	No	No
pre B Cell Receptor	No	Yes	No	No
mIgM	No	No	Yes	Yes
mIgD	No	No	No	Yes
CD25	No	Yes	No	No
CD23	No	No	No	Yes

TABLE 15.2
Comparison of B-1 and B-2 B Cells

Property	B-1 Cells	B-2 Cells
N regions	Few	Extensive
V region repertoire	Restricted	Diverse
Location	Peritoneum	Everywhere
Renewal	Self-renewal in situ	Bone marrow
Isotypes	IgM	IgM, IgD, IgA, IgG, IgE
Carbohydrate specificity	Yes	Rarely
Protein specificity	Rarely	Yes
Need T help	No	Yes
Somatic hypermutation	No	High
Memory development	No	Yes

Immature B cells having self-reactive mIgM on their membrane bind to self-antigens in the bone marrow and undergo the process of apoptosis. This was experimentally proven by D. A. Nemazee and K. Burki. They introduced a transgene specific for K^k, an H-2^k class I MHC molecule into H-2^d and H-$2^{d/k}$ mice. This transgene encodes the heavy and light chains of an IgM antibody. All nucleated cells have Class I MHC molecules expressed on their surface. Therefore, the endogenous H-2^k and H-2^d class I MHC molecules are present in the stromal cells of bone marrow in H-$2^{d/k}$ mice. In the H-2^d mice which do not express K^k on the stromal cells of the bone marrow have mature peripheral B cells population which secrete antibody with transgene-encoded anti-K^k. Their membrane-bound antibody also express heavy and light chains encoded by the same gene (Figure 15.4a).

In contrast, in the H-$2^{d/k}$ mice, there is an expression of K^k and no mature, peripheral B cells expressed the transgene-encoded antibody specific for H-2^k (Figure 15.4b).

The results of the above-mentioned experiment showed that immature B cells which express autoantibodies undergo negative selection and are removed by apoptosis. It is known that crosslinking of antibodies in immature B cells leads to their death. Therefore, if the immature B cells express antibodies against self-antigens, crosslinking of the antibodies on the B cell membrane takes place and they die or get eliminated from the body.

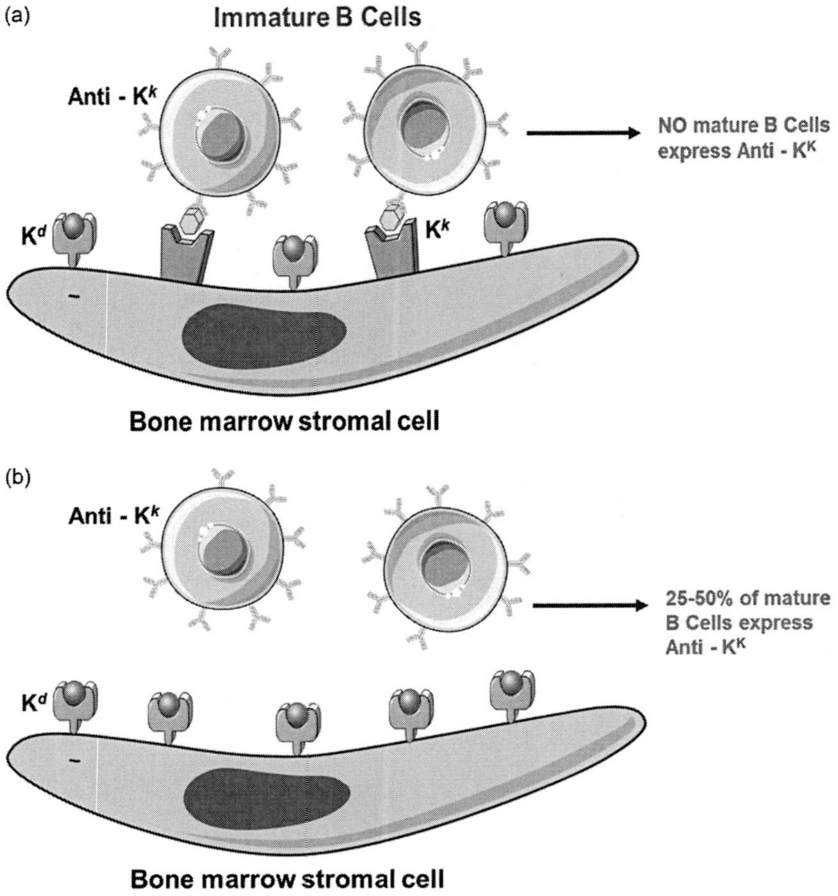

FIGURE 15.4 (a) H-$2^{d/k}$ transgenic mice: No mature peripheral B cells expressed the transgene-encoded antibody anti K^k (Author.) (b) In H-2^d transgenic mice mature peripheral B cells expressed the transgene-encoded antibody to H-2^k (Author.)

Some of the self-reactive B cells escape negative selection by editing their light chains. It has been found that the H-$2^{d/k}$ transgenics produced less number of mature B cells which expressed mIgM having the chain encoded by the transgene K^k, but different, endogenous light chains. But these B cells no longer remain self-reactive because they succeed in bearing light chains encoded by the endogenous chain gene regions. Consequently, these developing B cells begin to express an "edited" mIgM. Furthermore, mature self-reactive B cells are also negatively selected in the periphery.

15.3 ANTIGEN-DEPENDENT PHASE OF B CELL DEVELOPMENT

15.3.1 B Cell Activation and Proliferation

As B cells are released from the bone marrow, their further activation, proliferation, and differentiation occur in the periphery in presence of antigen. Interaction of the B cells with antigen results in their activation in the secondary lymphoid organs. Consequently, these activated B cells differentiate into antibody-secreting plasma cells and memory B cells. In absence of an interaction between B cells and the antigen, the life span of naïve B cells restricts to a few weeks only.

Depending on the antigen's nature, activation of B cell follows two routes - one dependent on T_H cells and the other that remains independent of T_H cells. Based on the need for T_H cells, the antigens are denoted as either Thymus-dependent (TD) antigen or Thymus-independent (TI) antigen.

Thymus-dependent antigens and thymus-independent antigens differ in their humoral response as illustrated in Table 15.3.

The response to Thymus-Independent (TI) antigens is generally weaker, memory cells are not formed, the predominant antibody secreted is IgM and low level of class switching. These differences between thymus-dependent and independent antigens responses, emphasize the important role played by T_H cells in the generation of memory B cells, affinity maturation and class switching. But in the case of Thymus-Dependent (TD) antigen which are soluble proteins, memory cells are formed, affinity maturation & class switching is observed and also they fail to act as polyclonal activators. Thymus-independent antigens are further divided into Type 1 and Type 2 (Table 15.4).

B cell activation requires two signals. In the case of Thymus-independent antigen, both signals are provided by binding of TI antigen while in Thymus-dependent antigen, signal 1 is provided by cross-linking of mIg, and signal 2 is provided by separate signal between CD40 on the B cell and its CD40L on an activated T_H cell (Figure 15.5). T_H cells play essential roles in most B cell responses (Figure 15.6).

TABLE 15.3
Summary of Thymus-Dependent & Thymus-Independent Antigens

Thymus-Dependent Antigen	Thymus-Independent Antigen
Activate B cells in the presence of T helper cells	Activate B cells in the absence of T helper cells
Activate mature B cells only	Activate both types of B cells: mature and immature
Soluble proteins	Bacterial cell wall components Lipopolysaccharide (LPS), Capsular polysaccharide, flagella etc.
Antigen is processed and displayed on the surface of B Cells in association with MHC-II.	Antigen processing is not needed
Immunogenic over wide range of dose.	Dose-dependent immunogenicity
Activate B cells monoclonally and there is no polyclonal activation.	High doses of Type-I (TI) antigens result in Polyclonal activation of B cells.
They are complex in nature, mainly proteins, hapten-carrier complexes and other antigens with diverse epitopes.	They have simple structures, mainly polysaccharides or proteins having repeating epitopes. Bacterial cell wall components including LPS can act as TI antigen
All classes of antibodies (IgG, IgM, IgA, IgD, IgE) are produced due to class switching	IgM is the predominant antibody as there is NO class switching
Memory B cells are produced	No memory B cells are produced
Affinity maturation present	Affinity maturation absent

TABLE 15.4
Distinguishing Features of Type 1 and Type 2 TI Antigens

Type 1 Thymus-Independent (TI) Antigen	Type 2 Thymus-Independent (TI) Antigen
Bacterial cell-wall components which include lipopolysaccharide (LPS)	Highly repetitious molecules such as polymeric proteins (e.g., bacterial flagellin) or bacterial cell-wall polysaccharides having repeating polysaccharide units.
TI-1 antigens are B cell mitogens and take action as polyclonal activators	They are not B-cell mitogens and so do not act as polyclonal activators.
TI-1 antigens activate both types of B cells: Mature and Immature	TI-2 antigens activate mature B cells and inactivate immature B cells.
	Though B-cell response to these antigens does not require direct interaction with T_H cells whereas cytokines secreted by T_H cells are required for efficient B-cell proliferation and class switching from IgM to other types.

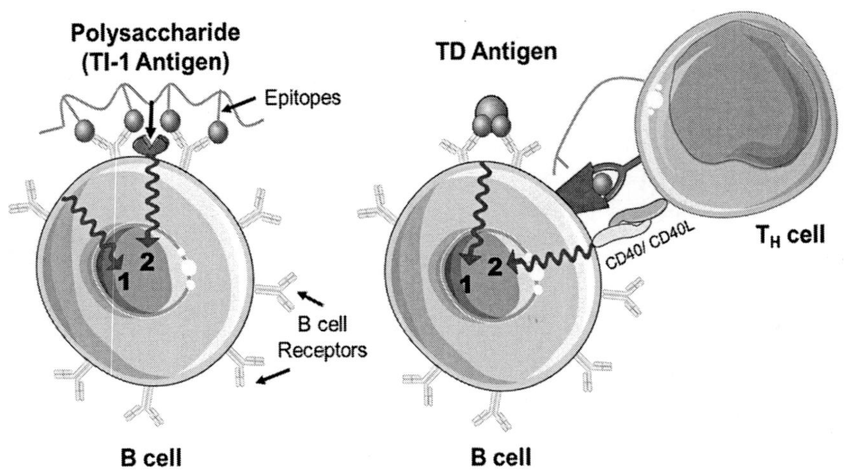

FIGURE 15.5 B cell activation in TI – Type 1 antigen and TD antigen (Author.)

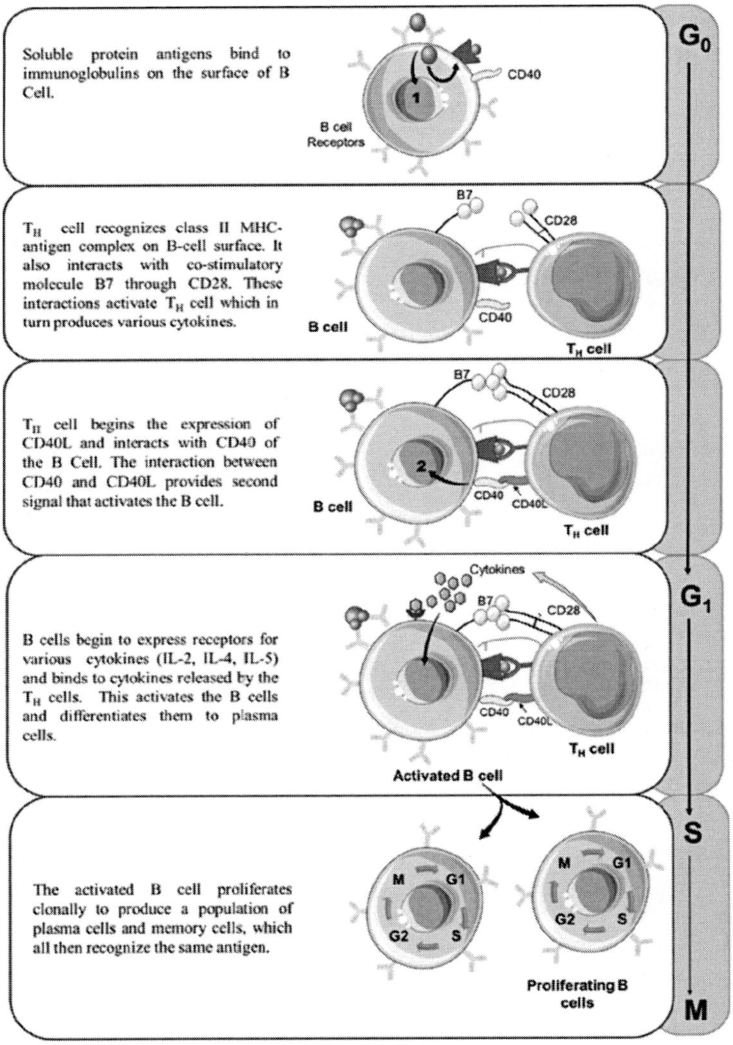

FIGURE 15.6 The sequence of events during activation of B cell by Thymus-Dependent Antigen and role of T_H cells in the activation of B cells (Author.)

15.4 HUMORAL RESPONSE

15.4.1 Primary Response

Humoral response which results from the activation of naïve lymphocytes is called primary response. The first contact of antigens with the individual results in the formation of antibody-secreting plasma cells and memory B cells. The primary response involves a Lag Phase during which B cells differentiate into memory cells or plasma cells. Lag phase is followed by the Logarithmic Phase which is marked by increased levels of serum antibody. The levels initially reaches a peak, attains a plateau for a variable time period, and then decline. IgM is the first antibody to be secreted followed by a class switch in IgG. The duration of primary response ranges from a few days to even several weeks depending on the persistence of the antigen (Figure 15.7).

15.4.2 Secondary Response

A later contact and activation of the memory cell with the same antigen is called secondary immune response. The secondary response has shorter lag phase with an heightened magnitude because the memory cells which interact with the antigen are high in number as compared to the naïve cells involved during the primary response. The classes of antibodies other than IgM predominates in secondary response. Unlike primary response, it involves affinity maturation too. A comparison between the primary and secondary responses can more clearly be understood by referring to Figure 15.8 and Table 15.5.

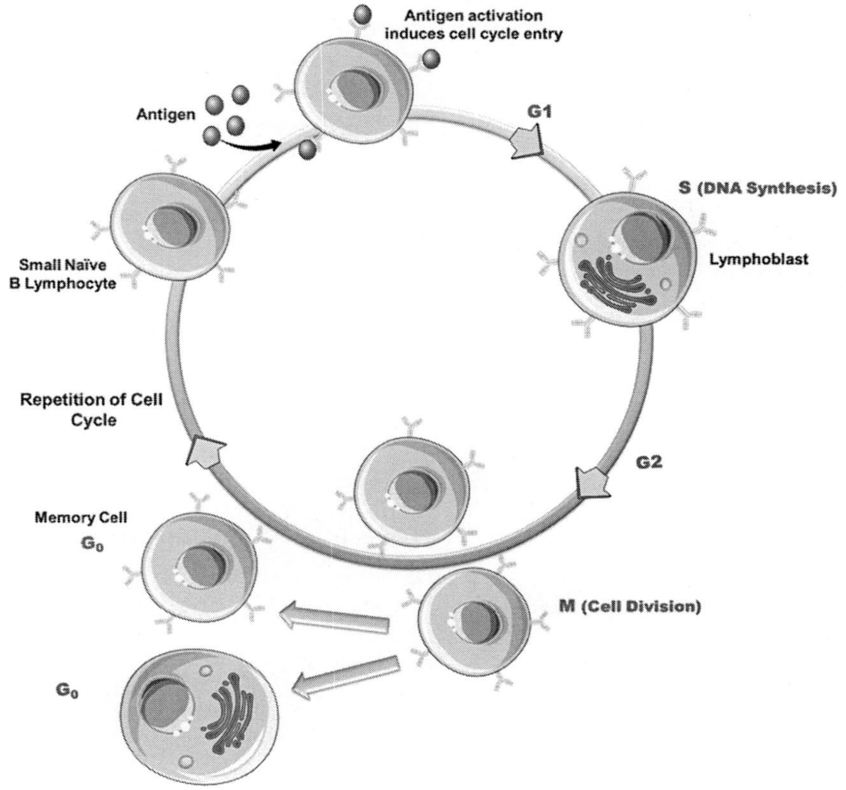

FIGURE 15.7 Primary Immune Response: Encounter of naïve B Lymphocyte with an antigen for the first time (Author.)

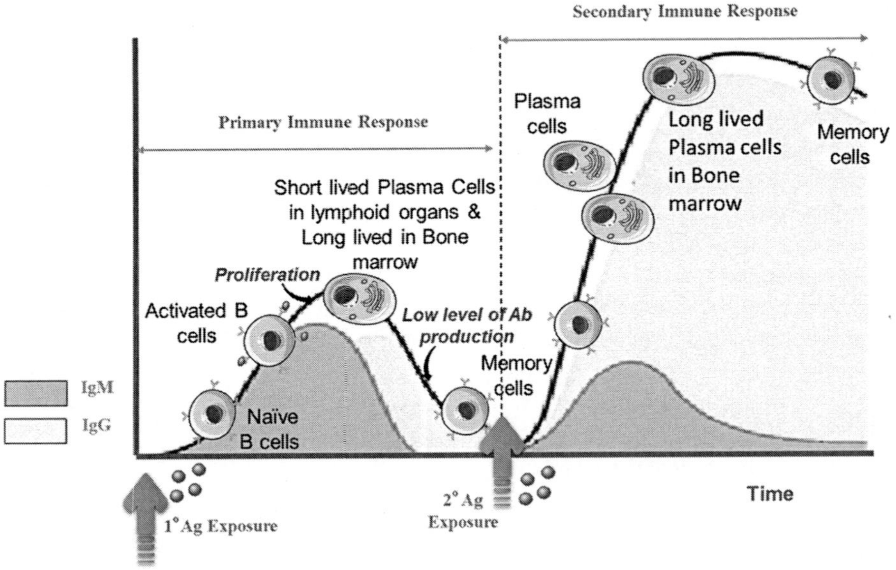

FIGURE 15.8 Differences between Primary and Secondary Antigen Response (Author.)

TABLE 15.5
Comparison of Primary and Secondary Humoral Response

Primary Humoral Response	Secondary Humoral Response
This is an immune response against primary contact with an antigen.	This is an immune response against second or subsequent exposure to the same antigen
Naïve B-cell and T-cell are responding cells	Memory cells are the responding cells
Lag phase is longer ranging from 4 to 7 days and sometimes may even last for weeks or months. The time is required for naïve B cell activation in response to the antigen and TH cells followed by proliferation and differentiation into plasma cells.	Lag phase is either absent or short ranging from 1 to 4 days due to the presence of clonally expanded memory B cells which show a more rapid response to the antigen than the naïve B cells.
The level of antibody peak at about day 14. It subsequently shows a drop as the plasma cells perish off.	The level of antibody reaches peak in just 3–5 days. This occurs because the memory cells are present in large numbers in comparison to the naïve cells for eliciting a response. This in turn ends up producing still large number of plasma cells and thus higher levels of antibodies (100–1,000 fold more).
The first antibody produced during a primary response is chiefly IgM though small amounts of IgG is also produced.	IgG antibody is predominantly produced with trace amounts of IgM being produced at times. It may involve production of other immunoglobulins such as IgA or IgE in the case of certain allergic reactions or situations.
Specific antibodies are produced which are less avid	Less specific antibodies are produced which are comparatively more avid.
It shows a rapid decline in the antibody levels.	For extended period of time, the Antibody level remains high.
Antibodies exhibits a lower average affinity towards the antigen.	Antibodies show a higher average affinity (affinity maturation) for the antigen.
It appears mainly in the lymph nodes and spleen.	It appears chiefly in the bone marrow, followed by the spleen and lymph nodes.
This response is shown by both TD and TI antigen.	This response is shown exclusively by TD antigen.

15.5 ANATOMICAL DISTRIBUTION OF B CELL POPULATION

B cell activation and differentiation occur in different anatomical sites. Antigens are concentrated in various peripheral lymphoid organs. Spleen filters the blood-borne antigens whereas regional lymph nodes filter the antigens drained by the lymphatic system from tissue spaces.

15.5.1 B CELLS IN LYMPH NODES

In the lymph nodes, Antigen (Ag) or antigen-antibody (Ag-Ab) complexes enter either alone or with antigen transporting cells and macrophages. There are three types of antigen-presenting cells (APCs) in the lymph node:

i. Interdigitating dendritic cells (in the paracortex)
ii. Macrophages (scattered in the node) or
iii. Specialized follicular dendritic cells (in primary and secondary follicles and germinal centers).

As antigen enters a regional lymph node by the lymphatic system, first of all in the paracortex, it is trapped, processed and presented along with class II MHC molecules present on the interdigitating dendritic cells. This results in the activation of T_H cells. The initial activation of B cells takes place in the paracortex region which has a large number of T cells. After activation, T_H and proliferating B cells form small foci at paracortex edges. Some of the B cells differentiate into antibody (IgM and IgG) secreting plasma cells in the foci. Plasma cells in these foci are mainly responsible for antibody production during primary humoral response. Similarly, initial B-cell activation takes place in the T-cell–rich periarterial lymphatic sheath, (PALS) of the spleen. Within 4–6 days of antigen challenge, these foci reach their maximum size and a less number of B cells and T_H cells migrate from PALS to the primary follicles of the cortex. Within a primary follicle, cellular interactions take place between B cells, T_H cells and follicular dendritic cells which lead to the development of a secondary follicle having a central germinal center from primary follicle. A less number of plasma cells generated in the germinal center of secondary follicles migrate to the medullary areas of the lymph node while many migrate to the bone marrow (Figure 15.9).

15.5.1.1 Antigen-Induced B-Cell Differentiation in Germinal Centres

After 7–10 days of exposure of thymus-dependent antigen, germinal centers arise. During this initial stage of germinal center formation, the activated B cells undergo intense proliferation to form centroblasts. Centroblasts are large in size with diffused chromatin, expanded cytoplasm and Ig are absent or nearly absent on their surfaces. Centroblasts then give rise to Centrocytes that are small, non-dividing B cells which express membrane Ig. Centrocytes eventually migrate to a region bearing follicular dendritic cells. Centrocytes make contact with the displayed antigen on the surface of follicular dendritic cells. Three important events of B cell maturation i.e. affinity maturation, class switching, and formation of plasma cells and memory B cells take place in these germinal centers (Figure 15.10).

15.5.1.1.1 Affinity Maturation by Somatic Hypermutation and Selection

Affinity of the antibodies produced during humoral response increases by the affinity maturation. Within the germinal centres, Ig genes in the B cells undergo extensive mutations. Various types of mutations including deletions, insertions and point mutations are introduced into the rearranged immunoglobulin genes too. The majority of these mutations occur within the three CDRs (complementarity-determining regions) throughout the V region when the centroblasts develop in the germinal center dark zone.

Since somatic mutations occur randomly, it leads to formation of few B cells with high-affinity receptors and many cells with unchanged or low-affinity receptors for a particular antigen.

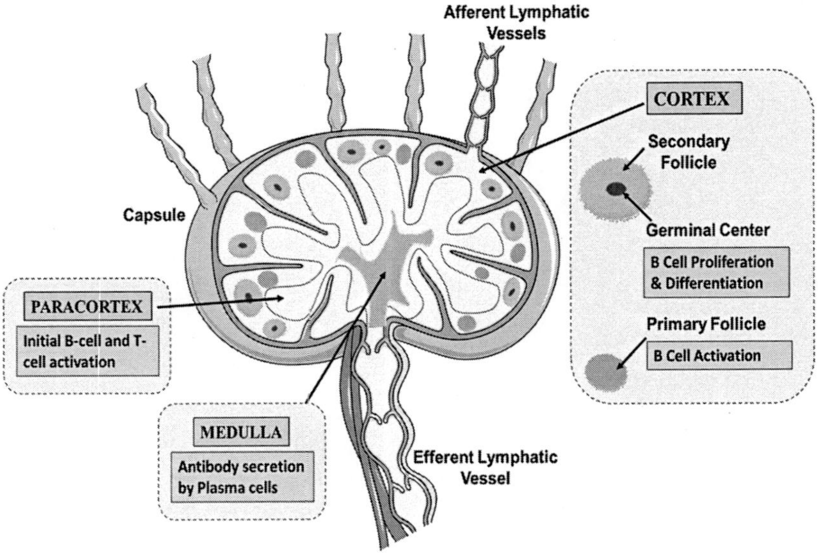

FIGURE 15.9 Lymph node: Lymph enters through the afferent lymphatic vessels, passes through the cortex, paracortex, medulla and finally exits from a single efferent lymphatic vessel. The cortex homes a large number of B cells, Follicular Dendritic Cells and macrophages within the lymphoid follicles, whereas the paracortex houses many lightly scattered T cells. The antibody-secreting plasma cells are stacked within the medulla. (Author.)

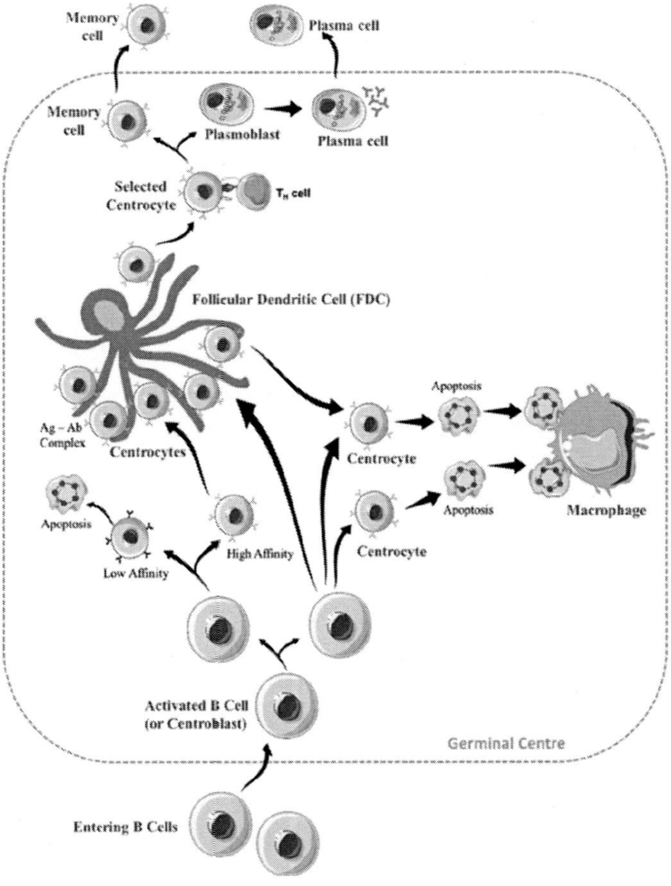

FIGURE 15.10 An overview of the events occurring within the germinal centers. (Author.)

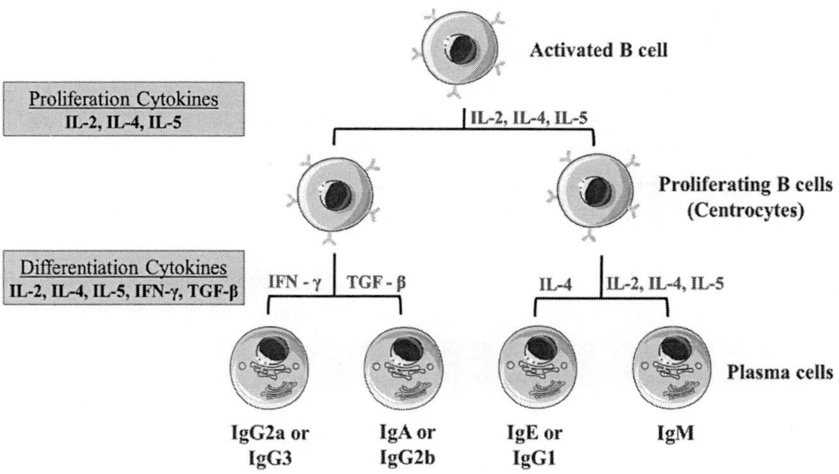

FIGURE 15.11 B cell interaction with cytokines elicit signals required for their proliferation and class switching as the B cells differentiate into plasma cells. (Author.)

This brings about a need to select a population of cells with high affinity. B cells showing high-affinity receptors for the antigen are selected positively and exit the germinal center, whereas cells exhibiting low affinity for the antigen are negatively selected and die in the germinal center. Selection process occurs in the light zone among the non-dividing population of centrocyte. The major factor which influences selection is the ability of centrocyte membrane-bound immunoglobulin (Ig) molecules to recognize and bind antigens which are displayed by follicular dendritic cells (FDCs).

15.5.1.1.2 Class Switching

Antibodies have V_H and V_L domains that bind specifically to an antigen. The biological effector functions of the antibodies are determined by the isotype of the heavy-chain constant domain. Class switching allows the antibody to be specific to an antigen through the biological effector activities of the antibody may vary. Cytokines also play an important role in class switching. Class switching occurs extensively in the humoral response due to thymus-dependent antigens. As mentioned earlier, isotypes other than IgM are present in the humoral response of TD antigens whereas IgM is found in TI antigen. Class switching occurs in TD antigen humoral response because of the interaction between CD40 and CD40 Ligand respectively on B and T helper cells along with the production of cytokines (Figure 15.10).

After selection of high-affinity mIg bearing B cells in the germinal center, some B cells differentiate into plasma cells and others move ahead to become the memory B cells. Cytokines play a significant role in this differentiation of B cells into plasma cells and memory cells (Figure 15.11).

SUMMARY

- While the development of B cells takes place in bone marrow, the antigen interaction occurs in the secondary lymphoid organs. This leads to formation of antibody secreting plasma cells and memory cells.
- Immature B cells are formed from pro B cell by Ig gene rearrangements which further develop into naïve cells.
- Self-reactive B cells with BCR that binds to the self-antigens are removed by apoptosis during negative selection.

- Antigenic interaction in secondary lymphoid organs leads to activation and further differentiation of B cells. T helper cells are required for B cell activation which in turn produces the antibody. Interaction between B and T helper cells takes place via CD40 and CD40 ligand which is not only important for the survival of B cells but also for the formation of germinal centres, memory cells and somatic hypermutation.
- Primary response is delayed as it is associated with a lag phase which occurs in response to the first exposure to antigen. IgM Antibody is produced, following which class switching occurs.
- The second exposure to the same antigen leads to secondary immune response which is short and has more logarithmic phase. Antibodies other than IgM such as IgG are produced.
- Affinity maturation and class switching occurs in germinal centres of secondary follicles. These are also the sites for formation of plasma cells and memory cells.

QUESTIONS

1. Define the following:
   a. Germinal center b. Class switching c. Affinity maturation d. Plasma cells e. Stromal cells
2. Differentiate between the following:
   a. Naïve B cells and Memory B cells
   b. Primary and Secondary immune response
   c. Primary and secondary follicle
   d. TI type 1 and TI type 2 antigens
   e. B1 and B2 B cells
3. Expand the following:
   a. HIM b. HSC c. NK d. VCAM e. VLA f. BCR g. TdT h. mIgM
4. Fill in the blanks:
   a. Interdigitating dendritic cells are present in ....
   b. Formation of plasms cells takes place in ....
   c. The classes of antibodies other than .... predominates in secondary response.
   d. Thymus-independent antigen act as B cells ....
   e. .... discovered the function of B cells while demonstrating the effect of surgical removal of *Bursa of Fabricius* in birds.
5. Long answer questions:
   a. Describe the structure and function of B cell co-receptor complex.
   b. Justify with an experiment that self-reactive B cells are eliminated.
   c. Give an account of the anatomical distribution of B cell population.
   d. Explain Somatic hypermutation.
   e. Give an overview of the cellular events of B cell differentiation occurring in the germinal center.

BIBLIOGRAPHY

Abul, A. A. and Andrew, L. H. 2003. *Cellular and Molecular Immunology*. V Edition, Saunders Publication, Philadelphia, PA, United States.

David, M., Jonathan, B., David, R. B. and Ivan, R. 2006. *Immunology*, VII Edition, Mosby, Elsevier Publication.

Kindt, T. J., Goldsby, R. A., Osborne, B. A. and Kuby, J. 2006. *Immunology*, VI Edition, W.H. Freeman and Company, New York, United States.

Murphy, K. and Weaver, C. 2017. *Janeway's Immunobiology*, IX Edition, Garland Science, New York, United States.

Discovery of B cells

Source:
https://en.wikipedia.org/wiki/Max_Dale_Cooper#/media/File:Max_Cooper_Royal_Society.jpg

In 1960s, Max Cooper discovered the function of B cells while demonstrating the effect of surgical removal of *Bursa of Fabricius* (B cell development site in birds from which the representation "B" cell was derived). Post-surgery irradiation of the chickens showed fully abrogated antibody production thereby confirming the distinct function of B cells in both adaptive and innate humoral immune responses.

16 Biology of the T Lymphocytes

Shailly Anand and Renu Solanki
Deen Dayal Upadhyaya College, University of Delhi

CONTENTS

16.1 INTRODUCTION

T lymphocytes, also called as thymocytes or T cells, are a type of white blood cells that play a key role in the body's immunity. From the site of their formation by hematopoiesis in the bone marrow, they migrate to the thymus for their maturation and differentiation. Hence, they derive their name from the site of their maturation as the T cells. There are two well-defined classes of T cells namely, T helper cells (T_H cells) and T cytotoxic cells (Tc) or cytotoxic T lymphocytes (CTL). They express T cell receptors (TCR) on their membranes which exhibit specificity and recognize MHC-bound antigen. T cells also express glycoprotein receptors such as CD2 and CD3 on their membranes. The presence of specific glycoprotein receptors (CD4 or CD8) serves as a basis for distinguishing the two types of T cells and also plays a role in defining their function. T cells with CD4 glycoproteins recognize antigen bound to MHC-II while those with CD8 recognize antigen bound to MHC-I. Also, there exists no T cell containing both CD4 and CD8 glycoproteins.

16.2 T CELL RECEPTOR (TCR)

16.2.1 Structure of T Cell Receptor (TCR)

T cells have membrane-bound TCR. TCR comprise two chains, either αβ or γδ, and accordingly T cell receptors are designated as αβ TCRs (Figure 16.1) or γδ TCRs. The two chains (α and β), each

DOI: 10.1201/9781003286424-16

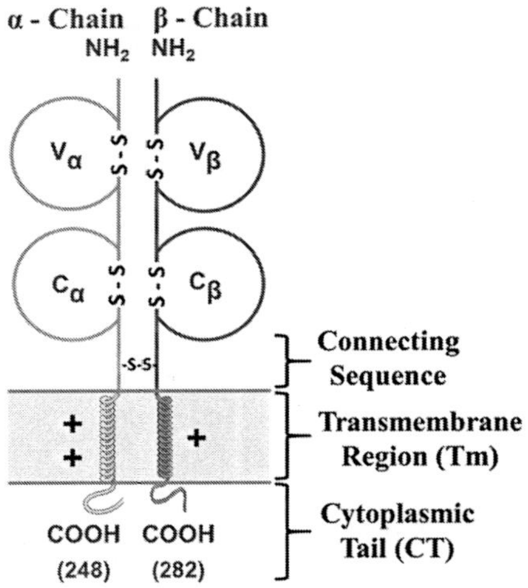

FIGURE 16.1 Structure of αβ T cell receptor. (Author.)

has two domains and an intra-chain disulfide bond spanning 60–75 amino acids. The sequence at the amino-terminal domain in both α and β chains show variations while the sequence of the remaining chain is conserved. The variable: V and constant: C domains of TCR are structurally homologous to the variable and constant domains of immunoglobulins. There are three hypervariable regions in the variable domain of TCR which are equivalent to the **C**omplementarity **D**etermining **R**egions (CDRs) of the heavy and light chains of immunoglobulin. There is an additional hypervariable region (HV4) in the β chain that does not form a contact with the antigen and therefore is not considered as CDR. Each TCR chain has a short connecting sequence in which a cysteine residue forms disulfide bonds thereby linking the two chains or heterodimers. Each chain that anchors α and β chains in the plasma membrane bears a transmembrane region of 21 or 22 amino acids. Transmembrane regions contain positively charged amino acids and these amino acids enable interactions between chains of TCR heterodimer and signal-transducing CD3 complex. Following the transmembrane domain is present a cytoplasmic tail of about 5–12 amino acids at the carboxyl-terminal end. In 95% of T-cells TCR is encoded by the αβ genes and in 5% by γδ genes (Table 16.1).

TABLE 16.1
Differences and Similarities between αβ and γδ T Cells

αβ T Cells	γδ T Cells
Abundant T cells about 95%	Only 5% of total T cell population
These cells interact with peptide antigens processed and presented with MHC molecules on the surface of cells. Self MHC restricted	These cells interact with antigen that is neither processed nor presented with MHC molecules. No self MHC restriction.
The long axes of the V and C regions of αβ consists of an elbow angle of 149°	The elbow angle of γδ TCR is 111°
These cells are positive for both CD4 and CD8,	A large number of cells are negative for CD4 as well as CD8. They mostly express a single-chain subtype.
Play an important role in adaptive immune response	Play an important role in innate immune response
Bind with peptide antigens	Bind with phospholipid antigen
There is MHC restriction	There is no MHC restriction

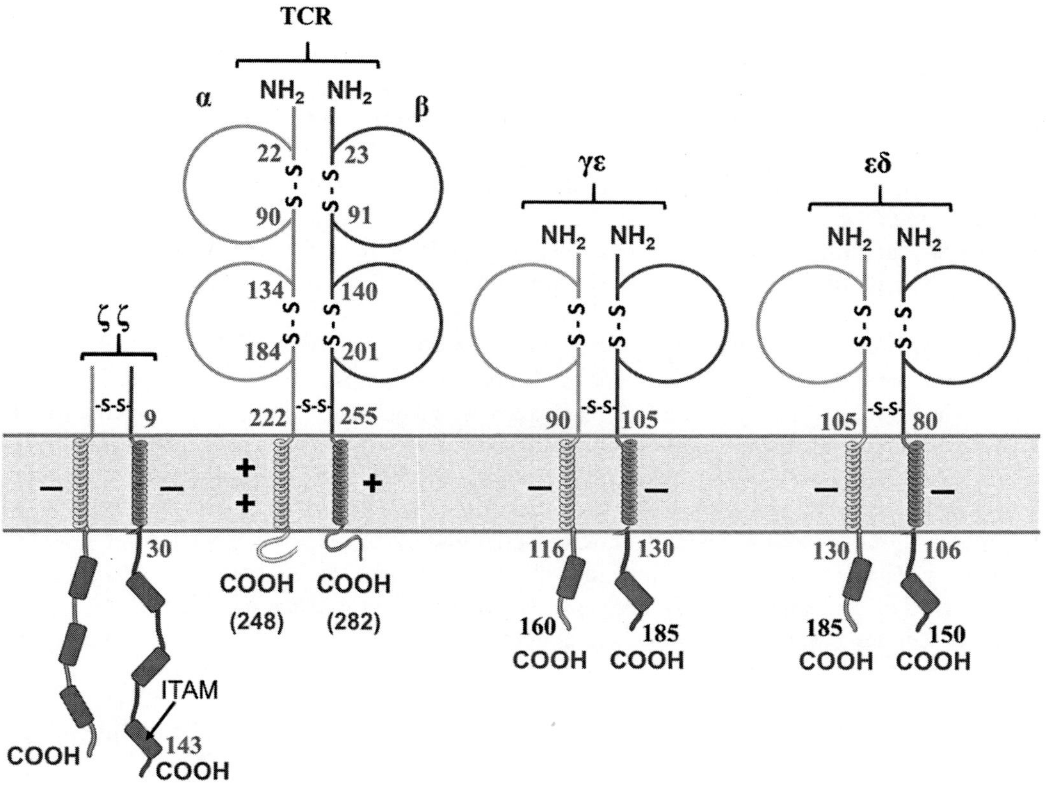

FIGURE 16.2 Diagram of TCR CD3 complex. (Author.)

16.2.2 T Cell Receptor Complex: TCR-CD3

T-cell receptor associates with CD3, forming the TCR-CD3 membrane complex (Figure 16.2). This complex is similar to the B-cell antigen receptor, in which the B cell membrane-bound immunoglobulins associate with the Ig-α/Ig-β heterodimer to form the B-cell antigen receptor. CD3 on T cells and Ig-α/Ig-β heterodimer on B cells function as an accessory molecule that helps in signal transduction after the interaction of T and B cells with the antigen, respectively. CD3 is a complex of three dimers composed of five invariant polypeptide chains:

i. a heterodimer of delta and epsilon chains (δε),
ii. a heterodimer of gamma and epsilon chains (γε),
iii. a homodimer of two zeta chains (ζζ)/a heterodimer of zeta and eta chains (ζη).

Therefore, the TCR complex has four dimers. The TCR heterodimer specific for the antigen and three dimers of CD3 is for membrane expression of the T-cell receptor followed by the signal transduction. **I**mmunoreceptor **T**yrosine-based **A**ctivation **M**otifs (ITAM) are present in the cytoplasmic tails of the CD3 chains, which are required for the signal transduction.

16.2.3 Self MHC Restriction of T Cell Receptor (TCR)

TCR is a membrane-bound receptor present on both types of T cells i.e. T_H and Tc cells. Unlike in B cells, TCRs are neither soluble nor specific for a particular antigen alone instead associate with antigen combined with self MHC (**Ma**jor **H**istocompatibility **C**omplex) molecules. TCR is a heterodimer composed of either αβ or γδ chains and this defines T cells specificity. T-cell receptor is associated with a signal-transducing complex, CD3, on the membrane as mentioned earlier. TCR is self-MHC restricted which has been proven experimentally by R. M. Zinkernagel and P. C. Doherty in 1974. They were awarded the Nobel Prize in 1996 for this work. Their studies demonstrated that:

i. T cells not only recognize specific antigens but also the antigen associated with an MHC molecule.
ii. T cells can only identify antigens presented by a self MHC molecule existing on the cell membrane. This attribute is called *self-MHC restriction.*

16.2.3.1 Doherty and Zinkernagel's Experiment to Show Self MHC Restriction of T Cells

H-2^k mice were immunized with the lymphocytic choriomeningitis (LCM) virus to produce cytotoxic T lymphocytes (CTLs) specific for this virus. Spleen cells containing T_C cells specific for the LCM virus were then isolated and kept under incubation with target cells of different H-2 haplotypes. Cells were radioactively labelled with ^{51}Cr and were infected either with the same or different virus. The number of target cells killed by cytotoxic T lymphocyte was measured by estimating the amount of ^{51}Cr in the culture supernatant. Radioactivity was observed only when the cells were of the same haplotype as the CTLs and infected with the same virus (Figure 16.3).

The results of the Doherty and Zinkernagel experiment can be explained in Figure 16.4.

The TCR in the given Figure 16.4 is specific for viral peptide A presented with H-2^k (MHC haplotype) molecule on the target cells.

A. An interaction of TCR with the viral peptide A presented with the self MHC molecule H-2^k resulted in the killing of the target cell.

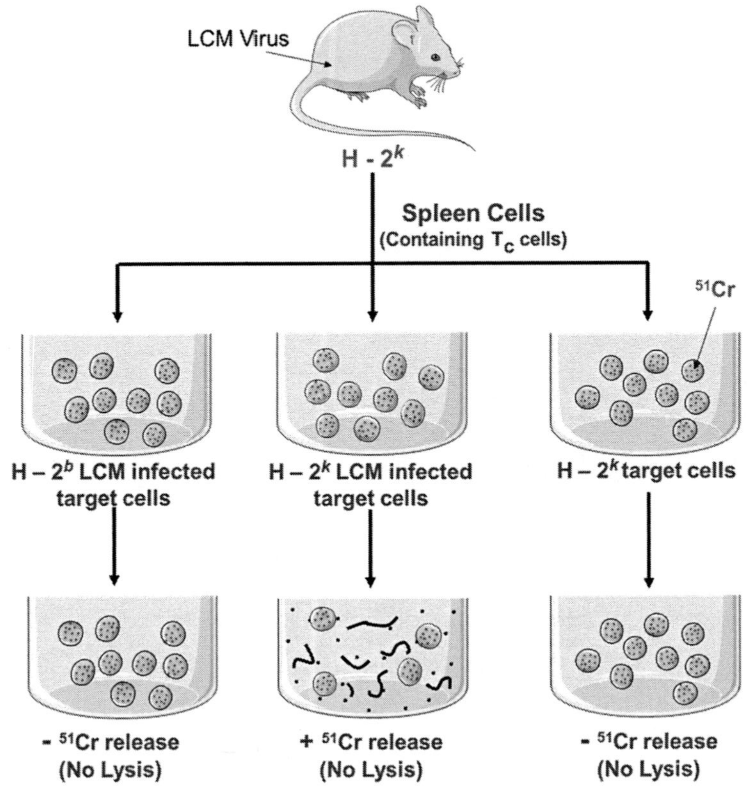

FIGURE 16.3 Doherty and Zinkernagel experiment. (Author.)

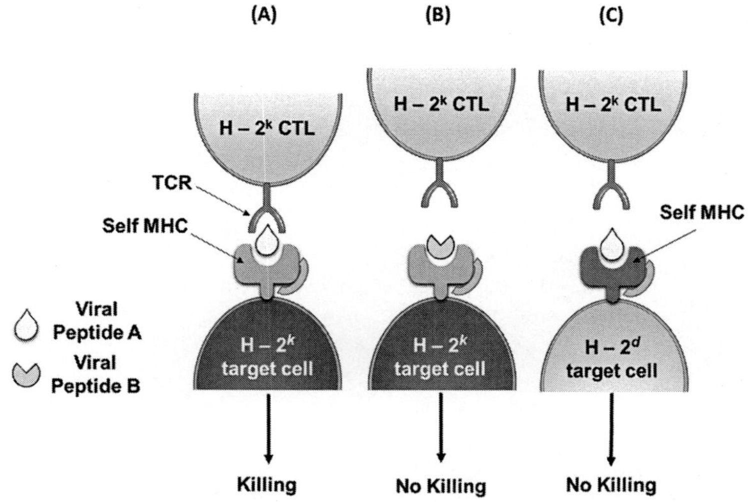

FIGURE 16.4 Self MHC restriction of T cell receptor (TCR). (Author.)

B. Since CTL had specificity only for peptide A, no antigen recognition occurred when the B antigen was displayed. Hence, the target cell was not killed.
C. Due to the display of a different haplotype of MHC, antigen recognition failed to occur resulting in no killing of the target cell.

The above experiment thus concludes that TCRs are not specific for either antigen or MHC alone but rather for both antigen and self MHC together.
It is known that TCRs are self-MHC restricted. The following two models were proposed by the scientists to explain the self-MHC restriction of the T-cell receptor.

i. **Dual-receptor model:** According to this model there are two separate receptors - one for interacting with the foreign antigen and another for the class I or class II self MHC.
ii. **Altered-self model:** This model proposed the presence of a single receptor. An alteration in the self MHC molecule because of binding of foreign antigen is recognized by the TCR. Altered-self model was accepted after the work of J. Kappler and P. Marrack.

16.3 T CELL ACCESSORY MEMBRANE MOLECULES

TCR-CD3 complex recognizes the antigen-MHC complexes while the other membrane molecules as mentioned in the table above play an important role either in antigen recognition or T-cell activation or both. Some of these accessory molecules help in the adhesion of T cells to the target cells or play an important role in the signal transduction or both (Table 16.2).

16.3.1 CD4 and CD8 Co-Receptors

T cells can be divided into two categories i.e. $CD4^+$ and $CD8^+$ T cells depending on the presence of either CD4 or CD8. $CD4^+$ T cells recognize the antigen presented with class II MHC molecules on antigen-presenting cells. $CD4^+$ T cells also designated as T helper cells (T_H). On the other hand, $CD8^+$ T cells, also called as Cytotoxic T cells recognize antigens presented with class I MHC molecules on nucleated cells.

CD4 and CD8 are classified as co-receptors of TCR as they are able to recognize the peptide-MHC complex in addition to their role in signal transduction and T cell activation.

TABLE 16.2
Accessory Molecules Present on T cells and Corresponding Ligands on Other Cells

T Cell Membrane Molecule	Ligand Present on Other Cells
CD4	Class II MHC on antigen presenting cells
CD8	Class I MHC on nucleated cells
CD2 (LFA-2)	CD58 (Lymphocyte Function Associated antigen-3 LFA 3) on antigen presenting cells mainly macrophages
LFA-1 (CD 11a/CD 18)	ICAM 1 (CD 54) Intercellular Adhesion Molecule 1 (endothelial cells or leucocytes)
CD28	B7 (B cells)
Cytotoxic T lymphocyte Antigen 4 (CTLA-4)	B7 (B cells)
CD45R	CD22 (Mature B cells)
CD5	CD72 (B cells)

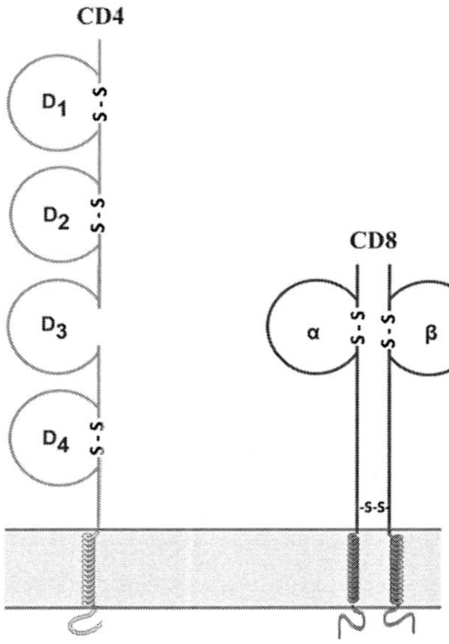

FIGURE 16.5 Structure of CD4 and CD8 co-receptors. (Author.)

Structurally, CD4 is a monomeric membrane glycoprotein with four extracellular domains (Dl–D4), similar to Ig domains. It also has a transmembrane region (hydrophobic in nature) associated with a long cytoplasmic tail as shown in Figure 16.5.

CD8 is a disulfide-linked dimer which is either a homodimer of αα chains or heterodimer of αβ chains. The α and β chains of CD8 are small glycoproteins bearing a single extracellular domain similar to that of the immunoglobulin domain. Both the chains are also associated with a transmembrane region and a cytoplasmic tail. The extracellular domain of CD8 interacts with the conserved region of MHC molecules by way of α2 and α3 domains of α chain by means of some contact with β_2 microglobulin.

On the other hand, CD4 interacts with the conserved region of Class II MHC molecules on antigen-presenting cells via the extracellular domain. CD4 binds to the β2 domain of Class II MHC molecule.

16.3.2 Role of Other Accessory Molecules

As compared to antigen–antibody interactions, the interaction of T cell receptor with peptide MHC complex is relatively weaker. Although multiple membrane molecules like CD2, LFA-1, CD28, and CD45R present on T cells interact with their corresponding ligands on the antigen-presenting cells

(APC), B cells or other target cells (Figure 16.6a and b); the expression of these adhesion molecules increase after activation. The interacting cells come closer and increase the efficiency of cytokines transfer and their functions. After activation, the expression of surface molecules declines followed by a detachment of B cell, antigen-presenting cell (APC) or target cell from T cells.

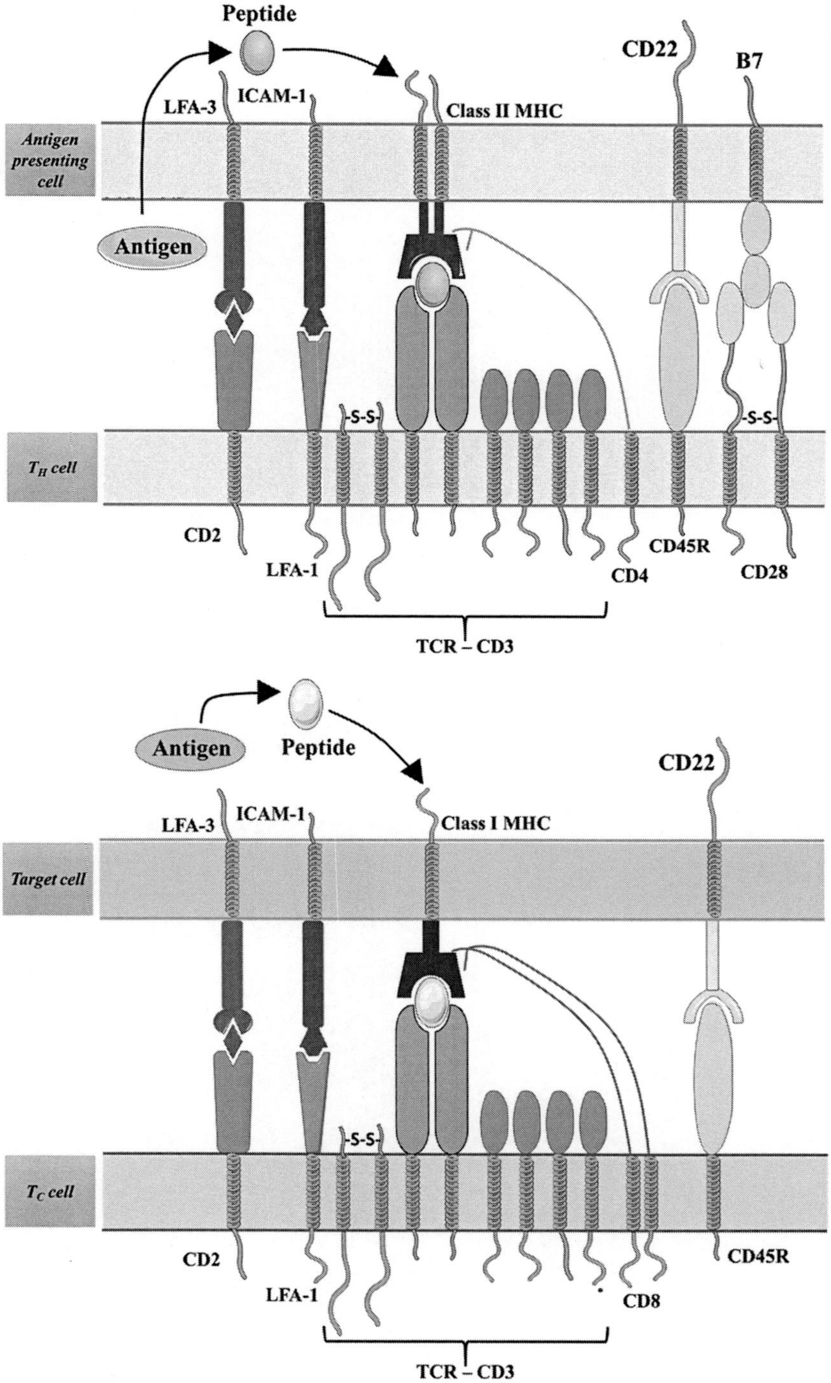

FIGURE 16.6 (a) Interaction of T_H cell with Antigen Presenting Cell (APC) via accessory surface molecules (Author.). (b) Interaction of Tc cell with target cell via accessory surface molecules (Author.)

16.4 T-CELL DEVELOPMENT

T cell development takes place in the thymus (Figure 16.7). After hematopoiesis, progenitor T cells migrate to the thymus for their maturation. T cell maturation includes germline TCR gene rearrangements and expression of markers on the cell surface. The developing T cells in the thymus, called as thymocytes proliferate and differentiate by way of different developmental pathways to generate functionally distinct sub-populations of mature T cells.

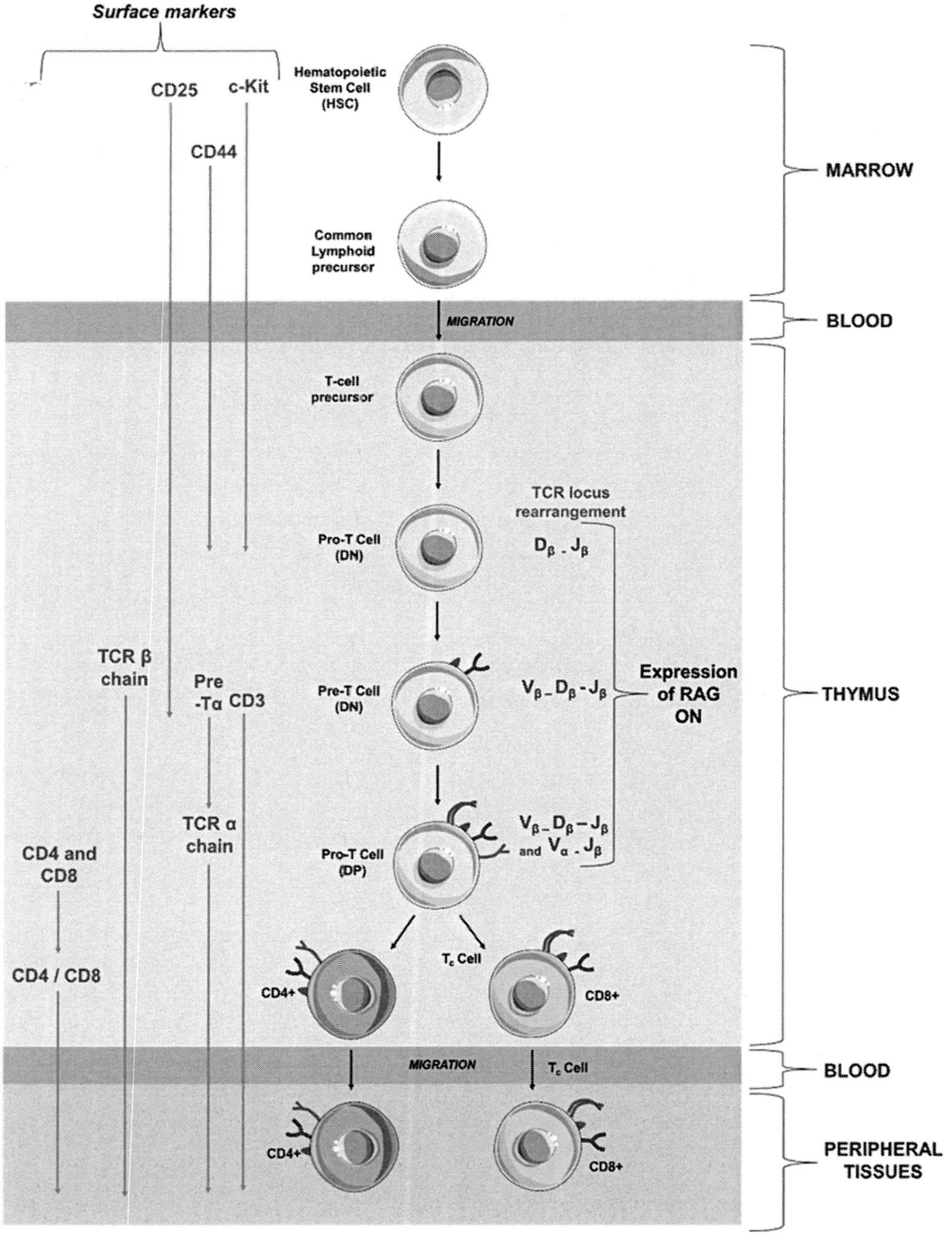

FIGURE 16.7 Development and differentiation of α and β T cells in the mouse: Through the bloodstream, T-cell precursors reach the thymus from the bone marrow and develop into mature T cells. These mature T cells exit to the periphery where they differentiate into effector and memory cells through antigen-induced activation. Stage-specific intracellular events and cell surface markers characterize each stage of the development. (Author.)

The selection of thymocytes takes place in the thymus. There are two types of selection processes:

i. **Positive selection:** T cells whose TCRs are able to recognize self-MHC molecules are selected leading to the formation of self-MHC-restricted repertoire of T cells.
ii. **Negative selection:** T cells which exhibit high affinity for self-MHC or with self-MHC along with self-peptides are removed leading to the formation of self-tolerant T cells.

T-cell precursors which arrive thymus lack T-cell receptors, CD3 complex, or the co-receptors CD4 and CD8 (specific surface markers of the T cell). In absence of RAG-1 and RAG-2 protein expression, TCR genes do not undergo any rearrangements. Thymocytes early in the developmental stage lack detectable CD4 and CD8, therefore, they are called as double negative (DN) cells ($CD4^-CD8^-$). The expression of surface markers including c-Kit, CD44, CD25, the β-chain of the IL-2 receptor, differ at different developmental stages of T cells. Non-expression of c-Kit and CD44 causes recombinase genes RAG-1 and RAG-2 to express resulting in subsequent rearrangements of TCR genes.

Most of the double-negative thymocytes ($CD4^-CD8^-$) progress down the αβ developmental pathway and a very few develop into γδ T cells. In response to the halt in cell proliferation, TCR β chain genes begin rearrangements and finally express the βchain. In case the β chains express even in the absence of rearrangement, the cells perish. The newly synthesized β chains associate with pre-T chain (glycoprotein of about 33-kDa) and the CD3 group to form a novel complex called the pre-T cell receptor or pre-TCR.

It has been suggested by some scientists that intra-thymic ligands are recognized by pre-TCR and a signal is transmitted through the CD3 complex which has several effects. It presents an indication about the productive TCR β-chain rearrangement and signals for the further proliferation and maturation of T cells (Figure 16.8). Rearrangement of TCR β chain genes is further suppressed (allelic exclusion) thereby, signaling the rearrangement of TCR α chain genes. Consequently, cells are progressed to the $CD4^+CD8^+$ double-positive state.

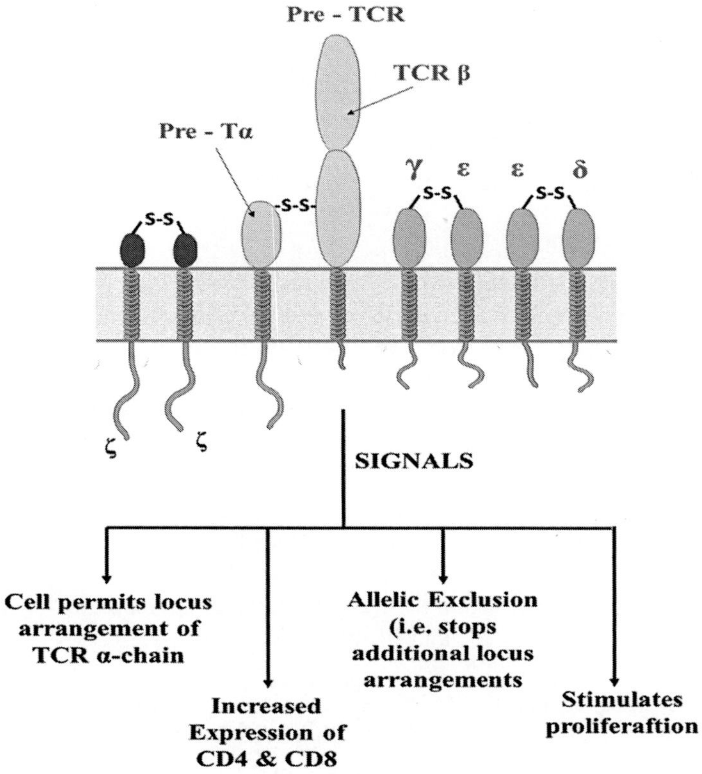

FIGURE 16.8 Pre-T cell receptor structure and function. (Author.)

After the expression of both CD4 and CD8 co-receptors, the thymocytes begin to proliferate. αTCR gene shows no rearrangements during this proliferative phase. Although, *RAG-1* and *RAG-2* genes are transcriptionally active in proliferating cells, but the RAG-2 protein is degraded rapidly. Therefore, rearrangement of the α-chain genes fail to occur. The α-chain genes undergo rearrangement only when the levels of RAG-2 protein increase and the double-positive thymocytes fail to proliferate. The diversity of the T cell repertoire increases first by the TCR β-chain rearrangement and later on by β-chain rearrangement.

During T cell development, the majority of the thymocytes do not mature and suffer death by apoptosis as they are unable to complete TCR rearrangement or could not survive thymic selection. Thymocytes which are double positive with the expression of TCR-CD3 complex and sustain thymic selection either develop into immature single positive CD8 or CD4 thymocytes. These cells move from cortex to medulla after facing one more negative selection. Subsequently, the cells enter the circulatory system.

16.4.1 Selection of T Cells in Thymus

A diverse population of T cells arise because of gene rearrangements that occur randomly within the TCR germ-line DNA along with junctional diversity. TCR genes are rearranged in such a way that their products do not exhibit any inherent affinity for foreign antigens presented with a self-MHC molecule. Cells then undergo two selection processes i.e. positive and negative selection in the thymus as mentioned earlier. This leads to the establishment of mature T cells which are self-MHC restricted and have self-tolerance.

16.4.1.1 Positive Selection

Positive selection takes place in the cortex of the thymus. During positive selection, immature thymocytes interact with cortical epithelial cells. According to some researchers, the interaction of T cell receptors on thymocytes with the MHC molecules on the cortical cells sends a protective signal to the immature thymocytes and prevents their death by apoptosis. Consequently, the immature T cells which can interact with thymic epithelial cells do not undergo cell death (Figure 16.9).

During positive selection, RAG-1, RAG-2 and TdT (Terminal deoxynucleotidyl transferase) proteins are expressed. So, immature T cells are capable of rearranging different TCR α-chain genes. These cells are then selected for self-MHC recognition. Therefore, only the T cells whose αβ TCR heterodimer is capable of recognizing a self-MHC molecule survive while the rest of the cells die.

16.4.1.2 Negative Selection

Positive selection results in two kinds of MHC-restricted thymocytes (i) low-affinity receptors and (ii) high-affinity receptors for self-antigen presented by self-MHC molecules. Negative selection of thymocytes takes place through interaction with thymic stromal cells. During negative selection, thymocytes interact with class I and class II MHC molecules of the dendritic cells and macrophages. Thymocytes showing high-affinity receptors either for self-antigen presented with self-MHC molecules or for self-MHC molecules alone are negatively selected and removed by apoptosis. Therefore, T cells that are reactive to self-antigens are eliminated and tolerance to self-antigens is achieved in the thymus after negative selection (Figure 16.9).

Double-positive $CD4^+CD8^+$ thymocytes either become $CD4^-8^+$ T cells (class I MHC–restricted $CD8^+$ T cells) or $CD4^+8^-$ T cells (Class II MHC-restricted $CD4^+$ T cells). Two different models have been proposed independently that explain the transformation of double positive $CD4^+CD8^+$ T cells into two different types i.e. $CD8^+$T cells and $CD4^+$T cells.

i. According to *instructive model*, if MHC I interacts with CD8, there occurs a down-regulation of CD4 and the cells end up becoming $CD8^+$T cells. Similarly, if MHC II interacts with CD4, a low expression of CD8 results in the cells to become $CD4^+$T cells (Figure 16.10).
ii. As per the *Stochastic model,* CD4 or CD8 expression is switched off at random without any correlation with TCR specificity (Figure 16.10).

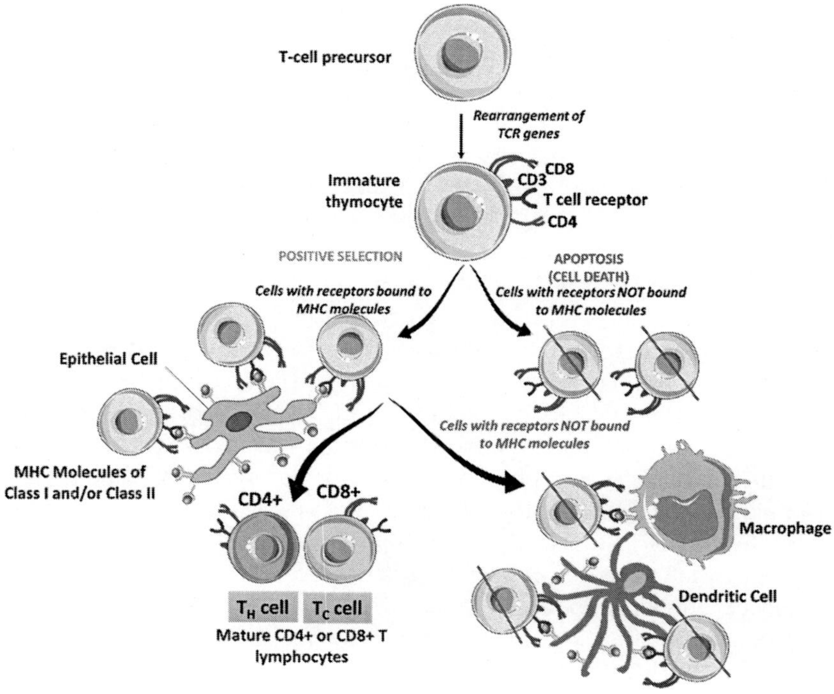

FIGURE 16.9 Selection of thymocytes in the thymus: Positive and Negative. (Author.)

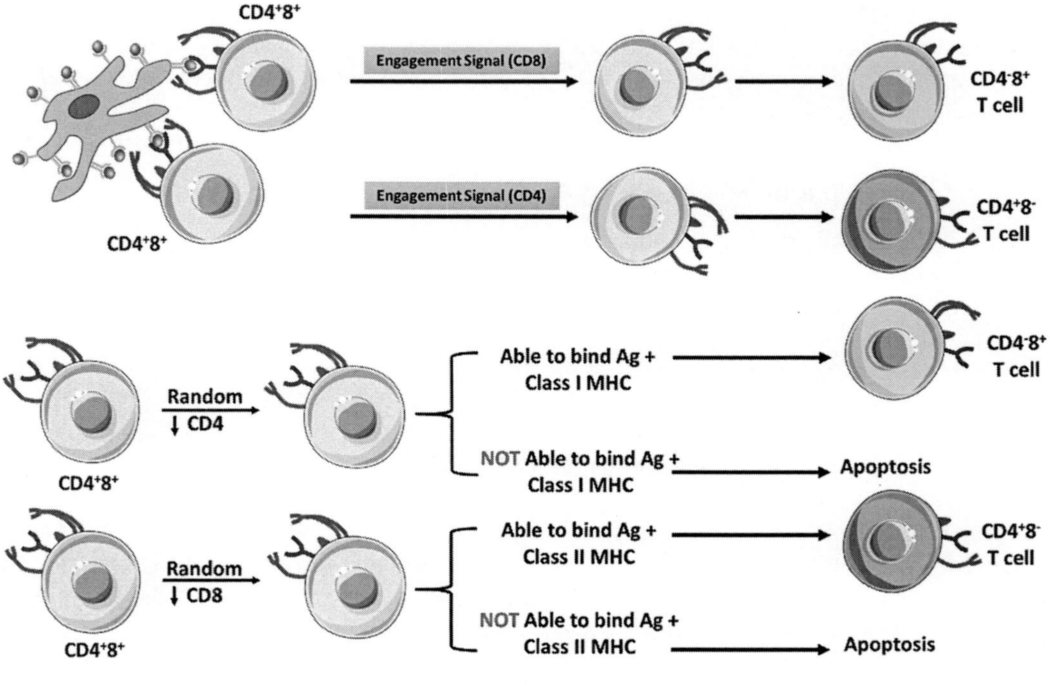

FIGURE 16.10 Proposed models for explaining the transformation of double positive $CD4^+CD8^+$ T cells into $CD8^+$ and $CD4^+$T cells (a) Instructive model (b) Stochastic model. (Author.)

16.5 T_H CELL ACTIVATION

Both humoral and cell-mediated immune responses include a key step of the activation of T_H cells followed by their clonal expansion.

Activation of T_H cells started after the interaction of the TCR-CD3 complex with the antigen presented on Class II molecules of antigen-presenting cells. For the activation of T_H cells, the interaction of various accessory membrane molecules which are present on the surfaces of T_H and antigen-presenting cells are required. After interaction with antigen, T_H cells enter into cell cycle and further proliferate and differentiate into effector and memory cells.

The signal mediated by the T cell receptor elicits an activation of several signal transduction pathways. These include pathways that generate transcription factors such as NF-AT (nuclear factor of activated T cells), NF-κB and activated MAPK (mitogen-activated protein kinase). For instance, phosphorylation of tyrosine activates ZAP-70 which in turn creates PLC gamma 1 (Phospholipase C gamma1). A cascade of events proceeds when activated PLC gamma 1 hydrolyzes PIP_2 (membrane phospholipid, Phosphatidylinositol 4,5-bisphosphate) to create second messengers including IP_3 (Inositol 1,4,5-triphosphate) and DAG (Diacylglycerol). The second messengers in collaboration with calcium ions stimulate PKC (Protein kinase C) which phosphorylates IκB (inhibitor NF-kappa B) thereby releasing NF-κB. Calmodulin also interacts with the calcium ions to evoke a phosphatase, calcineurin which dephosphorylates the cytoplasmic NF-AT. Upon dephosphorylation, the NF-AT move to the nucleus to trigger the transcription of numerous genes. ZAP-70 is also involved in the activation of Ras (a small G protein) pathway which in turn stimulates the MAP kinase pathway. A cascade of phosphorylations result in MAPK activation which then migrates to the nucleus to induce a transcription factor. As a consequence, gene encoding Fos is transcribed. The resulting product Fos interacts with Jun to activate yet another transcription factor, AP-1 (Figure 16.11).

CD28, a co-stimulatory molecule also engages with B7 to provoke signal transduction pathways such as MAP kinase. While Rac (a small G protein) initiates this cascade, the termination is marked by phosphorylation of Jnk (Junk or Jun kinase). This activated Jnk further phosphorylates and activates Jun by translocating into the nucleus. Hereafter, Jun collaborates with Fos in AP-1. The end result of all the transcription factors activated by the signal transduction pathways is to collectively increase multiple gene expressions which in turn induces expression of many more genes including the gene encoding T cell growth promoting IL-2 cytokine.

16.5.1 Co-Stimulatory Signals for T Cell Activation

Interactions of multiple surface molecules are required for T cell activation. Naïve T cells which have not interacted with antigens need two different signals to proceed on the path of activation and proliferation to finally mature into effector cells.

Signal 1: Arise because of interaction between MHC presented antigenic peptide and TCR-CD3 complex.

Signal 2: Co-stimulatory signals because of interaction between CD28 on the T cell and B7 family members on the antigen-presenting cells (APC).

There are two B7 family members namely, B7-1 and B7-2. B7 family members belong to the immunoglobulin superfamily in which the organization of extracellular domains is the same, whereas cytosolic domains vary. B7-1 and B7-2 are expressed on activated macrophages and B cells but constitutively expressed on dendritic cells. B7 molecules interact with the T cell membrane molecules CD28 and CTLA-4 (also known as CD152). CD28 and CTLA-4 are disulfide-linked homodimers. CD28 is expressed on both resting and activated T cells whereas CTLA-4 is expressed on activated T cells. The structures of CD28 and CTLA are similar but their functions are opposite. While CD28 gives a positive signal, the CTLA-4 signal is inhibitory for the T cell activation. CTLA-4 plays an important role in the regulation of T lymphocyte activation and proliferation. Expression of CTLA-4 increases after interaction of CD28 and B7 family members. CTLA-4 competes with CD28 for B7 binding site and inhibits the co-stimulation of CD28 (Figure 16.12).

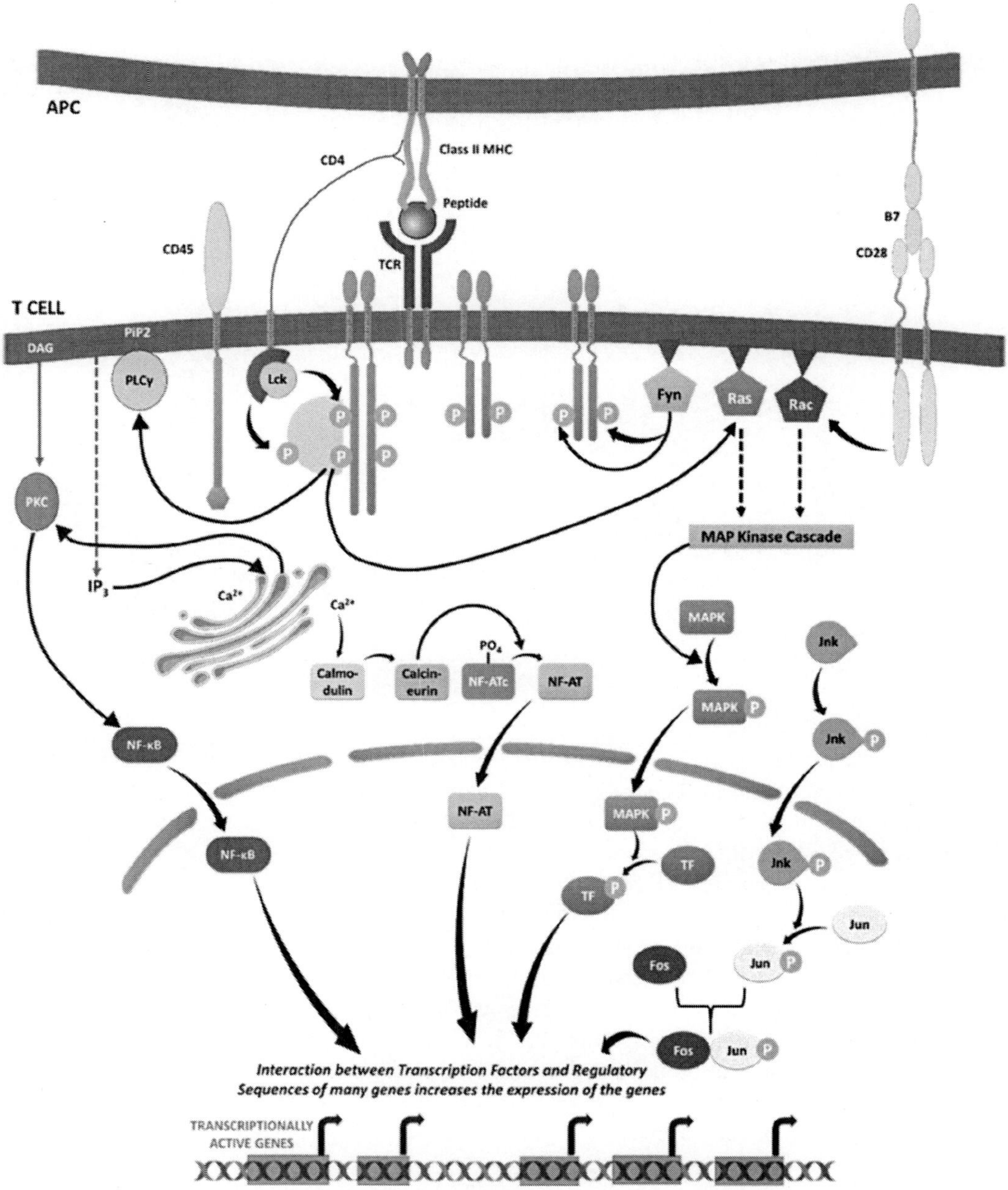

FIGURE 16.11 Signal transduction during T helper cell activation. (Author.)

In absence of any co-stimulatory signal when T cells receive only signal 1 by the interaction of TCR CD3 complex with peptides presented on MHC molecules, T cells enter into a state of non-responsiveness called clonal anergy. Cells are unable to proliferate in response to just one signal and no co-stimulatory signal.

16.6 T CELL DIFFERENTIATION

The two types of T cells i.e. $CD4^+$ and $CD8^+$ T cells are in the G_0 Phase of the cell cycle when they leave the thymus and enter the blood. Naïve T cells have condensed chromatin, very little cytoplasm and very less transcriptional activity. If the Naïve T cells don't encounter antigen in the secondary

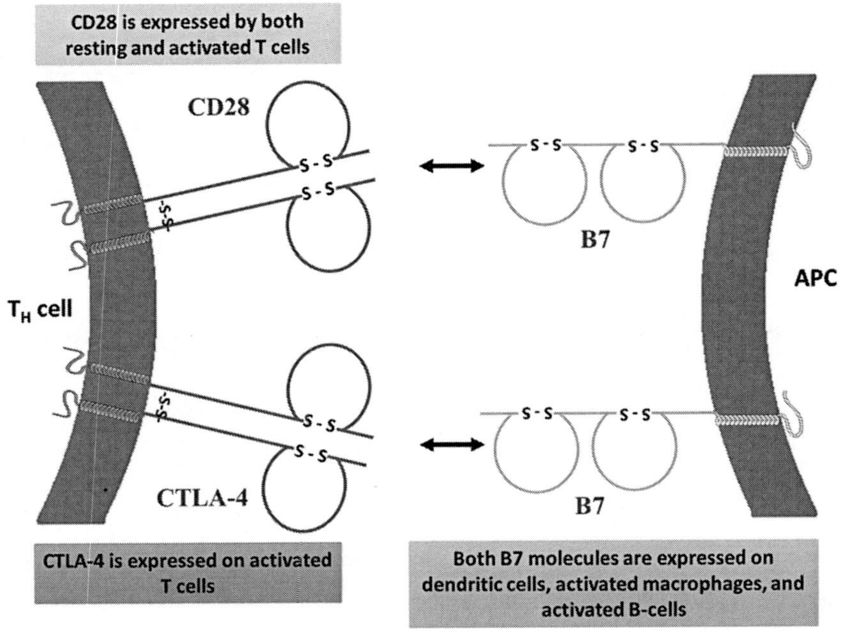

FIGURE 16.12 Co-stimulatory signals due to interaction of CD28 and CTLA-4 with B7 family members. (Author.)

lymphoid organs, they come back to the blood via efferent lymphatics. Every 12–24 hours, the naïve T cells circulate from the blood to the lymph nodes and back to the blood again. This recirculation of T cells is important for their interaction with appropriate antigens as about 1 in 10^5 naive T cells is specific for any given antigen.

If naïve T cells get an opportunity to interact with self MHC-presented foreign peptide, they get activated. It gets two signals - one from the interaction of the TCR-CD3 complex with peptide and other by co-stimulatory signals. In response to these two signals, the T cells enter the G1 phase of the cell cycle. In the activated T cells, the expression of IL-2 increases which then binds to the IL-2 receptor and culminates in the proliferation and differentiation of naïve T cells. Activated T cells give rise to multiple clones of progeny cells which further differentiate into memory or effector T cells (Figure 16.13).

There are two types of effector T cells:

i. Activated $CD4^+$ T_H cells whose main function is secretion of cytokines and B cell activation. These cells form two sub-populations based on cytokines secretion. T_H1 subset secretes cytokines like Interferon γ, Tumor Necrosis factor β and IL-2. This class play an important role in the delayed type of hypersensitivity (Hypersensitivity Type 4). T_H1 or T_H2 subset plays an important role in B cell activation.
ii. $CD8^+$ Cytotoxic T lymphocytes exhibit cytotoxic activity and produce few cytokines. They play an important role in the elimination of self-altered cells i.e. cancerous cells and virus-infected cells.

After antigen activation both naïve and memory cells differentiate into effector cells which have a life span of a few days to a few weeks. On the other hand, naïve and effector cells both give rise to memory cells in response to antigen interaction. Since these cells are long-lived, a heightened response is generated in case of subsequent exposure to the same antigen.

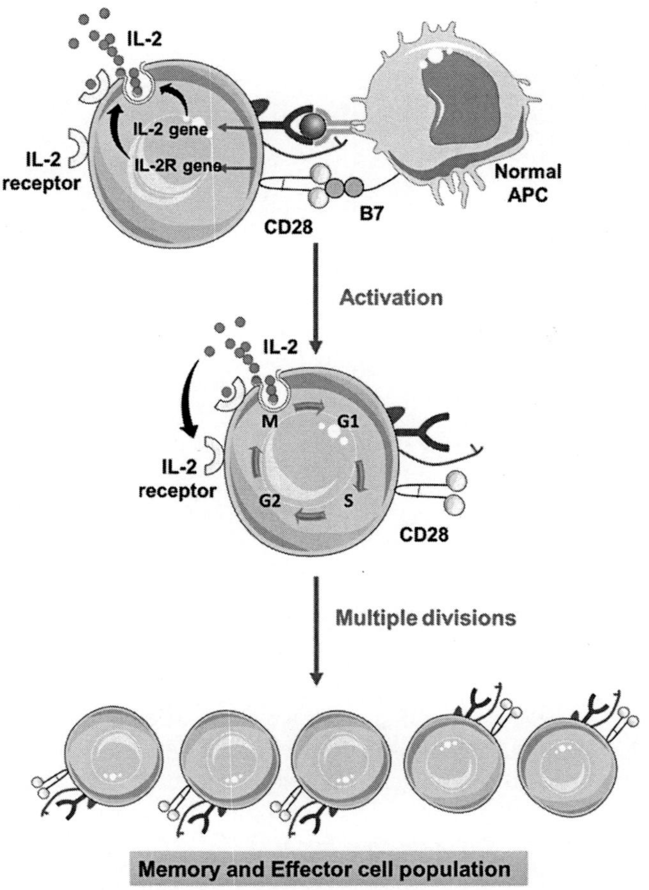

FIGURE 16.13 Up-regulated expression of IL-2 and IL-2 receptor leading to T_H cell activation causing multiple rounds of division of T_H cell followed by differentiation into memory and effector cells. (Author.)

16.7 SUMMARY

- Progenitor T cells are formed through hematopoiesis in the bone marrow following which further development occurs in the thymus where rearrangement of TCR genes take place.
- Majority of the T cells have αβ chains and very less number has γδ chains. The thymocytes in the early development are double negative ($CD4^-CD8^-$) and during development form either $CD4^+CD8^-$ cells or $CD4^-CD8^+$ T cells.
- Within thymus, the developing T cells are subjected to negative selection that removes the self-reactive T cells and also positive selection which eliminates those that cannot recognize self MHC molecules.
- T cells are activated by interaction of TCRCD3 complex with MHC bound antigen. Additionally, an interaction of accessory molecules is also required.
- $CD8^+$ T cells recognize the antigen bound to Class I MHC molecules on nucleated cells and $CD4^+$ cells interact with Class II MHC bound molecules on antigen presenting cells. On the other hand γδ T cells which play an important role in innate immunity are not MHC restricted.
- T cells are arrested in Go phase of the cell cycle before antigen interaction and thereafter proceed further in the cell cycle to differentiate into effector T cells.

QUESTIONS

1. Define the following:
   a. Dendritic cells
   b. Calcineurin
   c. TH cells
   d. TC cells
   e. Allogenic
2. Differentiate between the following:
   a. αβ and γδ T cells.
   b. CD4+ and CD 8+ T cells
   c. Instructive model and Stochastic model
3. Expand the following:
   a. CTLA 4
   b. TCR
   c. MAPK
   d. RAG
   e. MHC
   f. ITAM
   g. APC
   h. NF-AT
4. Fill in the blanks:
   a. T cells which are reactive to self-antigens are eliminated by ……………………
   b. Positive selection of the T cells takes place in the……………………….of the thymus.
   c. B7 family members which are required for T cell activation are present on …………………
   d. TCR of majority of T-cells is encoded by the …………………… genes.
   e. CD3 is a complex of ………………… dimers.
   f. Knockout mice which has no class I MHC molecules do not produce thymocytes having ……………………
5. Long answer questions:
   a. Explain the selection process during T cell development.
   b. Describe self MHC restriction of T cells with the help of an experiment.
   c. Give an account of T cell accessory membrane molecules
   d. Explain the sequence of events in TH cell activation.
   e. What are the components of T cell receptor complex?

BIBLIOGRAPHY

Abul, A. A. and Andrew, L. H. 2003. *Cellular and Molecular Immunology.* V Edition, Saunders Publication, Philadelphia, Pennsylvania, United States.

David, M., Jonathan, B., David, R. B. and Ivan, R. 2006. *Immunology*, VII Edition, Mosby, Elsevier Publication, Philadelphia, PA, United States.

Kindt, T. J., Goldsby, R. A., Osborne, B. A. and Kuby, J. 2006. *Immunology*, VI Edition, W.H. Freeman and Company, New York, United States.

Murphy, K. and Weaver, C. 2017. *Janeway's Immunobiology*, IX Edition, Garland Science.

17 Cell-Mediated Cytotoxic Responses

Charu Tripathi
CMP College, University of Allahabad

CONTENTS

17.1 INTRODUCTION

Immune responses belong to two categories, depending on the type of response generated and cells involved, namely, humoral and cell-mediated branches of the immune system. The branch that involves the action of B cells or B lymphocytes (which mature in bone marrow) which harbor membrane-bound antibodies or secrete soluble antibodies in the serum is known as humoral immune system. Whether the antibodies are membrane-bound or secreted, humoral response can cater only to extracellular antigens. Antibodies fail to recognize pathogens residing within cells (intracellular pathogens). The cell-mediated immunity fulfils the function of neutralizing intracellular pathogens. Cell-mediated immunity is responsible for eliminating intracellular pathogens and works by the action of T cells or T lymphocytes. T cells mature in the thymus and harbor T-cell receptors (TCRs) as surface molecules. T cell receptors, however, cannot recognize sole antigens, but recognize antigens only when they are processed and then presented by antigen-presenting cells (APCs) or target cells. APCs present antigens in conjugation with MHC class II molecules and target cells present antigens on MHC class I molecules. The cell-mediated branch thus eliminates intracellular pathogens which are processed and presented on APCs or target cells, viruses and manipulated self-cells like tumor cells. Interestingly, both the branches of the immune system are so intricately and extensively interdependent on each other that the full potential of one cannot be realized in the absence of the other. For example, the cell-mediated immunity destroys and eliminates the antigen-antibody complexes formed after the activity of the humoral branch. Also, naïve T cells get activated on facing appropriately presented antigen, leading to the formation of T_H (T helper) and T_C (T cytotoxic)

DOI: 10.1201/9781003286424-17

cells. T_H cells start cytokine production, activating both T_C cells as well as B cells which then become plasma cells and start secreting increased amounts of antibodies.

The cell-mediated response encompasses not only antigen-specific cells such as $CD8^+$ cytotoxic T lymphocytes (T_C cells or CTLs) and $CD4^+$ T_H cells but also non-specific cells, namely NK (natural killer) cells, macrophages, eosinophils, neutrophils and NKT cells. NKT cells exhibit properties of both T cells and NK cells. Antibody-dependent cell-mediated cytotoxicity (ADCC), as evident from the nomenclature, comprises the formation of antigen-antibody complexes which trigger cell-mediated responses, after which the complement system is recruited, which in turn releases chemotactic peptides that signal the cell-mediated components to come into action. Cell-mediated cytotoxic responses thus refer to the immune responses which involve both direct killing of cells by cells of the immune system along with indirect action involving the non-specific cells and complement system activation. The cells involved in these responses recognize host cells infected with pathogens and altered host cells, including tumor cells. T cells, macrophages, dendritic cells and NK cells secrete cytokines. Cytokines are very important immune signaling molecules, playing a crucial part in the induction of cell-mediated effects through the effector cells. In children with DiGeorge syndrome, the thymus is not present since birth and cell-mediated immunity is lacking. While they can fight extracellular bacterial pathogens, they cannot fight intracellular pathogens, viruses and fungi. Even mild infections such as attenuated viruses present in vaccines can prove fatal for them.

17.2 EFFECTOR T CELLS

17.2.1 Properties

Effector T cells include $CD4^+$ T_H cells (T_H1 and T_H2 cells) and $CD8^+$ cytotoxic T cells. T_C cells mount direct cytotoxic activity leading to cell killing when activated and the T_H cells are involved in delayed-type hypersensitivity (DTH) reactions. Briefly, the effector cells have the following three properties:

1. The effector T cells are different from naïve T cells because they have been activated. The need for a secondary co-stimulatory signal is not present in the case of effector T cells as required in the case of naïve T cells. The primary signal for the activation of naïve T cells is the interaction of the T cell receptor along with CD8 or CD4 with the MHC-peptide complex. However, apart from this, a co-stimulatory signal is also required for the activation of naïve T cells. Once activated, effector T cells are formed. Effector T cells' interaction with the MHC-peptide complex does not require secondary molecules. A costimulatory molecule B7 is present on APCs and target cells which is very important for in naïve T cell activation. However, even cells lacking the B7 molecule can relay activation signals to effector T cells, but the same is not true for naïve T cells.
2. Effector T cells express an increased amount of cell adhesion molecules like CD2 and the integrin LFA-1 on their surface.
3. They produce effector molecules which are responsible for downstream signaling and action. These effector molecules are both soluble and membrane-bound. They include cytotoxins (perforins and granyzymes), cytokines, IFN-γ, TNF-β, IL-2, IL-3, and GM-CSF among many others.

Apart from these three important properties of effector T cells, there are also variations in surface molecules between naïve T cells and effector T cells. Two different isoforms of CD45 are present on their surfaces, namely, CD45RA on the surface of naïve T cells and CD45RO present on the surface of effector T cells.

TABLE 17.1
Functions Performed by the Effector Molecules Secreted by Effector T Cells

Cell Type	Effector Molecule	Function
CTL	FASL, perforins, granzymes	Target cell destruction
T_H1 cells	Membrane-bound: TNF-β Soluble: IFN-γ, GM-CSF	Macrophage activation
T_H2 cells	Membrane-bound: CD40L Soluble: IL-4, IL-5, IL-6	B cell activation

17.2.2 Role of Cell Adhesion Molecules in TCR Interactions

Cell adhesion molecules (CAMs) CD-2 and LFA-1 are expressed on the surface of T cells. Cell adhesion molecules are responsible for the binding of T cells with APCs and target cells. The target cells and APCs express LFA-3 and intercellular cell adhesion molecules (ICAMs) on their surface. These bind to the TCR of the effector cell. This binding of LFA-3/ICAMs with TCR occurs when the TCR detects a peptide-MHC complex on the APC. Detection of such a complex increases the affinity of LFA-3 for CD2 on T cells and ICAMs for LFA-1 on T cells. LFA-1 and CD2 receptors are present two to four times more on effector T cells as compared to naïve T cells. It is for this reason that effector T cells can even interact with APCs and target cells expressing low amounts of ICAMs and LFA-3.

17.2.3 Effector Molecules Produced by T Cells

Effector T cells produce effector molecules which perform crucial functions like target cell destruction, macrophage activation and B cell activation (Table 17.1). These effector molecules may be membrane-bound or soluble. The membrane-bound molecules include molecules belonging to the tumor necrosis factor (TNF) family, namely, FASL (Fas ligand) on $CD8^+$ CTLs, TNF-β on T_H1 cells and CD40L (CD40 ligand) on T_H2 cells. Soluble or secreted molecules include cytotoxins (perforins and granzymes) and cytokines (IFN-γ and TNF-β) secreted by cytotoxic T lymphocytes. The functions mediated by these molecules are summarized in Table 17.1.

17.3 CYTOTOXIC T LYMPHOCYTES (CTLs)

17.3.1 Generation of Effector CTLs

Naïve T cytotoxic cells (T_C cells) have to be activated to give rise to cytotoxic T lymphocytes. After activation, the CTLs bring about the destruction of infected and altered cells. CTLs are generally $CD8^+$ which are restricted by class I MHCs. As MHC class I molecules are present on all nucleated cells of the body, therefore virtually all infected cells of the body can be lysed once encountering effector CTLs. There are two phases through which naïve T_C cells get activated. In the first phase, naïve T_C cells are activated. Naïve T_C cells are also known as CTL precursors (CTL-Ps). After activation, the naïve T_C cells become activated and cytotoxic T lymphocytes are formed (Figure 17.1). Now, these CTLs go through the second part of development. In this, the activated CTLs on being engaged with peptide-MHC I complex become effector CTLs which bring about the destruction of infected cells.

The activation of CTL-Ps is a complex and regulated process requiring the action of three sequential signals as mentioned below:

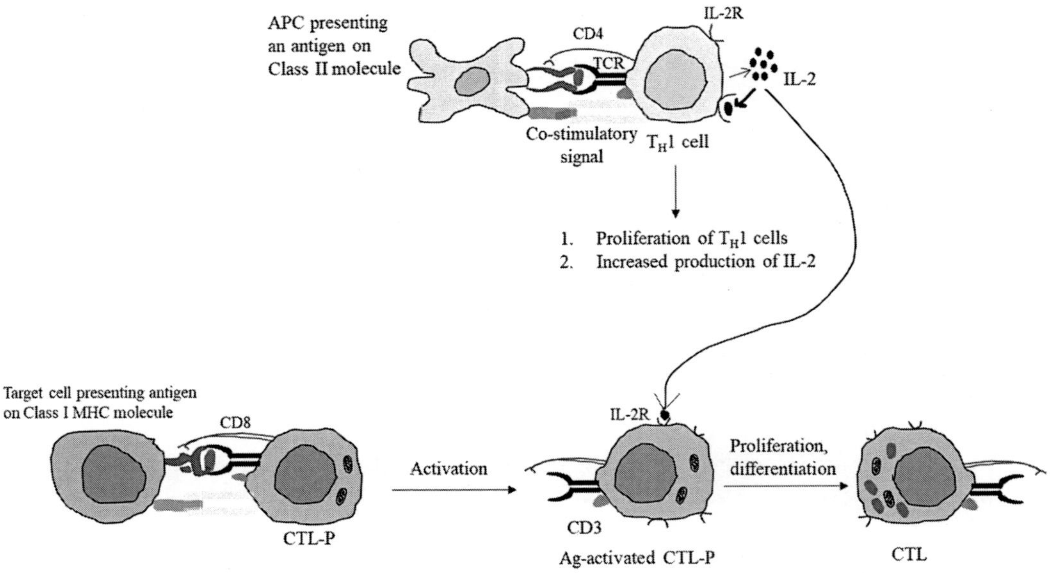

FIGURE 17.1 Schematic diagram demonstrating the activation of CTL-Ps leading to the formation of effector CTLs. CTL-P activation also requires the activation of T_H cells and subsequent production of IL-2.

1. The first signal is the interaction of peptide-MHC class I complex presented by a "licensed" APC with the CTL-Ps.
2. The second signal is the co-stimulatory signal relayed when CD28 present on T_C cell interacts with B7 present on antigen-presenting cells.
3. The third signal is the association of IL-2 and IL-2 receptor present on the CTLs, which initiates the killing action of the CTL.

Licensed APCs are generally dendritic cells which have acquired the property to activate CTL-Ps. In the process of licensing of APCs, a dendritic cell expressing the peptide in conjugation with an MHC class II molecule interacts with a $CD4^+$ T_H1 cell. After this, a co-stimulatory signal is needed for the completion of licensing of the APC. This involves an interaction between CD40 present on dendritic cells and CD40 ligand (CD40L) present on T_H1 cell. It must be noted here, that dendritic cells are unique in that they can process the antigen via both exogenous and endogenous pathways and thus present the antigens on their surface via MHC class I and class II molecules, hence fitting suitably for the process of activation of CTL-Ps. The process of licensing can also occur via the expression of toll-like receptors (TLRs) on infected APCs. Once the APCs are licensed, they can interact with CTL-Ps via class I MHC-peptide complex. Whether the T_H1 cell is still engaged with the APC during the process of CTL activation is not very clear. This whole process of activation of CTLs is meant to safeguard unnecessary activation of self-reactive T_C cells. Thus, CTL activation first requires the APCs to become licensed and then the APCs to engage with naïve T_C cells to activate them.

Activated CTLs differ from CTL-Ps in many respects. CTL-Ps are non-proliferative, non-cytotoxic cells which do not produce the cytokine IL-2. In addition, they do not have IL-2 receptors on their surface. After activation, CTLs begin expressing IL-2 receptors as their surface molecules, along with secreting low amounts of IL-2. High amount of IL-2 is secreted by T_H1 cells that are engaged with APCs. IL-2 binds to its receptor on the activated CTLs. This induces the proliferation of CTLs, after which activated CTLs differentiate to form effector CTLs. Similar to this process of activation of CTLs, memory CTL-Ps also undergo activation, but their activation requirements are not as complex as those of CTLs.

17.3.2 CTL Mediated Response

The killing process of target cells by effector CTLs begins when effector CTL engages with peptides that are presented by MHC class I by a target cell. Briefly, the following steps are involved (Figure 17.2):

1. A conjugate is formed between cytotoxic T lymphocyte and target cell.
2. Degranulation of CTL occurs following which the membrane of the target cell is attacked.
3. CTL dissociates from the target cell.
4. Target cell is destroyed.

The formation of a conjugate between the target cell and CTL is a calcium (Ca^{2+}) requiring, energy-dependent process, which not only involves the primary interaction between peptide-MHC I on the target cell and TCR-CD3 on CTL membrane, but also a secondary interaction between the integrin LFA-1 on CTL membrane and ICAM present the target cell. It must be noted here that after the process of CTL activation, LFA-1 gains a high-affinity state wherein it readily binds ICAM harbored by the target cell and leads to conjugate formation. When the conjugate is formed, the membrane of the target cell is attacked by two proteins, namely perforin and granzymes (fragmentins). Electron dense vesicles have been observed by electron microscopy near the membrane of CTLs where conjugate formation has occurred. Extraction and purification of these granules have led to the identification of their contents as monomers of a pore-forming protein perforin (molecular weight 65 kDa) and a serine protease fragmentin or granzyme. The cytoplasm of CTL near conjugate junction can be visualized to contain a large number of Golgi stacks and vesicles which move to the membrane and deliver their contents outside by exocytosis. When perforin monomers come in contact with the membrane of target cell, there occurs a rearrangement in their conformation. The monomers acquire a domain that is amphipathic in nature, thus it inserts into the membrane of the target cell. Many such monomers of perforin get lodged, forming tunnels passing through the target cell membrane. The pores formed have diameters of ~5–20 nm. Through these pores, granzyme gains entry into the target cells. Another mode by which granzymes enter target cells is by attaching to mannose-6-phosphate receptors present on the target cell. The receptors are internalized after the binding of granzyme B to the mannose-6-phosphate receptor. Thus, granzymes enter target

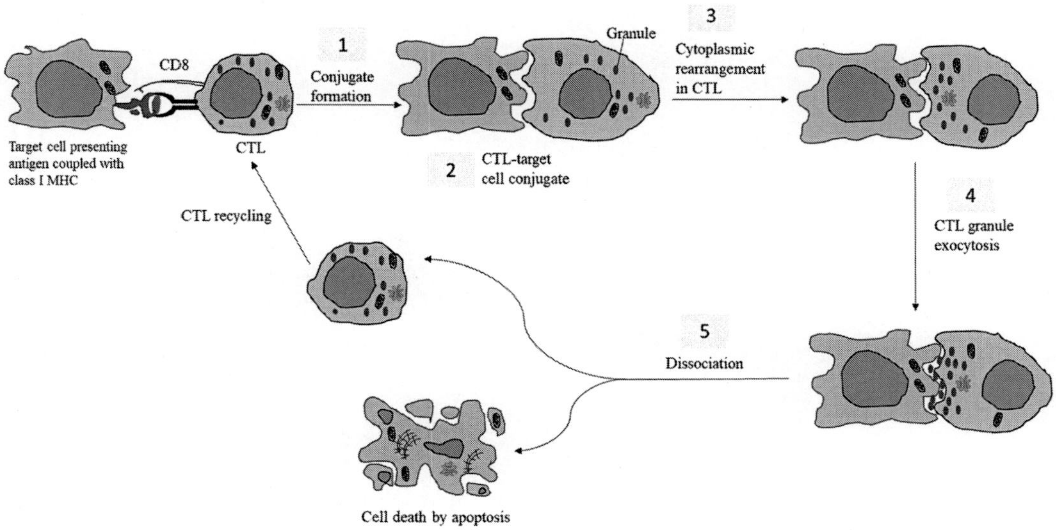

FIGURE 17.2 Step-wise depiction of CTL mediated lysis of target cells by conjugate formation, movement of granules near target cell membrane, degranulation and lysis of target cell.

cells along with the vesicles. Perforin, also internalized at the same time forms pores in granzyme vesicles and releases these proteases. Granzyme B within 5 minutes of entering, starts the apoptotic process leading to DNA fragmentation into 200 bp oligomers and cell death. In CTL cell lines lacking perforins and granzymes also bring about lytic function. This occurs through the interaction of Fas ligand (FasL) on CTLs and Fas on the target cell. Fas-FasL interaction triggers apoptotic pathway and induces cell death. Apoptosis initiates with the conversion of an inactive procaspase into an active caspase which in turn activates other caspases in a cascade of events finally leading to programmed cell death. Both Fas-FasL interactions and granzyme act by initiating the formation of the initiator caspase.

17.4 NATURAL KILLER CELLS

Natural killer (NK) cells belong to the lymphoid lineage of cells. They are large granular lymphocytes which constitute 5%–10% of circulating lymphocytes. NK cells are non-specific killer cells that can destroy viruses, intracellular pathogens and tumor cells even before the CTLs activate and proliferate. NK cells, thus, belong to the innate immune system which responds immediately to the rise in the concentration of IFN-α, IFN-β and IL-2, molecules produced by cells infected with viruses. Immediately after infection, the number of NK cells is seen to rise in just 3 days, whereas the activation of CTL cells takes about 7 days. NK cells carry out the non-specific cytotoxic activity by producing important cytokines like IFN-γ which have multiple functions. IFN-γ activates macrophages to start phagocytosis and anti-microbial activity. It also brings about the differential differentiation of T_H1 and T_H2 cells. It promotes the differentiation of T_H1 and inhibits the differentiation of T_H2. The molecules present on the membranes of NK cells share some similarities with T cells (Figure 17.3), however, there are a lot of variabilities even within the group of NK cells which

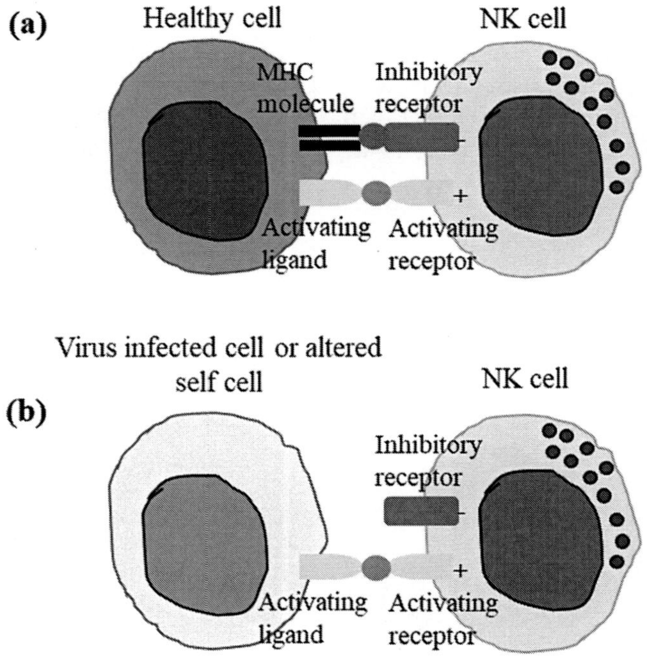

FIGURE 17.3 The NK cells make distinctions between a healthy cell and a target cell by means of the presence of MHC molecules. MHC molecules are expressed by healthy cells but not by tumor and infected cells. The detection of MHC molecules on cell surface, thus acts as an inhibitory signal, which when present, even overrides the activating signal. (a) The killing activity of NK cell is inhibited; (b) NK cell is activated, leading to the killing of the target cell.

might depend on different NK cell subpopulations or on the developmental stages of these cells. Some of the key molecules present on their surfaces are CD2 (β subunit of IL-2 receptor) and CD16/ $F_{c\gamma}$RIII (receptor for F_C region of IgG).

17.4.1 Mechanism of NK Cell-Mediated Cytotoxicity

The mode of action of NK cells through which they destroy target cells is similar to CTL-mediated cytotoxicity except for the fact that NK cells need not be activated like CTL cells and NK-mediated response does not involve antigen-specific MHC expressed antigens. NK cells have FasL receptor on their membranes through which they interact with target cells bearing Fas on their membranes and kill them. They also express perforins and granzymes inside cytoplasmic granules. These granules are present constitutively in their cytoplasm, as opposed to CTLs in which they appear only after CTL activation. When NK cells encounter an infected cell, they engage with it and at this time degranulation of cytoplasmic granules takes place, releasing perforins and granzymes which mediate cell killing by activating the apoptotic pathway. Another distinction from CTLs is that NK cells being cells of the innate immunity, do not bear the property of immunologic memory. Thus, there will be no difference in the intensity of NK cell-driven response when exposed to the same antigen a second time.

The pertinent question regarding the mechanism of NK cell activity is how do these cells differentiate between self-cells and non-self or infected cells. NK cells are stimulated to become active when cytokines IFN-α and IFN-β, TNF-β, IL-12 and IL-15 are present. The signal that activates NK cells is expressed by virus-infected cells, tumor cells and other stressed cells in the form of activating receptors present on their surface. Apart from this, NK cells are inhibited by inhibitory receptors. Earlier it was thought that there is just one set of activating and inhibitory receptors in what was proposed as "two receptor model". However, with ongoing research, it became clear that the activation and inhibition of NK cells is a complex phenomenon involving simultaneous action of both activating and inhibitory receptors. This model, known as "opposing signals model" holds that at a time, both activating and inhibitory signals might be getting delivered to NK cells via the involvement of many simultaneous receptors, however, the one which predominates brings about the respective effect, i.e., either activation or inhibition. This is a key safeguard mechanism which prevents NK cells from attacking self-cells. It is the minute balance of membrane receptors that determines whether the NK cell will carry out cytotoxic activity or not. Structurally, NK receptors are classified as lectin-like receptors and immunoglobin-like receptors. This structural classification, however, does not designate whether functionally they are activating or inhibitory. Both lectin-like and immunoglobin-like receptors may have activating as well as inhibitory receptors. Immunoglobin-like receptors consist of an important group of receptors called killer immunoglobin-like receptors (KIRs). Similarly, lectin-like receptors include CD94:NKG2. Interestingly, CD94:NKG2 has two forms, one of which delivers activating signals and the other delivers inhibitory signals.

17.5 NKT Cells

NKT cells are so named because they exhibit properties of both NK cells and T cells, having some surface markers common with T cells and some common with NK cells. Similar to T cells, NKT cells harbor T-cell receptors on their surfaces, however, their TCR is invariant. It is encoded by specific genes $V_\alpha 24$-$J_\alpha 18$ and $V_\beta 11$ which encode TCRα and TCRβ chains, respectively. Another differentiating point between T cell TCR and NKT TCR is that NKT TCR does not recognize peptide associated with MHC molecules, rather, it recognizes glycolipids presented on CDld molecules. The property of memory is also not associated with NKT cells, as no NKT memory cells have been known. Thus, due to their non-specific nature and absence of memory cells, these cells are a part of innate immunity. Recognition of lipid antigens present on tumor cells by NKT cells has also been observed. They are known to be activated by glycosyl ceramides (similar to lysosomal glycosphingolipid iGb3 required for their activation in the thymus) that are present in many bacterial species.

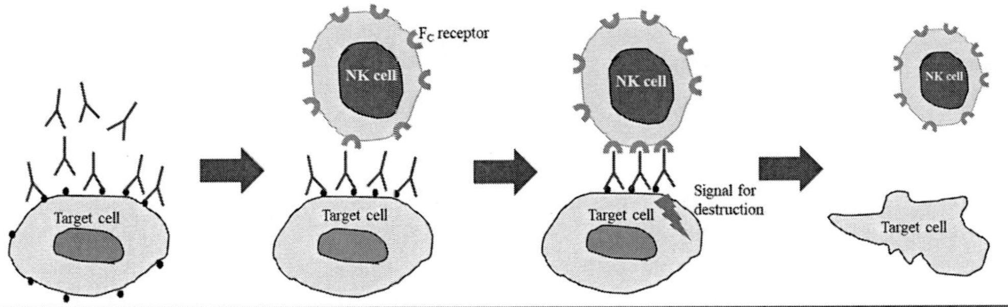

Binding of antibody to cell surface antigens present on the target cells	Antibodies bound to the target cells are recognized by Fc receptors present on NK cells	Binding of Fc receptors to antibodies signals NK cells to kill target cells	Target cells are destroyed by activation of apoptosis

FIGURE 17.4 Schematic representation of steps involved in ADCC, involving binding of antibodies to the antigen, binding of the antigen-antibody complex to Fc receptors present on NK cells activating NK cells to release lytic enzymes which finally destroy the cells.

17.6 ANTIBODY-DEPENDENT CELL-MEDIATED CYTOTOXICITY

Many cytotoxic cells also harbor receptors which recognize the Fc region of antibodies. When antibodies bind to their specific antigens, such antigen-antibody complexes can be recognized by cells bearing Fc receptors, leading to the destruction of target cells (Figure 17.4). This kind of immunity in which antigens are recognized via specificity of the antibody but cytotoxic cells bring about cell destruction is called antibody-dependent cell-mediated cytotoxicity (ADCC). Cells which harbor Fc receptors, viz., NK cells, macrophages, monocytes, neutrophils and eosinophils are the mediators of ADCC. When such an association between cytotoxic cells and antigen-antibody complex is formed, the level of lytic enzymes like perforins and granzymes rises in the lysosomes and granules of these cells, finally degranulating and causing the destruction of the target cell. NK cells, macrophages and monocytes secrete TNF when their Fc receptors bind antigen-antibody complexes which kill the target cell. Similarly, NK cells and eosinophils produce perforins and NK cells produce granzymes to neutralize the target cells captured via Fc receptors.

17.7 SUMMARY

The cell-mediated immune system neutralizes antigens via the action of cells such as $CD8^+$ T cells, $CD4^+$ T cells, NK cells, NKT cells, macrophages, eosinophils and neutrophils. T cell-mediated cytotoxicity is brought about by cytotoxic T lymphocytes (CTLs). CTLs are effector T cells that need to get activated before they can exhibit a cytotoxic response. CTLs are differentiated from naïve T_C cells by their lower activation requirements and a high number of cell adhesion molecules. CTLs produce effector molecules which are soluble as well as membrane-bound. Naïve T_C cells are known as CTL precursors (CTL-Ps), which are activated in two steps to become CTLs. The first step involves activation through the interaction of MHC class I-antigen complex expressed on a licensed APC with TCR of T cell. The second step involves attachment of IL-2 on an enhanced number of IL-2 receptors being expressed on activated CTLs, now converting them into effector CTLs. CTL-mediated killing involves conjugate formation with the target cell, movement of granules carrying perforins and granzymes towards target cell membrane and degranulation causing the release of perforins and granzymes. Perforins mediate killing by forming pores or tunnels in the plasma membrane of the target cell and granzymes activate the apoptotic pathway leading to target cell death. NK cells and NKT cells, on the contrary, exhibit non-specific, non-MHC restricted cytotoxic responses and fall under the category of innate immunity. NK cells get activated in response to viral infections, intracellular infections and tumor cells, causing lysis of these cells via the release of perforins and granzymes, similar to CTLs. NK cell receptors are classified structurally as lectin-like receptors and immunoglobulin-like receptors, both of which can deliver both activating and inhibitory signals. NKT cells are so named because they harbor a TCR, however, this receptor is invariant and non-specific, responding to glycolipid antigens. Most other membrane molecules harbored on NKT cells are common with NK cells. ADCC is carried out by macrophages, NK cells, eosinophils, monocytes and neutrophils. These cells harbor Fc receptors that bind to the Fc region of immunoglobulins, thus also capturing antigens bound to antibodies. After this, they release lytic enzymes like perforins, granzymes and TNF to kill target cells.

QUESTIONS

1. Expand the following:
    i. TCR
    ii. CTL
    iii. NK
    iv. KIR
    v. TNF
    vi. ADCC
    vii. IFN
2. Differentiate between naïve Tc cells and effector CTLs.
3. What is the role of cell adhesion molecules on the membrane of effector T cells?
4. Describe the process of activation of CTL-Ps.
5. Describe the process of CTL-mediated cytotoxicity.
6. What are NK cells? How do NK cells get activated?
7. Explain how NKT cells are considered as hybrids of NK cells and T cells.
8. Describe how target cell killing is brought about by ADCC.

BIBLIOGRAPHY

1. Kindt, T.J., Goldsby, R.A. and B.A. Osborne. 2006. *Kuby Immunology.* Sixth edition. W.H. Freeman and Company, New York.
2. Delves, P.J., Martin, S.J., Burton, D.R. and I.M. Roitt. 2011. *Roitt's Essential Immunology.* Twelfth edition. Wiley-Blackwell, New York, United States.
3. Janeway, C.A. Jr., Travers, P. and M. Walport. 2001. *Immunology: The Immune System in Health and Disease.* Fifth edition. Garland Science, New York, United States.

Index

Note: **Bold** page numbers refer to tables and *italic* page numbers refer to figures.